The Industrial Revolutions

VOLUME 3

The Industrial Revolution in Britain
II

THE INDUSTRIAL REVOLUTIONS

General Editors R. A. Church and E. A. Wrigley

Volume 1
Pre-industrial Britain
Edited by J. A. Chartres

Volumes 2 and 3
The Industrial Revolution in Britain
Edited by Julian Hoppit and E. A. Wrigley

Volumes 4 and 5
The Industrial Revolution in Europe
Edited by P. K. O'Brien

Volume 6
Industrialization in North America
Edited by Peter Temin

Volume 7
The Industrialization of Japan
Edited by W. J. Macpherson

Volume 8
The Textile Industries
Edited by D. T. Jenkins

Volume 9
The Metal Fabrication and Engineering Industries
Edited by Sidney Pollard

Volume 10
The Coal and Iron Industries
Edited by R. A. Church

Volume 11
Commercial and Financial Services
Edited by R. C. Michie

The Industrial Revolution in Britain II

Edited with an introduction by

Julian Hoppit and E. A. Wrigley

The Economic History Society

BLACKWELL
Oxford UK & Cambridge USA

First published 1994

Blackwell Publishers
108 Cowley Road, Oxford OX4 1JF
UK

238 Main Street
Cambridge, Massachusetts 02142
USA

British Library Cataloguing in Publication Data

A CIP catalogue record for this book is available from the British Library.

Library of Congress Cataloging-in-Publication Data

The Industrial Revolution in Britain/edited, with an introduction, by Julian Hoppit and E. A. Wrigley.
p. cm. — (The Industrial revolutions : v. 2-3)
"The Economic History Society."
Includes bibliographical references.
ISBN 0-631-18071-0 (v. 1). — ISBN 0-631-18072-9 (v. 2)
ISBN 0-631 17486 9 (11-volume set).
1. Great Britain—Industries—History. 2. Technological innovations—Great Britain—History. 3. Great Britain—Economic conditions. I. Hoppit, Julian. II. Wrigley, E. A. (Edward Anthony), 1931- . III. Economic History Society. IV. Series.
HD2321.I533 1994 vol. 2-3
[HC253]
338.09 s—dc20
[338.0942]

93-40549
CIP

Introduction typeset in 11 on 12 point Plantin by Typesetting Solutions, Slough, Berks., UK.
Printed in Great Britain by T. J. Press Ltd, Padstow, Cornwall.

Table of Contents

VOLUME 2

VOLUME 3

Part 4. Land: Agriculture and Raw Material Supply

Part 5. Technology, Invention, Innovation, Education

Part 6. Supply and Demand, Trade Cycles

General Editors' Introduction

R. A. CHURCH and E. A. WRIGLEY

The aim of this series of volumes is to bring together journal articles and chapters from collections of essays that deal with some aspect of the industrial revolution and may be regarded as 'classics'. Some of the articles were published recently, but writings from each decade since the Second World War are well represented and there is a scattering of articles that appeared in the interwar period. Very few libraries contain all the journal series, collections of essays, and research monographs from which the material reproduced in these volumes was drawn.

As well as indicating the latest trends in research, these volumes provide an insight into the development of thinking about the issues that have informed work on the industrial revolution over the last half-century. Each piece is reproduced with its original pagination, but with continuous pagination added for ease of reference. In every case but one the original was reproduced photographically. In the one remaining case (volume 4, article 7), the text was reset. Each volume contains an introduction which explains the choice of material reproduced, describes the main debates to which they relate, and provides a guide to further reading that should make it possible to pursue particular questions effectively.

The series consists both of topical volumes dealing with an individual industry or area of economic activity (for example, textiles or metal fabrication and engineering industries), and of volumes that cover the occurrence of the industrial revolution in a particular country or continent (Britain or North America).

By 'industrial revolution', a loose and imprecise term, we mean principally what is often referred to as the first industrial revolution, the period of the cotton mill and the steam engine, rather than that of electrical engineering, fine chemicals, and the internal combustion engine (though the first volume deals with the period immediately before the conventional dating of the industrial revolution in Britain). The industrial revolution was an event of such profound and continuing importance for all societies, both those initially affected and those which were remote from the main centres of early growth, that achieving a better understanding of its nature should be a prominent element in the reflections of all those concerned about society today as well as in the past. We hope that this series may prove of use to those who are persuaded of the truth of this assertion.

Part 4

Land: Agriculture and Raw Material Supply

The Growth of Labor Productivity in Early Modern English Agriculture*

ROBERT C. ALLEN

Department of Economics, University of British Columbia

The distinctive feature of English agricultural development was the rapid growth in labor productivity. This paper explains that growth in terms of the rise in grain yields and reductions in employment due to the increase in farm size. Estate surveys and the land tax assessments are used to measure changes in farm size over the 17th and 18th centuries. Data collected by Arthur Young are used to assess the implications of farm size for employment. The reasons large farms were able to economize on labor are explored. The estate surveys, Young's data, and estimates of crop yields are used to simulate labor productivity in English agriculture from 1600 to 1800. The implied growth in efficiency is consistent with independent estimates of the increase and with estimates of English productivity in comparison with French and Russian performances.

> Improvements in agriculture are of two kinds: those which increase the productive powers of the land, and those which enable us to obtain its produce with less labour.
>
> David Ricardo, *On the Principles of Political Economy and Taxation* (1817), p. 80

Several indices can be used to measure agricultural productivity. One of the commonest is grain yields. Between the Middle Ages and 1800, English yields approximately doubled. For instance, wheat, which had produced about 10 bushels per acre in the 13th and 14th centuries, produced 20 or more in the early 19th (Allen, 1986a). This increase did not make England unusually productive, however, for wheat yields ca. 1800 were similar in Ireland, northeastern France, Holland, and probably in western Germany and Belgium.[1]

* I thank Chris Archibald, Greg Clark, Nick Crafts, Stan Engerman, Joanne Innes, Peter Lindert, and Keith Snell for helpful discussions. I am grateful to Nancy South, Don Andrews, and Susan Ghan for research assistance and to Frank Flynn for assistance with programming and data analysis. I thank the Social Science and Humanities Research Council of Canada for financial support.

[1] For yield estimates see Allen and Ó Gráda (1986) for England, Ireland, and France, Van Zanden (1985) for Holland, and Chorley (1981, p. 83) for Germany. For a European-wide perspective using yield–seed ratios, which are indicative of yields per acre, see Slicher van Bath (1963).

0014-4983/88 $3.00

TABLE 1
Agricultural Labor Productivity, 1840

Country	Output[a]
U.K.	17.5
France	11.5
Belgium	10.0
Germany	7.5
Sweden	7.5
Switzerland	8.0
Italy	4.0
Russia	7.0

Source. Bairoch (1965, p. 1096).
[a] Net output in millions of calories per male worker.

It was unusually high labor productivity that distinguished English agriculture from farming in the rest of northwestern Europe (Crafts, 1985, p. 121). Bairoch (1965, p. 1096) compared productivity in several European countries in 1840 (see Table 1). He found it highest in the United Kingdom, next highest in Belgium and France, and lowest in the other countries. French productivity was 66% of the British, and Russian productivity only 40%. O'Brien and Keyder (1978, pp. 102–145) have used a purchasing power exchange rate to compare labor productivity in England and France in the first half of the 19th century. Their work confirmed that the productivity gap was of the order Bairoch measured. Wrigley (1985, p. 720) has pushed the measurement of the English–French productivity gap back to the 16th century by estimating the ratio of the total population to the agricultural population for the two nations. For countries like these that were largely self sufficient in food, this ratio is indicative of agricultural labor productivity. Wrigley's results (in Table 2) show that productivity was similar in England and France in 1500 and 1600 and grew little in the intervening century. Between 1600 and 1800 output per worker increased 17% in France but 73% in England with the result that in 1800 labor productivity in France was only 69% of the English level. Wrigley's calculations thus corroborate Bairoch's and O'Brien and Keyder's estimates of the labor productivity gap in the early 19th century and show that it emerged in the 17th and 18th centuries.

The purpose of this paper is to explain England's unusually rapid growth in labor productivity. There are two points of departure. The first is the identity that output per worker equals output per acre multiplied by acres per worker. We can understand the growth in labor productivity by partitioning it into these effects. Thus, crop yields did double in early modern England and that increase raised real farm revenue (output) per

TABLE 2
English and French "Agricultural Labor Productivity," 1500–1801

	England	France
1500	1.32	1.38
1600	1.43	1.45
1700	1.82	1.58
1750	2.19	1.63
1801	2.48	1.70

Note. For each country the ratio of the total population to the agricultural population is shown. The English figure shown for 1500 is more properly for 1520.

Source. Wrigley (1985, p. 720).

acre but (as we shall see) not by enough to account for the growth in labor productivity.

O'Brien and Keyder's work provides a second point of departure. They discovered that output per acre was similar in England and France. England's higher labor productivity was due to lower employment per acre. As they point out, there is a long tradition that attributes this difference to the English system of enclosed, large scale farming in contrast to the French system of small peasant proprietorships. That tradition derives originally from the critics of enclosures and large farms who argued that those changes depopulated the countryside. Addington (1772), Davies (1795, pp. 51–57), and Price (1792, pp. 283–284) were important 18th century proponents of that view, but its roots run back to the 15th century (Tawney and Power, 1924, Vol. III, pp. 12–81; Beresford, 1961).

This view was by no means universally accepted. Defenders of enclosure and large farms often argued that these changes *increased* employment by creating new jobs (hedging and ditching) and led to convertible husbandry, a farming system that increased yields and employment at the same time. Fitzherbert (1539) may have been the first writer to advance these arguments, but they reached their fullest exposition with Arthur Young. He developed the theory of the "capital intensive farmer" as the agent of agricultural improvement. The theory presumes that only large farmers were rich and only rich farmers could afford to increase livestock and undertake improvements like marling, draining, and hoeing their turnips the many times that Young recommended. All of these tasks expanded employment, so a movement to large farms led to more output and more jobs. Young also believed enclosure encouraged the same process. Chambers (1953) repeated these contentions, and they have become academic orthodoxy. No new evidence, however, has been adduced in support of this position since the work of Young. We will see that, in fact, the evidence collected by Young supports an assessment of enclosures and large farms more in tune with Davies and Price than with his own opinions.

What I shall show in this paper is that the increase in English labor productivity between 1600 and 1800 had two causes. The first was the rise in crop yields. This cause operated across northwestern Europe causing labor productivity to rise throughout the region. The second was the shift to large, capitalist farms. This reorganization increased labor productivity by reducing employment per acre. Moreover, that increment to efficiency was about equal to the Anglo–French productivity gap. The English superiority noted by Wrigley, Bairoch, and O'Brien and Keyder was due to England's peculiar agrarian institutions—in particular, large farms operated with wage labor.

The Growth in Farm Size

English farms have increased in size through several stages since the Middle Ages.[2] While demesnes have always been large—100 acres or more—villein farms seldom exceeded 30 acres (one virgate) in the 12th and 13th centuries and were usually 15 acres or less (Kosminski, 1956, pp. 216, 223; Postan, 1972, pp. 127–131). These farms could support a family (Titow, 1969, pp. 64–96) but did not exhaust the family's labor.

Farm sizes increased after the Black Death. Farms of several hundred acres were the norm in the new enclosures (Dyer, 1980, pp. 17–21; Finch, 1956, pp. 14–20, 41, 73–74, 104, 114, 138). In open field villages, customary holdings increased to 30–60 acres but rarely exceeded 100 acres. Customary holdings averaged 34 acres in the 16th century surveys collected by Tawney. Leasehold farms, which were mainly demesnes or enclosed farms, were much larger: they averaged 276 acres.[3] Hoskins (1950, p. 146) used probate inventories to estimate farm sizes in Leicestershire: "Though the average farm was one of roughly 40 or 50 acres in sixteenth century Leicestershire, half of the farms of the county were below this size. On the other hand, about four per cent of the farms were 100 acres and upwards in extent. . . ." He showed that the same pattern continued into the early 18th century (Hoskins, 1951, pp. 12–15). Havinden (1961)

[2] I call a holding a farm only if it contained at least five acres of agricultural land. I call a rural holding with a house and less than fives acres of agricultural land a "cottage."

[3] Tawney (1912, pp. 64–65, 212.) The data were retabulated in accord with the size categories of Tables 3 and 4. The calculations in the text ignore holdings of less than 5 acres. Farms in Lancashire and Northumberland were excluded. For farms between 5 and 100 acres, the total acreage of each category was obtained by multiplying the mean size of the retabulated category by the number of farms in the category. I assumed the farms between 100 and 120 acres had an average size of 110 acres and the 17 farms over 120 acres had an average size of 150 acres.

confirmed that Oxfordshire was little different in this regard. Thus, enclosed farms and demesnes in open field villages were typically several hundred acres in early modern England, while other farms in open field villages averaged about 50 acres. Their farmers were the English yeomen.

Farm size remained constant into the early 18th century. To follow farm sizes in this period, I reconstructed the farm size distribution from a sample of surveys of south midlands estates taken between 1595 and 1850. The results are shown in Tables 3 and 4. Enclosed farms continued large, although the small size of the samples for the early 17th and 18th centuries means that the distributions for those years are not precisely ascertained. Open farms averaged 59 acres in the "early seventeenth century" (based on surveys taken between 1595 and 1650), which is consistent with the findings of Tawney, Hoskins, and Havinden.[4] Average farm size in the open fields was still only 65 acres in the early 18th century.

In the 18th century, there was a revolution in farm size—by 1800, the average size of an open field farm had exploded to 145 acres. Average farm size was almost the same (147 acres) in enclosed villages at that time, so the 18th century shift to "capital" farms in the open fields represented a convergence to the farm size distribution previously characteristic of enclosures. The conclusion that the 18th century witnessed a great increase in farm size is consistent with Mingay's (1962, pp. 480–483) investigation of the Kingston estates in Nottingham and the Bagot and Gifford estates in Straffordshire as well as Wordie's (1974) study of the very large Leverson-Gower estates in Shropshire and Staffordshire.

Estate surveys suffer from two disadvantages as a source for reconstructing the farm size distribution. First, estates that were well surveyed tended to be large and so may not have been representative. Second, we cannot be certain that the tenants (whose holdings were tabulated in Tables 3 and 4) were actually farmers or (more important) that tenancies equalled farms. Since most tenancies before the late 18th century were held on long-term agreements, they were often sublet (Harrison, 1979), and it is possible that enterprising leasees built up large farms by leasing many small tenancies. We know that happened occasionally. If it were widespread, Tables 3 and 4 would not be an accurate representation of the farm size distribution.

Fortunately, the land tax assessments provide an independent basis for estimating the farm size distribution ca. 1800.[5] Their advantage is

[4] Exact comparison with Tawney's results is impossible since the 17th-century surveys often failed to label demesnes as such. Consequently, Tables 3 and 4 tabulate demesnes and customary holdings together. The inclusion of demesnes means that the average size of early 17th-century farms described in Tables 3 and 4 (59 acres) exceeded the average size of Tawney's customary holdings (34 acres).

[5] Probate inventories, which indubitably apply to integral farms, perform an analagous

TABLE 3
Distribution of Farm Acreage from Estate Surveys

Farm size (acres)	Acres			Percentage of acres		
	Early 17 cen.	Early 18 cen.	About 1800	Early 17 cen.	Early 18 cen.	About 1800
Open farms						
5–10	174	189	6	0.9	0.7	0
10–15	180	393	40	0.9	1.5	0.2
15–30	1,123	1,769	256	5.8	6.8	1.2
30–60	5,018	4,063	1,304	25.9	15.7	6.2
60–100	6,623	5,634	1,596	34.2	21.7	7.6
100–200	4,233	10,712	5,959	21.9	41.3	28.3
200–300	704	1,984	6,476	3.6	7.7	30.8
300–400	301	300	3,704	1.6	1.2	17.6
400–500	492	886	0	2.5	3.4	0
500–1000	513	0	1,691	2.7	0	8.0
1000+	0	0	0	0	0	0
Total	19,361	25,930	21,032			
Average size	59	65	145			
Enclosed farms						
5–10	18	8	11	0.4	0.1	.0
10–15	10	72	47	0.2	0.9	0.1
15–30	23	207	845	0.5	2.5	1.8
30–60	254	764	2,976	5.3	9.1	6.2
60–100	135	1,270	3,086	2.8	15.1	6.4
100–200	465	3,687	12,248	9.6	43.9	25.5
200–300	651	964	12,689	13.5	11.5	26.5
300–400	350	337	9,590	7.2	4.0	20.0
400–500	0	427	4,017	0	5.1	8.4
500–1000	1,214	645	2,438	25.1	7.7	5.1
1000+	1,716	0	0	35.5	0	0
Total	4,836	8,401	47,947			
Average size	210	100	147			

Sources. Some surveys underlying this table are from Gray (1915, p. 444) and Lennard (1916). Most, however, are from the following manuscript sources:

Bedford Record Office:

R Box 792, CRT 100/34, LA 1/14–27;

Bodleian Library, Oxford:

Ms. Top. Berks e 21, Ms. Top. Berks d 26, Ms. DD Bertie c. 18/3, Ms. DD Bertie c. 18/4, Ms. DD Bertie d 1/27, Ms. Top. Oxon b 121, Ms. Top Oxon c. 381, Ms. DD Harcourt b. 37, Ms. DD Harcourt b. 34, DD Harcourt e. 7;

Huntingdon Record Office:

dd M b 10/1-7 Hinch /5/70, C4/2/2/13, C4/2/5/5a, C4/2/6/11, C4/2/6/13, C4/2/7/5;

Northampton Record Office:

Brudenell ASR 95, Brudenell ASR 96, Brudenell ASR 138, Brudenell B.ii.49, H.xi.26, H.xi.31, O.xxii.6, D(CA) 211, D(CA) 213, D(CA) 215, D(CA) 306, D(CA) 444, G. 1654, G. 3916, G. 3898, C(A) 5739, F(M) MISC. VOL. 201, F(M) MISC. VOL. 555.

TABLE 4
Distribution of the Number of Farms from Estate Surveys

Farm size (acres)	Number of farms			Percentage of farms		
	Early 17 cen.	Early 18 cen.	About 1800	Early 17 cen.	Early 18 cen.	About 1800
Open farms						
5–10	24	26	1	7.3	6.5	0.7
10–15	15	34	3	4.6	8.5	2.0
15–30	53	80	10	16.2	20.1	6.9
30–60	114	96	29	34.8	24.1	20.0
60–100	84	73	22	25.6	18.3	15.2
100–200	32	77	38	9.8	19.3	26.2
200–300	3	9	28	0.9	2.3	19.3
300–400	1	1	11	0.3	0.3	7.6
400–500	1	2	0	0.3	0.5	0
500–1000	1	0	3	0.3	0	2.0
1000+	0	0	0	0	0	0
Total	328	398	145			
Enclosed farms						
5–10	2	1	2	8.7	1.2	0.6
10–15	1	6	4	4.3	7.1	1.2
15–30	1	11	39	4.3	13.1	11.9
30–60	6	18	69	26.1	21.4	21.1
60–100	2	16	39	8.7	19.0	11.9
100–200	4	25	81	17.4	29.8	24.8
200–300	3	4	52	13.0	4.8	15.9
300–400	1	1	28	4.3	1.2	8.6
400–500	0	1	9	0	1.2	2.8
500–1000	2	1	4	8.7	1.2	1.2
1000+	1	0	0	4.3	0	0
Total	23	84	327			

Sources. Same as for Table 3.

that they include all the land (not just that in large estates) and allow one to add up all of a farmer's land, no matter who owned it. The disadvantages of the assessments are, first, that one must adopt some procedure to convert tax liability to acreage,[6] and second, that it is usually

role for the 16th and 17th centuries. Since Hoskins' and Havinden's studies of farm size were based on inventories, and since those studies point to the same conclusion as the estate surveys, we can be confident that the surveys accurately indicate the size of farms even if the manorial tenants were not the farmers themselves.

[6] There has been prolonged controversy among historians as to whether it is possible to convert tax liability to acreage. See Turner and Mills (1986) for the most recent round and a summary of the earlier debates. In this study, the tax liability of each occupier was

impossible to identify and thence remove nonfarm property like woodland and deer parks. In the event, however, the assessments tell a similar story to the surveys.

Tables 5 and 6 show the distribution of farm acreage and numbers distinguishing enclosed farms by their period of enclosure.[7] The tables support three important conclusions. First, villages open at the time of the assessments and those enclosed in the 18th century were indistinguishable in terms of farm size. Second, although farm sizes in early enclosed villages may have been somewhat larger than in 18th-century enclosures, the differences were modest. Notice, however, that early enclosed villages had distinctly fewer cottages, i.e., properties tallied as 0–10 acres in these tables.[8] Third, although Tables 5 and 6 support the conclusion that most of the land ca. 1800 was organized in farms exceeding 100 acres, these tables moderate the conclusion somewhat, for they show a bit more of the land and definitely more of the farms to have been less than 100 acres than Tables 3 and 4 suggest. As a result, average farm size in open and recently enclosed villages was about 120 acres according to the land tax assessments, while it was about 145 acres according to the estate surveys.[9] The discrepancy probably reflects the fact that surveys survive disproportionately for large estates, and they had larger farms than average.[10] Nonetheless, the conclusion that the eighteenth century marked a widespread shift to "capital" farms is confirmed.

This conclusion is consistent with some accounts of the decline of the English peasantry (e.g., Johnson, 1909; Gray, 1910; Habakkuk, 1965), but it is at odds with Thompson's (1966, 1969) and Mingay's (1968) influential[11] reassessments of the aggregate evidence. They deny that the 18th century saw the end of peasant agriculture in England. Thompson, however, was primarily concerned with the ownership of land rather

first computed. Then, acreage was estimated by apportioning the total acreage of the village across all the occupations in proportion to their tax liabilities. In other words, an "acreage equivalent" specific to each parish was used to convert tax liability to acreage.

[7] The tables are derived from a sample of land tax returns for 636 villages in the counties pf Bedford, Buckingham, Cambridge, Huntingdon, Oxford, Northampton, Leicester, and Warwick. The return for 1790 (or the nearest suitable year) was used.

[8] The apportioning method in this study tends to overestimate the acreage of cottages since for them the ratio of the value of the structure to the value of the land was exceptionally high. Rough calculations suggest that occupations with up to 10 computed acres were in reality cottages with less than 5 acres.

[9] In these comparisons, we compare farms of 5–100 acres in Tables 3 and 4 to farms of 10–100 acres in Tables 5 and 6. The aim in both cases is to exclude cottages at the lower end.

[10] Subsidiary calculations show that the discrepancy does not affect the conclusions we will draw from the estate surveys.

[11] Beckett 1984, p. 5.

TABLE 5
Distribution of Farm Acreage from Land Tax Assessments, ca. 1790

Farm size (acres)	Acreage				Percentage of acreage			
	EEE	OEE	OOE	OOO	EEE	OEE	OOE	OOO
0–5	122	418	2,721	5,881	0.2	0.5	1.0	1.2
5–10	215	636	3,368	8,739	0.4	0.7	1.2	1.7
10–15	263	1,115	3,026	7,260	0.4	1.2	1.1	1.4
15–30	1,102	2,918	9,524	17,097	1.9	3.2	3.5	3.4
30–60	3,151	4,794	15,826	32,135	5.3	5.3	5.8	6.3
60–100	4,480	9,122	26,952	51,837	7.6	10.0	9.9	10.2
100–200	12,965	21,561	68,022	129,732	21.8	23.7	25.0	25.6
200–300	9,651	17,783	46,409	97,065	16.3	19.6	17.0	19.2
300–400	8,498	11,061	34,927	51,146	14.3	12.2	12.8	10.1
400–500	3,985	8,973	18,368	29,292	6.7	9.9	6.7	5.8
500–1000	7,707	9,485	25,892	58,157	13.0	10.4	9.5	11.5
1000 +	7,230	3,106	17,444	18,562	12.2	3.4	6.4	3.7
Total	59,369	90,972	272,479	506,903				
Average farm size	157	122	119	111				

Note. The columns refer to the period of enclosure. EEE—before the middle of the 16th century. OEE—between the middle of the 16th century and 1676. OOE—between 1676 and the date of the land tax assessment (ca. 1790). OOO—open at the time of the land tax assessment. Average farm size is computed for holdings of 10 acres or more.

TABLE 6
Distribution of the Number of Farms from Land Tax Assessments, ca. 1790

Farm size acres	Number of farms				Percentage of farms			
	EEE	OEE	OOE	OOO	EEE	OEE	OOE	OOO
0–5	47	150	1162	2050	—	—	—	—
5–10	31	88	472	1237	—	—	—	—
10–15	21	88	247	593	5.6	12.0	11.0	13.4
15–30	49	134	442	800	13.0	18.2	19.7	18.0
30–60	69	110	360	753	18.3	15.0	16.1	17.0
60–100	57	115	338	656	15.1	15.6	15.1	14.8
100–200	94	148	470	928	24.9	20.1	21.0	20.9
200–300	38	72	191	397	10.1	9.8	8.5	8.9
300–400	25	33	101	148	6.6	4.5	4.5	3.3
400–500	9	20	41	67	2.4	2.7	1.8	1.5
500–1000	12	13	40	87	3.2	1.8	1.8	2.0
1000+	3	3	11	12	0.8	0.4	0.5	0.3
Total	455	974	3875	7728				

Note. The percentages are computed only for farms greater than 10 acres on the grounds that computed acreages up to that level are in reality cottages with 5 acres or less of land. These should be excluded in this table to allow comparison with Table 4 in which holdings of less than 5 acres are also excluded. The columns refer to the period of enclosure. EEE—before the middle of the 16th century. OEE—between the middle of the 16th century and 1676. OOE—between 1676 and the date of the land tax assessment (ca. 1790). OOO—open at the time of the land tax assessment.

than its occupation, so his conclusion is not necessarily inconsistent with that advanced here. Mingay's views, on the other hand, are relevant since he was undoubtedly concerned with farm size. However, his conclusion that the number of small farmers did not decline over the 18th century is unpersuasive since it is based on Gregory King's 1688 social table of England and Wales. This has been shown to be exceptionally unreliable in the cases of freeholders and farmers (Cooper, 1967; Holmes, 1977). Aggregate statistics like King's are seductive, but in this case they are unreliable and preference must be given to estimates based on the traditional documentary sources.

FARM SIZE AND PERFORMANCE

What was the significance of the 18th-century growth in farm size? To answer the question, it is necessary to compare the operations of small and large farms. The most systematic source for these comparisons is the information collected by Arthur Young in his tours of England in the late 1760s. For several hundred "representative farms," he recorded the cropping pattern, the numbers of livestock, and the number of employees broken down by age and sex. From this information, we can

TABLE 7
Farm Size and the Ratio of Hired to Family Labor

	Constant	Acres	Acres squared	R^2
1. Pasture	2.34 (14.43)	−0.01 (−7.04)	1.28×10^{-5} (4.82)	0.42
2. Arable	1.48 (13.75)	−0.01 (−7.49)	7.79×10^{-6} (5.68)	0.41

Note. There are 111 pasture farms and 128 arable farms in the sample. *t*-ratios are in parentheses. The dependent variable is the ratio of the cost of full-time hired labor to family labor.

estimate the prices and quantities of all inputs and outputs on the farms.[12] Small and large farms differed in several important ways.

One classic difference was in the system of labor organization—small farms were operated mainly by family labor, while large farms used mainly hired labor. To measure the shift to hired labor, I computed the ratio of the cost of hired labor to the cost of family labor (valued at market wages) and investigated how the ratio varied with farm acreage. Table 7 reports results of regressing the ratio on a quadratic function of farm size for arable and pasture farms.[13] The R^2's are substantial and the coefficients highly significant. The ratio of family dlabor to hired labor declined with size. For arable farms, the ratio reached 1 at 77 acres. For pasture farms, the unit ratio was reached at 150 acres.[14] If we define "peasant" farms to be family farms while "capitalist" farms are operated mainly by wage labor, then we can call arable farms of less than 60 acres "peasant," while farms of 60–100 acres were "transitional," and farms over 100 acres were "capitalist." The 18th-century movement

[12] Allen (1982) provides a description of the data. I have changed the way they are analyzed in two ways. In Allen (1982), farm specific input and output prices were used, but it was shown that the variation in those prices explained little variation in surplus. In this paper, revenue, cost, and surplus are computed using average prices. Also harvest labor is estimated at the cost of mowing and reaping rather than as a proportion of the wage payments of laborers, which was the method Young (1771, IV, p. 356–357) used. The results are similar on average, but the procedure adopted here seems appropriate since we are investigating the effects of farm size on employment and since the mix of servants and laborers varied with size.

[13] Initially farms were assigned to the "arable" and "pasture" groups if the share of arable was greater or less than 50%. However, inspection of the data showed that 45% was a much more natural division, so the few farms with 45–50% of their land arable were included with the arable farms in subsequent analysis.

[14] The greater size is the result of the assumption that the family labor supply was the same for both types of farms in conjunction with the fact that pasture farms employed less labor per acre than arable farms.

TABLE 8
Revenue and Cost in Arable Farms
(Pounds per Acre)

Farm size (acres)	Total revenue per acre	Total cost per acre	Labor cost per acre	Ricardian surplus per acre	Number of farms
0–50	4.0399	3.0615	1.5016	0.9783	8
50–100	4.3094	3.1146	1.3023	1.1948	45
100–150	3.7694	2.4290	1.0215	1.3404	16
150–200	4.4774	2.3931	0.9598	2.0842	22
200–250	4.0605	2.5283	0.6228	1.5322	4
250–300	3.5078	2.0145	0.7641	1.4934	12
300–350	5.0522	2.7088	0.9025	2.3434	4
350–400	3.2252	1.8165	0.6191	1.4087	2
400–450	4.9065	2.4322	0.6422	2.4743	2
450–500	5.3128	2.3392	0.6781	2.9736	3
500–550	—	—	—	—	—
550–600	4.8578	2.2600	0.6199	2.5478	6
600–650	3.3538	2.0028	0.4878	1.3510	1
650–700	4.0183	2.1921	0.5910	1.9162	3

Source. This table was computed from the sample of farms surveyed by Arthur Young ca. 1770 and analyzed in Allen (1982). As noted in the text, the analysis in this paper introduces two changes—a uniform set of prices is used in the computations and harvest labor is estimated by piece rates. Arable farms are more than 45% arable and pastoral farms less than 45% arable. Initially a 50% division was used, but inspection of the data showed that there was a more natural division at 45%. Farms operated by gentlemen are excluded in this and all other calculations as are farms of more than 700 acres.

to large farms in open field villages was, thus, a transition from peasant to capitalist agriculture.

Large and small farms also differed in terms of economic efficiency. Ricardian surplus, the difference between farm revenue and the opportunity cost of all inputs other than land, is a convenient measure of efficiency (Allen, 1982). Tables 8 and 9 show how surplus varied with size for arable and pastoral farms. Expansion in size was accompanied by a continuous rise in Ricardian surplus but at a diminishing rate and reaching a peak at about 200 acres. Further increases of surplus with size look to have been unlikely for arable farms. For pastoral farms, the data indicate decreasing returns to scale above 200 acres. It is difficult, however, to be precise given the random fluctuations around the trend.

To test for increasing returns to scale, Ricardian surplus per acre was regressed on a quadratic function of farm size. The results are shown in Table 10. All the equations show surplus increasing with size (the positive coefficient of acres) but at a diminishing rate (the negative coefficient of acres squared). Equations (2) and (4) include a variable to

TABLE 9
Revenue and Cost in Pasture Farms (Pounds per Acre)

Farm size acres	Total revenue per acre	Total cost per acre	Labor cost per acre	Ricardian surplus per acre	Number of farms
0–50	3.8464	3.5715	1.4518	0.2749	16
50–100	3.3481	2.6911	0.9687	0.6570	35
100–150	3.2639	2.2647	0.7394	0.9992	19
150–200	3.1823	2.1576	0.5989	1.0246	23
200–250	2.8764	2.1280	0.7463	0.7484	1
250–300	2.4017	1.6609	0.4958	0.7408	8
300–350	1.8683	1.2689	0.5396	0.5993	1
350–400	2.1593	1.5232	0.5178	0.6361	5
400–450	2.4385	1.7417	0.4639	0.6968	2
450–500	—	—	—	—	—
500–550	3.4157	2.5452	0.2789	0.8705	1
550–600	2.9644	2.3241	0.2863	0.6403	1
600–650	—	—	—	—	—
650–700	2.8003	1.6181	0.5526	1.1822	1

represent the stint of farms with commons.[15] The coefficient of commons is positive and significant, an important result in its own right since it establishes that the management of commons was effective enough to prevent rent dissipation. The inclusion of commons improves the estimation of the coefficients of acres and acres squared. The low R^2's show that many things besides size influenced farm efficiency.

The equations show that estates organized in large farms generated more income than those divided into small farms. Equation (1) predicts for arable farms that an increase in size from 50 to 200 acres would raise Ricardian surplus per acre 64%. Amalgamating twenty 50-acre farms into four 250-acre farms would increase the annual value of a 1000-acre estate by 64%. For pasture farms, regression 3 predicts a 48% rise in Ricardian surplus per acre with an increase in size from 50 to 200 acres. Financial considerations of this sort prompted the extinction of peasant farming.

Why were large farms more efficient than small farms? Their advantage was not due to greater output. According to the tables, revenue per acre was independent of size. While that was true, Young's data do not really prove the point. The problem is that Young did not report yields for individual farms. Instead, he reported average yields for villages. In computing revenues for each farm, I assigned the average yield for the

[15] For farms without common rights, the commons variable was assigned a value of zero. For farms with common rights, the number of sheep on the farm was taken as the measure of commons. See Allen (1982).

TABLE 10
Scale Economies in Farming

	Constant	Acres	Acres squared	Commons	R^2
(1) Arable	0.77309	0.00592	-0.57053×10^{-5}		0.09
	(2.587)	(2.319)	(−1.504)		
(2) Arable	0.51296	0.00621	-0.67250×10^{-5}	0.51788	0.18
	(1.756)	(2.560)	(−1.860)	(3.795)	
(3) Pasture	0.45569	0.00292	-0.41591×10^{-5}		0.04
	(2.903)	(1.929)	(−1.586)		
(4) Pasture	0.09571	0.00485	-0.64839×10^{-5}	0.23579	0.28
	(.646)	(3.0601)	(−2.817)	(6.138)	

Note. The dependent variable in the regressions is Ricardian surplus per acre. *t*-ratios are in parentheses.

village to all farms in the village, whatever their size. This procedure would obscure a positive relationship between size and yield if such existed. However, it has been possible to test directly for a relationship between size and yields using a data set constructed from probate inventories. That test confirmed that there was no correlation between size and yield for farms in the range of 20–250 acres (Allen, 1986a). The lack of correlation between size and revenue in Tables 8 and 9 is, therefore, not misleading.

The advantage of large farms lay in lower costs, principally the costs of implements, draught animals, and labor. The economies in implements and draught animals probably reflected each farmer's wanting his own complement of equipment and horses so as not to be dependent on renting these crucial inputs. The savings on these inputs, however, were small compared to those arising from economies in labor.

LABOR SAVING AND THE ECONOMIES OF LARGE SCALE FARMING

Most of the advantage of large farms came from lower labor costs. As Table 11 shows, employment per acre declined with size for men but especially for women and boys. Savings were effected in both family labor and hired labor.

The rapid decline with size in the employment of women and boys compared to men meant that the 18th-century shift to large farms changed the sex balance of rural employment. We saw that farms of less than 50 acres were family farms in the sense that a family could operate the farm without much if any hired labor. Table 11 shows a second sense in which these small farms were family farms—they employed men, women, and boys in equal numbers, so they offered reasonably full employment to all family members. The shift to large farms meant that only the husbands in laborers' families were employed in agriculture.

Inspection of the data suggests that large farms used less labor than

TABLE 11
Employment per Acre (Workers per Acre)

Farm size (acres)	Arable farms			Pasture farms		
	Men	Maids	Boys	Men	Maids	Boys
0–50	0.0364	0.0329	0.0392	0.3300	0.0407	0.0420
50–100	0.0398	0.0233	0.0225	0.0254	0.0252	0.0215
100–150	0.0309	0.0179	0.0175	0.0210	0.0170	0.0142
150–200	0.0316	0.0142	0.0119	0.0179	0.0139	0.0095
200–250	0.0197	0.0105	0.0095	0.0250	0.0125	0.0125
250–300	0.0268	0.0096	0.0082	0.0160	0.0084	0.0084
300–350	0.0328	0.0091	0.0105	0.0187	0.0094	0.0062
350–400	0.0182	0.0079	0.0104	0.0180	0.0095	0.0065
400–450	0.0199	0.0079	0.0092	0.0167	0.0044	0.0067
450–500	0.0228	0.0065	0.0071	—	—	—
500–550	—	—	—	0.0091	0.0055	0.0055
550–600	0.0225	0.0048	0.0048	0.0117	0.0033	0.0033
600–650	0.0185	0.0031	0.0077	—	—	—
650–700	0.0211	0.0063	0.0068	0.0214	0.0071	0.0043

small farms and that for men, at least, the conversion of arable to grass reduced employment. We also wonder whether enclosure had any additional affect on employment over and above the effects it had in converting arable to pasture. The regressions in Table 12 were estimated to explore these possibilities. Employment per acre of men is the dependent variable in Eq. (1). The negative coefficient of acres shows that employment per acre decreased with size while the positive coefficient of acres squared shows that the savings were greatest in amalgamating small farms. The coefficient of the share of arable is positive, which confirms that male employment was highest on arable farms. All of these coefficients are highly significant. The coefficient of the dummy variable for enclosed farms is tiny and insignificant, which indicates that enclosure played no effect (beyond increasing the share of grass) in reducing employment. The coefficient of the partly enclosed dummy variable is slightly puzzling since it is negative and significantly so, but its magnitude is not particularly large—a 150-acre partly open farm that was 75% arable employed 16% fewer men per acre than an open farm.

The employment patterns of women and boys differed in several respects. Equations (2) and (3) show their employment per acre declined with farm size but at a continuously diminishing rate. The coefficients of acres in both cases were more negative than the coefficient of acres in the regression for men. This result shows the more rapid fall off in the employment of women and boys as farm size increased. In Eq. (2), the coefficient of the share of arable was significant but negative, a result which indicates

TABLE 12
Employment per Acre Regressions

	Equation			
	(1) men	(2) women	(3) boys	(4) total labor
Constant	0.02929 (8.668)	0.03838 (17.728)	0.03592 (14.754)	1.30405 (16.623)
Acres	-9.256×10^{-5} (−5.728)	-1.458×10^{-4} (−14.083)	-1.642×10^{-4} (−14.105)	-4.793×10^{-3} (−12.775)
Acres sq.	9.399×10^{-8} (3.725)	1.578×10^{-7} (9.757)	1.889×10^{-7} (10.389)	5.139×10^{-6} (8.772)
Arab.	0.02370 (8.688)	−0.00375 (−2.149)	0.00082 (0.417)	0.55392 (8.748)
Encl.	-9.541×10^{-4} (−0.377)	1.290×10^{-4} (0.080)	9.099×10^{-4} (−0.498)	−0.03406 (−0.579)
Part	-5.495×10^{-3} (−1.938)	-9.814×10^{-5} (−0.054)	-7.522×10^{-4} (−0.368)	−0.13636 (−2.072)
R^2	0.37084	0.60821	0.56482	0.59728

Notes. Acres sq. acres squared; arab., share of arable; encl., 1 if the farm is enclosed, 0 otherwise; part, 1 if the farm is partly enclosed, 0 otherwise. The dependent variable in Eq. (1)–(3) is the number of people, whereas the dependent variable in Eq. (4) is labor cost in pounds. *t*-ratios are in parentheses.

that female employment rose when arable was converted to pasture. Given the usual sexual division of labor, this increase is not surprising. The magnitude of the increase, however, was small: Enclosing a 150 acre farm that was 75% arable and reducing the share of arable to 25% increased female employment by only 12%. Equation (3) shows that the employment pattern of boys was independent of the share of arable. The enclosure dummy variables are negligible for both women and boys.

Equation (4) shows how overall employment changed with farm size, the conversion of arable to pasture, and enclosure. In this equation total employment equals total labor cost; i.e., the numbers of men, women, and boys are weighted by their wages. The equation shows the same decline in employment with size as was shown in Eqs. (1)–(3). Equation (4) also shows that total employment, in the sense of labor cost, declined as land was converted from arable to grass. The independent effect of enclosure on employment is again negligible, although the lower level of employment in partly open villages compared to either open or fully enclosed villages is again present.

The results deduced here on farm size and employment would have surprised Young, for he thought his data showed large farms increased population. He convinced himself of that by first tabulating the average number of servants, maids, boys, and laborers according to farm size (Young, 1771, IV, p. 246). There is nothing peculiar about this table and if it is analyzed as I have analyzed Young's data it yields the same answer—employment declined with size. Young obtained a contrary result by investigating the population supported by agricultural employment. He assumed that 90% of the laborers were married and had families of five people on average, while five-sixths of the farmers were married and had families of four. (Servants, maids, and boys were unmarried and so had families only of one.) He then computed the total population, including dependents, supported by farms of each size. Since larger farms employed a higher proportion of laborers to servants than small farms, the population supported by agriculture increased with farm size.[16] So he concluded "that the farms most advantageous to population, without exceptions, are those from five hundred acres upwards; and of such, those above a thousand acres are the superior; those under five hundred acres much inferior" (Young, 1771, IV, p. 254).

This conclusion does not contradict our analysis of Young's data, for he was not measuring the relationship between size and employment. Instead, he was confounding that relationship with the form of the employment contract—whether men were hired as day laborers or servants. Young (1771, Vol. IV, pp. 253–254) recognized the arithmetic of this: "great farmers do not keep near the proportion of servants, maids, and boys, that smaller ones do. Their superiority in population lies totally in labourers." He did not, however, distinguish clearly between the employment effects and the population effects of different size farms.[17]

His confusion between employment and population is also apparent by his use of his theory of the capital intensive farmer to explain the positive correlation between population and farm size.

> Great farmers are generally rich farmers; and it requires no great skill in agriculture to know that they who have most money in their pockets will, upon an average, cultivate the soil in the most complete manner; *good* culture, in most cases, is but another word for *much* labour. (Young, 1771, IV, p. 253)

The right test of this theory is to look at aggregate employment per acre as a function of farm size, and, as we have seen, there was a

[16] See Kussmaul (1981) for a discussion of farm service and especially pp. 18–20 for a discussion of Young's data.

[17] Young's conclusion that large farms supported more people becomes less optimistic when it is recognized that the larger population was in greater poverty. Each male servant earned (on average) 171. 9s. 9d per year, while each male laborer earned 181. 8s. 4d. (Young, 1771, IV, p. 356.) The servant, however, had to support only himself while the laborer also had four dependents.

declining relation between them. Indeed, Young should have been suspicious of his capital intensive farmer theory on the basis of the characteristics of the "average" farm he observed on the northern tour. He remarked:

> The article of labour is much below what it ought to be; five men and one boy are too few for cultivating such a farm [287 acres—the average of the tour] in a complete manner, or, indeed, upon the improved system of several counties. . . The husbandry that is conducted with this strength cannot be good; and it is much to be regretted that the average of the kingdom should lie under this deficiency of labour. . . Very few farmers employ the hands they ought. (Young, 1771, Vol. IV, p. 204, 195)

When Young extolled the large, rich farmer who employed many workers, he was not describing "best practice"—let alone "average practice" as minuted on his tours—but instead an ideal that was scarcely if ever realized.[18] Why Young confused his ideal farmer with the ones he met on his tours is a tantalizing question.

The results reported here are also inconsistent with Chambers' (1953) conclusions. As is well known, he claimed that enclosure and 'improved agriculture" led to greater employment. In contrast, our results show that enclosure reduced labor demand when it led to conversion of arable to pasture; otherwise it had no effect. To give Chambers his due, he collected much evidence on the growth of population in Nottinghamshire, and he may be right that population growth was the main source of the industrial work force. But it needs to be emphasized, in view of the great influence of his pronouncements on enclosure and employment, that he gathered no new information on that subject. All he did was repeat the opinions of Arthur Young.

Our findings also differ from those of Snell (1985), a recent critic of Chambers. Snell studied the pattern of seasonal unemployment, as measured by the date of application for poor relief, rather than labor demand as we have done. He found that enclosure in regions where there was considerable conversion to pasture (e.g., Leicestershire and Nottinghamshire) had little impact on male relief applications, while enclosure led to an increase in male winter unemployment relative to male summer unemployment in regions that continued to specialize in grain. This result suggests that enclosure had the biggest seasonal impact on the labor market where it had the smallest effect on farming. However, the male labor demand equation in Table 12 establishes that enclosure reduced employment most in places where it led to widespread conversion to pasture.

[18] It should also be noted that large farms kept no more livestock than small farms, but, indeed, used less capital per acre.

Snell (1985, pp. 15–66, 155–158) discovered that women's wages declined relative to men's after the middle of the 18th century. He attributed the decline to enclosure. Our labor demand functions, however, show enclosure actually improved the employment prospects of women when it led to a shift to pasture, a situation Snell (1985, p. 40–46) also noted. What reduced female labor demand and led to declining women's wages was the pronounced reduction in female employment that resulted from the transition to large farms. This reorganization occurred during the 18th century and accounts for the trends in Snell's wage data.

Farm Size and the Management of Labor

Why did large farms employ less labor than small farms? The answers are simplest for boys and women. The immediate reason is that their total employment increased slowly as farm size increased. Thus, the average arable farm of less than 50 acres employed on average 1.4 maids and 1.7 boys, including family labor. Even very large farms on average did not employ more than 3 of each. As a result, employment per acre of maids and boys declined with size.

The failure to expand the use of women and boys proportionately to increases in farm size is not mysterious. Even though boys were cheap, they were difficult to supervise, so their use was limited on large farms. The employment of maids did not increase much with size because farmers did not increase the size of their dairies a great deal. Dairy cows yielded little profit (Young, 1771, IV, p. 167) and were kept only because they provided a steadier source of cash income than corn. Large farmers made do with little more "insurance" than small farmers. Thus arable farms of less than 50 acres had on average 5 cows while farms of 150–200 acres had 12 cows. Larger farms had only 10–12 cows. Consequently, employment per acre of women declined with farm size.

The employment of men did expand continuously with farm size, but economies were effected on larger farms that reduced cost per acre. Specialization and division of labor were the source of the savings.

On small farms without much hired labor, the farmer had to do all the jobs—ploughing, sowing, harvesting, threshing, carting, hedging, ditching, and tending the sheep, horses, and cattle. Large farms employed specialists to do some of these tasks. This reduced costs in two ways.

First, specialization allowed the employment of less-skilled labor in place of the farmer's more valuable time. The possibilities are suggested by a Rutland wage assessment of 1563. It distinguished three kinds of farm servants:

> A chief servant of husbandry of the best sort, which can eire [plough], sow, mow, thrash, make a rick, thatch, and hedge the same, and can kill and dress a hog, sheep, and calf, may have in wages by the year 40s., and for his livery 6s.
>
> A common servant in husbandry, which can mow, sow, thrash, and load a cart,

> and cannot make a rick, hedge and thatch the same, and cannot kill and dress hog, sheep, and calf, may have in wages by the year 33s. 4d. and for his livery 5s.
>
> A mean servant in husbandry which can drive the plough, pitch the cart, and thrash, and cannot expertly sow, mow, nor make a rick, nor thatch the same, may have for his wages by the year 24s. and for his livery 5s. (Rogers, 1892, IV, p. 120, *V.C.H., Ruts,* I, p. 241)

A farmer of 50 acres approximated a chief servant in husbandry since he had to perform all those tasks himself. A farm with more employees could hire less skilled—and cheaper—people to perform the simpler tasks. On 19th-century farms "plough men are the staple class of labourers."

> His principal skill consists in the management of the plough; and his other and aggregate skill must comprehend a practical knowledge of everything connected with the care of his horses, and with the uses and working of the implements they draw. (Wilson, 1847, p. 234, 874)

The important point about this description is how limited the skills were—the 19th-century ploughman was comparable to the "mean servants" in the 1563 assessment, and their substituion for the more skilled yeoman was one reason for the lower costs of larger farms.[19]

The cow-herd on the 19th-century farm was another example of substituting cheap labor for expensive labor.

> The cow-herd or cattle-man is generally a man past middle life, and sometimes an old and almost superannuated ploughman. He receives considerably lower wages than a ploughman. (Wilson, 1847, p. 234)

Farms had to be large, however, for there to be enough work for this substitution to be feasible. Savings of this sort were imperfectly incorporated into the analysis of Young's data by valuing the farmer and the first servant hired as "first class" servants and other servants as "ordinary" servants.

Second, even though the yeoman had to perform a variety of tasks, he was likely to have been "a jack of all trades and a master of none." Large farms could employ men specialized in single lines of work who could perform it more effectively than the family farmer. Thus large midlands farms employed shepherds to manage the sheep. Hedging was a task performed by specialists on large farms.

> A superior hedger ranks with a master-ploughman, requires to be a person of considerable intelligence, and has charge of planting, pruning, plashing, and repairing hedges, of pruning orchard and forest trees, and of superintending all the other

[19] Reducing the number of small farms and reducing their occupiers to ploughmen amounted to the deskilling of a considerable fraction of the rural population.

shrubby and dendritic plants of the farm. . . Only a farm of large extent. . . employs a superior hedger. (Wilson, 1847, p. 234)

The employment of specialists allowed reductions in employment per acre.

Third, the large farm could employ gangs of men to perform tasks at high efficiency. The corn harvest was often done by gangs. In this case, small farms were not at a disadvantage when mobile gangs of reapers contracted to do that work. Such contractors, however, were not commonly available for mowing hay, which was also done by gangs. In his *General View of . . . Bedfordshire,* Batchelor (1808, p. 110) reported that "Cocking and dragging [required] five men to ten acres. Carrying, ten men at six acres per day. . . ." Indeed, any activity involving the transporting of bulky materials like hay, grain, or manure was done by teams. Thus, of the carriage of corn, Young (1805, p. 423) observed:

In a farm-yard where there are teams enough, carting the wheat crops requires three wagons: one loading in the field, one unloading, and one upon the road going backwards and forwards: five or six horses are sufficient for them, and two men to pitch, two to load, one to drive, and two to unload; in all seven: which make good dispatch.

Batchelor (1808, p. 109) made a similar point about building ricks:

A field of fiteeen acres, when laid on one stack, appears to occupy two men about one day in covering it securely with thatch, and these, together with the four *yelmers* and *servers* cost about 20s. per day.

Later in the winter when the corn was taken from the rick yard to the barn for threshing, Batchelor (1808, p. 111) observed that "six men are employed in this business." Similarly, "the carriage [of manure] occupies either two or three drivers . . . and four men are employed in filling and spreading" (Batchelor 1808, p. 106)

These quotations point to the widespread utilization of gangs on large farms. There was obviously latitude in the size of the work crews, but the numbers mentioned in these examples are not arbitrary. There are 6 to 10 men mentioned, and that was the range of men employed full time on rent maximizing farms. The "specialists" like the ploughmen spent much of the year doing their peculiar task and then worked together in a team during hay-making, corn harvesting, and when produce and manure had to be moved about the farm.

Further confirmation of the efficiency-raising effects of the gang system comes from an unlikely source—Arthur Young's *Political Arithmetic*. In the course of arguing that large farms increased employment, he considered the counter arguments. In this case, he had to admit some merits:

> It is said, that large farms are in fact machines in agriculture, which enable the cultivators of the soil to do that with few hands which before they did with many; resembling a stocking-loom, for instance, which enables the master manufacturer to turn off half his hands, and yet make more stockings than ever. A lively argument but false in almost every particular; indeed the resemblance holds no further than the capacity of performing in some operations much more with ten men in one farm, than with the same number divided among five farms; of which there can be no doubt: But I appeal to all persons conversant in husbandry, if this holds true through one-tenth of the labour of a farm; witness ploughing, harrowing, sowing, digging, mowing, reaping, threshing, hedging, ditching, and an hundred other articles in which one man, separately taken, performs the full tenth of ten men collected. The saving of labour is but in few articles, such as carting hay or corn; carting dung or marle; keeping sheep, &c. (Young, 1774, p. 294)

The economies of scale in carrying bulk commodities are conceded. It should be noted that some of the items, like hedging and ditching, where there were not economies from the gang system were subject to economies through the employment of specialists. The issue is how much labor, in total, was saved by large farms from all of these sources. As Table 12 shows, it was much more substantial than Young was willing to concede.

So in addition to being an agronomist, a large-scale farmer had to administer a complex system of employment. This is a recurring theme in the early 19th-century agricultural handbooks. Thus, Loudon (1831, p. 548) says

> The grand point to be aimed at by . . . the occupier of a large farm, is to hit on the proper number of sub-managers; and to assign each his distinct province, so that the one may never interfer with the other. Having attained this, the next thing is to keep the whole machine in regular action; to keep every man, from the lowest operator to the highest, strictly to his duty.

Orwin and Whetham (1964, p. 82), in describing employment conditions in the middle of the 19th century, observed that

> On large farms the head horseman, the head cowman and the shepherd held positions of real responsibility, often stayed on the same farm for most their lives, and identified themselves completely with its fortunes. Their knowledge of and pride in their charges were such that they would refer to "*my* sheep," "*my* horses" and so on.

The effectiveness of specialization and the work crew system in raising labor productivity depended on the foremen being this committed. In addition, the contemporary farm management literature anticipated 20th-century scientific management in urging farmers to keep time books in which every employees' exact whereabouts and activities were carefully monitored. These records allowed the deployment of labor to be rationally planned. (Loudon, 1831, p. 550). By exploiting the possibilities of an internal division of labor, large farms economized on adult male labor and generated more Ricardian surplus per acre.

Farm Amalgamation and the Growth of Labor Productivity

We can now confront the main question: Why did labor productivity rise in early modern English agriculture? Two developments were working in the same direction. First, English farm sizes increased substantially in the 18th-century, and larger farms employed less labor per acre than smaller farms. Second, English grain yields increased in the early modern period, especially in the 17th century (Allen 1986a). Since the labor requirements of most tasks in cereal production depended on the acreage involved and not on the volume of grain harvested, an increase in yield also caused labor productivity to rise (Parker and Klein, 1966, p. 528). What was the relative importance of these two developments? Were they significant enough to account for the rise in labor productivity as measured by Wrigley?

We can use the details of Young's representative farms to perform an approximate accounting. We distinguish arable from pasture farms. We estimate how output per acre and labor per acre varied with changing crop yields and farm size and then combine those results to simulate labor productivity.

Consider, first, output per acre, i.e., total farm revenue divided by total farm acreage. Table 13 shows the results of computing output per acre using average medieval yields, open field yields and enclosed yields ca. 1800. The values in Table 13 are averages across all farm sizes and thus incorporate the assumption that changes in the farm size structure did not affect yields. Further, the weighting of the changes in yields reflects the cropping patterns and prices ca. 1770. No allowance is made for changes in livestock efficiency.[20]

Next consider labor per acre. Table 14 uses a variant of Eq. (4) in Table 12 to estimate average employment per acre for the distributions of farm size drawn from estate surveys and shown in Table 3. Employment per acre was computed for each range of farm size using the acreage of the midpoint of the range and an overall average was computed by weighting these values with the acreage of land in each range. Table 14 shows a continuous decline in employment per acre in open, arable farms with most of the decline occurring in the 18th century when average farm size increased dramatically. The calculations for enclosed, pasture farms were erratic due to the small number of such farms tabulated for the early 17th and early 18th centuries.

Table 15 combines results from Tables 13 and 14 to estimate labor

[20] For this reason output per acre rises considerably more on arable farms than on pasture farms.

TABLE 13
Output per Acre with Various Corn Yields
(Pounds per Acre)

		ca. 1800	
	Medieval yields	Open yields	Enclosed yields
Pasture farms	2.6654	3.0105	3.1689
Arable farms	2.5451	3.4933	3.9187
Based on the following yields:			
Wheat	10.7	18.6	22.1
Barley	16.8	26.3	32.1
Oats	11.7	30.0	38.5
Beans/peas	10.0	20.4	22.9

Note. The medieval yields are defended in Allen (1986a) and the 1800 yields are from preliminary calculations for the pasture district as discussed there. These yields are quite close to Turner's (1982, p. 500) yields for the grains. As explained in Allen (1986a), the difference between open and enclosed yields ca. 1800 shown in this table probably overstates the usual difference. Using a smaller difference would reduce the already small contribution that enclosure made to the growth in labor productivity. Output per acre equals revenue per acre for arable and pasture farms computed by substituting the assumed yields for the actual yields in the computer program that generated Tables 4 and 5. These calculations thus embody the production patterns and price structures of the arable and pasture farms in Young's sample.

TABLE 14
Labor per Acre with Various Farm Size Distributions
(Pounds per Acre)

	ca. 1600	ca. 1700	ca. 1800
Arable (open)	1.241021	1.174268	0.911618
Pasture (enclosed)	0.421973	0.739729	0.584714

Note: Employment was computed with the following equation:

$$\text{employment} = 1.25928 + 0.54416\, S - 0.00479\, A + 0.51590 \times 10^{-5} A^2,$$

where S is the share of arable and A is farm size. Arable farms were presumed to be 75% arable and pasture farms 20% arable. The distribution of the acreage of open farms in Table 3 was interpreted as the distribution of the acreage of arable farms. Likewise, the distribution of enclosed farms in Table 3 was interpreted as the distribution of pasture farms. Labor per acre was computed for the midpoint of each size category in Table 3. For farms of less than 30 acres, which lay outside the range of farm sizes for which the employment equation was estimated, employment per acre was set to the value for farms of 30–60 acres. The value of employment per acre for farms of 400–500 acres was also assigned to larger farms since the employment equation began increasing above 500 acres, a result that seems spurious on inspecting the data.

TABLE 15
Labor Productivity in Grain Farming

	ca. 1600 open	ca. 1700 open	ca. 1800 open	ca. 1800 enclosed
Output per acre	2.55	3.49	3.49	3.92
Labor per acre	1.24	1.17	0.91	0.91
Output per worker	2.05	2.97	3.83	4.30
Index	1.00	1.45	1.87	2.10

Source. Row 1—Table 13. Values for arable farms are used. Using the ca. 1800 open value for ca. 1700 presumes that open field farmers had accomplished the 1800 yields by 1700. Row 2—Table 14. Values for arable farms are used. This presumes that the labor requirements for grain growing were the same in open and enclosed villages. This assumption is reasonable in view of Eq. (4) in Table 12. Row 3—Row 1 divided by Row 2. Row 4—Row 3 divided by 2.05.

productivity in grain farming over the early modern period. Comparing open field farms ca. 1600 with enclosed farms ca. 1800 shows that labor productivity increased by a factor of 2.1. This is the same order of magnitude as Wrigley's estimate (1.9) of the growth in labor productivity in English farming over the same period. Moreover, Table 15 shows that half the proportional increase occurred in the 17th century. That growth in labor productivity was almost entirely due to the increase in grain yields since farm size increased little. Of the rise in labor productivity in the 18th century, most was due to the amalgamation of farms. Not much was due to yield increases that followed enclosure.

Table 15 embodies many assumptions. The most controversial is probably the assumption that all of the increase in grain yields that occurred in open field villages between the Middle Ages and the 19th century occurred in the 17th century. If some of the increase occurred earlier, the only consequence is that the first value in Table 15 should be given an earlier date. If some of the yield increase occurred in the 18th century, the estimate of labor productivity ca. 1700 should be reduced with the result that more of the growth in labor productivity should be assigned to open field farms in the 18th century. This change would not assign any greater importance to enclosure.

Table 16 extends the calculation of labor productivity to the south midlands as a whole.[21] The evolution of labor productivity in Table 16

[21] This table embodies four more assumptions than the previous table. First, on the basis of subsidiary estimates of enclosure chronology, rough estimates of the acreage of arable and pasture in the south midlands were made for 1600, 1700, and 1800. These estimates presumed that enclosure before 1800 was to convert arable to pasture, while enclosure after 1800 was to improve arable farming. (See Allen, 1986b, for a graph of the chronology of enclosure in the south midlands that distinguishes districts where the aim

TABLE 16
Labor Productivity in South Midlands Farming

	ca. 1600	ca. 1700	ca. 1800
Output per acre	2.57	3.31	3.37
Labor per acre	1.13	0.96	0.71
Output per worker	2.27	3.44	4.75
Index	1.00	1.52	2.09

Source. The ca. 1600 value is a weighted average of the values shown in Table 13 where the weights are estimates of the shares of land open and enclosed. The ca. 1700 value is similar except that it uses the ca. 1800 open values. This calculation presumes, therefore, that open field farms had realized the 1800 yields by 1700. The ca. 1800 value is a weighted average of the ca. 1800 open, enclosed, and pastoral values in Table 13, where the weights are estimated shares of open field land, enclosed land devoted mainly to arable farming, and enclosed land converted to pasture. Row 2—Each entry is a weighted average of the corresponding value of labor per acre in open fields shown in Table 14 and 0.58714, the ca. 1800 value of labor per acre, which was assumed to apply to ca. 1600 and ca. 1700 for which the underlying samples were quite small and erratic (see Table 4). The weights are estimated proportions of open field and enclosed land in 1600 and 1700. In 1800 the open field labor value is weighted by the estimated share of open field land plus enclosed land devoted mainly to arable farming. The 1800 enclosed labor per acre value is weighted by the estimated share of land that was enclosed and converted to pasture. Row 3—Row 1 divided by Row 2. Row 4—Row 3 divided by 2.27.

is not very different from the history in Table 15. The overall increase in productivity was, again, similar to Wrigley's estimate for English farming as a whole. Productivity increased in about equal proportional amounts in both the 17th and 18th centuries.

Table 15 suggests a reconciliation between English and continental history. Farmers in Belgium and northern France realized crop yields ca. 1800 of the same order as English farmers, but farms in Belgium and France were much smaller than English farms. Table 15 indicates that if the size distribution of continental farms was similar to the distribution of English farms ca. 1700—before the transition to capitalist agriculture—then Belgian and French labor productivity should have been two-thirds of English productivity early in the 19th century. And that was indeed the size of the gap. Moreover, Russian farmers in the 19th century reaped yields like English medieval yields and had small, peasant farms, so one might expect labor productivity in 19th-century Russian agriculture to have been of the same order was English medieval productivity. Indeed,

of enclosure was conversion to pasture from districts where the aim was to improve arable farming.) Second, employment per acre in pasture farming was assumed to equal the 1800 value shown in Table 15 for all years. Third, the three values shown in Table 14 for output per acre in pasture farming were assumed to apply to 1600, 1700, and 1800. Fourth, output per acre in arable farming was taken from Table 16, and the 1800 values were weighted roughly in proportion to the acreages of open and enclosed arable land at that date.

Bairoch found Russian productivity to have been 40% of English productivity, while Table 15 indicates that medieval English labor productivity was 48% of 19th-century English productivity. These rough calculations suggest that our analysis of the growth in English agricultural labor productivity also explains the international differences in productivity at the beginning of the 19th century.

Capitalist Agriculture and English Economic Development

There are two theories about the effect of enclosure and large farms on employment and production in English agriculture.

Arthur Young's capital intensive farmer theory, through its restatement by Chambers, is the most influential. It contends that enclosures and large farms increased employment and output. The latter increased more, so labor productivity also rose. In Allen (1986a) I measured the impact of enclosures and large farms on crop yields to test the predictions about output. They were false. Indeed, the increase in English corn yields in the 17th and 18th centuries was mainly accomplished by open field farmers before the general shift to very large farms in the 18th century. In this paper, we have examined the predictions about employment. Contrary to Young's theory, employment declined with farm size. Enclosure also reduced employment when it led to conversion to pasture; otherwise it had no effect. This should not have been a surprise to Young since he had observed that employment was generally less than his theory predicted. Since there is no evidence that supports Young's capital intensive farmer theory and much evidence opposed, historians ought to abandon that theory.

The second theory about the effects of large farms and enclosures derives from the critics of enclosure who argued that it depopulated the countryside. Before the 18th century, enclosure did lower population or stifle its increase, but after 1700 it no longer had that effect (Allen 1986b). Throughout the 18th century, however, enclosure and the increase in farm size reduced agricultural employment. That reduction raised labor productivity above continental levels.

We can conceive of productivity growth in early modern English agriculture as a two part development. First was the rise in corn yields in the 17th century. Enclosure and capitalist agriculture made no contribution to this advance, which also raised labor productivity. Second was the further rise in labor productivity in the 18th century due to the shift to large farms. The release of labor from agriculture was the *sole* contribution of large farms to England's economic development.

REFERENCES

Addington, S. (1772), *An Enquiry into the Reasons for and Against Inclosing Open Fields*. Coventry: J. W. Piercy.

Allen, R. C. (1982), "The Efficiency and Distributional Consequences of Eighteenth Century Enclosures." *Economic Journal* **XCII,** 937–53.

Allen, R. C. (1986a), *Enclosure, Capitalist Agriculture and the Growth in Corn Yields in Early Modern England.* U.B.C. Department of Economics Discussion Paper 86–39.

Allen, R. C. (1986b), *Enclosure, Depopulation, and Inequality in the South Midlands, 1377–1801.* U.B.C. Department of Economics Discussion Paper 86–36.

Allen, R. C., and Ó Gráda, C. (1988), *On the Road Again with Arthur Young: English, Irish, and French Agriculture during the Industrial Revolution. Journal of Economic History* **48,** in press.

Bairoch, P. (1965), "Niveau de développement économique de 1810 à 1910." *Annales: Économics, Sociétés, Civilisations* **20,** 1091–1117.

Batchelor, T. (1808), *General View of the Agriculture of the County of Bedford.* London: Phillips.

Beckett, J. V. (1984), "The Pattern of Landownership in England and Wales, 1660–1880." *Economic History Review* 2nd series, **37,** 1–22.

Beresford, M. (1961), "Habitation versus Improvement: The Debate on Enclosure and Agreement." In F. J. Fisher (Ed.), *Essays in the Economic and Social History of Tudor and Stuart England.* Cambridge: Cambridge Univ. Press.

Chambers, J. D. (1953), "Enclosure and Labour Supply in the Industrial Revolution." *Economic History Review* 2nd series, **5,** 319–343.

Chorley, P. (1981), "The Agricultural Revolution in Northern Europe, 1750–1880: Nitrogen, Legumes, and Crop Productivity." *Economic History Review* 2nd series, **34,** 71–93.

Cooper, J. P. (1967), "The Social Distribution of Land and Men in England, 1436–1700." *Economic History Review* 2nd series, **20,** 419–440.

Crafts, N. F. R. (1985), *British Economic Growth during the Industrial Revolution.* Oxford: Clarendon.

Davies, D. (1795), *The Case of Labourers in Husbandry Stated and Explained.* Bath and London: C. G. & J. Robinson.

Dyer, C. (1980), *Lords and Peasants in a Changing Society.* Cambridge: Cambridge Univ. Press.

Dyer, C. (1981), *Warwickshire Farming, 1349–c. 1520.* Oxford: Dugdale Society Occasional Paper 27.

Finch, M. E. (1956), *The Wealth of Five Northamptonshire Families, 1540–1640.* Oxford: Northamptonshire Record Society. **XIX.**

Fitzherbert (1539), "Surveyenqe." In R. H. Tawney and E. Power (Eds.), *Tudor Economic Documents,* 1924. London: Longmans, Green. Vol. III, pp. 22–25.

Gray, H. L. (1910), "Yeoman Farming in Oxfordshire from the Sixteenth Century to the Nineteenth." *Quarterly Journal of Economics* **24,** 293–326.

Gray, H. L. (1915), *English Field Systems.* Cambridge, MA: Harvard Univ. Press.

Habakkuk, H. J. (1965), "La disparition du paysan anglais." *Annales: Économies, Sociétés, Civilisations* **20,** 649–663.

Harrison, C. J. (1979), "Elizabethan Village Surveys: A Comment." *Agricultural History Review* **27,** 82–89.

Harvey, B. (1977), *Westminster Abbey and Its Estates in the Middle Ages.* Oxford: Clarendon.

Havinden, M. A. (1961), *The Rural Economy of Oxfordshire, 1580–1730.* Oxford, B. Litt. Thesis.

Holmes, G. S. (1977), "Gregory King and the Social Structure of Preindustrial England." *Transactions of the Royal Historical Society* 5th series, **27,** 41–68.

Hoskins, W. G. (1950), The Leicestershire Farmer in the Sixteenth Century." In W. G. Hoskins (Ed.), *Essays in Leicestershire History.* Liverpool: University Press.

Hoskins, W. G. (1951), "The Leicestershire Farmer in the Seventeenth Century." In W. G. Hoskins (Ed.), *Provincial England,* 1963. London: Macmillan & Co. Pp. 149–169.
Johnson, A. H. (1909), *The Disappearance of the Small Landowner*. Oxford: Clarendon.
Kosminski (1956), *Studies in the Agrarian History of England in the Thirteenth Century*. Oxford: Basil Blackwell.
Kussmaul, A. (1981), *Servants in Husbandry in Early Modern England*. Cambridge: Cambridge Univ. Press.
Lennard, R. (1916), "Rural Northamptonshire Under the Commonwealth." In P. Vinogradoff (Ed.), *Oxford Studies in Social and Legal History*. Oxford: Clarendon. Vol. 5.
Loudon, J. C. (1831), *An Encyclopedia of Agriculture*. London: Longman, Rees, Orme, Brown, and Green. 2nd ed.
Mingay, G. E. (1962), "The Size of Farms in the Eighteenth Century." *Economic History Review* 2nd series, **14,** 469–488.
Mingay, G. E. (1968), *Enclosure and the Small Farmer in the Age of the Industrial Revolution*. London: Macmillan & Co.
O'Brien, P. K., and Keyder, C. (1978), *Economic Growth in Britain and France, 1780–1914*. London: George, Allen & Unwin.
Orwin, C. S., and Whetham, E. H. (1964), *History of British Agriculture, 1846–1914*. London: Longmans.
Parker, W. N., and Klein, J. L. V. (1966), "Productivity Growth in Grain Production in the United States, 1840–60 and 1900–10." In *Output, Employment, and Productivity in the United States after 1800*. New York: National Bureau of Economic Research, Studies in Income and Wealth. Vol. 30, pp. 523–580.
Pinchbeck, I. (1930), *Women Workers and the Industrial Revolution, 1750–1850*. London: Routledge.
Postan, M. M. (1972), *The Medieval Economy and Society*. London: Weidenfel & Nicolson.
Price, R. (1792), *Observations on Reversionary Payments*. London: T. Cadell. 5th ed.
Ricardo, D. (1817), *On the Principles of Political Economy and Taxation*. 1st ed. In P. Sraffa (Ed.), *The Works and Correspondence of David Ricardo,* Cambridge: Cambridge Univ. Press, Vol. 1, 1951.
Rogers, J. E. T. (1866–1902), *A History of Agriculture and Prices in England*. Oxford: Clarendon.
Slicher van Bath, B. H. (1963), *Yield Ratios, 1810–1820*. A. A. G. Bijdragen. Vol. 10.
Snell, K. D. M. (1985), *Annals of the Labouring Poor*. Cambridge: Cambridge Univ. Press.
Tawney, R. H. (1912), *The Agrarian Problem in the Sixteenth Century*. London: Longmans, Green.
Tawney, R. H., and Power, E. (1924), *Tudor Economic Documents*. London: Longmans, Green.
Thompson, F. M. L. (1966), "The Social Distribution of Landed Property in England in the Sixteenth Century." *Economic History Review* 2nd series, **19,** 505–517.
Thompson, F. M. L. (1969), "Landownership and Economic Growth in England in the Eighteenth Century." In E. L. Jones and S. J. Woolf (Eds.), *Agrarian Change and Economic Development*. London: Methuen. Pp. 41–60.
Titow, J. Z. (1969), *English Rural Society 1200–1350*. London: Allen & Unwin.
Turner, M. (1982), "Agricultural Productivity in England in the Eighteenth Century: Evidence from Crop Yields." *Economic History Review* 2nd series, **35,** 489–510.
Turner, M., and Mills, D, (1986), *Land and Property: The English Land Tax 1692–1832*. New York: St. Martin's Press.
Van Zanden, J. J. (1985), *De economische ontwikkeling van de Nederlandse landbouw in de negentiende eeuw, 1800–1914. A. A. G. Bijdragen* **25,** Wageningen.
Wilson, J. M. (1847), *The Rural Encyclopedia*. Edinburgh: A. Fullarton.

Wordie, J. R. (1974), "Social Change on the Leveson-Gower Estates." *Economic History Review* 2nd series, **27**, 593–609.
Wrigley, E. A. (1985), "Urban Growth and Agricultural Change: England and the Continent in the Early Modern Period." *Journal of Interdisciplinary History* **XV**, 683–728.
Young, A. (1770), *The Farmer's Guide in Hiring and Stocking Farms*. London: W. Strahan.
Young, A. (1771), *A Six Month's Tour Through the North of England*. London: W. Strahan, 2nd ed.
Young, A. (1774), *Political Arithmetic*. London: W. Nichol.
Young, A. (1805), *The Farmer's Kalendar*. London: Richard Phillips.

Harvest Technology and Labour Supply in Britain, 1790-1870[1]

By E. J. T. COLLINS

> "Manual labour is superior to any application of unconscious power, on account of its intelligence, as shown in selection, variation in force, delicacy of touch, and adaptation to every minute difference in the material acted upon... It is therefore well to pause before substituting machines, when men are available and wages are not too high."
>
> J. WRIGHTSON (*c.*1910)

IMPROVEMENTS in output per worker in farm operations such as ploughing and harvesting are nowadays usually associated with mechanization. Historically, however, it is doubtful whether economic growth in agriculture has been at all closely correlated with either machines or other high-cost technological inputs. In Britain and Japan the most successful developments in raising farm output were secured initially via the gradual penetration of inexpensive biological innovations. As late as 1870 roots and legumes were still the bases of high-output farming over much of Britain and most operations were then still performed using labour-intensive hand-tool methods.

The lesson for developing economies might be that cheap and reproducible biological inputs plus an efficient deployment of manpower are likely to prove more effective engines of growth than expensive machines. Yet costly and sophisticated technologies are often recommended, as the panacea for under-development, to poor peasant agricultures with small reserves of capital and large and increasing reserves of underemployed labour.[2] This does not argue that the machine has no place in such economies, nor that there does not exist within them a need for saving labour. On the contrary it has often been observed how, even under extreme conditions of seasonal underemployment, the peak activity periods of the farming year are characterized by serious labour shortage.[3]

Thus it is that, both historically and at the present day, the small grains harvest has created the exceptional demand for labour which has fully drained the local pool and has perhaps required the importation of many temporary workers. In

[1] Version of a paper read at the British Agricultural History Society Conference, Norwich, 1968. I am indebted to Prof. A. W. Coats, C. A. Jewell, and Dr E. L. Jones for their many helpful comments on earlier drafts of this article. In another article, 'Labour Supply and Demand in European Agriculture, 1800–1880', in E. L. Jones and S. J. Woolf, eds. *Agrarian Change in Economic Development, the Historical Problems* (1969), I have carried this analysis a stage further to include other areas of Europe and to examine in more detail the role of labour in proto-industrial agricultures.

[2] See, for example, R. Bradfield, *Towards More and Better Food for the Filipino People* (Agricultural Development Council, New York, 1966); *A Report on Agriculture and Agricultural Engineering in China* (Chicago, 1949) (cf. J. L. Buck, 'Farm Implements and Machinery for China', *Economic Facts*, L (1945), 757–66); B. Higgins, *Economic Developments: Problems and Policies*, p. 344; *Implement and Tractor*, LXXX, no. 18 (1965), 68; *World Crops*, Oct. 1967, pp. 10–11; *Marchés Tropicaux et Méditerranéens*, 17 Feb. 1968, p. 325.

[3] *Bulletin Agricole du Congo Belge* (1960), pp. 603–15; *Agricultural Development in Nigeria, 1965–1980* (F.A.O., Rome, 1966), pp. 205–6; P. Leurquin, *Rice in Colombia*, Ford Foundation Institute Studies, VII (1967), 266–9, 284.

seventeenth-century England interregional harvest labour flows were already making good local deficiencies. Migrations on a grander scale were observed by Redford in the Balkans during the First World War. Today they are integral to many agricultures in the Near East, the Mediterranean, and South America.[1]

It has been asked how low-income agricultures are to offset seasonal labour shortages when perhaps only marginal factor savings are required, and when the more drastic remedy of mechanization may not only reduce employment opportunities in a key earning period but also, more seriously, lead to massive redundancies which the non-farm sector cannot absorb. American experience in the inter-war years confirms that even in highly industrialized countries the social costs of over-rapid farm mechanization can be uncomfortably high.[2]

One answer may lie in the development of an "intermediate" or "middle-way" technology of improved hand-tool methods or simple man- and animal-powered machines which, at lower cost and with less social disruption, can provide acceptable gains in work output commensurate with the necessarily slow growth of capital investment in urban housing and employment. The suggestion is by no means a new one. But the curious fact remains that its possibilities have so far been little exploited.[3]

The primary concern of this article, therefore, is to demonstrate from historical experience, that of the British small grains harvest in the period 1790–1870, that in one proto-industrial agriculture the demand for labour-saving methods was successfully met by an intermediate technology, developed within or alongside an existing framework of traditional hand-tool techniques.

I

Customarily it is assumed that within recent times no changes of consequence occurred in harvesting technique in Britain before the arrival of American-type reaping machines early in the third quarter of the nineteenth century. Thus has been echoed the nineteenth-century view that agricultural progress was dependent on the rapid closing of the technological gap between the farm and factory, with its predilection to measure progress by the most conspicuous indicators, of which machinery was among the most important. By the 1860's the railway imagery in the shape of steam-ploughs and reaping machines had come to dominate the agricultural imagination. "We must all appreciate steam power," insisted Macdonald, "there is nothing like it."[4] "Unquestionably one of the greatest benefits which the application of Mechanics to farm purposes

[1] A. Everitt, 'Farm Labourers', in Joan Thirsk, ed. *Agrarian History of England and Wales*, IV (Cambridge, 1967), 434. For earlier examples see G. C. Homans, *English Villages of the Thirteenth Century* (New York, 1960), p. 136. For later and present-day examples: J. P. Arles, 'Manpower Mobilization and Economic Growth: an Assessment of Moroccan and Tunisian Experience', *International Labour Review*, XCIV, no. 1 (1966), 3; A. A. Pepelasis and P. A. Yotopoulos, *Surplus Labour in Greek Agriculture, 1953–1960* (Athens, 1962), p. 158; A. Redford, *Labour Migration in England, 1800–1850* (2nd edn, Manchester, 1964), XIII, 4–6; J. Tack, *Die Hollandsgänger in Hannover und Oldenburg* (Leipzig, 1902); *Contrasts in Emerging Societies* (1965), ed. D. Warriner, pp. 16, 247–8, 264–5, 308; E. Wolf, *Peasants* (New Jersey, 1966), p. 46.

[2] C. McWilliams, *Ill Fares the Land* (1945 edn). Between 1926 and 1933 it was estimated that 150,000 workers had been displaced by combine harvesters alone (ibid. pp. 82–6 and *passim*).

[3] There are a few important exceptions, however, in particular H. J. Hopfen, *Farm Implements for Arid and Tropical Regions* (F.A.O., Rome, 1960).

[4] D. G. F. Macdonald, *Hints on Farming* (1868 edn), p. 53.

has conferred upon agriculture" was how Prof. J. Wilson described the reaping machine.[1]

On the other hand the neglect by the more influential contemporary agricultural writers of routine hand-tool processes has implied that this constituted the least dynamic sector of British agriculture. Nor have historians of farm mechanization studied hand-tool technologies in any depth.[2] Such attitudes towards the routine, so-called "traditional", farm practices have led them to overlook the possibility that, between different hand-tool methods, there may have been important efficiency differentials. One extreme view of inertia would have us believe that in Britain, "sickle, scythe and hook existed side by side from antiquity which would seem to negative the idea that one was better than the other", and that "most often the machine took over from each in its place".[3] Many, no doubt, would disagree with I. P. Roberts, the American historian, in believing that the cradle scythe compared with the sickle was a more significant step forward than was the reaping machine compared with the cradle scythe.[4]

Thus it has been assumed that prior to the reaping machine farmers were technically constrained from introducing greater efficiency into the harvest field. But, this apart, the widely held belief that up to the early 1870's farm labour was oversupplied presupposes that previously there had been little incentive to labour-saving innovation. Alternatively, the introduction of the reaping machine after 1850 has been seen as the predictable response to an initial but not too drastic drying up of the harvest labour market.[5]

Against these assumptions must be set the following facts. First, that as late as 1870 much the greater part of the British corn harvest was still cut by hand tools, even though other farm operations such as threshing and livestock feed preparation, which mostly occupied the off-peak periods of the farming year, were already highly mechanized.[6] And second, that if there was a failure to initiate change within harvest hand-tool technology in Britain, this was in remarkable contrast to much of western Europe, Scandinavia, and the United States where between 1750 and 1850 there was a massive substitution of the higher working scythe for the sickle as the standard corn-harvesting tool.[7]

This article will go on to advance the counter-arguments: first, that between 1790 and 1870 there occurred in Britain an important transformation in hand-harvesting technology, secured for the most part through the intermediate technology of improved hand-tool methods; second, that this was primarily a response to a deteriorating harvest labour market; and third, that for some time

[1] J. Wilson, *Our Farm Crops* (1859), I, 55.

[2] Leo Rogin, *The Introduction of Farm Machinery* (Univ. of California, 1931), being the outstanding exception. The most detailed ethnographic historical survey of harvest hand tools at present available is Axel Steensberg, *Ancient Harvesting Implements* (Copenhagen, 1943).

[3] F. G. Payne, 'The Retention of Simple Agricultural Techniques', *Gwerin*, II (1959), 129–31.

[4] Cited in Earle D. Ross, 'Retardation in Farm Technology before the Power Age', *Agricultural History*, XXX (1956), 11.

[5] C. S. Orwin and E. H. Whetham, *British Agriculture, 1846–1914* (1964), p. 113; J. D. Chambers and G. E. Mingay, *The Agricultural Revolution, 1750–1850* (1966), pp. 72, 190.

[6] Contemporary estimates suggest that as late as 1871 barely one-quarter of the British corn area was mechanized.—P. A. David, 'Mechanization in Nineteenth Century Agriculture—Britain and America' (unpublished paper read at the Centre for the Advanced Study of Italian Society, Feb. 1968).

[7] The European and North American evidence is summarized in Collins, loc. cit. *passim*.

after 1850 many farmers preferred to deepen their investment in this range of seemingly obsolescent techniques rather than to exploit the more worth-while factor-saving alternative of mechanization.

II

In 1790 the national stock of harvesting techniques was essentially that which obtained in the sixteenth century.[1] Almost everywhere wheat was hand reaped with the toothed sickle or more occasionally with the smooth reap-hook. Reaping high, a foot or so from the ground so as to leave a long stubble, was the common mode over much of southern and eastern England, particularly on the heavier soils, while elsewhere reaping low close to the ground mostly prevailed. In southern Britain, south of a line from Chester to York, barley and oats were generally mown with the scythe, but in northern Britain and Ireland these crops were reaped, the scythe being applied only to grass and forage crops. Most cereals were cut at the dead-ripe stage, early cutting being practised in only a few localities such as Middlesex and Wiltshire.[2] The *status quo* was first threatened in the mid-eighteenth century, at about which time the practice of "bagging" with the heavy hook began to replace hand reaping in the London area, Devon, and the English Marcher counties, the smooth reap-hook began to supersede the serrated sickle, while in the East Riding the scythe was adapted into a wheat-harvesting tool.[3] Yet by 1790 the impact of these innovations was confined to just a few districts of southern Britain. The corn scythe, which had become established in southern England in the medieval period, had by this late stage failed to penetrate much beyond the Humber.

The chief features of the revolution in hand-harvesting methods which subsequently was to take place were, first and most important, a switch from low into higher working capacity tools; second, improved hand-tool designs and constructions; and, third, earlier cutting of grain.

The changes in methods can be summarized as follows: in southern Britain mowing or bagging replaced hand reaping as the standard procedure for wheat;

[1] Compare the following account with the descriptions of late medieval and sixteenth-century harvesting methods described in 'Fitzherbert's Book of Husbandry' in *Ancient Tracts concerning the Management of Landed Property* (1747), pp. 27–8; Thomas Tusser, *Five Hundred Pointes of Good Husbandrie* (English Dialect Society, 1878), p. 37; Lord Ernle, *English Farming Past and Present* (3rd edn, 1922), p. 12. The distinguishing characteristics of the harvesting methods available in the period under review are outlined as follows.

(*a*) *Hand reaping*. Performed with the serrated-edged sickle or the slightly heavier and broader smooth-edged reap-hook (weight 8–15 oz., blade length 15–24 in.). From a kneeling or stooping position the reaper would clutch a handful of straw, cut with a sawing or drawing action, and lay each cut on to a ready-prepared straw band until enough had accumulated to form a sheaf.

(*b*) *Bagging*. Performed with a large and heavy smooth-edged hook (weight 2–4 lb., blade length 20–30 in.). With the help of a stick, or second hook, or "small rake" to tension the corn the bagger would cut the standing corn by a series of low slashing strokes. Sometimes he would do his own gathering and binding, but more usually this would be undertaken by followers.

(*c*) *Mowing*. Performed with the scythe to which was often attached a bow (a looped length of twig or wire) or a cradle (a wooden fork-like construction) to assist in the even laying down of the cut. Gathering and binding were invariably done by specialist followers.

[2] J. Middleton, *General View ... Middlesex* (1807), pp. 216, 241. T. Davies, *General View ... Wiltshire* (1811), p. 66.

[3] In 1771 mowing wheat was described as "a new method invented to save labour [and] a subject much discussed within a few years and with great warmth".—Mr Stillingfleet, *The Farmers Kalendar* (1771), pp. 234–6. See also D. Henry, *The Complete English Farmer* (1771), pp. 216–18.

in northern Britain and Ireland mowing displaced reaping of all grains; while, as an interim measure, and over a broad front, low reaping replaced high reaping and the serrated sickle gave way to the smooth reap-hook.

Regional and local chronologies of change are inevitably complex, but for present purposes we can distinguish four more or less discrete phases of innovating activity: 1790–1815, 1815–35, 1835–70, and 1870–90.[1]

The period 1790–1815 may be viewed as the experimental and "early adoption" phase of the revolution, characterized by a sharply increasing rate of innovation and, through the new farming literature, the wider dissemination of previously little-known techniques. "Formerly," it was said, "provincial practices seemed to be entirely locomotive; they continued unnoticed where they began . . . now a mutual adoption of each other's practices takes place to the greater expedition of harvest labour."[2] Nevertheless, change was selective and regionalized, but although only a small minority of farmers broke sufficiently with tradition to introduce new methods, many more had progressed from a point of indifference to one of mild awareness. Indicative of the anxiety in some quarters to widen the range of available techniques was the attempt by a local agricultural society in Hampshire in 1814 to popularize the Hainault scythe, Flemish prisoners of war being specially recruited to instruct local workmen in its use.[3]

The second phase, 1815–35, was almost one of hiatus. Exceptional activity in the east Midlands and in north-east Scotland, plus a rapid adoption of the reap-hook in southern Scotland, combined to maintain the overall rate of innovation at wartime levels.[4] Elsewhere inertia mostly prevailed while, in some areas, there was even a tendency for the improved wartime techniques to be abandoned in favour of slower, more traditional, methods.[5] The case for hiatus is supported most strongly, however, by later retrospective testimony which insists that the hand-tool revolution gained decisive momentum only from the mid-1830's.[6]

The years 1835–70 constituted the most active or, as might be termed, the "majority adoption" phase of the revolution. Interest in the heavy tools revived suddenly from about 1835, and already by the mid-1840's rapid change had been reported over a broad front.[7] From the later 1850's it was becoming almost

[1] The following section pretends only to provide a brief summary of the nature and chronology of harvesting change during the period 1790–1870. For a rather more detailed and better documented account see Collins, loc. cit.

[2] W. Davies, *General View . . . South Wales* (1815), pp. 1, 427.

[3] *Farmer's Magazine* (hereafter referred to as *F.M.*), Aug. 1825, p. 340; *Minute Books of the Christchurch Agricultural Society*, in the hands of the South Avon and Stour Agricultural Society.

[4] *British Farmer's Magazine*, Oct. 1840, pp. 350–2; *British Husbandry* (1837), II, 188; Rev. Farquharson, 'On Cutting Grain Crops with the Common Scythe', *Prize Essays and Transactions of the Highland Society*, new ser. IV (1835).

[5] In the Whitehaven area of Cumberland the method of mowing wheat introduced by Welsh militiamen during the war was abandoned on their withdrawal.—W. Dickinson, 'The Farming of Cumberland', *Journal of the Royal Agricultural Society of England* (hereafter referred to as *J.R.A.S.E.*), XIII (1852), 232. In Devon in the 1820's the sickle was displacing the heavy "hewing hook".—T. Moore, *The History of Devonshire* (1829), I, 429.

[6] E.g. "Only within the last few years that this system [mowing wheat] has received anything like general acclaim. . ."—*F.M.* Sept. 1845, p. 213.

[7] R. Baker, 'On the Farming of Essex', *J.R.A.S.E.* V (1844), 28–9; G. Buckland, 'On the Farming of Kent', ibid. VI (1845), 226, 261; C. S. Read, 'On the Farming of South Wales', ibid. X (1849), 133; M. M. Milburn, 'On the Farming of the North Riding of Yorkshire', ibid. IX (1848), 501; *British Husbandry*, op. cit. II, 189; Henry Stephens, *Book of the Farm* (1844 edn), III, 1080; *F.M.* Sept. 1840, pp. 163–4, Oct. 1841, pp. 280–1, Dec. 1841, p. 453, Aug. 1845, pp. 100–1.

a matter of pride to be able to report the extinction of time-honoured practices, the sickle in one case being described as so obsolete that it might as well be "hung up in the British Museum". Over large areas of southern Britain the sickle had by the 1870's been relegated to smallholdings and "common yards and allotments".[1] The most significant feature of this phase of the revolution was that between 1850 and the mid-1860's the bagging-hook and scythe appear to have gained ground as fast as, if not faster than, the reaping machine. In many areas of Britain large-scale mechanization did not even begin until the early 1870's.

In 1870 the hand-tool revolution was complete over much of arable Britain, the conspicuous laggards being confined to the midland clay vales, the lower Cotswolds, the Border Counties, and the north and west of Ireland. In these areas, and in Ireland in particular, change went on through and beyond the 1870's,[2] although the rapid introduction of reaping machinery in the early 1870's may sometimes have pre-empted a really large-scale adoption of the scythe and bagging-hook. Yet where mechanization took place it was frequently only partial, with the result that new hand methods were subsequently introduced on to non-mechanized sections of the crop.[3] Significantly, the Irish scythesman was to have his day in England where previously he had handled only the sickle.[4]

The evidence does not permit accurate quantification of long-run changes by affected crop areas. Yet given the assumptions, first, that the national corn area and regional crop distributions in 1790 were roughly similar to those of 1870 and, second, that prior to 1870 only a small proportion, say less than 10 per cent, of the national grain area was affected by direct switching from sickle to reaping machine, then it can be calculated that between these two dates new hand-tool methods had at some time or other been introduced on to upwards of five million acres.[5] Whereas in 1790 perhaps 90 per cent of wheat, 75 per cent

[1] *F.M.* Aug. 1869, p. 122; J. G. Cornish, *Reminiscences of Country Life* (1939), pp. 82–3; G. E. Evans, *Ask the Fellows who Cut the Hay* (1956), pp. 117–18. By 1900 the sickle had more or less disappeared from the folk memory, although in the more remote parts of Ireland it is still in use today. In 1905 one of Sturt's informants told him of a tool called the sickle which he had heard of but never seen and of the reap-hook which he had used but very occasionally. At this stage the bagging hook was still common in west Surrey.—K. Mackerness, ed. *The Journals of George Sturt* (Cambridge, 1967), II, 455–6.

[2] "As late as 1868, when the third edition of this work was being prepared for the press, the sickle was considered to be still employed so extensively as to warrant the retention of the detailed account given in former editions of the manner of reaping with the sickle. Since that time, however, the work of the harvest has been completely revolutionized, in most parts the scythe supplanted the sickle and now the reaping machine has driven both into disuse."—Stephens, op. cit. (1891 edn), III, 56; E. Estyn Evans, *Irish Folk Ways* (1957), pp. 155–66; A. W. Boyd, *A Country Parish* (1951), p. 41; J. Wrightson, 'Comparative Economy of Different Methods of Harvesting Corn Crops', *J.R.A.S.E.* LXVII (1906), 99–100.

[3] In some areas the wheat scythe was first used to make "roadways" around the field for the reaping machine. This practice was general up to the coming of the combine harvester.

[4] E. E. Evans, op. cit. pp. 155–61.

[5] The method of computation is, briefly, as follows. Areas under crop for each county were obtained from *Agricultural Statistics, 1870*. In the absence of adequate quantitative data, crop/method distributions were based on qualitative evidence from the literary sources, i.e. *General Views* and *Statistical Accounts*, *et al.* (for 1790); the national survey of harvesting methods in *Gardener's Chronicle and Agricultural Gazette*, 17, 24 Aug. 1867 (hereafter referred to as *Ag. Gaz.*); *J.R.A.S.E.* Prize essays (1840–80), *Ag. Gaz.*, *F.M.*, *Mark Lane Express*, *et al.* (for 1870). The shortcomings of this method are obvious yet the degree of error is probably not large. In most cases it has been possible to narrow considerably the areas of uncertainty. For example, it can be taken that in 1790 the sickle was the exclusive tool in northern Britain and Ireland, that in southern Britain wheat was invariably reaped and spring corn mown, with the deviant areas small and easily pinpointed. Similar sorts of assumptions can be made for 1870, and although the overall pattern

of oats, and 25 per cent of barley were reaped with small tools, by 1870 the respective proportions had fallen to 20 per cent, 40 per cent, and 10 per cent. The stream of technological change was in fact much greater than these values suggest. Over large areas of southern England, Scotland, and Northern Ireland the reap-hook had replaced the sickle before 1850[1] while on almost a majority of farms there had occurred not one but a succession of changes involving three or more tools; the commonest progressions being sickle→reap-hook→scythe, and sickle→reap-hook→bagging-hook.[2]

Changes in hand-tool methods were complemented by a number of important modifications to hand-tool design and construction. Between 1830 and 1850 several improved tool forms appeared on the edge-tool market. New thick-back and ridged designs of hook and sickle were developed during the 1830's. Serratures were made finer and blades became broader, heavier, and more rigid—developments which did much to standardize the design of the bagging-hook, which was a tool requiring above all a well-defined forward balance and a heavily weighted blade to do its horizontal slashing job efficiently.[3] In about 1840 Sorby of Sheffield secured a major breakthrough with his cast-steel swedged blades,[4] which combined lightness with greater blade strength and provided the smooth hook with a more durable edge requiring less frequent sharpening in the field.

Equally important advances were made in scythe-making technology. Some time before 1835 manufacturers had modified traditional designs to meet the demand for a blade to cut grains which was more robust, heavier, and stiffer than that of the grass scythe.[5] The demand was better served by the appearance in 1840 of the "patent" riveted scythe blade while, soon after, cast steel replaced blister steel in the construction of the cutting edge.[6] There were parallel improvements in sned (handle) design. The straight or poled sned was often preferred to the conventional S-shape for corn,[7] but rather more important was the short two-helved sned, a specialist cradle-scythe mounting, developed in Aberdeen around 1830, and which spread after 1835 into other areas of Scotland and then to northern England and County Down.[8] Some attempts may also have been made to improve the design of the "cradle" attachment.

The appearance of improved tool forms early in the "majority adoption"

---

here is more complex than in 1790 the evidence is sufficiently comprehensive to allow most important regional variations to be detailed with some precision.

[1] *British Husbandry*, op. cit. II, 188; T. Sullivan, 'On Reaping and Harvesting Grain Crops', *F.M.* Aug. 1845, p. 101.

[2] In west Berkshire there was even a four-stage progression. After 1900 the scythe sometimes took over from the bagging-hook.—Private communications: Mr H. Bew, Padworth, Berkshire; Mr J. Wyeth, Swallowfield, Berkshire.

[3] Stephens, op. cit. (1844 edn), III, 1079; Sullivan, op. cit. p. 101.

[4] Stephens, op. cit. (1844 edn), III, 1079.

[5] R. W. Dickson, *General View . . . Lancashire* (1815), p. 182; *Sketches of Rural Affairs, The Sickle* (S.P.C.K. 1845), pp. 32–3; *British Husbandry*, op. cit. II, 189.

[6] *Country Life*, Dec. 1952; J. M. Wilson, *The Farmer's Dictionary* (n.d.), II, 518. For descriptions of tool forms see *Official Descriptive and Illustrated Catalogue of the Great Exhibition* (1851), II, Class 22.

[7] G. H. Andrews, *Modern Husbandry* (1853), p. 373; Wilson, *Dictionary*, II, 230.

[8] Stephens, op. cit. (1844 edn), III, 1081–2. Private communications: Mr S. Fenton, National Museum of Antiquities of Scotland; Dr A. Gailey, Ulster Folk Museum. See also Martin Doyle, *Practical Husbandry* (Dublin, 1839), p. 266; J. C. Morton, *Cyclopedia of Agriculture* (1851), V, 10–11; *F.M.* Jan. 1841, p. 30.

phase of the revolution was of special significance and their contribution to higher working efficiency is self-evident. But quite as important was their effect in accelerating the breakdown of local hand-tool preferences. The transformation of the scythe and bagging-hook into more efficient and more reliable tools overruled a number of the technical objections which had been raised against the traditional designs.

The third element in the hand-tool revolution, that of time of cutting, can be dismissed briefly. It became a usual practice during the first half of the nineteenth century to cut wheat, oats, and non-malting varieties of barley at an earlier stage of maturity. The advantages here were, first, that yield and quality of sample might be improved and, second, that the front end of the harvest could be advanced, giving an eight to ten days' longer cutting season and greatly reducing the risks of shedding usually associated with dead-ripeness.[1]

III

That the slashing and stroking tools were faster than the hand-reaping tools is well known. In practice, however, their labour-saving advantages were much less spectacular than simple cutting rates suggest. For the sickle and reap-hook required only very few, while the bagging-hook and scythe required large numbers of ancillary workers for the connected operations of gathering, binding, and raking. Making certain assumptions about these linkage labour demands we arrive at the following notional labour requirements per acre for the six competing methods.

Work Rates and Labour Requirements for Competing Harvest Methods in Britain

A = Cutting rate per worker-day (in acres); B = Ratio of linkage workers to cutters; C = Worker-days per acre

	Wheat			*Barley and Oats*		
	A	*B*	*C*	*A*	*B*	*C*
High reaping (sickle)*	0·35	1:7	3·60	—	—	—
Low reaping (sickle)	0·25	1:5	4·80	0·33	1:5	4·00
Low reaping (reap-hook)	0·33	1:5	4·00	0·40	1:5	3·30
Bagging	1·00	2:1	3·00	1·30	2:1	2·20
Mowing (binding and sheaving)	1·10	2:1	2·70	2·00	2:1	1·50
Mowing (leaving loose in swathe)	—	—	—	2·00	0·5:1	0·75

*Excludes autumn clearance of stubble.

Thus in the cutting, binding, and stooking of wheat the labour-saving potential of the reap-hook, bagging-hook, and scythe relative to the sickle was of the crude order of 15 per cent, 35 per cent, and 40 per cent. In the longer run high reaping proved to be more labour intensive than low reaping, although in the primary operations, with which we are here chiefly concerned, it enjoyed parity with the reap-hook.

Some of the gains secured by bagging and mowing in the primary field opera-

[1] Stephens, op. cit. (1844 edn), III, 1056–7, 1072–3; *Sketches of Rural Affairs*, op. cit. pp. 18–19; R. Low, *Elements of Practical Agriculture* (1838), p. 212. A Hampshire farmer argued that because "a week was often of importance in this critical period earlier cutting allowed him to recruit many labourers who afterwards would be unobtainable".—*Farm Reports ... of Select Farms*, bound as vol. III, *British Husbandry* (1840).

tions were lost in the subsequent operations of carting, stacking, and threshing, chiefly because their sheaves were bulkier, less regular, and more difficult to handle than those of reaped corn.[1] But there were other compensations. Bagging and mowing permitted a very flexible deployment of the work-force in that female and child labour could often be substituted in the linkage operations, while for any farmer who cared to sacrifice thoroughness to expediency there were many opportunities for small savings in time and labour. The labour-saving advantages of mowing were greatly enhanced when spring corn was left loose in the swathe and not bound and stooked.[2] Additional, if indirect, savings stemmed from the fact that sheaves of bagged and mown corn dried out faster than those of reaped corn, which made for earlier carting and, with the field cleared, enabled an earlier start on autumn cultivations.

Owing to the extreme regional variations in work organization and wage rates it is difficult to arrive at precise cost differentials. Such costings as are available suggest that in the cutting and stooking operations, mowing was rather cheaper than bagging, 20–25 per cent cheaper than low reaping, and 25–30 per cent cheaper than high reaping.[3] Actual costs would vary not only with the wage rates but also with the age and sex composition of the harvest work-force and the relative price elasticities of cutters and ancillary workers. But as a general rule the price of baggers and mowers was rather higher than for hand reapers, and female linkage workers were cheaper than male.[4] The cost-saving advantages of bagging and mowing over hand reaping in the primary field operations would in practice have been partially offset by higher costs in the secondary operations.[5] Probably a key incentive to adopt the scythe and bagging-hook was the increased volume of their straw cut, which at off-farm prices often represented a substantial gain on this one item. Assuming one inch of stubble to produce 56 pounds of straw per acre, a gain of up to five hundredweight per acre could result from the switch from high reaping to mowing and bagging. The scythe was reckoned to cut two inches lower and the bagging-hook four inches lower than the sickle and reap-hook.

IV

Proponents of change argued that mowing and bagging allowed substantial savings in labour and time, and therefore labour costs, while adherents to the reap-hook and sickle justified their inertia on the ground that the opportunity

[1] Threshing, particularly, was reckoned a more difficult and costly operation, though less so with the threshing machine than the flail. The argument was that considerable time (up to 30 per cent) was wasted rearranging up-ended straws to meet the flail.

[2] Over most of southern and eastern England barley was seldom put into stooks, but a certain amount of labour was required to turn or "cock" the swathe to allow even ripening and thorough drying out. In northern and western Britain all grain was stooked, while in the wetter areas where the need for weather-proofing was greatest, the process was sometimes elaborate and very labour-consuming.

[3] For detailed costings see P. Love, 'On Harvesting Corn', *J.R.A.S.E.* XXIII (1862), 223.

[4] Thus on one Roxburghshire farm the substitution of high-priced scythesmen for relatively low-priced female reapers made mowing a more expensive mode than hand reaping.—*Ag. Gaz.* 28 Sept. 1867, p. 1011; G. Robertson, *Rural Recollections* (Edinburgh, 1829), pp. 247–8. Conversely in Norfolk the substitution of women binders and stookers for men resulted in a 15 per cent cost saving in these operations.—J. C. Morton, *Hand Book of Farm Labour* (1868 edn), p. 105.

[5] Love, op. cit. p. 223, suggests, in fact, that over the whole harvest operation, cutting to threshing, hand reaping was actually cheaper than mowing or bagging.

costs of change were too high. Essentially, the issue was that of economic incentive versus technical constraint.[1]

As late as the 1860's there was some measure of agreement among practical agriculturists that in technical performance the sickle was a more versatile and meticulous tool than either the scythe or the bagging-hook. In 1853 the Royal Agricultural College still taught reaping as the cleanest and most "expert" and mowing as the most "slovenly" method of cutting wheat.[2]

The bases of antipathy towards the bagging-hook and scythe are not always easy to establish. On the one hand it was unquestionably true that their wastage rates were much higher than those of the sickle, particularly in badly lodged and over-ripe stands of corn. In west Surrey some sixty years after its introduction, it was reckoned that with the bagging-hook "a man leaves his wages on the ground", and even though the scythe was rather tidier the wastage of corn and straw was still regarded as "something terrible".[3] It was also true that the choice of technique was to a large extent determined by actual physical conditions. Where field surfaces were uneven, heavily ridged, or thick with rocks or tree stumps, the scythe was virtually inoperative except in the hands of highly skilled workmen. The prolific straw growth and extreme lodging associated with highly nitrogenous soils, such as those of the Fens and the Lothians, imposed such severe constraints on the use of the scythe that despite sometimes serious labour shortages the sickle and reap-hook continued as the predominant hand tools for wheat and oats up to and beyond 1780.[4] Even in areas where the mowing of wheat was already well established a common practice in the 1860's was still to reserve the heaviest stands for the sickle, notwithstanding the much higher costs involved. In 1866, for example, an East Riding farmer mowed his lightest wheat at 8s. to 11s. per acre and reaped the rest at 14s. to 16s. per acre.[5] In south midland and central southern England farmers often preferred to bag or reap rather than mow their heavier crops of spring corn. Farm account evidence reveals that on many farms sickle, reap-hook, bagging-hook, and scythe were viewed, for a time at least, as interchangeable tools, as complementary elements within the spectrum of available hand-tool techniques, the exact tool mix in any one season being determined by labour supply and crop condition.

Some of the objections to new methods were less rationally conceived. Traditionally there was always a reluctance to apply brutal methods to the bread corns. The regional synonyms for bagging—cuffing, slashing, and hacking—significantly connote violence of treatment as against the gentle meticulousness of hand reaping. Even the reap-hook did not entirely escape censure,[6] while it was only in the 1830's that bagging became generally recognized as a procedure distinct from orthodox hand reaping, with its own specialized tool and individual *modus operandi*. Aesthetic preference was a further constraining element among

[1] See Stephens, op. cit. (1844 edn), III, 1048–1105; J. Wilson, *Cyclopedia of Agriculture* (1851), IV, 26–7; *F.M.* Aug., Sept., Oct. 1845, for detailed treatments of comparative efficiencies.

[2] *Ag. Gaz.* 22 Jan. 1853, pp. 58–9.

[3] G. Jekyll, *Old West Surrey* (1904), p. 184; P. McConnell, *Diary of a Working Farmer* (1906), p. 253.

[4] *F.M.* July 1845, pp. 64–5; *Ag. Gaz.* 24 Aug. 1867, pp. 888–92; Stephens, op. cit. (1891), III, 57; A. D. Hall, *Pilgrimage of British Farming* (1913), p. 75; P. McConnell, *The Practical Farmer* (1911), pp. 391–2.

[5] *Ag. Gaz.* 24 Aug. 1867.

[6] *F.M.* June 1811, p. 181; *Dreweatt, Watson & Barton, 1759–1959* (privately printed, Newbury, 1959), p. 27.

men who worshipped at the shrines of even stubbles, tidy symmetrical sheaves, and clean, unlittered fields.[1]

Resistance to change was most apparent where innovation involved the retraining of the harvest labour-force in a range of new and unfamiliar techniques. The failure of the Hainault scythe to establish itself in Scotland, even after field trials which had demonstrated convincingly its superiority over the sickle, may be ascribed to the disinclination of local workmen to learn what appeared to require of them "some peculiar and indescribable dexterity". Potential employers of the corn scythe were warned that "strong and able workmen are necessary who must possess dexterity as well as strength, otherwise they will only make confusion in the field." In the Lothians and the Border Counties, for want of skilled local workmen, much of the mowing and sometimes even the gathering and binding were let out to migrant harvesters from Aberdeen.[2] Wherever the bulk of the harvest was taken by women physically incapable of using the heavier tools, by part-time industrial workers lacking the basic skills of mowing or bagging, or by Irish migrant harvesters whose preference for the sickle was for a long time all-pervading, the difficulties of introducing the scythe and bagging-hook were especially obvious. For example, it was in the Scottish Lowlands and the Border Counties, where these categories of workers were most abundant, that the reaping machine was more often able to take over from the reap-hook without the intervention of the scythe.

Such was the effect of workers' hand-tool preferences and farmers' unwillingness to risk the adoption of untried techniques that we can observe the tendency for new methods to spread outwards in wave fashion from just a few concentrations; for example, the wheat scythe from the East Riding, Dorset, west Norfolk, and the Vales of Clwyd and Carmarthen, the two-helved scythe from Aberdeen, and the bagging-hook from Devon, Hereford, and Middlesex. The "splash" effect, that is simultaneous innovation at a large number of widely scattered locations, was a less common phenomenon.

V

That in the long run bagging and mowing did replace hand reaping must suggest that farmers came to rate labour saving more highly than grain saving and technical performance. Several considerations are suggestive here. First, that between the 1830's and 1860's harvest wages rose faster than corn prices, thus raising the marginal costs of saving ears to the point at which labour-intensive, low-loss methods were no longer profitable. Second, that from the late 1830's average corn yields rose quickly while corn prices fell only slightly, which increase in net income per acre largely offset any crop loss incurred by method switching.[3] Moreover, as the economic importance of corn in the national

[1] One farmer, for example, condemned the "slovenly higgledy-piggledy work" of the wheat scythe, with its "shapeless and unsightly" stacks and its "dead robbery... Robbing Peter to pay Paul".—*British Farmer's Magazine*, Oct. 1843, pp. 430–5.

[2] Wilson, *Cyclopedia*, II, 337; *F.M.* Aug. 1825, pp. 339, 349; *Sketches of Rural Affairs*, op. cit. p. 31; *F.M.* July 1845, pp. 64–5, Sept. 1845, p. 211.

[3] Between 1830–9 and 1860–9 wheat prices fell by 8 per cent while barley and oats prices increased by only 11 per cent and 6 per cent.—B. R. Mitchell & P. A. Deane, *British Historical Statistics* (Cambridge, 1962), pp. 448–9. Over the same period average weekly earnings rose by 17 per cent in England and 57 per

economy diminished, and as livestock farming became more prominent, so many of the non-economic constraints on innovation, in particular the folk attitudes towards the harvest, of which the sickle was a symbol and through which the harvest was regarded as a sacred and unalterable ritual, became less operative. We may conclude, therefore, that the primary incentive to change was provided by a deterioration in the supply of harvest labour relative to demand.

The close correlation between labour supply and harvest technology is at most points easy to establish. The case was put most cogently in 1814 by the *Irish Farmer's Journal*, which, arguing against the introduction of the Hainault scythe, concluded that "we would not, in this country, where the population is great and without resource, except in agriculture, for employment, recommend the too great abridgement of manual labour. What is gained in money is often lost in sustaining those who are thrown into distress from poverty, the result of inoccupation."[1] The presence of an abundant resident supply of labour usually carried with it the social obligation to employ it. As Morton pointed out, farmers could not employ just a few labourers and feed them well; they had to maintain all in the parish, either in the field or in the workhouse.[2] In Joseph Ashby's Tysoe the sickle was still extensively used in the early 1870's because, it was said, men were cheap and plentiful.[3] Farmers invariably delayed the adoption of faster working tools until hard experience and the failure of all other shifts had convinced them that only a substantial improvement in output per worker could remove the risks of serious crop losses in difficult seasons, and provide insurance against the blackmail of exorbitant wage demands in years of labour scarcity. In South Wales, for example, "much corn and too few hands" first suggested bagging, which practice was subsequently introduced from the West Midlands in the Napoleonic Wars. Even at this early stage proponents of the sickle were qualifying their enthusiasm with the reminder that hand reaping was practicable only where labour was well enough supplied. In the 1860's depopulation in Inverness had reached a point where there was no longer sufficient labour for the reap-hook, which had resulted in the general adoption of mowing, "a rough and ready way of getting through the work [which] involved a considerable sacrifice of grain".[4]

Conversely, farmers were cautious about adopting methods which threatened to displace workers or to reduce their earning capacity. Attempts to introduce the mowing of wheat into Middlesex around 1800 were frustrated by labourers' reactions to a tool which they believed would deprive them of work. In Kent in the 1830's the scythe "on all occasions excited the ill-will of the labourers to a very dangerous extent, for it is a most powerful and efficient implement and it is thought that if brought into use it would extinguish the usual harvest earnings". Indicatively, when mowing was recommended to Scottish farmers, it was on the

cent in Scotland.—Ibid. pp. 349–50. Harvest wage data are notoriously scarce but between the late 1840's and the late 1850's hiring rates rose by approx. 30 per cent.—*Ag. Gaz.* 27 Apr. 1850, pp. 266–7, 30 Apr. 1850, pp. 392–3. For corn yields see above, p. 466.

[1] *Irish Farmer's Journal*, 6 Aug. 1814. [2] J. C. Morton, *Handbook of Farm Labour* (1868 edn), p. 76.

[3] M. K. Ashby, *Joseph Ashby of Tysoe* (Cambridge, 1961), p. 25; see also Flora Thompson, *Lark Rise to Candleford* (Oxford, 1954 edn), pp. 256–8.

[4] W. Davies, op. cit. I, 424–8; *Ag. Gaz.* 1 April 1861, p. 296.

understanding that it should not harm the resident labour force but affect only the casual and migrant workers hired at the public markets. Farmers were averse to displacing at one point labour which might otherwise not be available to them earlier or later in the season. Some at least were prepared to create work or to forgo labour-saving innovations so as to retain labour on the farm.[1]

We will now examine in greater depth the long-run movements in the demand and supply schedules for harvest labour during the most active phase of the revolution, 1835–70.

The corn harvest required, in the cutting and binding operations, a large and extremely elastic supply of labour to meet different conditions of weather and crop, and variable speed of ripening. Time was the critical factor. In a dry season, failure to cut wheat and oats within eight to ten days of ripening could result in heavy losses by shattering and shedding, while in a wet "catchy" season irreversible grain-wetting was likely to occur if cutting was suspended for too long.[2] Labour requirements varied also with crop yield and condition, it being estimated that a heavy, badly laid crop required roughly double the labour input of a light standing crop, and more if the scythe had to be replaced by the sickle.[3]

This special demand could be met only by a massive recruitment of temporary workers and by offers of wages considerably in excess of normal weekly rates, along with attractive perquisites ranging from food and beer to coals, livestock keep, and the free use of a cottage or potato ground. Although harvest wages tended to be highest in the labour-scarce districts of northern Britain, there was no straightforward correlation between weekly and harvest wage rates. In East Anglia, for example, weekly wages in 1850 ranged from 6s. to 8s. per week, almost the lowest in England, while harvest wages stood at over 20s. per week, considerably higher than in any northern county.[4] In many corn-growing districts the additional demand prior to mechanization may have been as high as 70 or 80 per cent. As late as 1921 the summer casual labour-force in English and Welsh agriculture still numbered 207,000.[5] In 1851 it would have been much greater. In Kent in the 1830's, temporary workers outnumbered the regular hands by three to one, while even with the occasional help of reaping machines the summer work-force on ten Berkshire farms in the 1860's was 60 per cent higher than at any other season.[6] The sources of this supply were many and various. On Tom Strong's "Stubble Farm" the harvest company was composed of "all sizes and ages, men and women, Irish and English, strollers and

[1] J. Middleton, op. cit. p. 216; *Report on the State of the Irish Poor in Great Britain* (1835), App. G, p. 165; Stephens, op. cit. (1851 edn), II, 347. Thaër, too, warned that innovations would "often damp, if not destroy, that alacrity and cheerfulness on the part of the men which cannot be too diligently fostered during harvest time."—Albert Thaër, *The Principles of Agriculture*, trans. W. Shaw and C. W. Johnston (1844), II, 380–1; A. H. Savory, *Grain and Chaff from an English Manor* (Oxford, 1920), pp. 80–2.

[2] During the period 1835–70 southern Britain had eleven years with July and August rainfall sufficiently below normal to promote rapid crop ripening, and eight years with August rainfall in excess of four inches, the generally accepted threshold for irreversible wetting.—Calculated from rainfall data and other information kindly provided by L. P. Smith, Meteorological Office, Bracknell, Berkshire.

[3] Wilson, *Cyclopedia*, IV, 27. [4] *Ag. Gaz.* 27 Apr. 1850, pp. 266–7.

[5] *A Century of Agricultural Statistics* (H.M.S.O. 1968), p. 62.

[6] *Report on the State of the Irish Poor*, op. cit. App. G, p. 165; *Royal Commission on the Employment of Children, Young Persons, and Women in Agriculture (1867)* (hereafter referred to as *R.C. Employment*), *Second Report* (1869), App. 2, p. 368.

neighbours, reapers and faggers, good workmen and bad, grandmothers and children, kettle boilers and tiers, married and single". In the early nineteenth century weavers and industrial workers were essential components of many harvest gangs in northern England and the Scottish Lowlands. Rural craftsmen, gypsies, male servants, and soldiers on furlough were recruited by Kent farmers. As late as 1918 Pembrokeshire farmers still engaged unemployed colliers, quarrymen, and fishermen.[1] Many areas of Britain were, of course, very dependent on migrant workers.

The long-run movements in national harvest labour requirements were determined first by the extent of the corn area and second by crop yields. The national corn area may have grown slightly between 1835 and 1855; from the mid-1850's it fell gently away but with the major decline occurring in northern and western Britain, leaving the specialist corn-growing areas of the south and east relatively unaffected. Contemporary estimates suggest that between 1835 and 1870 wheat yields rose by at least 20–25 per cent, barley by 10 per cent, and oats by 15 per cent.[2] Assuming that labour input was a direct function of yield we can estimate on a basis of constant techniques a 20–25 per cent growth in harvest labour requirements between 1835 and 1870, the rate of increase being rather greater under conditions of rapidly improving yields, increasing crop area, and rising wheat values in the first half of the period than subsequently. The real increase in demand would have been considerably higher. The more extensive cultivation of roots and vegetables added an important and perhaps often critical element of on-farm competition for labour during the harvest period.

A reliable long-run supply curve for harvest labour cannot be constructed from the available statistical data. It must be assumed that the occupational census, because it was taken in the spring, would exclude a majority of would-be casual and temporary harvest workers, who would be described under some category other than agricultural labourers, or else not enumerated as gainfully employed. It would exclude, too, almost all Irish migrant harvesters, whose presence on the mainland was chiefly confined to the period June to September. Broadly, the problem is how to isolate the several categories of harvest worker: the full-time, the part-time, the casual, and the long-distance migrant, all with their different supply curves. The complex and highly sensitive inter- and intra-sectoral labour flows which characterized this early phase of economic growth cannot be detected by the static census, and thus can be explored only qualitatively.

[1] Anon. *Stubble Farm* (1880), II, 17–18; T. S. Ashton, *An Economic History of England: The Eighteenth Century* (1955), p. 202; W. Hasbach, *A History of the English Agricultural Labourer* (English edn, 1908), p. 82; I. Pinchbeck, *Women Workers and Industrial Revolution* (1930), pp. 55–9; *Report on the State of the Irish Poor in Great Britain*, op. cit. App. G, p. 165; *Ag. Gaz.* 24 Aug. 1867, p. 890; *Select Committee on Wages and Conditions of Employment in Agriculture* (1919), II, 489.

[2] For a review of contemporary yield estimates see P. G. Craigie, 'Statistics of Agricultural Production', *Journal of the Royal Statistical Society*, XLVI (1883), 1–47. Healy and Jones have suggested, however, an improvement in wheat yields of 40 per cent from 1815–24 to 1850–9, with a very marked upturn from the late 1830's, coinciding with the start of Ernle's "High Farming" period, the rapid step-up of artificial manure inputs shown by Thompson, and the opening of the majority adoption phase of the hand-tool revolution.—M. J. R. Healy and E. L. Jones, 'Wheat Yields in England, 1815–1859', *Jnl. R. Stat. Soc.* ser. A, CXXV (1962), 574–9; F. M. L. Thompson, 'The Second Agricultural Revolution, 1815–1880', *Economic History Review*, 2nd ser. XXI (1968), 62–77.

On the likely assumption that full-time workers in agriculture constituted the largest single category of harvest workers, probably to the extent of 50–60 per cent of the total work force, it is clear that even small proportional changes in their numbers would have had an important and immediately perceptible effect on the overall harvest labour-supply situation. That their numbers declined after 1850 is well known. That they increased over the 1830's and 1840's is less easily demonstrated, but for present purposes we will assume the conventional estimate of a 17 per cent increase between 1831 and 1851 as approximately correct.[1] There is some basis, however, for believing that the greater part of this growth occurred in the early 1830's and the later 1840's, between which points the farm labour-force grew only slowly, while briefly in 1835–9 it may even have declined. It seems likely, then, that between 1835 and 1845 the growth in the harvest labour-force was insufficient to satisfy increased crop demand.

From 1834 railway building, a new phase of industrial expansion, and agricultural revival combined to dry up much of the conspicuous rural labour surplus. In 1836 an anxious Sussex landowner forecast that before many years farmers would experience difficulty in getting the harvest in. In 1839, even before the commencement of work on the Great Eastern Railway in the area, apprehension was rife in north Essex. "Where the men are we do not know. Month after month may pass by without a single application of any of them for work, where formerly we sometimes had hundreds." Similar reports of full agricultural employment and harvest shortages came from areas as far apart as Bedfordshire, Cumberland, Glamorgan, Hertfordshire, Kent, Lincolnshire, Nottinghamshire, Oxfordshire, and Yorkshire.[2]

Although the supply position eased somewhat in the early 'forties, until 1846 the extraordinary demand for unskilled men in railway building and its linkage industries exercised a selective and locally very disruptive influence on the agricultural labour market. In that year (1846) there were noisy complaints from north-eastern England about the great scarcity of harvest hands.[3] After 1846 the collapse of the railway boom, industrial recession, and massive Irish immigration provided agriculture with many temporary recruits, slowed down rural migration, and set in motion a reverse flow of labour back to the countryside.[4] Significantly, there are suggestions of a slowing down of the rate of harvesting innovation during these years.

[1] Numbers in the agriculture, forestry, and fishing group of industries are computed to have risen from 1·8 million in 1831 to 1·9 million in 1841 and to 2·1 million in 1851, and subsequently to have fallen to 1·8 million by 1871.—P. Deane and W. A. Cole, *British Economic Growth, 1688–1959* (2nd edn, Cambridge, 1967), p. 143. This and other evidence is summarized in E. L. Jones, 'The Agricultural Labour Market in England, 1793–1873', *Econ. Hist. Rev.* 2nd ser. XVII (1964), 327–33.

[2] *Select Committee of H. of L. on the State of Agriculture* (1836), *Third Report*, pp. 88, 108, 269, 291, 422–3; *Report on the State of the Irish Poor*, op. cit. App. G, pp. 165, 197–8; *Poor Law Commission, Second Annual Report* (1836), p. 543; *F.M.* June 1839, pp. 480–1; *Mark Lane Express*, 16 Sept. 1839, p. 10, 14 Oct. 1839, p. 10, 24 Dec. 1938, p. 10.

[3] A. D. Gayer, W. W. Rostow, and A. J. Schwarz, *The Growth & Fluctuations of the British Economy, 1790–1850* (Oxford, 1953), I, 300–3; *F.M.* Sept. 1846, p. 278; Jones, op. cit. pp. 327–8; T. Tooke, *A History of Prices* (1926 edn), IV, 44–72.

[4] Gayer, Rostow, Schwarz, op. cit. II, 337–40; J. P. Lewis, *Building Cycles & Britain's Growth* (1965), p. 85; Redford, op. cit. pp. 42, 125–31, 156–8. For local examples of labour abundance see *F.M.* Oct. 1848, p. 374, Nov. 1848, pp. 466–7, Sept. 1850, p. 266; J. D. Marshall, 'The Lancashire Rural Labourer in the Early Nineteenth Century', *Transactions of the Lancashire and Cheshire Antiquarian Society*, LXXI (1961), 92.

The harvest labour market deteriorated again, and this time more permanently, in the early 1850's. The smart upturn in the trade cycle, heavy emigration, and the Crimean War soon taught farmers the wisdom of Caird's warning "that as labourers begin to withdraw" they would soon discover that the surplus was "not so great as they had led themselves to believe".[1] In the late 1850's rustic grumblers were predicting for the first time the collapse of intensive mixed farming as the labour-supply position deteriorated further. "The labour market will beat us all," forecast one Kent farmer.[2] Gone was the time when, as on one Cambridgeshire farm, a whole crop of wheat had been sickled in four or five days and 300 men at trifling cost set reaping in a single day.[3]

Farmers discovered that in years of high industrial activity their supplies of harvest labour could fall to dangerously low levels as the younger men of the parish and would-be migrant and casual harvest workers took their labour elsewhere. They also discovered that the supply price elasticity of harvest labour was low and falling, and that offers of high wages no longer guaranteed them a labour-force sufficient to avert heavy crop losses by shedding and spoiling. Rapid crop-ripening coincided with high-activity phases of the trade cycle in nine different years between 1852 and 1870 to make harvesting a hazardous and uncertain process.[4]

At such times, as the market advantage shifted away from the farmer, wage bargainings were characterized increasingly by strikes, disturbances, and much-advanced wages. On a sample of 42 farms harvest wages stood 30 per cent higher in the labour-scarce year of 1859 than in the labour-flush year of 1849. However, the chief complaint was less about the price of labour than its scarcity. In 1860 Derbyshire farmers, although still under pressure, were actually congratulating themselves that in contrast to previous years "men could be got for money".[5]

The pattern that emerges shows clearly that the short-run fluctuations in harvest labour supply were more consequential than the long-run deterioration. The latter would have been met by a smooth and regulated uptake of labour- and labour cost-saving inputs, but in practice a farmer's decision to change his method was more closely correlated with the short-run convergence of the labour supply and demand schedules; for in most cases he knew his exact labour

[1] J. Caird, *English Agriculture in 1850 & 1851* (1852), p. 518. For the Crimean War situation see Olive Anderson, 'Early Experiences of Manpower Problems in an Industrial Society at War', *Political Science Quarterly*, LXXXII (1967).

[2] *Quarterly Journal of Agriculture*, XXII (1859–69), 294; *Ag. Gaz.* 12 Dec. 1859, p. 1000; *F.M.* Feb. 1859, p. 174, March 1860, p. 220; *R.C. Employment, First Report* (1868), pp. 123, 167, 170–1; and in the 1870's, F. Clifford, 'The Labour Bill in Farming', *J.R.A.S.E.* 2nd ser. XI (1875), 67 ff.

[3] S. Jonas, 'On the Farming of Cambridgeshire', *J.R.A.S.E.* VII (1846), 49.

[4] In the years 1852–4, 1857, 1859–60, 1864, 1866–7. 1859 was perhaps the most critical year, "an event foreseen for some years ... but which has taken some unobservant farmers by surprise—Corn will not be got in in some areas".—*F.M.* Sept. 1858, p. 190. For regional evidence see *F.M.* Aug. 1859, p. 132, Sept. 1859, p. 265, Oct. 1859, pp. 313–14; *Reading Mercury*, 6 Aug. 1859, 13 Aug. 1859, 27 Aug. 1859. There is a noticeable correlation between rural unemployment (as indicated by numbers of adult able-bodied males receiving outdoor relief in the *Pauperism Returns*) and the state of the trade cycle. The relative over-supply position in slumps was much enhanced if, during the previous phase of labour scarcity, farmers had switched permanently into faster methods.

[5] *Mark Lane Express*, 8 Aug. 1853; *F.M.* Sept. 1859, p. 160; *Reading Mercury*, 21, 28 July 1866; Sample made up largely from national wage surveys in *Ag. Gaz.* 27 April 1850, pp. 266–7, 30 April 1860, pp. 392–3; *F.M.* Oct. 1860, p. 370; and, in Norfolk, *F.M.* Nov. 1866, pp. 367–8.

requirements only on the eve of the harvest, while fluctuations in labour supply could be sharp and sometimes unpredictable.[1] Thus in Berkshire the resumption of labour flows in the economic downswing of 1861 enabled a large proportion of oats to be bagged and bound which previously had been mown and carted loose, while in 1866, a year of labour scarcity, the procedure was reversed. Similarly, bagging was first introduced on to one Oxfordshire farm in the high-wage years of 1856 and 1857, then abandoned in favour of reaping, until taken up again in the next crisis year, 1866.[2]

VI

Notwithstanding the extremely vigorous adoption of labour-saving factors after 1835, contemporary accounts suggest that even with the reaping machine, harvest labour shortages in the 1860's were as acute as at any other time in the previous three decades. This cannot be attributed, except perhaps locally, to the post-1850 decline in the numbers of full-time workers. Numbers fell only by 15–20 per cent between 1851 and 1871. Assuming that in 1851 permanent farm workers made up half of the total harvest work-force, the resultant 7–10 per cent shortfall would easily have been offset by technical innovation. The explanation therefore must lie elsewhere, in the more rapid and infinitely more drastic reduction in the numbers of casual and temporary harvest workers, whose supply schedules tended to be less predictable and more vulnerable to extreme fluctuation than those of full-time workers.

Two factors were operative here. First, as urban employment opportunities improved and as the gap between farm and non-farm incomes widened, so there was less incentive for non-agricultural workers and their families to supplement their incomes by harvest work; and second, and for much the same reasons, potential casual and migrant workers sought seasonal work elsewhere than in agriculture—in mining, building, and engineering where wages were more attractive and the season longer than in the harvest fields.

The decline of cottage industry in areas in which traditionally it had been a major source of supply of casual harvest labour was to have serious repercussions. At an earlier point in the century Lancashire and West Riding weavers had left their looms to assist in the harvest. By 1860 this supply, and with it large numbers of full-time agricultural workers, had been swallowed up by the factories. In the 1830's south of England farmers had more or less resigned themselves to the fact that town help would soon cease entirely as labour became more exclusively "town bred" while, in the Scottish Lowlands, Irish harvesters were filling the ranks left vacant by mechanics, weavers, and colliers. "There was never a difficulty with farm labour", remembered one Lancashire farmer, "as long as hand loom weaving was kept up. Our cottages were largely filled with hand loom weavers . . . these men were always available for a

[1] The operation of this two-way labour flow in the late 1860's is evidenced in: *F.M.* Aug. 1866, p. 170; *R.C. Employment, First Report* (1868), p. 399; *Second Report* (1869), App. 2, p. 401; H. Evershed, 'Report of the Farming of Kent, Sussex and Surrey', *Journal of the Bath and West of England Agricultural Society*, 3rd ser. II (1871), 17.

[2] *Reading Mercury*, 14 Sept. 1861, 8 Sept. 1866; Reading University Library, *Farm Records Collection*, pp. 295–7. See also loc. cit. KEN 4/7/2, BER 13/5/3, 4 for a similar correlation.

push."[1] In the short run the decline of rural domestic industry may have meant a flow of new recruits into agriculture. In the longer run their labour was transferred to the towns and the factories. The work discipline of the new industrial capitalism succeeded eventually in destroying those "older collective rhythms" which until late on in the second quarter of the nineteenth century had still been able to "break through the new".[2] In 1870 the only substantial, purely urban labour contribution to agriculture was in the hop fields and market gardens.

From 1850 there is also evidence for a decline in the size of the female agricultural labour force, and although this was less marked in the harvest than in other operations, clearly it was becoming increasingly difficult to recruit married women for outdoor farm work. Unmarried girls preferred the superior status and more genteel environment of domestic service to the bucolic atmosphere of the harvest field. By the mid-1860's arable farmers were apprehensive at the prospect of a further withdrawal of female labour, since this labour was reckoned essential for the timely performance of summer operations. Somerset women were reported as "above working out now, it makes their hands dirty"; in parts of Kent only the wives of farm workers could be persuaded to help in the harvest; while in the north-west women were so scarce that even light repetitive farm tasks were performed by men. The situation was particularly acute where farmers had to compete directly with market gardening and local industry for female workers.[3] In such areas farmers waited hopefully for a trade recession to bring them a temporary increase in supply.

Most critical, however, was the post-1850 decline in the volume of migrant harvest labour flows. A fall in this category of harvest worker had quite disproportionate effects on the overall supply position, first because most migrants exploited the phased crop ripening between highland and lowland, and light and heavy soils, to work two or even three harvests a season, and secondly because as young men and as piece-workers their work output was extremely high.[4] As regional labour shortages developed so interregional labour flows became more complex. By the 1850's it had in some areas become almost a tradition that ambitious young men should take to the roads in the summer, selling their labour to the highest bidder. In others the flow was more regularized. North Hampshire men went into Sussex and Wiltshire, Stour Valley scythesmen into Foulness and the Marsh Hundreds, Vale of Taunton men on to the Mendips,

[1] D. Bythell, 'The Hand Loom Weavers in the English Cotton Industry during the Industrial Revolution: Some Problems', *Econ. Hist. Rev.* 2nd ser. XVII (1964), 344–7; Redford, op. cit. pp. 129–31; *Report on the State of the Irish Poor*, op. cit. App. G, pp. xlv–xlvi, 166; *Royal Commission on Labour*, V (1894), pt II, § IIB, p. 291.

[2] E. P. Thompson, 'Time, Work Discipline and Industrial Capitalism', *Past & Present*, XXXVIII (1967), 62–3. Workers in domestic industry often resorted to agriculture in slump years; Marshall, op. cit. p. 92; *R.C. Employment, Second Report* (1869), App. II, pp. 225, 522, 526. The eventual decline of the domestic textile industry in Furness was held directly responsible for the scarcity of farm labour.—J. D. Marshall, *Furness in the Industrial Revolution* (Barrow, 1958), pp. 237–8. Conversely, when trade was brisk, as in Forfar in 1825, and in the straw-plait districts of Essex, 1850–70, industrial workers were virtually unobtainable.—*F.M.* Nov. 1825, p. 490; *V.C.H. Essex*, II, 376–7.

[3] *R.C. Employment, Second Report* (1869), App. I, pp. 14, 35, 54, 123, 136, 154, App. II, pp. 27, 112, 215, 434–7, 477, 489, 518–21, 548.

[4] In 1867, for example, the beginning of the harvest varied from 5 Aug. on Foulness to 20 Aug. at Dunmow (Essex); 5 Aug. at Woking to 13 Aug. at Carshalton (Surrey); 20 Aug. at Wetherby to late September at Whitby (Yorks.).—*Ag. Gaz.* 24 Aug. 1867, pp. 888–92.

Vale of White Horse men on to the Wiltshire Downs, Blackmoor Vale men on to the Dorset Downs, Dorset and Somerset men into the Isle of Wight, Dalesmen into the Yorkshire Wolds, Farnham baggers down to Brighton, and Aberdeen mowers to the Lowlands and Border Counties.[1]

Some old-established flows dried up after 1850; others became so irregular that farmers could no longer rely on them. While in some source areas the outflow was checked by permanent migration to the towns and rising local demand for summer labour, in others, and particularly where wages were low, seasonal migration may have increased, but with the growing risk that instead of returning for their local harvest young men would prefer to stay abroad all summer. Better, it was said, to spend £100 extra on wages in winter than to risk losing £200 in shed and spoilt grain in the summer.[2] Rather more worrying was the growing tendency for the direction of migrant labour flows to vary with the timing of the harvest in the receiving areas, and for their volume to fluctuate with the level of urban and industrial demand for unskilled labour.[3]

The drying up of the Celtic labour flows, historically the largest and most reliable source of seasonal workers for English and Scottish arable farmers, was perhaps more pertinent than the slowing down of internal circulations. In central Scotland the supply of crofter labour began to tail off from the 1820's with the result that by the 1870's the supply of "solitary highland lassies" had dwindled to insignificant proportions. The Welsh flows to the English Midlands had already largely dried up by the mid-1850's, as the population of the source areas drifted to the South Wales coalfield and the Lancashire and Birmingham conurbations. By the late 1860's Welshmen were found only in the Shropshire wheatlands, their presence in Herefordshire having ceased abruptly during the Crimean War.[4]

The numbers of Irish seasonal workers rose so dramatically over the first half of the century that in some areas, such as the Scottish Lowlands and Yorkshire, they entirely displaced some of the already existing seasonal flows, which subsequently were never renewed. But while up to the mid-1830's their entry into local harvest labour markets had often been the signal for violent disturbances,

[1] W. Clift, *Reminiscences* (Basingstoke, n.d.), pp. 64–6; A. Young, *General View . . . Essex* (1807), I, 304; *F.M.* July 1845, pp. 64–5, Sept. 1845, p. 211; Rev. Warne and A. & W. Driver, *General View . . . Somerset* (1798), p. 97; G. Legard, 'Report on the Farming of the East Riding of Yorkshire', *J.R.A.S.E.* IX (1848), 117. In southern England, the Irish, the most mobile of the migrant workers, harvested in Kent, Hampshire, and Surrey before moving on to the hop fields.—*Reading Mercury*, 1 Aug. 1868. In the north Midlands they took a hay harvest in north Derbyshire, went south for an early corn harvest, returned to Derbyshire for a late corn harvest, and then went to Lancashire for the potatoes.—*R.C. Employment, Second Report* (1869), App. II, pp. 25, 29. In Scotland they took a harvest in the Lowlands and Border Counties, moved north into Stirling, Fife, and the Carse of Gowrie for a later harvest, and finished up in the market gardens around Glasgow.—J. E. Handley, *The Irish in Scotland* (Cork, 1945), pp. 43, 44.

[2] Morton, *Handbook*, pp. 72–3; *Ag. Gaz.* 2 Jan. 1869.

[3] The synchronized flow of migrant workers between areas was possible only when ripening was an even, moderately slow, soil- and climatically induced process. Flows were seriously disrupted when as a result of fine dry weather grain ripened simultaneously over a wide area, or if the three harvests ran together. Very late or very early harvests were equally disruptive. In the early harvest of 1868 the Irish arrived in Lincolnshire too late, in the late harvest of 1869 too early.—*Reading Mercury*, 21 Aug. 1869.

[4] *R.C. Labour*, V (1894), pt II, § IIB, p. 291; A. H. John, *The Industrial Development of South Wales* (1950), p. 32; *Hereford Journal*, 21 Oct. 1857; H. Tanner, 'The Agriculture of Shropshire', *J.R.A.S.E.* XIX (1858), 17.

thereafter, apart from a passing spell after the Famine, the ill feeling between Irish and native labour evaporated under conditions of growing labour scarcity.[1]

Irish migrants estimated at 58,000 in 1841 probably reached their peak some time between 1846 and 1850. With the massive emigrations of the early 1850's, and helped by improved work opportunities at home, their numbers fell away rapidly to level off in the 'eighties at around 20,000–30,000.[2] The decline was quickly felt. Complaints about the non-appearance of the Irish were already loud in 1852–5. By the mid-'sixties the Irish element was "nearly extinguished" in Middlesex and much reduced in Cambridgeshire, Lincolnshire, Yorkshire, Wiltshire, and Sussex. In one area of Derbyshire the decline was estimated at over 80 per cent.[3] A further aggravating factor was that the Irish gravitated more and more towards the highest wage areas, the Fens, the Lincolnshire and Yorkshire Wolds, the London area, but particularly Cheshire, Lancashire, and the Glasgow region where potatoes and market-garden crops guaranteed a long working season and good wages within easy distance of the quays.[4] As their mobility decreased so the Irish took only one or two harvests where at one time they would have taken two or three. For many farmers the supply price of Irish labour, formerly reckoned the lowest in the market, had by the 1860's become too high. As they faded out, so the sickle was quickly replaced by locally wielded scythes and bagging-hooks, or, in a few areas, by reaping machines.

VII

In conclusion, I hope to have demonstrated the success of intermediate hand-tool technology in raising output per man in one farm operation. As an alternative to mechanization it had the advantages first, of simplicity, cheapness, and easy maintenance; second, of operational flexibility; third, of being able to offer satisfactory labour savings without creating serious technological redundancy; and fourth, of reversible method changes.

The usefulness of this brand of technology to modern underdeveloped agricultures is at once apparent. In the case of harvesting, an intermediate technology in the form of scythes, more efficient sickles, heavier hooks, or perhaps simple animal- or hand-powered machines may prove a viable alternative to power mechanization. This may be either when labour begins to leave the land in the process of economic growth or when the demand for harvest labour increases as a result of improved farming practices. Mechanization is not only a costly process which ties the poor farmer to fixed payments on capital in good

[1] *Report on the State of the Irish Poor*, op. cit. App. G, pp. xlv–xlvi; J. H. Clapham, *An Economic History of Modern Britain* (Cambridge, 1964 edn), pp. 56–66; Handley, op. cit. *passim*; Redford, op. cit. pp. 141–9; *F.M.* Sept. 1838, p. 378; Caird, op. cit. p. 251.

[2] *R.C. on the Land Acts (Ireland)* (1886), pp. 393, 424 ff; Mitchell and Deane, op. cit. p. 52; See also Redford, op. cit. *passim*; Handley, op. cit. *passim*; *R.C. Labour*, v (1894), pt II, §IIIA, 335–40. Evidence from the 1890's suggests for the earlier period marked annual fluctuations in the numbers of Irish migrants.—*Journal of the Board of Agriculture*, II, 245; III, 374, 407; VI, 409; VII, 408; VIII, 547; IX, 546.

[3] *Ag. Gaz.* 10 Nov. 1856, 15 Jan. 1853; *F.M.* Oct. 1859, pp. 283–4; J. C. Clutterbuck, 'On the Farming of Middlesex', *J.R.A.S.E.* 2nd ser. v (1869), 10–11; *R.C. Employment, Second Report* (1869), App. II, pp. 112, 222, 412; S. G. Kendall, *Farming Memoirs of a West Country Yeoman* (1944), p. 228; H. Robinson, *A South Down Farm in the 'Sixties* (1947), p. 14; *Ag. Gaz.* 11 May 1867, p. 500.

[4] *R.C. Employment, First Report* (1868), p. 539; *Second Report* (1869), pp. 214, 562; *Ag. Gaz.* 23 April 1870, p. 576.

seasons and bad, but it also invariably requires a complementary and often heavy infrastructural investment in technical education, maintenance depots, and fuel and replacements services.

Yet what might be done for harvesting might obviously be applied elsewhere in agriculture, in village industry, and in many other activities requiring low-cost improvements in work output to remove structural labour bottlenecks. The historical precedents may be found, *inter alia*, in the development of threshing machines, hoes, animal harnessings, axes, saws, spinning wheels, weaving looms, and milling equipment.

However, the present environment for the success of such a strategy is unpromising. Like Kipling's lascar stoker praying before his pressure boiler, policy-makers have often put their faith in the products of Western technology. A few economists, disturbed at the failure of so many underdeveloped economies to "take off", are now arguing that our economic principles, because they are culture-bound and too much modelled on the twentieth-century experience of the advanced countries, blind us to the real problems of the other half-world where neither production nor consumption is studded with gadgets, and where the chief demand is for simple and inexpensive sources of income.[1]

British historical experience may be very relevant here. Pre-1850 advances in both agriculture and industry were characterized by a highly sensitive use of the resources of labour and capital. This study of change in harvesting methods stresses the indivisibility of the cultural, social, economic, and technical elements within a complex process of technological change. It has attempted to demonstrate the importance of the short-run convergence and the fiction of the long-run trend as factors in farmers' decision-making. Above all, it has tried to show that there is a phase of development in which a scythe can be economically and socially more useful than a reaping machine.

University of Reading

[1] L. Dabasi-Schweng, 'The Problem of Transforming Traditional Agriculture', *World Politics*, XVII (1965); R. Dumont, *African Agricultural Development* (United Nations, New York, 1966); Polly Hill, 'A Plea for Indigenous Economics', *Economic Development and Cultural Change*, XV (1966), 10–20; M. Lipton, *The Listener*, 28 March, 4 April 1968; H. A. Luning, *Economic Aspects of Low-Income Farming* (Wageningen, 1967), pp. 5–6; T. W. Schultz, *Transforming Traditional Agriculture* (Yale, 1964).

E. L. Jones

Agriculture and Economic Growth in England, 1660-1750: Agricultural Change

BETWEEN the middle of the seventeenth century and the middle of the eighteenth century, English agriculture underwent a transformation in its techniques out of all proportion to the rather limited widening of its market. Innovations in cropping took place on a wide, though not a universal, front and independently of any great expansion of demand, which was to stimulate the extension of improved methods during the classic agricultural revolution of the late eighteenth century. Except in the sphere of stock breeding, the remainder of the century really had little to offer in the way of techniques which were new in principle. Yet the initial introduction of the most important advanced techniques had come during the late seventeenth and early eighteenth centuries, when the slow and ultimately uncertain growth of population and the modest rise in per capita national income combined to produce only a gradual growth of demand.

The problems posed are of two kinds: one relating to the economic conditions which induced the developments within agriculture, the other relating to the economic consequences of agricultural change. First, how, in a restricted market, did substantial technical innovation, a rise in output, and an expansion of agriculture's capacity to meet the subsequent burdens of an expanding economy come about? Second, how serious was the fall in agricultural investment and incomes which is thought to have taken place during the second quarter of the eighteenth century, and what were the effects on the course of economic growth? What I shall attempt, therefore, is a thumbnail sketch of the timing and distribution of technical changes before 1750 and of their effect on output. I shall then offer two complementary explanations of these achievements and mention briefly the implications for the aggregate levels of investment and income in agriculture, matters which are germane to any incomes-effect explanation of industrial growth. A fuller discussion by Arthur John of the impact of the agricultural changes on the growth of the whole economy is the subject of the following article.

It will be plain that my aim is threefold: first, to give a brief overview of present knowledge on the topic,[1] although it must be insisted that limitations of space prevent any but the most cursory discussion of the evidence, and this must therefore be pursued through the footnote references; second, to relate the fragments of evidence so far available in a provisional and speculative explanatory scheme; and third, by the necessary tenuousness of any model constructed at this stage of research, to indicate the program needed to fill the most critical gaps in our information.[2]

I

What were the early husbandry changes and how were they inserted into the agriculture of preindustrial England? During the sixteenth and seventeenth centuries, the range of plants at the farmer's disposal had been extended beyond any precedent. The discovery of the Americas, for example, had greatly increased the vegetable species available to England, although most of the introductions to this country were made from or through the more intensive agricultures of the Low Countries, and most of them not until the seventeenth century.[3] By that time, as recent work

[1] Especially as revealed in the following key sources: H. J. Habakkuk, "English Landownership, 1680-1740," *Economic History Review*, X (Feb. 1940), 2-17; A. H. John, "The Course of Agricultural Change, 1660-1760," in L. S. Pressnell (ed.), *Studies in the Industrial Revolution* (London: Athlone Press, 1960), pp. 122-155; G. E. Mingay, "The Agricultural Depression, 1730-1750," *Economic History Review*, 2d Series, VIII, No. 3 (Apr. 1955), 323-338. The present article is a slightly revised version of a paper given to the joint conference of the Economic History Society and British Agricultural History Society, at the University of Reading, April 10, 1964. Comments by Gordon Mingay and Donald Whitehead, and by B. H. Slicher van Bath and his staff at a seminar in the Department of Rural History, University of Wageningen, Netherlands, have been most helpful.

[2] In order to extend our knowledge of agricultural output, an analysis of eighteenth-century farm accounts has now been started at Nuffield College, Oxford, under the direction of R. M. Hartwell and the writer.

[3] The account of crop innovations which follows is drawn from references

has shown, farm organization even in the common fields was sufficiently flexible to utilize the new crops. The most influential newcomers were all fodder crops—two legumes (sainfoin and clover) and a root (the turnip).

The most rapid survey of published sources alone shows that the introduction of these plants as field crops came earlier and over a much wider area of England than was once thought. Sainfoin, for instance, was established at Daylesford, Worcestershire, in 1650, and about the same year at North Wraxhall, in northwest Wiltshire, by one Nicholas Hall of Dundry in Somerset.[4] The turnip, after being introduced as a vegetable by Dutch immigrants to Norwich soon after 1565, later spread until by the reign of Anne, according to Defoe, it was to be found "over most of the east and south parts of England."[5] Even in western counties, turnips were being grown during the seventeenth and early eighteenth centuries, and by the middle of the eighteenth century they were grown in Cornwall, the westernmost English county (by 1771, farm accounts for Menabilly Barton, Fowey, are referring to "the Turnip Oxen").[6] In the west, however, turnips were usually sown in small lots as supplementary feed for livestock, not as a full course in the rotation of the whole farm. Considering the high production costs and the unreliability of the turnip crop, and the year-round growth of grass in the southwest, there should be no surprise that it failed to sweep the board. But it was not absent.

Clover leys were sown during the seventeenth century up and down the country from the south coast to Sherwood Forest and across from East Anglia to the Welsh border. Clover was much sown in the west. For example, Andrew Yarrenton grew it at Astley, Worcestershire, and in 1662 wrote his *The Improvement*

listed below; a good contemporary summary is to be found in O. Lawson Dick (ed.), *Aubrey's Brief Lives* (Harmondsworth: Peregrine Books, 1962), especially pp. 28-29.

4 R. C. Gaut, *A History of Worcestershire Agriculture and Rural Evolution* (Worcester: The Worcester Press, 1939), p. 97; F. Harrison, *North Wraxhall, Co. Wilts* (London: Macmillan, 1913), p. 155.

5 K. J. Allison, "The Sheep-Corn Husbandry of Norfolk in the Sixteenth and Seventeenth Centuries," *Agricultural History Review*, V, No. 1 (1957), 27; Daniel Defoe, *A Tour Through England and Wales* (London: Dent & Sons, 1928), p. 58.

6 Rashleigh Farm Accounts, uncatalogued, Cornwall Record Office; J. Rowe, *Cornwall in the Age of the Industrial Revolution* (Liverpool: University Press, 1953), p. 221; on another western county, see E. L. Jones, "Agricultural Conditions and Changes in Herefordshire, 1660-1815," *Transactions of the Woolhope Club*, XXXVII (1961), 40.

improved . . . by clover. He engaged agents to sell the seed and his explanatory tract in many towns in the west Midlands. Similarly, Dutch woolen merchants living at Topsham on the Exe imported clover seed, which they distributed to markets throughout the southwestern peninsula during the second half of the seventeenth century.[7] In addition, the seventeenth century saw an immense spread of floated water meadows from their points of origin in Herefordshire, Dorset, and probably Shropshire during Elizabeth's reign.[8] By 1700, these irrigated meadows were widespread in the Wessex chalklands and the west Midlands, and were still extending.

From our vantage point these innovations may seem to have come in unhurriedly. As late as 1750, all of them, especially the turnip, still had much ground to conquer. Nevertheless, it is undeniable that they were by then widely established. Grafted onto an agriculture which was already being modified by the creeping consolidation of holdings, they forced open the bottleneck of too little fodder with which to over-winter a large stock of farm animals. The fodder crops and the forced grass of the water meadows had the same kind of impact: by increasing the supply of feed they enabled more stock to be kept and better fed, giving more dung per acre. This was a crucial advance for a society which lacked efficient artificial fertilizers. Heavier applications of manure would raise the yields of both cereal and fodder crops, more feed would permit still heavier stocking, and the whole slowly expanding circle would unfold once more. Total output would rise unless offset by a net withdrawal of land from cultivation.

The insertion of grass leys into farming routines, which has been acclaimed as a major advance during the sixteenth and early seventeenth centuries, was most unlikely to produce a *comparable* effect on total output. Ley husbandry may improve soil texture but not soil fertility, since on balance the grazing beasts will take out what they put in. While some improvement in yields may have

[7] Charles Wilson, *Holland and Britain*, (London: Collins, n.d.), p. 108.

[8] Rowland Vaughan, *Most Approved and Long Experienced Water Works* (London, 1610), no pagination; letter to Humphrey Weld from his tenants of Winfrith manor, 1598, Dorest Record Office, Weld Collection D10/E(103). A floated water meadow is one in which river water is conveyed for a series of short periods over adjacent, gently sloping meadow land by means of a complicated system of hatches and ditches. This is in contrast with the older and simpler system of catchwork meadows, watered by washing stream water over hillside grassland. The effects are irrigation, raising the temperature and thus facilitating grass growth, and fertilizing.

resulted, the biological constraints made any substantial rise in output per acre improbable. The achievement of the 1540-1640 period in this respect must be judged meager, its increases of output probably more the result of the extension of cultivation. Despite the early examples of floated meadows and a few cropping innovations, and despite many more studies of the farming of individual counties during the 1540-1640 century than have been made of the subsequent hundred years, no major technical transition in agriculture has been identified during the earlier period. That period looks to have been one in which inflationary profits were redistributed from the laborer to the yeoman and to the landowner, to be spent on the "Great Rebuilding" of farms and on the erection of London houses, rather than one in which productivity far overhauled the growth of population.

The true transformation of crop rotations was accomplished with the adoption on a significant scale in the latter half of the seventeenth century of the innovations in fodder cropping. Nitrogen-fixing by the legumes contributed directly to soil fertility; the improved permanent grass of the water meadows, washed by lime from the chalk streams, fed animals which were daily moved out to dung and thus to transfer nutrients and minerals to the arable fields. These changes were perhaps revolutionary in their effects although not in their pace.

Direct evidence of a resultant rise in total agricultural output is limited, but M. A. Havinden has provided us with some measures of the impressive increases which were secured in Oxfordshire—on the whole an open-field county which, although it grew the new crops, would not be expected to be in the van of the movement.[9] The figures, derived from large samples of probate inventory data, denote a marked rise in livestock populations, especially of sheep and especially on the light soils. They also show a shift in cereal cultivation in favor of the highest-priced cereal, wheat—a feature which is quite contrary to continental experience during the seventeenth- and early eighteenth-century depression. Unfortunately the evidence on total crop production is extremely slender, although the suggestion is of a rise.

There is evidence from elsewhere. The national sheep population responded to the expansion of feed supplies with a considerable

[9] "Agricultural Progress in Open-Field Oxfordshire," *Agricultural History Review* IX, No. 2 (1961), 73-83.

TABLE 1

		Limestone Uplands of Oxfordshire	*Thames Valley (Clays and Loams)*
Size of median sheep flock	1580–1640	14 animals	24 animals
	1660–1730	60 animals	51 animals
		Percentages	*Percentages*
Proportion of cattle herds containing over 5 animals (N.B.: not offset by a fall in the number of herds)	1580–1640	33	39
	1660–1730	46	45
Wheat as a proportion of total crops	1590–1640	14	25
	1660–1730	27	32

expansion during the seventeenth and early eighteenth centuries, and this was of course especially significant for the manuring of the wide expanses of potential new ploughland on the light-soiled uplands. On the other hand, the national cattle population seems hardly to have responded to the stimulus of more and better feed during the first half of the eighteenth century, although it seems to have done so earlier.[10] The output of tallow, hides, and soap remained in the doldrums, while a supply of dairy products was obtained from Ireland. After 1707, the Scots sent more and more store cattle to England, and the Channel Islanders found that they could send growing cargoes of "green hides" and of reexported French calfskins to this country, as well as more and more Guernsey and Jersey cattle.[11] Finally, grain prices slowly sank, and a struggle to encourage exports by means of the corn bounty went on (with intermissions like the lean years of the 1690's) from the 1660's to the 1750's.

This may indicate that supply did tend to pull just ahead of demand, despite the withdrawal from arable production of some land in the clay vales. At first sight more striking still, the notching up of agricultural output took place despite difficulties for the producers of grain. There was a high turnover of small farmers on heavy arable lands, especially in central and eastern England, during the extreme scarcity of the 1690's and the low prices of the

10 There is much scattered evidence on the size of the sheep and cattle populations and the output of their products, the most convenient summary being Phyllis Deane and W. A. Cole, *British Economic Growth* (Cambridge: University Press, 1962), pp. 68-74.

11 Southampton Civic Record Office, Petty Custom and Wharfage Books from 1723.

first years and second quarter of the eighteenth century. Arrears of rent were high on both heavy and light land during some years between 1702 and 1705, 1709 to 1712, and in the 1730's and 1740's.[12] During the last-mentioned decades, a fortuitous run of good harvests raised the supply of grain well above what the home market, the bounty on corn export, the distilling of gin, and the grassing down of the clays could clear. Both grain and livestock prices dropped. Yet the arable acreage continued to be extended and new farms to be created on the light-soiled uplands; fodder crops came more and more into use; and there was a noteworthy reequipping of farms by landowners.

II

First, we have to account for the improvement in husbandry techniques, chiefly the development of rotations involving fodder crops, during the second half of the seventeenth century and the first half of the eighteenth century. At the general level this may not be so difficult. European work shows the extension of fodder crops to be a regular feature of periods of poor cereal prices, when, *ceteris paribus*, relative prices favor a switch to livestock and therefore necessitate extra animal feedstuffs. The spread of the fodder plants as field crops would have been delayed during the late sixteenth century, when grain prices were relatively high. At that time, although many of the fodder plants were available, there was less incentive to sow them as field crops, so that in England they tended to remain localized and in use only as vegetables, whereas their field cultivation had already become established in the Low Countries during the depression of the late middle ages.[13] The mechanics of their subsequent colonization of England need to be more fully worked out, particularly as regards their widespread insertion into common field regimes which were until recently dismissed as totally inflexible; but it is worth noting that many country gentlemen (including many Royalists from western counties) made acquaintance with the Low Countries during the seventeenth century.

Second, and this is perhaps the surprising and uniquely English

[12] See especially G. E. Mingay, *English Landed Society in the Eighteenth Century* (London: Routledge, 1963), pp. 54-55.

[13] B. H. Slicher van Bath, "The Rise of Intensive Husbandry in the Low Countries," *Britain and the Netherlands*, I (London: Chatto and Windus, 1960), 130-53.

feature, we have to consider a rise not only in livestock production but also in cereal production at a time of poor prices for grain. It is certainly unlikely that net cereal output actually fell, while livestock output rose so that aggregate agricultural output would have increased. Two complementary explanations will be suggested for this phenomenon. But even if all the agricultural changes of the period were merely shifts in the locale of production within the framework of a static output of grain, it would still be necessary to postulate how, when cereal prices were falling, production could be so much extended on the light soils of the former sheep downs. The second part of our explanatory scheme could provide such an explanation.

We may first lay stress on English agriculture's built-in mechanism for minimizing differences in the level of investment between spells of rising prices and spells of falling prices. There were two distinct sources of agricultural investment—landowner and tenant. The shifting in their respective shares of estate-maintenance expenditure was (as has been observed of later periods) almost rhythmic. During the 1730's and 1740's, landlords wrote off arrears of rent and reshouldered burdens like the land tax which in somewhat better times they had begun to foist onto the tenantry. Freed from these charges, tenants were able to devote all available resources to productive ends. In addition, in order to compete for and to retain suitable tenants on their farms, landowners even felt obliged to subsidize productive outlays themselves. They provided tenants with grass seeds, rearranged farm layouts, repaired the existing farm buildings and put up new ones, and on the chalk uplands they undertook the heavy capital expenditures needed to float water meadows and to bring into cultivation commons and sheep downs.[14]

There may have been a shift toward investment in capital items by landowners: in buildings, hedges, and roads, where sheep down was being broken up for arable farms. The agricultural investment of the period, notably of the second quarter of the eighteenth century, was in part therefore the outcome of the estate system. The late seventeenth and early eighteenth centuries had seen the ex-

[14] E. L. Jones, "Eighteenth-Century Changes in Hampshire Chalkland Farming," *Agricultural History Review*, VIII, No. 1 (1960), especially 7-10; M. C. Naish, "The Agricultural Landscape of the Hampshire Chalklands, 1700-1840" (unpublished M.A. thesis, London University, 1960), *passim*.

tension of the estates belonging to great landowning families already established in the counties. This was at the expense of the smaller squires, who had been badly hit by war taxation from 1692 to 1715 and (once the hungry nineties had passed) by low grain prices, too. The great estates which had been built up were, in H. J. Habakkuk's phrase, "units of ownership not of production."[15] Their owners did not engage directly in anything so demeaning as husbandry; their incomes were lucratively supplemented, and sometimes half supplied, from nonagricultural sources such as army posts and pensions. The physical capital of most estates had been run down during the wars; but when high interest rates dropped after 1720, enclosures and improvements became profitable again. The larger landowners made it clear that they were in the market for good tenants and that, since they could let at lower rents than the lesser gentry, they would stay in the market even when times became less favorable. During the second quarter of the century, therefore, landowners, notably the great lords, were prepared to subsidize agriculture. For social reasons, and because of their other resources, they were comparatively insensitive to poor financial returns. They accordingly cushioned their tenants against the full force of falling prices and thus raised the level of investment above what it otherwise might have been.

III

A complementary explanation of the considerable innovation in agricultural practice which went on right through the low-price periods of the late seventeenth and early eighteenth centuries is to be derived from the dominant distribution pattern of types of farming. We must first discount the two dichotomies customarily used to organize English agricultural history: the distinctions between open-field and enclosed counties, and between a progressive east and a backward west of England. Research has diminished the sharpness of these divisions and proved them inadequate to explain the course of technical change or the economic experience of agriculturists. Examples have already been cited which make it clear that the east-west division is unhelpful: it is based on inappropriate indices of the rate of technical change—that is, on the adoption of a root break or of the full-blown Norfolk rotation. The

[15] Habakkuk, *Economic History Review*, X, 5.

link between this dichotomy and the rate of technical change in agriculture was in any case never very explicit. It depended mainly on the fact that East Anglia is nearer than the west country to Holland—the relevance of which seems less striking on a consideration of the actual spread of the new crops within England. Similarly, the divergence between open-field and enclosed areas has been blurred: in Oxfordshire, for example, it has been shown that the introduced fodder crops spread faster on the light soils of the limestone uplands, even in the open fields, than in the more enclosed clay vales like that of Thames.[16] In the case of the open-field–enclosure dichotomy, the link with innovation was supposedly through the "slowest ship in the convoy" thesis, whereby the most dilatory open-field farmer dictated a snail's pace of adjustment in the system of communal husbandry.[17] Modern work, showing that the new crops were early inserted into many common field routines, appears to indicate a different power structure, whereby the more pushing and articulate men often secured their neighbors' consent to changes. Both the east-west and open-field–enclosure distinctions do nevertheless refer to important factors influencing the detailed pattern and timing of agricultural innovation, but in no all-embracing or enormously revealing way.

A more fruitful division is offered by the scarp-and-vale topography of lowland Britain. This cuts boldly across county boundaries, dividing agriculture into two main groups of farming systems: on one hand, those on the free-draining, light soils of the chalk and limestone uplands, the lightest loams, and some of the more fertile sand lands; and on the other hand, those on the heavy loams and ill-drained clays. In the former category are the Cotswolds, the Wessex chalklands, the Norfolk "good sands," and so forth; in the latter, the Midland clay triangle and the clay vales of other parts.[18]

16 Havinden (cited in n.9), *passim.*

17 Compare the categorical dismissal of the ability of open-field farmers to utilize the new crops and the statements (culled from the biased "improving" literature) about the slowness of innovation, in Lord Ernle, *English Farming Past and Present*, (6th ed.; London: Heinemann, 1961), pp. 122, 134.

18 Many studies by historical geographers have attempted to measure agricultural differences between light-soiled uplands and heavy-soiled vales, and sometimes between finer units. Such work includes S. E. J. Best, *East Yorkshire: A Study in Agricultural Geography* (London: Longmans, 1930); D. R. Mills, "Enclosure in Kesteven," *Agricultural History Review*, VII, No. 2 (1959), 82-97; J. A. Sheppard, "East Yorkshire's Agricultural Labour Force in the Mid-Nineteenth Century," *ibid.*, IX, No. 1 (1961), 43-54; D. B. Grigg, "Changing Regional Values during the

Agricultural Change, England 11

The scope for innovating was not uniform across these two divisions. Although varying from crop to crop, the potential value of the innovations was, in total, greater on the light soils which had previously been too infertile for permanent tillage but which were dry enough for stock (especially sheep) to be kept on the land during winter to feed off the fodder crops, thus returning manure for a following cereal crop. The light land of the Wessex chalk and the Cotswolds benefited, too, from the floating of their valley meadows. Once it became possible to keep their thin soils fertile, cereals could be grown more cheaply on them than on the heavy clays, where, in contrast, traction costs were high, the working season was curtailed, and the land was too wet for stock to be folded on forage crops during the winter months. Because roots grow better in the fine texture of light soils and are difficult to feed off (or even to lift from) heavy, ill-drained land, the turnip was especially slow to penetrate clay districts. On the clays, therefore, less advantage could be taken of the contemporary innovations. The available studies do indeed suggest that such areas were usually very tardy in taking up the new crops.[19]

As a result, during the late seventeenth and early eighteenth centuries, some of the clays went down to grass and specialized more than ever in fattening and dairying, whereas there was a marked extension of cultivation on the former sheep downs of southern England, and as far north as the sand lands of Sherwood Forest. During the third quarter of the eighteenth century this distinction was emphasized, according to contemporaries, when the output per acre on light soils rose, for example on the Cotswolds, to "equal that of a like quantity of land in the vale, where the rents are double and treble the price and the land will not admit of proportionable improvements."[20] In other cases, there was

Agricultural Revolution in South Lincolnshire," *Transactions of the Institute of British Geographers* (1962), pp. 91-103. But these writers have hardly generalized their findings or pursued the economic implications, while historians are usually content to work with administrative units which often blur agricultural distinctions.

19 See, for example, G. H. Kenyon, "Kirdford Inventories, 1611 to 1776, with Particular Reference to the Weald Clay Farming," *Sussex Archaeological Collections*, XCIII (1955), 78-156; and C. W. Chalklin, "The Rural Economy of a Kentish Wealden Parish, 1650-1750," *Agricultural History Review*, X, No. 1 (1962), 29-45. The problems of the rates of assimilation of the new crops by different farming systems are discussed in E. L. Jones, *Transactions of Woolhope Club*, XXXVII, 32, 38-42, and in his "English Farming before and during the Nineteenth Century," *Economic History Review*, 2d Ser., XV, No. 1 (1962), 145-47.

20 Simon Moreau, *A Tour to Cheltenham Spa* (Bath, 1783), p. 62.

a reappraisal of rents in keeping with the altered values of the two kinds of land; for example, it was observed that "the dry grounds in Hertfordshire, which formerly were lett for a trifling rent, are now lett at twenty shillings per acre, since the introduction of clover & turnips into their poor & barren hills; while the low lying stiff grounds pay only ten shillings, which is the rent they gave near a century ago."[21]

Thus the centers of arable prosperity were shifting from the heavier lands to the lighter lands. As regards the overall economic condition of the Midlands and the heavy-clay vales, some compensation was probably found in two movements which may have tended to hide the problems of husbandmen on their arable open-field lands. First, cottage framework knitting was spreading into these areas, absorbing underemployed labor or hands ousted from farming as the vale countryside turned more to pastoral husbandry.[22]

Although the relative rates of population growth by land-use divisions have been assessed from nineteenth-century census data, next to nothing is known of their earlier history. Seemingly, there were greater opportunities and pressures toward additional settlement and nonagricultural employment in the vales and on the lowland heaths than on the light-soiled uplands. The conversion of the uplands to tillage and the establishment on them of isolated new farms may have needed some hands beyond those previously present but underemployed; but the new cereal culture could doubtless have been carried on without much of a permanent influx, since the work was seasonal and harvest hands, for instance, were obtainable from the vale and the clothing towns.

Second, the trends of animal production were less unfavorable to the heavier soils. Within the increase in the size of the national flock there was a noteworthy increase in the population of long-wooled sheep in the Midlands. In Defoe's day, Leicestershire was "a vast magazine of wool." That county, which in the first half of the seventeenth century had been notable for its tillage, had turned

[21] J. H. Smith, *The Gordon's Mill Farming Club, 1758-1764* (Edinburgh: Oliver and Boyd, 1962), p. 149.

[22] J. D. Chambers, "The Vale of Trent," *Economic History Review*, Supplement 3 (1957), p. 13, n.8; W. G. Hoskins, *The Midland Peasant* (London: Macmillan, 1957), p. 227; Joan Thirsk, "Industries in the Countryside," in F. J. Fisher (ed.), *Essays in the Economic and Social History of Tudor and Stuart England* (London: Cambridge University Press, 1961), pp. 70-88.

more and more to grass. Wool from Leicestershire, Northamptonshire, and Lincolnshire poured into Cirencester and Tetbury markets for the clothing trade of the west of England; more mutton was in demand as a cheapening loaf released spending power. This was the period which saw the rise of the gentleman graziers, building William and Mary and Queen Anne houses in the Midland shires; but it was a time which offered unsatisfactory compensation for sinking grain prices to those small farmers in the open fields whose fortunes still depended centrally on the cereal markets.[23]

Although stated very baldly here, this distributional model helps to explain much of the adjustment which was taking place within agriculture. The existence of this broad twofold grouping of farming systems was noted by contemporaries and continues to be recognized by farmers and naturalists. The topographic distinction is often sharp. But a mechanism whereby this could influence the course of innovation and output during spells of low prices and incomes has not hitherto been spelled out. The classification used here does not of course exhaust the ecosystems of lowland Britain, while any simple dichotomy runs the risk of violating some part of the evidence.[24] Nevertheless, the distinction seems to yield a useful categorization of farmers' economic fortunes and of the rates at which they assimilated new techniques. What the present discussion attempts to add is a brief account of the function of the dichotomy in terms of different physical constraints and costs of agricultural production.

Such an account offers a "pie-slicing" thesis. Producers on light soils were able to take a disproportionately bigger slice of the pie (the market) at the expense of the higher-cost clay-land cereal producers. Their comparative advantage was reinforced by their proximity to the markets: East Anglia to London and the east-coast ports; the Wessex chalk to London, Bristol, and Southampton (whence grain was sent to Iberia); the Cotswolds and the Ryelands of Herefordshire to Bristol and Monmouth, whence grain was also exported to Iberia. The Midlands farmers had to dispatch

[23] Defoe, II, 89; W. G. Hoskins, *The Making of the English Landscape* (London: Hodder & Stoughton, 1955), p. 124.

[24] The third main division in terms of area, the lowland heath, may however be ignored here as consisting of sandy soils also termed "light," but too infertile for cultivation except for a little temporary enclosure during periods of exceptionally high prices, such as the Napoleonic wars. See E. L. Jones and C. R. Tubbs, "Vegetation of Sites of Previous Cultivation in the New Forest," *Nature*, CXCVIII, No. 4,884 (June 1963), 977-78.

their grain farther and over worse roads to reach any of these outlets. Higher production and transport costs thus placed them in an inferior competitive position even given stable grain prices.

During the gentle subsidence of cereal prices during the late seventeenth and early eighteenth centuries, therefore, light-land farmers found that by innovating they could profitably expand their output. By putting a bigger volume of produce on the market, even at reduced prices, they could at least maintain their incomes. Technically, they had more room to maneuver than did their clay-land competitors. Except in some years between 1702 and 1712 and during the 1730's and 1740's, there is no reason to think that the total profits of light-land farmers fell. But the increased output of grain from the light soils necessarily tightened the screw of low prices on the higher-cost, technically less flexible, clay-land farms. The rapid turnover of tenants, the selling out of owner-occupiers, and the conversions to pasture on the clays suggest that, despite more modification of their open-field farming than was accepted by the older writers, the clays were being outcompeted in the grain markets. There is some evidence that clover was adopted by farmers on the Chilterns, where the chalk is capped with clay-with-flints, precisely because they were obliged to improve their yields if they were to remain competitive with the grain from the freshly ploughed up downland of the Salisbury Plain area.[25]

During the sharper price falls of 1702-1705, 1709-1712, and the 1730's and 1740's, not all light-land farmers escaped. Arrears of rent mounted up on some estates in those districts. In the last of these spells there are signs that output went up and the new husbandry spread faster than ever in Norfolk. It may be suggested that this was the result of a perverse reaction to lower prices. Faced with falling prices but with the costs in arable farming incurred well ahead of selling, farmers would find their incomes squeezed. The light-land men were the most likely to react to this situation by increasing output. On the light lands, the fixed, inescapable costs imposed by high rents on long leases (instead of the annual agreements and low rents of the clay-land open-field farms) and by the hiring of a high proportion of the labor input were greater than on the small wheat-and-bean farms of the clays. Costs were therefore "sticky," and although in the long run technical innova-

[25] John (cited in n.1), p. 147.

tion would reduce them and maintain profit margins, in the short run there were certain to be adjustment difficulties. Variable costs could not be cut down fast enough. The burden was to some extent thrown onto the landlord in the form of high arrears of rents, and in the long run it was eased by cost-reducing, yield-raising innovation, most of all on the light lands. The possibilities of cutting unit costs of production by introducing the new fodder crops and associated husbandry regimes were greater on the light soils than on the clays. Thus the price-cost squeeze on the light soils was most intense (the clay-land owner-occupier may at times have retreated into subsistence) and the prospects of eventual repayment for innovation, in the form of higher rents or at least by abolished rent arrears, were brightest. Certainly higher rents were sometimes a condition of the landowner's outlay on capital items on light-land farms—as an illustration, forty pounds was "allowed towards building a New Rickhouse 1738 as per Agreement on Farmer Morgans taking an advanced Rent" on a farm at Broughton, Hampshire.[26] And light-land farmers were willing to turn to the new crops specifically as a way out of their dilemma: for instance, in 1745 a farmer at Stoke Charity on the Hampshire chalk attempted to extricate himself from financial entanglements with his landlord by proposing to sow large acreages of "Cinque Foyle" in future years.[27]

Combining our explanations, some of the agricultural development of the period may be attributed to the rhythms of the landlord-tenant system and to the growing social motives for the ownership of agricultural estates, and some of it to the disparate technical and cost situations of light-land and heavy-land farmers. Light-land farmers, during the prolonged, gentle deflation of cereal prices, were persuaded or obliged to exploit their opportunities of taking a larger share of the market. During the sharper falls in agricultural prices, they inclined more energetically to expand their output to protect their incomes. This response was what the nineteenth century was to call high farming. In the first half of the eighteenth century, agriculture was even less free to contract if faced with a fall in prices than during the nineteenth century,

[26] Jones, *Agricultural History Review*, VIII, No. 1, 9, quoting Nottingham University Archives, Manvers Collection, Rentals.
[27] *Ibid.*, 9, citing Stoke Charity Papers, Hampshire Record Office: 18M54/Box E, pkt. A.

since in a preindustrial society there were comparatively few attractive alternative occupations, although there was a gradual drift of laborers into framework knitting and, as Arthur John shows below, into certain towns. A land-use inversion between the uplands and the vales lasting fifty or one hundred years was not, in the circumstances, a particularly sluggish adjustment. It is possible that the rise in national cereal output to be expected from the breaking up of new land could be held in check by a much more limited grassing down of the clays; for, after an initial spurt of fertility, yields per acre on the light lands were usually lower than in the vales. But there does seem to have been some pressure of excessive production because of the inability of resources to flow freely out of clay-land agriculture.

IV

What might the implications of this analysis be for the totals of agricultural investment and income? As regards investment, the scraps of evidence at hand do not point to striking changes in aggregate levels. Instead of producing more when his income was squeezed, the small clay-land owner-occupier may merely have worked harder himself to avoid buying inputs or hiring labor (he was, after all, likely to be underemployed much of the year), and to that extent cash spending on inputs may have fallen. But on larger holdings where labor-intensive fodder crops replaced a grain course or permanent pasture there was presumably a net increase in employment. On the light lands, farmers seem to have invested more vigorously in improved techniques in attempts to extricate themselves from the price-cost squeeze. And landowners certainly in many cases stepped up their agricultural investment during the 1730's and 1740's. The tentative conclusion (which is all that can be drawn) is that total investment, on balance, probably went up by some modest proportion.

When it comes to assessing changes in total agricultural income, we are making bricks quite without straw. I suggest as a working hypothesis that during the late seventeenth and early eighteenth centuries as a whole, when the small farmers of arable clay land were often embarrassed, any fall in their incomes may have been offset by the gains of the more successful innovators, predominantly light-land men. This perhaps implies the concentration of income in fewer hands, with implications for the ratio of basic to luxury goods purchased. During the 1730's and 1740's, total agricultural income may have dropped, for there were spells when on both heavy and light lands rents fell and farmers were in distress. Yet, as far as spending by agriculturists on the products of an incipient industrial economy is concerned, it has yet to be shown that this in itself had enormously depressing consequences. The period saw a sharper phase of the secular shift in resources and distributions in agriculture, whereby the clays tended to go down to grass and the light lands to be converted to tillage, with concomitant expenditures on new building, floating meadows, paring and burning the grass downs prior to their cultivation, and so forth. These undertakings were primarily located on the light lands, whereas during the third quarter of the eighteenth century, when grain prices were rising, landowner investment may have been lower on the light lands,[28] concentrating instead on Parliamentary enclosure (perhaps an attempt to create in the jumbled open fields of central England the physical setting of an improved agriculture which had earlier and more readily been provided on the empty sheep downs of the light soils) and on transplanting Norfolk-type rotations *en bloc* to less developed regions like Scotland. But the income generated by the undertakings of the second quarter of the century may have balanced any fall at that time in the purchase of consumer goods by the farm sector. In addition, most big landowners could bear their part in such enterprises from nonagricultural funds and without foregoing conspicuous forms of consumption. The checks and balances of landlord and tenant and of the two dominant groups of farming system, therefore, tended to circumscribe a fall in total investment or income.[29]

[28] See Naish, *passim.*

[29] English experience contrasts sharply with that of Europe, where the fall in population and depression in agriculture were very severe. This is of particular consequence with respect to the Low Countries, which had formerly been ahead of England in agricultural techniques but which failed to respond so energetically (or perversely) to the fall in prices. In the reclaimed areas of Holland, the drainage network remained a fixed charge on agriculture when farm-product prices were falling. The status of landowning was lower than in England (seignorial rights could be acquired without buying land) and did not attract investment in the same way. And when returns from agriculture fell after 1650, the one hundred or so Amsterdam merchants who owned land in the Beemster, North Holland—the reclamation of which they had financed—quit their estates and retreated to the city. On the European experience, see B. H. Slicher van Bath, *The Agrarian History of Western Europe, A.D. 500-1850* (London: Arnold, 1963), 206-20. I am also indebted to that author for a typescript of his lecture, "Die Europaischen Agrarven-

18 *E. L. Jones*

Any net decline in consumption in the farm sector might be expected to have been offset by the surplus real income created among the buyers of food, although the gain in income for the consumers of a cheaper loaf was spent partly on white bread as a replacement for rye bread, partly on more meat, partly on tea and sugar, partly on gin, and some of it surely was taken out in increased leisure. The second quarter century was not unreasonably described as a "golden age" for the laboring man.

Within agricultural production, systems were being regrouped, however much subsequent periods of high grain prices might rejuvenate clay-land arable farming and thus, temporarily, put the clock back. But by comparison with later periods (though not with earlier ones) resources were still choked up in agriculture as a whole. This, together with the near stagnation of population growth, meant that the rise in purchasing power brought by cheaper food was not yet strong enough to throw down the final barriers against large-scale industrial expansion. This was to come in the third and fourth quarters of the century, when prices rose because a long run of inclement weather depressed yields and (rather later) when the growth of population extended the market. At that time the upsurges of investment and income in agriculture seem to have swamped any damage done to expansion elsewhere in the economy by higher food prices.

Nuffield College, Oxford

hältnisse im 17. und der ersten Hälfte des 18. Jarhhunderts," and to Drs. A. M. van der Woude for showing me the Beemster.

The Second Agricultural Revolution, 1815–1880[1]

By F. M. L. THOMPSON

THERE are two sorts of manures, "enriching manures and exciting manures", according to the observation of an early nineteenth-century chemist, and it is here proposed to examine some of their possibilities for historical enrichment, if not excitement.[2] Manures, or their absence, are of course vital in every farming system, but this article is not concerned with the circumstances and purposes of their application, which must be left to those studies in agrarian history whose purpose has lately been rather unkindly defined as the plotting of "every rotation in every field in every parish in every county".[3] Nor is it concerned with farmyard manure, though efforts to conserve and utilize this, especially efforts to collect, store, and distribute the urine of farm livestock, played a not insignificant part in the process which is here termed the commercialization of British agriculture. Indeed the nineteenth-century discussions of the relative manurial merits of the urine of horses, cows, sheep, and pigs, the devices contrived for the measurement of the volume of urine voided in one year by examples of each of these beasts, and the solemn debates on whether it was more beneficial to use it warm, cold, fresh, or stale make interesting and, to the initiate, edifying reading.

Agricultural students of the later nineteenth century in fact had to read works dealing with such matters, and no doubt had to reproduce the percentage of phosphoric anhydrite in horses' urine in their examinations. They also learnt about the manurially significant chemical constituents of town sewage; and the effort to close the utilitarian circle of public health—from food, through excreta, flush-closets, egg-shaped sewer pipes, main outfall sewers, and back to food again—by inserting sewage irrigation systems into the end of this cycle to bring the product into profitable agricultural use, occupied quite a bit of the polemic energy of Edwin Chadwick and his disciples.[4] It also occupied some of the time and money of the more wealthy or bizarre agricultural improvers, who by and large seem to have lost heavily on this particular operation and to have been left with a lot of rusty old pipes and pumps as the sole reward for their attempts to pipe a supply of town sewage to their fields. Nowadays, a sewage farm is a place for purifying and getting rid of sewage, not a place for growing specially heavy food crops, though it is true that those who can stand the smell and cope with the bulky material involved purchase semi-dried sludge cake from municipal sewage works, and are thus the Chadwickian sewage-consumers in being.

Neither of these fascinating aspects of manurial history enter into this article, for the sufficient reason that they are not readily susceptible of measurement.

[1] Based on a paper delivered to the Anglo-Soviet Conference of Historians in Moscow, Sept. 1966.

[2] James Anderson, *Quick Lime as a Manure* (1797), quoted in Sir E. J. Russell, *History of Agricultural Science in Great Britain* (1966), p. 85.

[3] Lord Ernle, *English Farming Past and Present* (6th edn, 1961), introduction by O. R. McGregor, p. cxii.

[4] Edwin Chadwick, *Report on the Sanitary Condition of the Labouring Population of Great Britain, 1842*, ed. M. W. Flinn (1965), pp. 59–60, 120–4.

The purpose of this article is a simple one: to advance the proposition that there was not one single agricultural revolution but three, of which the second fell in the period 1815–80; and the method of proof involves reasonably reliable measurement. In the process of performing this elementary dissection of the concept of the agricultural revolution, some light may be thrown on the argument that landlords exploited and oppressed tenants, and generally blocked or made difficult the path of technical and economic progress.[1]

The notion that there was a single, unitary agricultural revolution has lately gained renewed currency and authority because it forms the title of a book, *The Agricultural Revolution 1750–1880*.[2] The inference is plain, that all agricultural history before 1880 was but a working out and an application of the techniques generated in the eighteenth century. It seems to me both more illuminating and closer to the facts of agricultural history to distinguish conceptually three different kinds of technical and economic changes which have led from the traditional open-field farming of medieval Europe to twentieth-century factory-farming. A sixteenth-century historian would undoubtedly make it four rather than three phases, and the first would be the shift from peasant subsistence-farming to cash-farming for the market, leaving the methods of cultivation and the physical organization of farms substantially unchanged.

The second change, and the agricultural revolution proper, involved the technical changes of crop rotations and livestock improvement; it generally involved the physical changes of enclosure, and it embodied the economic changes of increased intensity of cultivation, that is, production functions were altered by an increase in the amounts of both capital and labour that could be profitably employed in relation to land. Farmers clearly became more market-orientated than before, in deciding what crops to grow, when, and in what proportions. But the supply of the major part of the extra capital required came from landlords, even if it was the tenants who serviced it by paying increased rents, and the major part of the capital expansion was administered by the landlords also. Certainly the tenant-farmers had to mobilize extra working capital, but this was chiefly required to meet enlarged labour bills. Some extra capital, it is true, was required to stock the farms with the larger flocks of sheep required by the new husbandry. But it might be argued that the economies flowing from a rearrangement of strips or a consolidation of holdings furnished an opportunity for the internal accumulation of the necessary savings; or even that there was simply a more productive redistribution of tenants' capital which was already within the system.

From the tenant's point of view this agricultural revolution was mainly a managerial revolution, in the sense that the new husbandry required him to manage the resources of his farm in an orderly, competent, rational, and efficient manner in order to reap the rewards of larger yields, better stock, and greater income per acre farmed, which mixed farming offered. On the whole, these things did not cost him very much money, but they did cost him a great deal of intelligence in comparison with that displayed by his ancestors. Above all, the concept of the mixed farm, the spearhead of this revolution, was essentially a concept of a self-sufficient productive unit. Certain materials, like lime, chalk, or marl, might

[1] Lord Ernle, ed. op. cit. pp. cxxiv, cxxx–cxxxi.

[2] J. D. Chambers and G. E. Mingay, *The Agricultural Revolution, 1750–1880* (1966).

indeed be procured from outside and carted on to the farm, but these were largely intended to be permanent or long-term corrections of deficiencies in the soil of that farm, and the carting was looked on as almost a once-for-all operation. Fundamentally the production cycle of the mixed farm was, then, a closed circuit, and this was its whole beauty and symmetry. It produced for sale wheat, barley, meat, and some wool; the roots, clovers or other rotation grasses, and perhaps pulses, as well as the hay which it also grew, were consumed on the spot and furnished the richer and more abundant supplies of manure from which the larger cereal yields came, as well as supported the livestock production and the horse-power which worked the farm.

There were indeed several exceptions to this picture of individual farm self-sufficiency, most of minor importance. The exception of real significance was that the seeds for the new crops, particularly for turnips and clover but also for tares, sainfoin, lucerne, and in the nineteenth century Italian rye-grass, frequently had to be purchased off the farm and frequently were imported. In the course of the eighteenth century, if not earlier, certain regions or farmers established reputations as seed-raisers, for example parts of Kent which supplied the London market with high-quality turnip seed. Imports also grew. To take the one article of clover seed, a considerable import trade was established, drawing on Holland as its chief source of supply until the early nineteenth century, with Germany moving into the leading place in the 1820's, France of some significance by the 1830's, and the U.S.A. becoming an intermittent but occasionally a large supplier at the same time. By the closing years of the Napoleonic Wars 62,000 cwt were being imported annually, enough to sow about 438,000 acres of clover. On the possibly rash assumption that all this clover was used in four-course rotations, this implies an acreage of 1,750,000 under the new husbandry by 1815, quite apart from the acreage using indigenously raised seeds. The imported clover seeds were then costing British farmers about £200,000 a year, so that in this one item a regular habit of purchasing a key input had been established. Financially the annual outlay involved was small in relation to farmers' total outgoings. But the ground was being prepared for further developments in the purchasing of inputs, and a network for their distribution was already being established in the shape of seed-merchants' houses in the chief agricultural towns.

The essence of the second agricultural revolution was that it broke the closed-circuit system and made the operations of the farmer much more like those of the factory owner. In fact farming moved from being an extractive industry, albeit of a model and unparalleled type which perpetually renewed what it extracted, into being a manufacturing industry. The technical changes involved concerned fertilizers and feeding stuffs, and those will shortly bring me back to manures. The physical changes were largely associated with field drainage and with purpose-designed farm buildings, and these were mainly provided by the landlords, usually without very handsome benefit to themselves.[1] The economic changes involved a further increase in the intensity of cultivation, but this time the shift in the production functions was an increase in the amount of capital that could be profitably employed in relation to both labour and land, and the rise in the number of agricultural labourers first slowed down and then went into

[1] F. M. L. Thompson, *English Landed Society in the Nineteenth Century* (1963), pp. 248–50.

reverse. Though part of the fresh capital required was supplied by the landlords, as before, a rapidly rising proportion of it was supplied by the farmers themselves. It is the twin facts of a substantial rise in the amount of tenant-farmers' working capital and of the adjustment in outlook required to regard farming as an activity in which, at least in significant degree, purchased raw materials are processed in order to produce a saleable finished product that justify the view that it was in the second agricultural revolution that farming became properly commercialized.

In the third stage labour was replaced by machinery—the labour of men by mowers, reapers, binders, tedders, and threshers, and the labour of horses by tractors. In addition, fertilizers were applied in hitherto undreamt-of quantities per acre, and continuous cereal cropping became possible without incurring the disapproval meted out to soil-mining. This development has taken the floor from under mixed farming and the idea of a cereal–livestock balance, and extreme specialization by crop has become technically possible, and in certain situations economically profitable. Possibly a subdivision of this final stage, or even a fourth stage, ought to be distinguished: the stage when the function of livestock-farming came to be regarded as the provision of standing room, preferably covered, on which eating machines may consume imported feeding stuffs and convert themselves into eggs, oven-ready chickens, or beefsteaks as quickly as possible.

If we settle for three stages only, it is still apparent that although they may be conceptually distinct they were not chronologically distinct in the tidy fashion of political revolutions. They were, however, much more distinct in time in Britain than in many other countries. It may, indeed, be argued that in Germany, for instance, the first and second revolutions, and for good measure the preconditions phase of transition from subsistence to market-farming, all happened together in a great rush in the last quarter of the nineteenth century and the years before 1914, and that it is this bunching which explains the phenomenal growth in German agricultural productivity in that period, compared to the relative stagnation in Britain. It is true that at any given moment in the last hundred and fifty years in Britain it is easy to find farmers who were just applying the features of one of these stages at the very time that the most go-ahead farmers or regions were several jumps in front. In parts of Wales, indeed, it seems that the first agricultural revolution only arrived in 1939, and was immediately telescoped into the tractor revolution.[1] Nevertheless, it is roughly true to regard the first revolution as being completed in Britain by 1815, when the great wave of enclosure was over, apart from tidying-up operations, and when the lure of easy wartime profits, with their inducement to mark time technically, was withdrawn. Before 1815 the embryo of the second revolution was present, but it only developed markedly thereafter; by about 1880 the force of this revolution was spent, its hallmarks were widely spread over the farms of the country, and any further spread was checked by the great price fall. Already before 1880 many of the implements of the mechanization revolution had been developed, and were in fairly wide use in Britain, so that once more we have a mature revolution pregnant with a new one. But we cannot regard this offspring as being fairly born in Britain until 1914, when the

[1] J. G. Jenkins, 'Technological Improvement and Social Change in South Cardiganshire', *Agricultural History Review*, XIII.

tractors began to move, with some assurance of reaching the other side of the field. This phase still continues, or has moved without pause into the fourth stage; so that in all the years since 1650, when the turnip arrived, there is only the one short period from 1880 to 1914 when British agriculture may plausibly be held to have been free from profound technical changes.

In this article the second agricultural revolution is the centre of attention. In a technical sense, though not in the sense of commercial application on a significant scale, it began well before 1815: and it began with manures. Its beginning was a matter of bones and of rape cakes. Bones had been used agriculturally before 1815, but not long before: Loudon in his 1835 *Encyclopaedia of Agriculture* states that they were first introduced into Lincolnshire and Yorkshire about 1800 by a bone merchant of Hull.[1] Previously this merchant had presumably supplied his bones entirely to the glue industry. And there is evidence that in 1815 about 8,000 tons of bones were imported into Hull from the Continent.[2] The prehistory of oilseed cakes, of which rape cake is one, is both longer and more obscure. These are made from the residues left when oilseeds—linseed and rape seed being the principal ones used in Europe until joined by cotton seed in the 1860's—are crushed in order to produce oils. These oils have long been required for industrial uses, linseed oil chiefly for paint-making and rape oil chiefly for use in the cloth-dressing stage of woollen manufacture, and were presumably needed for these purposes even in the Middle Ages. Rape oil, indeed, a Select Committee of 1816 was told, was quite indispensable to the woollen industry, which used three gallons for every hundred yards of broadcloth produced.[3] This being the case, sizeable quantities of the residues must have been available for a very long time; but for a very long time they were apparently regarded as worthless waste products. Very slowly there developed an agricultural demand which conferred commercial value on these wastes when pressed and stamped into cakes. From 1667 comes a report that linseed cake was being burnt as a fuel in Lincolnshire, while 1716 is said to be the date of the first reported use of cake as cattle feed, in Northamptonshire. This was rape cake which, however, was and remained through the nineteenth century unpopular among British farmers as feed; it was used as a manure, again in Lincolnshire, in 1725, and by the time of the 1816 inquiry was regarded as of use solely as a manure. Meanwhile, the Agricultural Report on Lincolnshire, 1794, establishes that the use of linseed cake for feeding was by then not uncommon there.[4]

By the time of the French Wars, therefore, the oilseed cakes were well on their way to upgrading their status from that of wastes to that of useful industrial by-products. They became, in the economists' language, joint products with the oils, always junior to the oils in price but still essential determinants of the oil prices, only with the great expansion of the agricultural demand for them in the seventy years after 1815. The changing status of the cakes is reflected in their price history. Linseed oil and rape oil were equally valuable, and each sold at virtually the same price per gallon both in the 1760's and in 1816; but whereas linseed cake

[1] J. C. Loudon, *An Encyclopaedia of Agriculture* (1835 edn), p. 807.

[2] J. M. Wilson, *The Rural Cyclopedia* (1848 edn), I, 474.

[3] *S[elect] C[ommittee] on Seeds and Wool*, British Parl. Papers 1816, VI, evidence of Jervis Walker, clothmaker, pp. 24–5.

[4] Cited by H. W. Brace, *History of Seed Crushing in Great Britain* (1960), pp. 11, 28–9.

was worth only 1·36 times as much as rape cake in the 1760's, in 1816 it was worth over three times as much; and though rape subsequently became a little more popular, the ratio settled down through most of the nineteenth century at 2:1 in favour of linseed.[1]

Luckily for the historian, virtually the entire output of these cakes was derived from imported materials, and it is therefore possible to establish the quantities produced, and to assume that when averaged over groups of years these were the same as the quantities consumed in agriculture. Certainly the crops of linseed and rape were grown commercially, though not on a large scale, in the British Isles, but except for linseed in Ireland in the first couple of decades of the century, neither crop was grown for seed.[2] Linseed was raised for the flax, and rape was grown as green fodder for stock or for ploughing in as green manure. Hence no awkward and insoluble question arises as to the extent of home production of oilcakes from home-produced raw materials. The crops, when allowed to go to seed, were held to be peculiarly damaging ones in terms of soil exhaustion, so that as British cattle and sheep steadily munched their way through more and more tons of oilcakes, and as steadily dropped their enriched dung on British fields, so they inexorably impoverished the soils of the chief suppliers of the seeds, at first Prussia, then from the 1830's Russia, India from the 1850's, and Egypt with the cotton seeds from 1865.[3]

Annual consumption of oilcakes in the last five years of the French Wars was running at around 25,000 tons. There was a small but established import of cakes from Europe of about 1,000 tons a year, and about the same quantity was coming in from Ireland; the rest was produced at home from imported seeds. There were seed-crushing mills at or near many east-coast ports, but London and above all Hull were already the main centres of the industry. Hull, favourably situated in relation to the European suppliers of seed as the point of import, was naturally favourably regarded by Lincolnshire, the East Riding, and Nottinghamshire as the point of supply for oilcakes. Maybe it was this locational factor which helped to account for the primacy of this region in agricultural advance in the post-1815 decades; on the other hand, as the case of bones will also demonstrate, maybe it was the existence of a strong and localized agricultural demand in this region which helped to determine the industrial location. Be this as it may, by 1856 out of 150 hydraulic seed-crushing presses in use in Britain, 100 were in Hull.[4]

By that date, 1856, the home production of cakes from imported seed had risen from the 23,000 tons of 1815 to 190,000 tons, and the imports of cakes from 1,000 tons to 83,000 tons. Neither the rise in consumption nor the geographical spread of the practice of using cakes had proceeded evenly. Consumption remained stable until 1820, and then increased in the depression years 1821–4, not by very much but still by a perceptible 15 per cent. It seems quite likely that this was a response by Lincolnshire farmers to falling prices for their farm products. Consumption then practically doubled in the late 1820's, but fell back very slightly in the next

[1] Ibid. p. 28; *S.C. on Seeds and Wool*, 1816, pp. 18–24, 67–9, 70–2; *Trade and Navigation Returns*, annual, with current import prices from 1854.

[2] *S.C. on Seeds and Wool*, 1816, p. 102.

[3] All import statistics and other details are from annual *Trade and Navigation Returns* except where otherwise stated.

[4] Brace, op. cit. p. 49.

depression, 1832–6; then in 1837–42 it doubled again, more than held this new level through the 1840's, moved upwards again in the post-Repeal depression of 1851–3, and thereafter entered a continuously rising trend until it reached 740,000 tons a year in the 1880's. As to the geographical spread, the only farmers who mentioned oilcakes with any enthusiasm to the 1836 Select Committee on Agricultural Distress were from southern Scotland and Lincolnshire, and we may well infer that until the jump in total consumption of 1837–42 their use was localized to those areas, becoming more widespread among mixed farmers thereafter, and eventually among stock and dairy farmers as well.[1]

It was primarily in manurial terms, at least until the mid-1860's, that oilcakes were prized. Quite apart from the direct use of rape cake as a manure, usually in the ground form of rape dust, which we have already noted, the main economic justification of feeding stock with expensive purchased oilcakes was held by farmers to lie in the greatly increased value of cake-based dung over ordinary dung. The fanatic of this agricultural philosophy, Alderman Mechi, even held that there was absolutely no limit to the quantity of oilcake which it was profitable to feed to stock, beyond the physical limit of the quantity which a determined farmer could force down a beast's throat. More sober farmers argued that 7 lb. of cake per day was as much as a beast could reasonably be expected to digest; though Hudson of Castle Acre, one of the wealthiest and most successful farmers of the day, was said to feed up to 20 lb. a day.[2] If cake paid for itself in manure rather than in greater weight of meat raised, it followed that the returns were seen in the succeeding crops; in the circumstances these were largely, though not exclusively, cereal crops, and hence until the late 1860's this was a set of outlays which concerned arable or mixed farmers much more than livestock farmers.

Similarly, it was such farmers who were interested in the agricultural use of bones. Indeed, until the 1830's it is likely that it was only the Lincolnshire and eastern Scottish farmers who were thus interested. The use of bones, either crushed into half-inch pieces or ground to a fine dust, was generally held to have been the foundation of the startling growth of agricultural prosperity and output on the Lincolnshire Wolds, a growth which was proceeding vigorously in the years 1815 to 1835.[3] The bones supplied the phosphates and nitrogen required to enable these light lands to produce heavy crops of roots and grains. The fact that they had to be purchased by the farmers supplied the reason for the growth of the famous "Lincolnshire tenant-right" or set of customary rules for compensating outgoing tenants for their unexhausted improvements, which was an essential concomitant of agricultural advance financed in this way. The import trade in bones stagnated in the immediate post-1815 years, but then leapt forward in the second half of the 1820's to a level of 16,000 tons a year, increased practically fourfold in the period 1837–42, but then remained stationary until the next large increase occurred in the late 1850's. From the early 1840's, however, bones had a strong competitor to contend with in the field of purchased manures, in the shape of guano; and by the 1850's the fresh cargoes of imported bones were prob-

[1] *S.C. on Agricultural Distress*, B.P.P. 1836, VIII, QQ. 7474–7, 9907–18, 13607. In Bedfordshire and Cambridgeshire decreased consumption was mentioned, 2340, 8321–3.

[2] *Farmers' Magazine*, 3rd ser. XXIV (1863), 341–4.

[3] *S.C. on Agricultural Distress*, B.P.P. 1836, QQ. 1070–2, 1325–30, 7386, 7415–17; Loudon, op. cit. p. 807; D. Grigg, *The Agricultural Revolution in South Lincolnshire* (1966), pp. 49, 148.

ably destined for the sulphuric acid vats to be converted into superphosphate, rather than for direct application to the fields untreated. In the earlier period the primacy of Lincolnshire is reflected in the primacy of Hull as a port handling the imports, about two-thirds of the total imports passing through that one port. East-coast Scottish farmers, with similar conditions to face, followed suit, and in the 1840's Aberdeen became another large bone importer.[1] The bones at this time came largely from Germany. The Doncaster Agricultural Association, conscious of the great local importance of the matter, investigated the use of bone-manure in 1828: their saying was that "one ton of German bone dust saves the importation of ten tons of German corn."[2] A decade later the German chemist Liebig, aware of this situation from the other side of the fence, commented that in the insatiable search for bones, English merchants were rifling the battlefields of Europe and ransacking the catacombs of Sicily; and, he added sourly, the proceeds "she squanders down her sewers to the sea."[3] Soldiers' and martyrs' bones apart, the field of supply was extended by the 1830's to Russia, and most notably to South America; for long the cattle of the pampas of Uruguay and the Argentine were valued by Britain for their hides, their tallow, and their bones, and presumably it was only refrigeration which deprived the natives of inexhaustible supplies of free meat.

Besides the imports there was, of course, a large home bone production available for commerce. Just how large, it is very difficult to say. Evidence that an organized trade existed is easy to come by, and descriptions of barge-loads of bones leaving London or Birmingham are not particularly rare.[4] But the total quantity made commercially usable all depends on what view one takes of the persistence of the English in collecting old bones. The perennial optimist J. C. Morton, in his 1855 *Cyclopaedia of Agriculture*, estimated the home supply on a population basis. Taking the population at 20 million, estimating its meat consumption at a level to please Hartwell more than Hobsbawm at 50 lb. per head a year, he arrived at a total annual meat consumption of 1,000 million lb. Assuming that bones constituted 20 per cent of carcass weight, and that all of them were collected by the rag-and-bone men, he put the home supply at 100,000 tons a year.[5] I have adopted a more conservative approach and, working backwards from a credible home supply of 55,000 tons in 1913, by applying corrections for population growth and imports of meat on the bone, I arrive at a home supply of about 30,000 tons for 1855.[6] In fact the home supply and the imports seem to have been roughly equal in amount until 1850, and thereafter the imports were generally about double the size of the home supply. By the late 1840's the bones were costing farmers around £300,000 a year; and by applying the recommended rates of application per acre of bone-manure to the estimated available supply, one can calculate that there was sufficient to deal with about 60,000 acres in the early 1820's, 100,000 acres in the 1830's, and 150,000 acres by 1850. Since the effects of boning were expected to last for at least five years, five times these

[1] Wilson, op. cit. I, 474; J. C. Morton, *A Cyclopedia of Agriculture* (1855), II, 271.

[2] Quoted in Wilson, op. cit. I, 473.

[3] Quoted in J. Hendrick, 'The Growth of International Trade in Manures and Foods', *Trans. Highland and Agric. Soc. of Scotland*, 5th ser. XXIX (1917), 16.

[4] *S.C. on Agriculture*, 1833, v, Q. 12458; *S.C. on Agricultural Distress*, 1836, Q. 499.

[5] Morton, op. cit. I, 270. [6] Estimate of home supply in years before 1913 in Hendrick, op. cit. p. 17.

areas of land were actually affected, and the areas become sufficiently large for one to be confident that the great majority of farmers in Lincolnshire and the adjoining light-land districts, and in east Scotland, were practising the bone husbandry.

Similar calculations can be done for all the other purchased manures and feeding stuffs which came into use in the course of the period. Guano from the beginning of the 1840's, the first of the highly concentrated manures, was helped on its way to acceptance by farmers by the fortunate coincidence that it happened to smell just like "natural" farmyard manure, and taste like it, for one of the recommended methods of testing a sample of guano was to taste it.[1] Superphosphate introduced on a commercial scale at the end of the 1840's was from then until the late 1860's almost entirely made from a home-produced raw material, coprolites, the fossilized dung and bones of birds and animals which must have died in droves in prehistoric Cambridgeshire and Hertfordshire, where it was mined. Coprolite mining was a home industry with a neat life cycle, born in 1847, attained maximum virility in 1876 when it produced the equivalent of 446,000 tons of superphosphate worth £3·5 million, and stone dead in 1900.[2] It was killed partly by its own exhaustion, mainly by the advent of huge and cheaper supplies of imported phosphate rocks from Spain, France, Florida, the West Indies, the Pacific Islands, and finally North Africa. Nitrates came into agricultural use from the early 1850's in the form of "cubic nitre" imported from Chile; sulphate of ammonia, the other main source of farming nitrogen, was apparently allowed to run to waste in the ammoniacal liquors that formed the effluents from gasworks until the 1860's, when a sizeable home supply sprang up. Potash, the last of the three fertilizing chemicals, was traditionally procured from wood ashes, and still is obtained from their bonfires by many gardeners; it also chanced to be present in small quantities in ordinary farmyard manure. Though industrial potash was procured by timber-burning operations on a commercial scale, for example in Canada, there was no organized supply of potash to farming until the discovery of the potash mines at Stassfurt, Prussia, in 1861. From then until 1914, under the state-directed Potash Syndicate, Germany had a world monopoly in potash supplies. It does not appear that British farmers took very kindly to the German potash, and by 1892 it was estimated that they were using only 30,000 tons a year, against about 100,000 tons used by German farmers.

Again, among feeding stuffs, maize was added to the oilcakes; first imported on a massive scale in 1846–7 in an attempt to succour the starving Irish, maize remained on the import list to feed British poultry, cattle, pigs, and horses, who proved far more tractable than the Irish at becoming convinced of its palatability. By the 1880's farmers were spending about £10 million a year on maize alone (though it appears possible that some proportion of this was in fact being spent to feed non-farm animals). For all these items of purchased raw materials used as inputs by the farming industry it is possible to construct tables showing the

[1] MS agricultural student's notebook, *c.* 1885, in the author's possession, fo. 25 bis.

[2] On coprolites, see *Estates Gazette*, VII (23 July 1864), reporting a paper by J. C. Morton on agricultural progress since 1840; J. R. McCulloch, *Dictionary of Commerce* (1871 edn), p. 424; Hendrick, op. cit. p. 18; A. N. Gray, *Phosphates and Superphosphates* (1944 edn), p. 22. The statistics on the other materials are derived from *Trade and Navigation Returns*; the *Report of the Departmental Committee on Fertilizers and Feeding Stuffs*, P.P. 1892, XXVI; Gray, op. cit. and his 1930 edn, especially Table 10, p. 101.

quantities and the values consumed year by year, in a continuous series from 1815 onwards, and these are given in an Appendix.

Some general conclusions may be derived from these figures. In the first place, it must be emphasized that the costs of these purchased inputs were carried by tenant farmers, not by landlords. Adding on to the import or wholesale values an arbitrary allowance of 10 per cent as a gesture towards covering the costs of distribution to the farm gate, these costs aggregated in round figures £600,000 a year in the first decade after 1815, £3 million a year in the late 1830's, £10 million a year in the early 1850's, £17 million a year in the early 1870's, and £25 million a year in the early 1880's; by 1913 the same items, by Ojala's reckoning, cost £29 million a year and, by 1939, £35 million.[1] It is therefore fair to say that the period of rapid growth in farmers' financial outlay on these materials was over by 1880. It is true that in the period after 1880 there was a great fall in the prices of most of the items, and that the tonnage of both feeds and fertilizers used in 1939 was about double that of 1880. Nevertheless, in the fifty years before 1880 the tonnage of feeds had increased ten times, and the tonnage of fertilizers 27 times; a doubling of both the feed and fertilizer tonnages occurred in thirteen years between 1864 and 1877. With perhaps just a shade less conviction it is thus also possible to claim that in physical terms the main force of the second revolution was spent by 1880.

In the second place many, although not all, of these purchased inputs did not produce their full results in the immediately succeeding crop, or within the year of their purchase. The effects of bones, as we have seen, lasted at least over five years, and outlays on feeding stuffs, as well as on the quicker-acting artificial manures, were all subjects of valuation when a tenant was compensated for unexhausted improvements. At a conservative estimate, the increased crop and grass yields attributable to these outlays as a whole continued for at least two years. This is the same as saying that the amount of tenant-farmers' working capital tied up in these materials was double the annual outlay, and therefore between, say, 1820 and 1880 the amount of tenants' capital increased by £50 million, or approaching £3 for every farmed acre likely to have been affected. In a different way, by taking the usual rates per acre at which the different fertilizers were applied, we can say that by 1880 sufficient was being used to treat about 17 million acres, which is either virtually the whole of the arable acreage of Britain, or about half of the total cultivated acreage. Similarly the oilcakes and maize, if they had been exclusively fed to cattle, would have been enough for about three million of them, half the total British herd.

Quantitatively, therefore, this revolution was a most substantial affair. It also meant that something like half the farmland and half the farm output—though undoubtedly a good deal less than half the individual farmers—came under the sway of a system of commercialized farming, in which farmers regarded their activities as a business; a business that required them to purchase raw materials in the cheapest market, process them in their factories, and sell the final products in the dearest market, just like any cotton lord. Further, the whole process was one in which agriculture, mining, and manufacturing industry became increasingly interdependent and intertwined, not simply in the general sense of mutually

[1] E. M. Ojala, *Agriculture and Economic Progress* (1952), pp. 212–13.

beneficial development of sources of supply and markets, but in the highly specific sense that entire industries like oil-crushing and heavy chemicals were only able to develop as they did because there was an agricultural demand for their oil-cakes and their sulphuric acid, as well as an industrial demand for their oils and their alkalis. It has sometimes been said, most recently by Prof. Landes, that the consumption of sulphuric acid "is a rough yardstick of general industrial development" because of the great range of the technically advanced production processes in which it is used. It is therefore of some significance that the fertilizer industry used more than one-third of a total annual British output of around 900,000 tons of sulphuric acid in the early 1880's. In 1865, in relation to an acid output of 380,000 tons, fertilizers had only required about 11 per cent of total output; in 1850, for which there is unfortunately no estimate of total acid output, the 5,000-odd tons of acid needed in the production of superphosphates must have been a negligible proportion of the whole. Here was a major growth factor in heavy chemicals.[1] In addition, the shipping industry both stimulated, and was stimulated by, these imports of often bulky agricultural raw materials which by 1880 involved an annual import of around three million tons. The derived effects on shipping were also considerable, for each ton of sulphuric acid produced at home was liable to occasion the import of 18 cwt of pyrites. The agricultural demand for oilcakes at home enabled a thriving export trade in oils to develop, which had grown to between 65,000 and 75,000 tons a year in the 1870's and 1880's, worth over £1·5 million a year; while the fertilizer industry, on the basis of its large home market, established an export which had reached an annual level of over 300,000 tons, worth £2 million, by the end of the 1880's.

Finally, we must remember that this transformation, financed and managed by tenant-farmers, took place within an unregenerated institutional framework of landlord–tenant relations, before any statutory enactment of compulsory compensation for unexhausted improvements, which did not come until 1883, and while the great majority of tenants were tenants-at-will legally liable to eviction on six months' notice. In the face of the evidence of what actually happened to agriculture it seems idle to pretend that these legal and institutional arrangements discouraged enterprise by tenant-farmers, let alone that they made them into an oppressed and exploited class. The solution to the riddle is, of course, that appearances belied reality. Compensation in fact existed, and customs of the tenant-right variety spread and developed, even if no general law existed; and security of tenure obtained in practice for all the reasonably efficient and prompt rent-payers among the annual tenants, even if it did not do so in law. Institutional arrangements which were theoretically objectionable did in fact work in practice, and the cases of hardship whose importance was inflated for political reasons by radical polemic were in fact no more than evidence of friction at the margins of the system, and were of no economic importance.

[1] David S. Landes, 'Technological Change and Development in Western Europe, 1750–1914', in *Cambridge Economic History of Europe*, VI (1965), 501. Estimates of total sulphuric acid output from L. F. Haber, *The Chemical Industry during the Nineteenth Century* (1958), pp. 103–4. For amounts of sulphuric acid used in superphosphate manufacture, see Gray, op. cit. pp. 151–6; G. Lunge, *A Treatise . . . on Sulphuric Acid and Alkali* (1903), quoted by Haber, op. cit. p. 103. In the 1880's the production of soda by the Leblanc process was the major user of sulphuric acid; by the 1900's, with the spread of the Solvay process, the fertilizer industry was left as the largest single user.

The new husbandry of the second revolution, of course, required landlord participation in capital improvements, principally in drainage and farm buildings. In terms of capital provision, just as in terms of land management, farming became increasingly a joint enterprise of landlords and tenants. In a sense, as I have argued elsewhere, the leadership in this joint enterprise was increasingly in the hands of the tenants.[1] Whether the low returns enjoyed by landlords were a special consequence of their trailing role, or whether tenants also only obtained low rates of return on their growing investment, are questions which can only be answered when it is possible to relate the increased inputs dealt with in this article to estimates of agricultural output. It might then turn out that the enterprise shown by tenants was quite often equally unprofitable or misdirected.

University College London

APPENDIX

The Statistics of Agricultural Inputs, 1806–91

Annual averages have been calculated, mainly for five-year periods, but the precise periodization has been varied in order that the impact of agricultural depressions and other cyclical and secular trends may be judged.

Table 1. *Inputs of Selected Feeding Stuffs and Seeds: Quantities*

Among the more important purchased feeding stuffs, wheat offals and brewers' grains have been omitted, both because of the guesswork involved in estimating their amounts and because of the uncertainty as to what proportion of them were used on farms in any case—as distinct, for example, from town dairies; similarly no imported grains other than maize have been included, since their use as feed was probably very small before *c.* 1890. Buckwheat, which Peel thought to be of growing importance as a feed in the 1840's, was in the event never imported in any quantity except in 1846 and 1847. Among the imported seeds used for sowing (not pressing) only clover has been included; there were sizeable imports of other seeds, for instance tares, but these were not of great value. The actual inputs of feeding stuffs and seeds were therefore somewhat greater than those recorded in the table.

Table 1

Period	*Oilseed cakes (tons)* *Imported*	*Oilseed cakes (tons)* *Home-produced*	*Maize and maize-meal ('000 cwt)*	*Clover seeds (cwt) Imported*
1806–13	1,623	?		?
1810–14	2,803	24,866		62,649
1815–20	3,825	24,183	1·8	67,302
1821–4	10,643	24,437	7·7	69,258
1825–31	24,035	45,533	152	114,691
1832–6	23,509	37,597	2·6	65,443
1837–42	63,210	66,393	53	139,229
1843–6	72,890	84,118	799	139,339
1847–50	66,572	89,912	8,685	137,240

(*continued overleaf*)

[1] Thompson, op. cit. pp. 253-5.

Period	Oilseed cakes (tons) Imported	Home-produced	Maize and maize-meal ('000 cwt)	Clover seed (cwt) Imported
1851–3	57,830	122,066	6,555	161,341
1854–8	84,008	152,478	5,849	161,244
1859–63	101,496	201,789	9,619	208,744
1864–7	116,596	261,524	9,048	204,333
1868–71	160,722	310,455	12,756	248,121
1872–6	160,174	375,359	24,248	303,825
1877–81	208,874	402,910	35,834	311,291
1882–6	259,278	480,926	27,466	313,083
1887–91	288,600	486,546	32,598	335,239

Sources: For imported oilseed cakes, maize and maize-meal, and clover seeds, *Annual Trade and Navigation Returns*, supplemented by *S.C. on Seeds and Wool*, 1816, VI; *S.C. on Agricultural Distress*, 1836, VIII, pt I, app. 3; and A. & P. *Customs*, 1847/8, LVIII, articles which in 1841 paid customs duty of £1,000 and upwards. For home-produced oilseed cakes the procedure was as follows: for years prior to 1827 the recorded import values of linseed and rape seed were converted into quantities by applying the Official Value factors of 2s. 3d. per bushel of flax and linseed and 2s. 6d. per bushel of rape seed. From 1827 the quantities of imports are recorded. The quantities of imported seeds were converted into quantities of oilseed cake by applying an extraction factor, itself derived from *S.C. on Seeds and Wool*, 1816, pp. 69, 72; H. W. Brace, *History of Seed Crushing in Great Britain*, p. 49, for 1856; and *Report of Departmental Committee on Adulteration of Artificial Manures, Fertilizers, and Feeding Stuffs*, 1892, XXVI, for 1892. The amount of cake made from one-quarter of oilseed was taken as 300 lb. in 1816, 294 lb. in 1856, and 290 lb. in 1892, and an interpolation made to allow for the improvement in the oil-extraction rate. A separate extraction rate was calculated for cotton seed, of 0·523 tons of cake from 1 ton of cotton seed, from the 1892 Departmental Committee, Q. 400.

Table 2. *Inputs of Fertilizers: Quantities*

Lime is the most important omission from the fertilizers listed: it was used in considerable quantities, but no source has been found from which these might be estimated. Several minor substances, such as gypsum and woollen rags used for manure, in both of which there was a small import trade, have also been omitted. Substantial amounts of various potash compounds, sulphate and muriate, were imported; but these also have been omitted, on the assumption that they were for industrial, not agricultural, use. The further assumption has been made that the imports of agricultural potash from Germany, when they were established, were recorded in the category of "unenumerated manures" which appears in the *Returns* from 1854. On the other hand, the export of fertilizers from Britain poses a problem in measurement. Exports of guano were straightforward re-exports and were recorded as such; the amounts have been deducted from gross guano imports, and the figure in the table is of the amount retained for home consumption. But exports of all other fertilizers were recorded in a single item of "Manures (including Sulphate of Ammonia and other Chemical Manures)", which first appeared in the *Returns* in 1865. Besides the sulphate of ammonia, these exports clearly included superphosphates and manufacturers' compound fertilizers, in unknown proportions; and they were produced partly from domestic, partly from imported, raw materials, again in unknown proportions. Since it is thus impossible to establish the amounts of exports or re-exports of individual items, except for guano, the figures in this table are those of the gross amounts imported or produced; the necessary adjustments to establish net home consumption of fertilizers in general are carried out in the summary Tables 4 and 5.

Table 2

All figures are in tons

Period	*Bones* Imports	*Bones* Home produce	*Guano* net imports	*Superphosphate* From imported materials	*Superphosphate* From home-produced materials	*Nitrates* Imports	*Nitrates* Home-produced	*Un-enumerated manures* Imports
1810–14	8,000	18,000						
1815–20	?	19,500						
1821–4	4,400	21,000						
1825–31	16,100	23,000						
1832–6	16,900	25,000						
1837–42	46,390	27,000						
1843–6	36,860	28,000	119,940					
1847–50	29,730	28,500	88,540					
1851–3	39,540	29,000	165,350		17,000	13,000		
1854–8	68,340	30,000	231,610		35,000	15,000		1,380
1859–63	66,080	31,000	138,940		85,000	24,800		2,690
1864–7	68,150	32,500	166,440		180,000	38,700		5,440
1868–71	86,510	35,000	209,460	21,300	200,000	45,100	40,000	15,370
1872–6	86,690	37,000	114,750	142,900	270,000	101,900	46,000	69,280
1877–81	79,780	41,000	82,950	227,300	282,900	53,000	60,000	98,000
1882–6	72,700	44,500	39,790	390,000	46,000	77,900	97,000	83,900
1887–91	66,830	47,500	18,780	500,400	10,000	86,000	134,000	85,780

Sources: BONES. Imports: For 1810–14, Wilson, *The Rural Cyclopedia* (1848), I, 473–4; 1821–53, entered in the *Annual Returns*, but at the declared value only. However, for 1842–6 quantities as well as values were returned in A. & P. *Customs*, 1847/8, LVIII, at an average value of £4·74 per ton, at which rate the earlier value figures have been converted; 1854 onwards, quantities entered in the *Annual Returns*.

Home produce: These figures are highly conjectural, the only definite estimates being those of J. C. Morton that the home supply was 100,000 tons a year in 1855 (*Cyclopaedia of Agriculture* (1855), I, 270, derived from the assumption that the population of 20 million consumed 1,000 million lb. of meat a year), which has been rejected as too fanciful, and of Prof. J. Hendrick that the home supply in the years just before 1913 was between 50,000 and 60,000 tons a year ('The Growth of International Trade in Manures and Foods', *Trans. Highland and Agric. Soc. of Scotland*, 5th ser. XXIX (1917), 17). Hendrick's estimate has been accepted as credible, and it has been assumed that for 1909–13 the home supply averaged 55,000 tons a year, and that two-thirds of this derived from home-raised livestock and one-third from meat imported on the hoof or on the bone, an assumption which seems unlikely to exaggerate the contribution of home-raised livestock. Bone-equivalents were then calculated for the average meat imports of 1909–13, and for the average number of stock units kept by British farmers in 1909–13, on the basis of 1 stock unit = 1 cow = 7 sheep = 5 pigs. These bone-equivalents were then applied to the meat imports and stock units of 1867–71, the earliest period for which statistics of British livestock are available, yielding an estimated 35,000 tons of bones. Estimates for the period between 1867–71 and 1909–13 were then interpolated. The output attributable to home-raised livestock in 1867–71, 32,000 tons, was then divided by total population at that date, and the resultant quotient of 1·03 tons of bones per 1,000 people per annum was applied to total population from 1811 onwards, with appropriate additions for meat imports which became significant from 1851. Since this calculation rests on the assumption of a constant ratio between population and livestock for the period 1811–67, it probably results in an overestimation for the earlier part of the period, which may however be balanced by the probable decline in the zest for bone collection as living standards rose.

GUANO. Imports for 1841–50 in A. and P. *Customs*, 1852, LII, and thereafter in the *Annual Returns*. No exports recorded before 1854.

SUPERPHOSPHATE. From imported materials: Imports of phosphate rock were separately recorded in the *Annual Returns* from 1882. For the period from 1868, when rock imports are known to have started, until 1881, these imports were included in the category of "unenumerated manures". By taking the imports of "unenumerated manures" for each country from which an import of phosphate rock is known to have come, from the year in which it is known to have started, and by taking for each country the percentage which phosphate rock formed of imports of phosphate rock plus "unenumerated manures" in 1882, an estimate has been built up of imports of phosphate rock in the period 1868–81. Each ton of phosphate rock produced 1·73 tons of superphosphate.—A. N. Gray, *Phosphates and Superphosphates* (1944), p. 16; Gray also gives invaluable details of the chronology of the opening up of phosphate deposits.

From home-produced materials: Commercial exploitation of coprolites began in 1847, and figures of output have been obtained for the following years: 1853, 1854, 1860, 1861, 1863, 1874–8, 1876, 1877, 1884–8, 1887, and 1894–8—*Estates Gazette*, VII (23 July 1864), p. 1,840; J. R. M'Culloch, *Dictionary of Commerce* (1871 edn), p. 424; Hendrick, op. cit. p. 18; Gray, op. cit. p. 22.

NITRATES. Imports: 80 per cent of the imports of "cubic nitre" have been taken as destined for use as fertilizer, a proportion which accords with the *Report of the Departmental Committee on Adulteration of Artificial Manures, Fertilisers, and Feeding Stuffs*, 1892, XXVI, vi.

Home-produced: From the table of British production of sulphate of ammonia, 1870–1913, in Hendrick, op. cit. p. 11.

UNENUMERATED MANURES. Gross imports less estimated imports of phosphate rock in the period 1868–81.

Table 3. *Inputs of Feeding Stuffs and Fertilizers: Values*

In £'000

Period	*Oilseed cakes*	*Maize*	*Clover seed*	*Bones*	*Guano*	*Super-phosphate*	*Nitrates*	*Un-enumerated manures*
1810–14	220		187	130				
1815–20	223	9	201	115				
1821–4	280	4	207	125				
1825–31	556	83	343	195				
1832–6	488	1	196	205				
1837–42	845	29	417	365	1,285			
1843–6	1,333	423	417	325	932			
1847–50	1,329	4,603	411	290	1,798			
1851–3	1,618	3,343	483	340	2,628	119	221	
1854–8	2,099	2,456	515	548	1,284	280	259	5
1859–63	2,449	3,366	572	494	1,511	680	342	13
1864–7	3,005	3,166	428	550	1,811	1,440	487	39
1868–71	3,999	4,719	679	738	2,466	1,768	1,266	73
1872–6	5,189	8,971	739	824	1,244	3,304	1,925	193
1877–81	4,960	10,750	715	732	760	4,080	1,627	205
1882–6	5,372	7,965	697	737	342	3,052	1,890	159
1887–91	4,669	8,149	684	558	125	3,315	2,046	128

Sources: As in Table 2, the figures for guano relate to net retained imports, those for "unenumerated manures" relate to gross imports less estimated imports of phosphate rock in the period 1868–81, and for superphosphate and nitrates (imported element) to gross imports.

Values for 1854 onwards are average import values, from the *Annual Returns*, and do not include any allowance for internal distribution and merchanting costs. The home-produced product, in the case of oilseed cakes, superphosphate, and nitrates, has been arbitrarily priced at the declared import value. For the period before 1854 no series of prices is available, except for the price of maize which appears in the *Annual Returns*. There are scattered references to prices in the sources previously cited, which have been used, and supplemented by the comparison between Official and Declared Values presented in the 1854 *Return*. The figures for the period before 1854 must therefore be regarded as approximate only.

As mentioned in the text, the initial large import of maize was for human, not animal, consumption.

Table 4. *Summary of Inputs: Quantities*

In '000 tons

Period	*Feeding stuffs*	*Seed*	*Fertilizers (including gross imports)*	*Fertilizers (exports and re-exports) General*	*Fertilizers (exports and re-exports) Guano*	*Fertilizers (net retained)*
1810–14	27	3	26			26
1815–20	29	3	23			23
1821–4	35	3	25			25
1825–31	76	6	39			39
1832–6	61	3	42			42

Period	Feeding stuffs	Seed	Fertilizers (including gross imports)	Fertilizers (exports and re-exports) General	Guano	Fertilizers (net retained)
1837–42	132	7	73			73
1843–6	197	7	184			184
1847–50	590	7	146			146
1851–3	506	8	263			263
1854–8	528	8	424		41	383
1859–63	783	10	365		17	348
1864–7	830	10	498	20	8	470
1868–71	1,108	12	656	47	3	606
1872–6	1,747	15	899	87	31	781
1877–81	2,402	16	949	147	25	777
1882–6	2,113	16	864	275	13	576
1887–91	2,403	17	955	295	6	654

Table 5. *Summary of Inputs: Values*

In £'000

Period	Feeding stuffs	Seed	Fertilizers (including gross imports)	Fertilizers (exports and re-exports) General	Guano	Fertilizers (net retained)	Total
1810–14	220	187	130			130	537
1815–20	232	201	115			115	548
1821–4	284	207	125			125	616
1825–31	639	343	195			195	1,177
1832–6	489	196	205			205	890
1837–42	874	417	1,650			1,650	2,941
1843–6	1,756	417	1,257			1,257	3,430
1847–50	5,932	411	2,088			2,088	8,431
1851–3	4,961	483	3,666		358	3,308	8,752
1854–8	4,555	515	2,828		452	2,376	7,446
1859–63	5,815	572	3,218		178	3,040	9,427
1864–7	6,171	428	4,412	166	85	4,161	10,760
1868–71	8,718	679	6,346	374	35	5,937	15,334
1872–6	14,160	739	7,870	699	380	6,791	21,690
1877–81	15,710	715	7,703	1,180	299	6,224	22,649
1882–6	13,337	697	6,306	1,928	126	4,252	18,286
1887–91	12,818	684	6,222	1,903	50	4,269	17,771

The Supply of Raw Materials in the Industrial Revolution

By E. A. WRIGLEY

Any great increase in the output of industry, such as began in England towards the end of the eighteenth century, must have as its counterpart an equally great increase in the input of industrial raw materials at the other end of the process of production.[1] The problem of providing an adequate raw material supply had been acute in many branches of industry in earlier centuries. The removal of these constrictions is intimately connected with several important aspects of the rapid growth which occurred, and its study affords a vantage point from which they can conveniently be surveyed.

The most important change in raw material provision which took place was the substitution of inorganic for organic sources of supply, of mineral for vegetable or animal raw materials. This was a *sine qua non* of sustained industrial growth on a large scale, for when industrial growth is based upon vegetable and animal raw materials present success can usually be obtained only at the cost of future difficulties. England in the sixteenth and seventeenth centuries provides some typical examples of the dilemma which confronts industries when they use animal or vegetable raw materials. The iron industry of the Weald was able to expand without prejudice to its future prosperity only up to the point at which the annual cut of timber equalled the yearly increment of new growth. Any expansion beyond this point could take place only at the cost of contraction in the future. Expansion without prejudice to future supplies could, of course, have been secured if more land had been devoted to the production of timber, but in a country where the area of unused land was small more woodland meant less ploughland or pasture. Competition for the use of scarce land was a perennial problem in these circumstances and a permanent, radical increase of industrial raw material supply was very difficult to obtain. Those Tudor pamphleteers who complained that the sheep were eating up men

[1] Except, of course, in so far as technological changes permit raw material saving.

were directing attention to the central problem of industrial raw material supply in an age when organic materials were essential for most industrial processes. More land devoted to the production of timber or wool meant less land available to produce food. The price of raw materials was sure to rise because of competition for the use of land and so inhibit industrial growth even where the government refrained from direct political action to guarantee the supply of food. If the government did not intervene to restrict pasture in the interest of tillage, the play of the market would ultimately produce the same result. Moreover, industrial growth not only provoked problems of this type directly by competing for the use of land, but also indirectly by encouraging a growth in population which in turn increased the demand for land upon which food could be grown. Once the spread of settlement had brought all available land into use, the only way in which the supply of food and of industrial raw materials of vegetable or animal origin could be increased simultaneously was by a general rise in the productivity of the land.[1]

The view that the productivity of the land controls the growth of industry no less than that of agriculture is a recurring theme in the *Wealth of Nations*. Adam Smith wrote just before the dramatic changes in industrial raw material supply had become fully apparent and did not recognize their importance. In his chapter *Of the natural Progress of Opulence* he began by defining the exchange of products manufactured in the towns for agricultural produce and raw materials as 'the great commerce of every civilised society'.[2] Later in the chapter he enlarged upon the nature of this exchange.

> It is this commerce which supplies the inhabitants of the town, both with the materials of their work, and the means of their subsistence. The quantity of the finished work which they sell to the inhabitants of the country, necessarily regulates the quantity of the materials and provisions which they buy. Neither their employment nor subsistence, therefore, can augment, but in proportion to the augmentation of the demand from the country for finished work; and this demand can augment only in proportion to the extension of improvement and cultivation. Had human institutions, therefore, never disturbed the natural course of things, the progressive wealth and increase of the towns would, in every political society, be consequential, and in proportion to the improvement and cultivation of the territory or country.[3]

If the productivity of the land in the last analysis governed the wealth of any country, it is hardly surprising that Adam Smith claimed of investment in agriculture that 'Of all the ways in which a capital can be employed, it is by far the most advantageous to society'.[4]

The supply of mineral raw materials forms an interesting contrast with the supply of vegetable and animal raw materials. In the very long run the mineral

[1] Within any one national market area, of course, it was possible to expand the supply of food and raw materials without increasing pressure on the land by import from abroad. Cheap sea transport enabled England to make good some of her shortage of timber, for example, in this fashion.

[2] A. Smith, *An Inquiry into the Nature and Causes of the Wealth of Nations*, ed. J. R. McCulloch (Edinburgh, 1828), II, 171.

[3] *Ibid.* 175–6.

[4] *Ibid.* 150.

supply problem is insoluble in a sense which is not true of organic raw materials, since every mine is a wasting asset. It cannot be made to give a sustained yield in the way which is possible with a forest or a farm. A forest can yield indefinitely: a mine cannot. Nevertheless, in any but the very long run the difficulty of obtaining a large increase in supply is less pronounced with mineral raw materials. Given an adequate mining technique and the existence of rich deposits production can rapidly be built up to high levels. As the individual mine nears exhaustion the price of extraction must rise, of course, but as long as it is possible to sink other pits to tap equally rich deposits the price of the product need not rise, and may well fall if increasing production encourages the creation of larger and more efficient production units. Moreover, an increase in the production of mineral raw materials does not take place at the expense of the supply of food or of other industrial raw materials. There is no equivalent in the production of inorganic raw materials to the competition for land which accompanies an expansion in the production of organic raw materials.[1] In the half century after the publication of the *Wealth of Nations* the vital importance of the new sources of industrial raw materials became clear. The passage which McCulloch in his edition found 'perhaps the most objectionable' [2] in the book was that which concluded that capital was best employed in agriculture from the point of view of society as a whole. Adam Smith had argued that rent is created by those 'powers of nature' which give an added productivity to agriculture, but had added that in manufacture nature does nothing for man; he must do everything for himself. McCulloch objected to this definition of rent, but he also denied that the powers of nature favoured agriculture alone. His was a world in which the importance of manufacture had ceased to be regulated solely by agricultural productivity. The amount of capital which could profitably be invested in manufacture was no longer controlled by the agricultural surplus in the manner suggested by Adam Smith.[3] Mineral sources of raw material had given another dimension to the discussion. To McCulloch the 'powers of nature' revealed in the steam-engine were as remarkable as any that Adam Smith had noticed in the fields.[4]

There is a second difference of great importance between mineral production on the one hand, and vegetable and animal production on the other. Production of the former is punctiform; of the latter areal. The transport problems involved in moving a million tons of coal from pitheads scattered over an area of only a few square miles are quite different from those involved in moving the same weight of grain or timber from an area of several thousands of square miles. The former implies heavy tonnages moving along a small number of routeways, whereas the latter implies the reverse. A heavy capital investment

[1] Open-cast mining forms a minor exception to this rule.

[2] Smith, *Wealth of Nations*, II, 150 n.

[3] See, for example, *Wealth of Nations*, II, 214–15, where his argument leads him to the conclusion that the manufactures of Leeds, Halifax, Sheffield, Birmingham and Wolverhampton are 'the offspring of agriculture'. The general argument of the chaper *Of the different Employment of Capitals* is also interesting in this connexion.

[4] '... are not the pressure of the atmosphere and the elasticity of steam, which enable us to work the most stupendous engines, the spontaneous gifts of nature?' *Wealth of Nations*, II, 150 n.

in improved communications is unlikely to give a good return when the raw materials of industry are organic since the traffic density along any one route is usually low. A large volume of mineral traffic, on the other hand, makes such an investment necessary to cope with physical difficulties, and financially attractive because the total possible savings are so much greater.[1]

I

The decisive technological change which freed so many industries from dependence upon organic raw materials was the discovery of a way of using coal where once wood had been essential. The timing of the change varied a great deal between the several industries. It came earliest in industries like the boiling of salt in which the use of coal presented no problem of undesired chemical change in the product because the source of heat was separated from the object by a sheet of metal. Industries like iron smelting and hop drying in which contact was more intimate presented greater problems in a period when chemical knowledge was slight. A long period of trial and error commonly elapsed before a successful method of substitution was developed. Coal-fired salt pans were a commonplace before the end of the sixteenth century: coke-fired blast furnaces were not successfully operated until the first quarter of the eighteenth, and in some branches of the iron industry it was near to the end of the century before charcoal could be dispensed with. In spite of the rather slow spread of coal use from one industry or process to another, however, it was already an industrial raw material of the first importance by the beginning of the eighteenth century. At that time coal production in England and Wales had reached a level of about three million tons per annum, or roughly half a ton per head of population. The production of coal both absolutely and *per caput* was already of quite a different order of magnitude in England from that obtaining on the continent; and by the end of the century production had tripled. Much of the coal was used for domestic rather than industrial purposes, but all helped to relieve the pressure on timber supplies. Wherever it could successfully be substituted for wood its effect was to liberate production from the physical limits upon output imposed on industries requiring a source of heat in a country where the timber resources were very limited. Unlike timber, a substantial increase in coal consumption in any one period did not prejudice supplies in the next, nor did an expansion in coal use in one industry affect others adversely. Moreover, once the initial period of prejudice against coal had passed and the difficulties involved in its use had been overcome, many industries discovered that coal was better suited to their purposes than wood had been.

In the absence of coal the timber requirements of a country whose industries were as large as those of England at the beginning of the nineteenth century

[1] Occasionally the older system had coped with quite heavy tonnages. The grain and timber trade, especially to the London market, was on a substantial scale and had meant large outlays on North Sea and coastal shipping and on river barges, but it was the new problems of mineral traffic on a large scale which produced the canals and the railways.

would have been enormous. Much heat was needed in a wide range of industrial processes, and to have provided it with wood must have denuded the forests of England, indeed the forests of Europe, in a few decades. The classic case is perhaps that of the iron industry. Benaerts quotes an estimate, for example, that the production of 10,000 tons of charcoal iron required the felling of 40,000 hectares of forest.[1] The pig which resulted from a coal or coke melt was for many years unsatisfactory and commanded only a very low price. The prejudice against it remained for some time even after Darby had overcome the main difficulties in smelting ore with coke, yet coke pig was to prove essential to the rapid progress of industrial growth. Without it there could have been no great expansion in the scale of iron output or fall in its price, and the physical properties of iron were so essential to the age of machines that it is difficult to believe that any great changes were possible without cheap and abundant iron. Many machines which were first constructed in wood could be greatly improved when made in metal, and many of the great engineering constructions after the turn of the century could not have been made at all without cheap iron. The physical properties of iron permit great precision of working: the steam engine and the machine tool depend on this. They and the iron ships, iron rails and iron bridges of the new age required the successful supplanting of vegetable by mineral fuels.

Coal could not be used as a direct substitute for wood in the building industry where timber was used not as a fuel but as a building material, but indirectly it was important because its use in the brick industry meant that the production of bricks could be expanded without unit costs of production rising, so that brick became the prime building material of the new age. The engineering and construction industries, as producer goods industries of central importance, are points of great sensitivity in any period of rapid industrial growth. The output of producer goods must necessarily expand faster than that of consumer goods at such a time, and it is vital that it should be easily expansible, and if possible that the costs of the raw materials involved should show a secular tendency to fall as the volume of production increases. In the past these industries had been heavily dependent on wood, which tended to induce a secular rise in the costs of the raw material when the scale of production grew. After the changeover to mineral raw materials the possibility of a much easier and unrestricted expansion was always present.

Those consumer goods industries, such as the Staffordshire pottery industry or the glass industry, which required much heat in their manufacturing processes also benefited from growing independence of vegetable fuel. Brewing, the paper industry, and some sections of the textile industry made use of coal too, though in their case the fuel was used to process organic rather than inorganic raw materials.

A part of the great increase in the productivity of industrial workers which began in the later part of the eighteenth century and has continued down to the present day arose in the manner which Adam Smith described and analysed.

[1] P. Benaerts, *Les Origines de la grande industrie alleman de* (Paris, 1933), p. 454. One hectare is equal to 2½ acres.

Markets grew larger; production processes were subdivided; industrial skills became more specialized and workers nicer in their skills; new machinery was developed; real costs fell as productivity increased. But a part of the increase in productivity which took place, that part which became possible as a result of the rise to prominence of a class of industrial raw materials whose significance Adam Smith did not fully appreciate, that part which arose from cheaper and more abundant heat and mechanical power, could not take place as long as the land produced not merely the food of the nation but also its industrial raw materials. It is notable how frequently the industries in which expansion was marked in the years between Adam Smith and McCulloch were those which were gradually freeing themselves from dependence upon organic raw materials, especially wood. This is true of the industries making iron, non-ferrous metals, most types of machinery, glass, salt, pottery and bricks. The industry in which the most dramatic growth of all took place was, perhaps, cotton which provides an instructive contrast with most other quickly growing industries in that its raw material was vegetable. The cotton industry conformed quite closely to the picture of industrial growth envisaged by Adam Smith and will require further consideration at a later stage in this discussion.

II

Though the large-scale use of coal offered great opportunities, it also brought problems, in whose solution may be seen some of the most important economic and technological foundations of the Industrial Revolution. The prime difficulty was the great expense of transporting coal with the existing transport media. Coal could only come into general domestic and industrial use if it were cheap, and could only be cheap if it could be cheaply transported. As long as the price of coal taken overland doubled within five miles of the pithead it was not likely to be widely used. This, of course, is the reason why the first large-scale coal industry was on the Tyne in touch by sea with the London market and the smaller markets down the east coast. Sea transport had always been the cheapest form of transport, and moreover it had long been accustomed to dealing with the range of problems raised by punctiform production on a large scale. Large ports had for centuries concentrated a great bulk of goods at a single point and forwarded them to a limited number of similar points, often far away. The biggest ships were capable of moving hundreds of tons at a time, whereas on land loads measured by the hundredweight were normal. Exceptionally much larger weights might be moved (as with large building stones), but only over short distances and by making special arrangements. Before the era of large-scale mineral production there was little incentive to try to alter the capacity of overland transport systems since the areal production of vegetable and animal products seldom calls for the movement of a great bulk of material along a single route. Improvements to overland transport which are precluded when raw material production is areal may be both necessary and economically practicable when raw material production is punctiform. Once the Newcastle-London coal trade had shown the very real advantages of coal over wood for many purposes there was always latent the possibility that a radical improvement in overland communication might take place.

Prof. Nef has shown how important the growth of the coal trade was in developing more efficient methods of ship construction and working in this country.[1] By the end of the seventeenth century about half of the total British merchant fleet by tonnage was engaged in the coal trade. But the effect of the development of large-scale punctiform mineral production upon shipping was only to develop traits characteristic of the movement of goods by sea for a long time. Ports and pits had much in common as sources of cargo. It was otherwise with overland transport. When the production of the coal industry came to be measured in hundreds of thousands and even millions of tons annually a new solution was necessary if the coal was to reach inland centres of consumption. However inefficient to later eyes may have been the movement of grain, wood and wool on horseback or in carts along pitted roads, it was economically inevitable since the volume of the traffic was too small to warrant the investment needed to provide good roads or canals. Areal production meant poor communications. Minerals had, of course, also been moved on horseback before the eighteenth century in spite of the punctiform nature of their production, but they had moved in quite small absolute quantities and had not afforded any opportunity for substantial improvement. For a long time coal was moved in the same way whenever it was moved overland. It moved on pack animals from the Staffordshire coalfield to the Northwich salt pans and from the Yorkshire pits to the Bradford dyers, or on small river craft on the Severn and Thames. As long as the inland consumption of coal remained small, coal moved as other raw materials had moved for centuries. Large-scale consumption provided new opportunities. The demonstration that coal could be used so successfully where wood had been used in the past created a large potential demand for it which could not be met while communications were poor, but which provided a powerful incentive to improve them. The canalization of the Weaver to Northwich after 1720 in order to provide cheaper coal than could be brought overland from Staffordshire was an early example of the improvements in transport encouraged by the use of coal; and the work done on the Douglas from Wigan to the Ribble estuary dates from the same period. As coal consumption rose, more ambitious works became possible. Waterways were not merely improved; they were created. Forty years after the work on the Weaver and the Douglas the Worsley Canal was built, to be followed by many others in the next half century.

The art of building canals was not new to Europe. The Dutch had a long experience of making them, and several long canals were built in France in the seventeenth century, but the English canal network was constructed in response to an incentive of a new type. If mineral raw materials were to continue to grow in importance in English industry they required such a network. Canals are well suited to the movement of goods produced at a point, but not to areal production. The truth of this is well illustrated by the history of the many canals which were built in the first flush of enthusiasm for canals

[1] J. U. Nef, *The Rise of the British Coal Industry* (1932), I, 238–40, 390–4.

in purely agricultural areas of the country, and which were seldom successful financially.[1] The successful canals were those on which there was a heavy volume of mineral traffic, usually coal, but occasionally other minerals also: for example, the canal from the pottery district of Staffordshire to the Mersey carried a china clay traffic. Agricultural areas through which the canals passed benefited greatly from their presence, both because they made possible much cheaper movement of food to the market and because some of the essentials of good husbandry, such as lime and manure, were more easily obtainable after their construction. Agricultural traffic contributed significantly to the revenues of many canals built to cater chiefly for mineral traffic, but agricultural traffic was characteristically insufficient to sustain canal finances on its own, and canals which were built in agricultural areas in the hope that their presence would create sufficient traffic to make them profitable seldom fulfilled their promoters' expectations.

The development of railways, the other chief means of cheap internal transport created during the Industrial Revolution, was also closely connected with the switch to inorganic raw materials, and especially the transport of coal. From the seventeenth century there had been railways connecting pitheads with coal wharves on the Tyne, developed to deal with the problem of coal movement overland on a large scale. The laying of wooden planks along which the horse-drawn carts could move was a simple way of increasing the load which each horse could shift. When he no longer had to expend most of his energy in overcoming the mud in wet weather and the deep ruts in dry, a horse drawing a cart along planked ways was able to move two or three times as much coal. Once the volume of coal moving over the short roads to the wharves had become large the heavy expense of improving roads in this way proved well worthwhile. In time flanged wheels were introduced and it became profitable to cover the wooden tracks with metal plates in order to increase their life under constant heavy usage. Railways were peculiarly a mining development (even down to the track gauge), and were created to overcome the problems posed by large-scale punctiform mineral production, initially as feeders to waterways, but later as an independent network. Like canals, they also, of course, proved in time of great benefit to other forms of production and made easier the movement of the vegetable and animal raw materials. Moreover, they developed a great passenger traffic. Yet it is true of railways as of canals that most of those built in purely agricultural areas in England did not generate enough traffic to make them profitable.

As the eighteenth century progressed the volume of coal output rose steadily, from three million tons at the beginning to about ten millions at the end. The great coalfields near Newcastle no longer grew in output as quickly as some of the inland fields because the improvement in communications, especially the development of canals, made available to industrial and domestic consumers over a steadily increasing area the advantages which during the seventeenth century had been restricted to the east coast ports and to very small areas on

[1] Clapham noted this condition, though he cast his conclusions about it in a different form. J. H. Clapham, *An Economic History of Modern Britain* (2nd ed. Cambridge, 1950), I, 82.

the inland fields. Even at the beginning of the century probably no other mineral, vegetable or animal raw material approached coal in weight of production: by the end it had far outdistanced any rival. It was therefore peculiarly coal which provided at once the chief stimulus to the building of traffic arteries capable of dealing with the quantities of raw materials now used by the economy, and the main goods traffic on the canals and later the railways.[1]

III

Or the issue may be put in different terms. The importance of the changes in raw material supply and in the transport system can be illustrated from the writings of the economists of the period as well as traced in the narrative economic history, especially in the discussions of the limits of economic growth. The starkest discussion of the organization of economic life in a society bounded by the productivity of the land and the problems of transport is perhaps that of Thünen. When Thünen published his discussion of the pattern of land use which would be found upon a featureless plain surrounding a central city, he deliberately made a limited number of simple assumptions about the nature of its economic life. Upon these assumptions he was able to show that the steady rise in the cost of transporting produce as distance increased from the central market would cause the land to be divided up into a series of concentric rings each marked by a different type of land use. The innermost ring was devoted to the production of perishable commodities like milk and vegetables; the next ring to woodland to meet the city's need for fuel, charcoal, building materials, tools and furniture; and the remaining rings to agriculture of a gradually decreasing intensity, shelving off finally into pastoral activities at the point where the cost of transporting grain to the market made it unprofitable to plough and plant the land. Everything depended upon the demand in the central market, local changes in agricultural productivity, and the level of transport charges. As the price of raw materials and food rose in the central market, for example, or alternatively as the cost of transport fell, the whole system of rings would expand allowing a larger area of land to be used more intensively. Thünen himself illustrated the dramatic effect of falling transport costs on the intensity of land use by examining the effect of a river running across the series of rings to the central city along which transport costs were only a tenth of those overland. Thünen's model underlined many of the characteristics of the economic life of

[1] Paradoxically, although so many of the most important changes in transport and power were connected with the mining industry and specially coal, and although it was the adoption of mineral raw materials generally in industry which alone made possible the scale of expansion which occurred, the mining industry itself did not experience any revolutionary increase in manpower productivity. Output per man-year in a large coal pit in 1700 was about 150 tons (Nef, *The Rise of the British Coal Industry*, II, 136n.), a figure already about two-fifths as large as the peak figure in the 1880's. The coal industry, indeed, is sometimes referred to as an example of the impossibility of designing machines to perform all jobs previously done by hand, and is classed with, say, agriculture, in this respect. But the central difficulty of the production of coal has never been the winning of coal at the coal-face, hard and dangerous though this has always been, but its transport within the pit, up the pit shaft, and from the pithead to the point of consumption. The canal and the steam-engine solved the prime difficulties of the coal mines. Those at the coal-face were less pressing.

earlier centuries. His scheme makes it clear how crippling the high cost of transport can be; how, for example, local famine might well occur in a country enjoying a general sufficiency; how, though timber was essential to such an economy, its great bulk and weight made it difficult to deliver to a market at a reasonably low cost. When he published his book in 1826 his model still fitted the economic life of parts of Germany without excessive distortion, though by that date it was no longer appropriate to England. Thünen acknowledged Adam Smith as his chief mentor,[1] but his scheme embodied only a part of the world Adam Smith had studied. The *Wealth of Nations* describes a much more complicated world; it does not merely abstract from it some of its salient characteristics as *Der isolirte Staat* had done. Although Adam Smith maintained that a gradual rise in agricultural productivity alone made possible the development of cities and industry, and further maintained that in the last resort the size and wealth of cities must continue to be governed by the productivity of the land, he understood and entertained within his scheme of analysis the great modification and complication which arose out of the development of trade between city and city and nation and nation, not restricting himself simply to the consideration of trade between a city and its surrounding countryside. The wealth of nations could increase greatly when Thünen's limiting assumptions were relaxed in this way, and Adam Smith showed clearly how this might come about, and discussed which policies were likely to expedite the process or to frustrate it. The wealth of nations could not, however, increase without limit upon the assumptions within which Adam Smith worked since the productivity of the land set a ceiling to growth. In the half century following the course of events proved beyond doubt that the assumptions might be still further relaxed. The ceiling which Adam Smith had assumed to exist generally now applied only to food and to a limited range of organic industrial raw materials. The use of mineral raw materials removed the limit from industrial production in general, both directly by making it possible for an enormous increase in the physical volume of production to take place without prejudice to future supplies of raw materials, and indirectly by demonstrating that the 'powers of nature' were present just as abundantly in the mines as in the land, so that capital invested in industry could yield at least as good a return as investment in the land from the point of view of the community as a whole. There is thus a certain restricted sense in which Malthus has remained more relevant to the modern situation than Adam Smith, for in discussing the limitations imposed upon the growth of population by the size of the supply of food, Malthus was concentrating upon a part of the economy in which it has not been possible as yet to substitute inorganic for organic raw materials. He was mistaken about the rate at which the supply of food could be expanded, but the problem he posed still lives with us today. His argument in a modern setting appears wrong in degree, but right in kind. Adam Smith's discussion of the wealth of nations, on the other hand, appears wrong in the second sense because he was not aware of the new possibilities of increasing production just

[1] J. H. v. Thünen, *Der isolirte Staat in Beziehung auf Landwirtschaft und Nationalökonomie* (2nd ed. Rostock, 1842–50), pt. II, p. 1.

becoming apparent in his day when the substitution of mineral for vegetable and animal raw materials removed an ancient and important obstacle to industrial growth.

IV

The importance of minerals in general and coal in particular to the development of an industrial economy at the end of the eighteenth century extends beyond the improvement in communications and the possibility of escaping the close limits set to expansion as long as organic raw materials were essential to most industrial processes, for the development of the steam engine is peculiarly a product of the problems of the mining industry.

Although the expansion of mineral production was not subject to the same limitations as the expansion of the production of vegetable and animal materials, the technological problems of increasing production were nevertheless considerable. Perhaps the most intractable was that of draining pits when they were sunk to a depth which made impossible drainage by the cutting of adits to a point on the surface below the level of the seam. Horse gins were useful when the depths were moderate, but at the depths where some of the richest seams of coal and veins of tin and copper occurred a more powerful engine was required if the inflow of water to the workings was to be held in check.[1] The urgent difficulties of the mining industry were the means of turning the Newcomen from an ingenious but unpractical machine into a reliable piece of equipment without which the deeper pits could not have been maintained in production. The early engines of Savery and Newcomen were essentially pumping machines for which the only big market was the mining industry. A few Newcomen engines were used for pumping water from rivers to help with the supply of water to towns, but the majority were used in mines. They were at once essential to the continued expansion in coal production, and virtually unusable without a supply of coal. They are a product of a coal age rather than a wood age and could only be used extensively when mines rather than woodlands were their source of fuel. Otherwise they would have devastated an area of timber as quickly as the early iron industry had done, for the early engines were extremely inefficient and required very large quantities of fuel. Theoretical knowledge of the power of steam had long existed, but the coal industry's problems first provided the catalyst to convert this into workable machinery rather than engaging toys. The first use of steam engines in industries other than mining reflects their background as pumping engines in mines, for they were used initially simply to complete the cycle of water movement from tail-race to mill-pond and so to render water wheels independent of ordinary stream flow and prevent those interruptions to their operation which had previously been inevitable in prolonged dry weather. Watt's improvements to the steam engine (or rather his invention of a steam in place of an atmospheric engine) and his development of methods of gearing which gave rotatory rather than

[1] It has been claimed that three-quarters of the patents issued in England between 1561 and 1668 were connected with the coal industry, either directly or indirectly, and that a seventh were concerned with the drainage problem. S. F. Mason, *A History of the Sciences* (1953), p. 217.

reciprocating motion represent radical improvements upon the earlier Newcomen engines and gave the steam engine great importance in a wide range of industries as time passed. The steam engine more than any other single development, perhaps, made possible the vast increase in individual productivity which was so striking a feature of the Industrial Revolution by providing a source of power which dwarfed human, animal and even hydraulic sources. Yet the machine Watt improved was already widely used in the mining industry, which had fostered its development for several decades.

V

The cotton industry has attracted more attention than any other in discussions of the Industrial Revolution, and since it grew vastly while continuing to use a vegetable raw material, it merits further consideration in the context of this essay. No other major industry grew as quickly as cotton in the late eighteenth century. There was a series of inventions in both spinning and weaving which led to a marked rise in the output per worker. Cotton spinning was the first industrial activity to become organized in factories in the fashion which became normal during the next century. Lancashire became the area which first acquired the full range of features characteristic of the new industrial scene; large urban manufacturing populations living a life divorced from the rhythm of the countryside about them, working in factories, caught up in a web of exchange which connected their livelihood with events throughout the world. Cotton has for long been treated as the *type par excellence* of the new manufacturing industry, the lead-off industry in the take-off into sustained industrial growth.

The cotton industry fits well into the pattern of industrial growth which Adam Smith described. The decisive importance of the size of the market is well illustrated in its history. The attempt to expand production to keep pace with demand caused new mechanical devices to be seized upon eagerly and developed rapidly into reliable manufacturing machinery. There was a steep rise in productivity as workers became more specialized and turned to cotton manufacture as a full-time employment rather than a useful subsidiary source of income. The price of the finished article fell and still further enlarged the market. Once Whitney's gin had proved its worth, the demands of the industry for raw materials could be met by breaking in new land in the southern United States. There was no bottleneck in raw material supply and raw cotton tended to fall rather than rise in price. There is nothing in the story to call in question the assumptions of Adam Smith about the part played by the growth of industry in promoting the wealth of nations. Some coal was used in the preparatory and finishing sections of the industry, but only after a generation of expansion had caused the need for power to outstrip the capabilities of the human arm and the water wheel was the steam engine brought into use, so that the problem of raw material provision was very different in the cotton industry from those industries in which organic raw materials were replaced by inorganic. The example of the cotton industry makes it clear that industrial expansion could go far and fast in some directions without provoking difficulties in raw material supply.

For this very reason, however, the cotton industry, in spite of its importance in the Industrial Revolution, cannot be regarded as a microcosm of the whole process. In particular the great changes in inland transport and in power were not closely connected with the cotton industry. The movement of cotton presented no great difficulties to the methods of goods transport which had been in use for centuries. The movement of raw cotton was measured by the million pound rather than the million ton and bore a far higher value per unit weight than, say, coal. In consequence it was able to support relatively high transport charges without a crippling increase in price. The fact that many early mills were built in quite remote Pennine valleys close to a head of water underlines this point. Similarly, the well tried sources of industrial power, initially the worker's arms and later the water wheel, sufficed to move the machinery used in the cotton industry during the first two decades of rapid growth at the end of the eighteenth century. The cotton industry benefited substantially from the opening of canal communication between Manchester and Liverpool, both in that transport charges were lower than by river or on horseback and in that delivery was more reliable and quicker: but the cotton industry did not create a large enough tonnage of traffic to justify the construction of canals. 1800 was the first year in which the import of raw cotton exceeded fifty million pounds in weight, which is only some 23,000 tons, and no more than the annual output of perhaps 150 coal miners. Even though cotton might produce a much greater revenue per ton-mile than mineral freight it is clear that the cotton industry in itself could offer little inducement to spend capital on the scale necessary to build canals. In so far as the presence of the cotton industry in Lancashire did hasten the construction of canals it was perhaps rather as a consumer of coal and as an employer of labour which consumed coal domestically that it exerted an influence. Again, the cotton industry became an important market for Watt's new engines. Without them it might have lost impetus as the most suitable heads of water were harnessed one by one, leaving only the inaccessible or insufficient to be brought into use. But the history of the development of the steam engine lies outside the cotton industry.

It might indeed even be argued that there is a sense in which the cotton industry was exploiting old lines of development with a new intensity rather than striking out in a radically new direction. There had been periods of technological innovation in the textile industry in the past. When, for example, the spinning wheel superseded the distaff there had been a marked rise in the productivity of the individual spinner. The invention of the knitting machine and the introduction of the Dutch swivel-loom had also in their time brought about important changes in productivity and in the amount of fixed capital per worker. Water power had been used in the fulling of wool and the throwing of silk for centuries. Even the bringing together of many textile workers under one roof was not unknown before the eighteenth century. During the late eighteenth century the cotton industry brought development along old lines to a new pitch of perfection, evolving better machinery in both spinning and weaving and extending the use of water power into spinning. There was a vast increase in the quantity of cotton manufactured. It was produced more cheaply and, after early difficulties with the new machinery had been overcome, the quality of both yarn and cloth was higher than anything achieved in England in the past. In contrast with this the range of industries in which coal replaced wood as the main source of heat might be held to have struck out along a new line of development. The change from organic to inorganic sources of raw material supply led to no sudden or dramatic change in the quantity or quality of production in the industries concerned, yet latent in this change were new possibilities for improvement in the transport of goods and the supply of power to industrial processes. To borrow a biological analogy the raw material bottleneck produced a mutation in raw material supply which proved intensely favourable and led to changes which helped to transform industrial production and ultimately society as a whole.

The temptation to treat the cotton industry as a microcosm of the whole Industrial Revolution has proved difficult to resist. The signs of the new age were first readily apparent in cotton, but its history is not therefore typical of the whole. The fact that the cotton industry was singularly free from raw material supply problems marked it out from many others, at once facilitating its expansion at an early date and isolating it from a range of problems faced and solved by many other industries. Cotton benefited from the new sources of power and better transport facilities, but these were available because of the successful struggle against difficulties of raw material supply which had taken place in other sectors of the economy.

VI

It is natural that special attention should be paid by all those interested in the Industrial Revolution to any aspect of English life which was different from its continental equivalent. Hence the great interest shown in such questions as social mobility and capillarity, the organization of financial affairs and the capital market in England, English agriculture and systems of land tenure, the forms of English government and law, and in exploring any differences between English and continental demographic patterns. The supply of industrial raw materials may, of course, be treated in the same way. It is not difficult to show that the continent clung longer to the traditional types of organic industrial raw material, nor is it difficult to suggest reasons for this. Wiebe's price series,[1] for example, suggests that the supply of some types of timber was causing much less difficulty on the continent than in England in the seventeenth century. Yet it is perhaps more illuminating to dwell on the occasional similarities between England and the continent than on the general dissimilarity.

The area most like the new English industrial areas was central Belgium. This was the only area on the continent in which the production of coal in the

[1] G. Wiebe, *Zur Geschichte der Preisrevolution des XVI. und XVII. Jahrhunderts* (Leipzig, 1895). See esp. tables 522, 524 and 528.

eighteenth century reached a level at all comparable with that achieved in English coalfields at the same time. Both at Liège and near Charleroi the coal seams outcropped to the surface near the good water communication afforded by the Sambre-Meuse system, and this made possible the development of a wide market for coal at an early date. The Mons coalfield was not so well endowed with natural water communication, but the same problems and opportunities which produced a burst of canal building in England led to a similar development in Mons. The Mons-Condé canal, for example, was completed in 1814, linking the coalfield area with the industries of Nord. The history of the successful search for coal in the concealed Nord field which resulted in the sinking of the famous mine at Anzin is evidence of how useful even in the middle of the eighteenth century the discovery of coal was judged to be in this part of the continent. At that time traffic along the valley of the Sambre-Meuse resembled that along the valley of the Severn in England. The metal communes of Maubeuge and Valenciennes, for example, were dependent upon primary iron imported from the *pays de Liège* in the late eighteenth century. Belgian pits were quick to adopt British devices. Already before the turn of the century Belgian pits had reached considerable depths and were using Newcomen engines extensively. Soon after the turn of the century other British developments were absorbed into Belgian practice: the use of the steam engine for winding up coal and men as well as for pumping water, the Davy safety-lamp, and so on. There were independent local contributions to coal-mining and other industrial problems in Belgium at this time. Joseph Chaudron, for example, discovered an improved method of protecting main shafts by using a revetment of iron (the *cuvelage en fer*). Dony succeeded in extracting zinc from calamine. Gas lighting, a Belgian invention,[1] was in use in factories by 1810. The remarkable achievements of the Cockerill family in engineering and textiles at Seraing and Verviers during the first quarter of the nineteenth century form an industrial epic on a scale worthy of comparison with English equivalents of the period. John Cockerill constructed successful coke-fired blast furnaces at Seraing in the 1820's, and built excellent marine steam engines. The advance of central Belgium, in short, was very rapid, and it is perhaps symbolic of this that Neilson's hot blast was widely used there at a time when it was still a novelty in British iron centres outside Scotland.

The history of the rapid acceptance of British industrial advances in the valleys of the Meuse and Sambre is not vastly different from the history of their acceptance, say, in the valleys of South Wales. There was, it is true, usually a time lag, though not always longer in the case of Liège than in the case of the less active British areas. In view of the fact that Belgium during the eighteenth and early nineteenth centuries was ruled by other countries and was frequently disturbed by the passage of armies, it is striking how swiftly the areas whose problems and opportunities were similar to those of the new industrial areas of Britain followed the British lead and occasionally improved upon it.

Central Belgium was not unique among continental industrial areas. The

[1] Minckelers lighted his lecture room with gas in 1785, though it was William Murdoch who in the 1790's showed the commercial possibilities of this method of illumination.

supply with the rapid change in England. When Thünen published his book the Ruhr was still an area of agriculture and marshland. The example of central Belgium, however, where change came so rapidly on the heels of developments in similar English areas, stresses in the context of raw material supply two points which are perhaps true of the Industrial Revolution as a whole: first that western Europe was a single economic community within which like circumstances might give rise to similar results; and secondly that industrial growth was essentially a local rather than a national affair. In this regard it is perhaps unnecessarily inexact to talk of England and the continent rather than, say, of Lancashire and the valley of the Sambre-Meuse. Each country was made up of a number of regional economies. Within Belgium, for example, the Flemish domestic linen industry was in great straits because of its St. Etienne region, for example, showed some similar traits, but in the main it is fair to contrast the conservatism of the continent over industrial raw material failure to adopt English methods at just the period of brilliant advance on the Belgian coalfields.

Perhaps the chief advantage in looking at the Industrial Revolution from the standpoint of raw material supply is that it makes it easier to understand the nature of the gap which separates Adam Smith and his world from the world which McCulloch knew. The ordinary categories of economic analysis do not pick up the differences very well. In the world which Adam Smith described there could well exist a technically perfect capitalism, with a developed money market, extensive international trade, many intermediaries between the producer and the ultimate consumer, a divorce between the workers and the ownership of the means of production, and so on. But in this world there was a ceiling to the possible size of industrial production set by the difficulty of expanding raw material supply at constant or declining prices as long as most industrial raw materials were organic. When this was no longer true this ceiling disappeared. Adam Smith's world and that of his Physiocratic predecessors was a world bounded by the fertility of the soil. This was the backcloth to any examination of industrial development. Ninety years later when Jevons published his great work on the coal industry he was prepared to assert firmly that the greatest single factor governing the industrial prospects of any nation must be its wealth in coal.[1] If McCulloch was impatient with Adam Smith it was at least in part because the world in which he lived was different. The weaknesses in his argument would not have escaped Adam Smith if he had the benefit of being able to observe a further half century of economic history. If he had been privileged to do so, what he would have seen would have convinced him that the use of inorganic raw materials in industry on a vast scale had revealed the existence of 'powers of nature' whose potentialities he had not suspected. *Peterhouse, Cambridge*

[1] 'Coal, in truth, stands not beside but entirely above all other commodities. It is the material energy of the country – the universal aid – the factor in everything we do. With coal almost any feat is possible or easy; without it we are thrown back into the laborious poverty of early times.' W. S. Jevons, *The Coal Question* (London and Cambridge, 1865), p. viii. See also the chapter *Of the Comparative Coal Resources of Different Countries.*

It is interesting to note that Jevons's main concern was that supplies of coal must soon run short in Britain; that mineral raw materials, being exhaustible, were a dangerous basis of national wealth and power.

Part 5

Technology, Invention, Innovation, Education

Technological Change and Industrial Development in Western Europe, 1750-1914

D. S. Landes

I. *The Industrial Revolution in Britain*

IN the eighteenth century, a series of inventions transformed the manufacture of cotton in England and gave rise to a new mode of production—the factory system. During these years, other branches of industry effected comparable advances, and all these together, mutually reinforcing one another, made possible further gains on an ever-widening front. The abundance and variety of these innovations almost defy compilation, but they may be subsumed under three principles: the substitution of machines—rapid, regular, precise, tireless—for human skill and effort; the substitution of inanimate for animate sources of power, in particular, the introduction of engines for converting heat into work, thereby opening to man a new and almost unlimited supply of energy; the use of new and far more abundant raw materials, in particular, the substitution of mineral for vegetable or animal substances.

These improvements constitute the Industrial Revolution. They yielded an unprecedented increase in man's productivity and, with it, a substantial rise in income per head. Moreover, this rapid growth was self-sustaining. Where previously, an amelioration of the conditions of existence, hence of survival, and an increase in economic opportunity had always been followed by a rise in population that eventually consumed the gains achieved, now for the first time in history, both the economy and knowledge were growing fast enough to generate a continuing flow of investment and technological innovation, a flow that lifted beyond visible limits the ceiling of Malthus's positive checks. The Industrial Revolution thereby opened a new age of promise. It also transformed the balance of political power, within nations, between nations, and between civilizations; revolutionized the social order; and as much changed man's way of thinking as his way of doing.

In 1760 Britain imported some 2½ million pounds of raw cotton to feed an industry dispersed for the most part through the countryside of Lancashire and existing in conjunction with the linen manufacture, which supplied it with the tough warp yarn it had not yet learned to produce. All of its work was done by hand, usually (excluding dyeing and finishing) in the homes of the workers, occasionally in the small shops of the master weavers. A generation later, in 1787, the consumption of raw cotton was up to 22 million pounds; the cotton manufacture was second only to wool in numbers employed and value of product; most of the fibre consumed was being cleaned, carded, and spun on machines, some driven by water in large mills, some by hand in smaller shops or even in cottages. A half-century later, consumption had increased to 366 million pounds; the cotton manufacture was the most important in the kingdom in value of product, capital invested, and numbers employed; almost all of its employees, except for the still large number of hand-loom weavers, worked in mills under factory discipline. The price of yarn had fallen to perhaps one twentieth of what it had been, and the cheapest Hindu labour could not compete in either quality or quantity with Lancashire's mules and throstles. British cotton goods sold everywhere in the world: exports, a third larger than home consumption, were worth four times those of woollens and worsteds. The cotton mill was the symbol of Britain's industrial greatness; the cotton hand, of her greatest social problem—the rise of an industrial proletariat.

Why did this revolution in the techniques and organization of manufacture occur first in Britain? A few theoretical considerations may help us to organize the argument. Technological change is never automatic. It means the displacement of established methods, damage to vested interests, often serious human dislocations. Under the circumstances, there usually must be a combination of considerations to call forth such a departure and make it possible: (1) a need or opportunity for improvement due to inadequacy, present or potential, of prevailing techniques;[1] and (2) a degree of superiority such that the new methods pay sufficiently to warrant the costs of the change. Implicit in the latter is the assumption that, however much the users of older, less efficient methods may attempt to survive by compressing the costs of the human factors of production, entrepreneurial or labour, the new techniques are enough of an improvement to enable progressive producers to outprice them and displace them.

The technological changes that we denote as the 'Industrial Revolution' implied a far more drastic break with the past than anything since the invention of the wheel. On the entrepreneurial side, they necessitated a sharp redistribution of investment and a concomitant revision of the concept of risk. Where before, almost all the costs of manufacture had been variable—raw materials and labour primarily—more and more would now have to be sunk in fixed plant. The flexibility of the older system had been very advantageous to the entrepreneur: in

[1] The criterion of adequacy would, for my purposes, be marginal costs. Steeply rising costs per unit under conditions of growing demand would imply a need for and incentive to technological improvement.

time of depression, he was able to halt production at little cost, resuming work only when and in so far as conditions made advisable. Now he was to be a prisoner of his investment, a situation that many of the traditional merchant-manufacturers found very hard, even impossible, to accept.

For the worker, the transformation was even more fundamental, for not only his occupational role, but his very way of life was at stake. For many—though by no means for all—the introduction of machinery implied for the first time a complete separation from the means of production; the worker became a 'hand'. On almost all, however, the machine imposed a new discipline. No longer could the spinner turn her wheel and the weaver throw his shuttle at home, free of supervision, both in their own good time. Now the work had to be done in a factory, at a pace set by tireless, inanimate equipment, as part of a large team that had to begin, pause, and stop in unison—all under the close eye of overseers, enforcing assiduity by moral, pecuniary, occasionally even physical means of compulsion. The factory was a new kind of prison; the clock a new kind of jailer.

In short, only the strongest incentives could have persuaded entrepreneurs to undertake and accept these changes; and only major advances could have overcome the dogged resistance of labour to the very principle of mechanization.

The origins of the entrepreneurial interest in machines and factory production must be sought in the growing inadequacy of the older modes of production, an inadequacy rooted in internal contradictions, themselves aggravated by external forces.

Of these pre-factory forms of organization, the oldest was the independent craft shop, with master often assisted by one or more journeymen or apprentices. Fairly early, however—as far back as the thirteenth century—this independence broke down in many areas, and the artisan found himself bound to the merchant who supplied his raw materials and sold his finished work. This subordination of the producer to the intermediary (or, less often, of weak producers to strong ones) was a consequence of the growth of the market. Where once the artisan worked for a local clientele, a small but fairly stable group that was bound to him personally as well as by pecuniary interest, he now came to depend on sales through a middleman in distant, competitive markets. He was ill-equipped to cope with the fluctuations inherent in this arrangement. In bad times he might be completely idle, with no one to sell to; and when business improved, he usually had to borrow from his merchant the materials needed to get started again. Once caught on a treadmill of debt—his finished work mortgaged in advance to his creditor—the craftsman rarely regained his independence; his work sufficed to support him—no more—and he was in fact if not in principle a proletarian, selling not a commodity, but labour.

Aside from his pecuniary difficulties, the local artisan was in no position to know and exploit the needs of distant consumers. Only the merchant could respond to the ebb and flow of demand, calling for changes in the nature of the final product to meet consumer tastes, recruiting additional labour when necessary, supplying tools as well as materials to potential artisans. It was largely in this way that the rural population was drawn into the productive circuit. Very early, urban merchants came to realize that the countryside was a reservoir of cheap labour: peasants eager to eke out the meagre income of the land by working in the off-season, wives and children with free time to prepare the man's work and assist him in his task. And though the country weaver, nail-maker, or cutler was less skilled than the guildsman or journeyman of the town, he was less expensive, for the marginal utility of his free time was, initially at least, low, and his agricultural resources, however modest, enabled him to get by on that much less additional income. Furthermore, rural putting-out was free of guild restrictions on the nature of the product, the techniques of manufacture, and the size of enterprise.

The above description of a long and complex historical process inevitably oversimplifies. If it seems reasonable to assert that, taking Europe as a whole, most putters-out came from the mercantile side, it is important to note the many exceptions: the weavers who became clothiers by hiring their less enterprising neighbours; the fullers and dyers who had accumulated capital in the finishing processes and integrated backwards by contracting directly for yarn and cloth. In some areas, most notably the region around Leeds in the West Riding of Yorkshire, rural artisans organized their own small weaving sheds, joined when necessary to create common facilities, and sold their pieces as independent clothiers in the weekly cloth halls. But even in Yorkshire, this fragmentation of enterprise was characteristic primarily of the woollen trade; in the worsted manufacture, where capital requirements were greater, the productive unit was larger and the merchant putter-out more important.[1]

The English textile industry built its fortune in the late medieval and

[1] In his discussion of the shift from urban to rural industry, P. Mantoux, *The Industrial Revolution in the Eighteenth Century* (London, 1928), pp. 64–6, conveys the impression that the putting-out system was the result of the decay of what he describes as 'domestic manufacture', that is, dispersed independent cottage industry of the kind found in Yorkshire. Often, as we noted, this was true, but even more often, probably, putting out was the product of mercantile initiative seeking new sources of labour and drawing the rural population into the commercial circuit.

early modern periods on rural manufacture. No centre of production, except perhaps Flanders, was so quick to turn from the towns to the countryside; it is estimated that as early as 1500 over half the output of wool cloth was accounted for in this manner. The trend continued: by the mid-eighteenth century, the great preponderance of the British wool manufacture was cottage industry; of all the towns immemorially associated with the wool trade, only Norwich remained as an important urban centre, and it was rapidly declining in relative importance. Allowing for such regional variations, moreover, and for occasional pauses, the industry as a whole had prospered impressively. In the late seventeenth and early eighteenth centuries, at a time when the Italian manufacture was a shadow of its former self, when Dutch cloth output was shrinking steadily, and when France was in the throes of a prolonged depression, British consumption of raw wool was growing at the rate of about 8 per cent a decade; and from about 1740 to 1770, the decennial increase was 13 or 14 per cent.[1]

This growth merits detailed attention, for it was the principal precipitant of the changes we denote by the Industrial Revolution, and understanding it may help us understand the reasons for British precedence in technological and economic development. In part the wool industry grew because of favourable conditions of production. Thus no country had so abundant a supply of raw wool, particularly the long wool required for the lighter, harder, worsted fabrics. And rural manufacture, largely unhampered by guild restrictions or government regulation, was in a position to make the most of this resource advantage by suiting its product to demand and changes in demand. In particular, it was free to develop cheaper fabrics, perhaps less sturdy than the traditional broadcloths and stuffs, but usable and often more comfortable. This freedom to adjust and innovate is particularly important in light industry, where resources and similar material considerations often are less important as locational factors than entrepreneurship. A good example from within the British wool industry is the rapid growth of the Yorkshire worsted trade, to the point where it passed the older centre of East Anglia in the course of the eighteenth century; compare Clapham's explanation: 'the ordinary case of a pushing, hard-

[1] P. Deane, 'The Output of the British Woollen Industry in the Eighteenth Century', *J. Econ. Hist.* XVII (1957), 220. These figures are derived from informed contemporary guesses and are therefore gross approximations. But it is the trend that interests us here. On this, compare the much slower growth of the Verviers-Hodimont area near Liège, one of the most enterprising centres of wool manufacture on the Continent. P. Lebrun, *L'industrie de la laine à Verviers pendant le XVIIIe et le début du XIXe siècle* (Liège, 1948), pp. 518–19. Note also the difference in size of output between Yorkshire alone (aulnage returns in T. S. Ashton, *An Economic History of England: the Eighteenth Century* (London, 1955), pp. 249–50) and the Verviers area.

working locality with certain slight advantages, attacking the lower grades of an expanding industry'.[1] We shall have occasion to remark comparable examples of the advantages of entrepreneurial freedom when we turn to the continental countries. In the meantime, we may note that the British wool manufacture profited the more from its liberty because some of its most dangerous competitors across the Channel were being subjected in the seventeenth and early eighteenth centuries to increasing regulation and control.

Finally, one should cite the relative freedom of British industry from the disturbance and destruction of war, the uneven but long and often rich inflow of skilled foreign artisans, and the access of the producing centres to water transport, hence distant markets—all factors conducive to lower costs of manufacture and distribution.

On the demand side, the British wool manufacture was comparably favoured. The population of the kingdom was not large, but it was growing, faster probably by the middle of the eighteenth century than that of any of the countries across the Channel. From not quite 6 millions around 1700, it rose to almost 9 millions in 1800; 70–90 per cent of the gain came in the second half of the period.[2] What is more, the absence of internal customs barriers or feudal tolls created in Britain the largest coherent market in Europe. This political unity was confirmed by the geography of the island: the land mass was small; the topography, easy; the coastline, deeply indented. By contrast, a country like France, with more than three times as many people, was cut up by internal customs barriers into three major trade areas, and by informal custom, obsolete tolls and charges, and, above all, poor communications into a mosaic of semi-autarkic cells.

Moreover, what nature bestowed, man improved. From the mid-seventeenth century on, there was a continuous and growing investment of both public and private resources in the extension of the river system and the construction of new roads and bridges. By 1750 there were over a thousand miles of navigable streams in Britain; and Parliament had been passing turnpike acts at the rate of eight a year for half a century. Impressive as this development was, it was inadequate to the needs of the economy, and the pace of investment increased markedly

[1] J. H. Clapham, 'The Transference of the Worsted Industry from Norfolk to the West Riding', *Econ. J.* XX (1910), 203. Eric M. Sigsworth, *Black Dyke Mills: a History: with Introductory Chapters on the Development of the Worsted Industry in the Nineteenth Century* (Liverpool: University Press, 1958), p. 17, subscribes to this point of view.

[2] For different but roughly concordant estimates of this increase, see Phyllis Deane and W. A. Cole, *British Economic Growth 1688–1959: Trends and Structure* (Cambridge, 1962), p. 5, n. 3. In the same period, the population of France went from about 20 to 27½ millions. E. Levasseur, *La population française* (3 vols.; Paris, 1889), I, 201–6, 215–18.

in the fifties and sixties. These years saw the first canals (Sankey Navigation, 1755–9; Duke of Bridgewater's canal, 1759–61) and turnpike acts at the rate of forty a year. In two decades (1760–80), navigable water and solid roads linked the major industrial centres of the North to those of the Midlands, the Midlands to London, and London to the Severn basin and the Atlantic.

Within the market of Britain, purchasing power per head and standard of living were significantly higher than on the Continent. We have no precise measures of national income for the eighteenth century,[1] but there is an abundance of impressionistic testimony by travellers from both sides of the Channel to the greater equality of wealth, higher wages, and greater abundance to be found in Britain. Thus one of the best signs of comfort in Europe is the consumption of white bread; in the nineteenth century, one can almost follow the rise in *per capita* income and the diffusion of higher living standards among the poorer sections of the population, into rural areas, and into central and eastern Europe by the wheat frontier. In the eighteenth century England was known as the country of the wheaten loaf. This was an exaggeration: in large areas, particularly in the Midlands and North, rye and barley were the staple grains, especially in the early part of the century. Even there, however, the bread grew whiter over the years, and nowhere was there anything like the reliance one found across the Channel on coarser cereals like buckwheat and oats. Similarly, there was much myth in the image of John Bull, beefeater. Yet when Arthur Young sat down to soup in the Pays Basque—'what we should call the farmer's ordinary'—he received 'ample provision of cabbage, grease, and water, and about as much meat for some scores of people, as half a dozen English farmers would have eaten, and grumbled at their host for short commons'.[2] Even workhouse menus, hardly designed to make life agreeable for the residents, provided for meat daily or at least several times a week.[3]

The English labourer not only ate better; he spent less of his income on food than his continental counterpart, and in most areas this portion was shrinking, whereas across the Channel it may well have risen during much of the eighteenth century.[4] As a result, he had more to spare for other things, including manufactures. The Englishman was

[1] See, however, the article of P. Deane, 'The Implications of Early National Income Estimates for the Measurement of Long-Term Economic Growth in the United Kingdom', *Economic Development and Cultural Change*, IV (1955), 3–38.

[2] Young, *Travels during the Years 1787, 1788 and 1789* (2 vols.; Dublin, 1793), I, 87 f., 93.

[3] Cf. Dorothy Marshall, *The English Poor in the Eighteenth Century* (London, 1926), p. 268.

[4] This is the position of C. E. Labrousse, *Origines et aspects économiques et sociaux de la Révolution française (1774–1791)* ['Les Cours de Sorbonne'] (Paris, n.d.), pp. 54–8.

reputed for wearing leather shoes where the Fleming or Frenchman wore clogs. He was dressed in wool where the French or German peasant often shivered in linen, a noble fabric for table or bed, but a poor shield against the European winter. Defoe vividly and proudly described the importance of this demand for British manufactures in his *Plan of the English Commerce* in 1728:[1]

> ...for the rest, we see their Houses and Lodgings tolerably furnished, at least stuff'd well with useful and necessary household Goods: Even those we call poor People, Journey-men, working and Pains-taking People do thus; they lye warm, live in Plenty, work hard, and [need] know no Want.
>
> These are the People that carry off the Gross of your Consumption; 'tis for these your Markets are kept open late on *Saturday* nights; because they usually receive their Week's Wages late...in a Word, these are the Life of our whole Commerce, and all by their Multitude: Their Numbers are not Hundreds or Thousands, or Hundreds of Thousands,but Millions; 'tis by their Multitude, I say, that all the Wheels of Trade are set on Foot, the Manufacture and Produce of the Land and Sea, finished, cur'd, and fitted for the Markets Abroad; 'tis by the Largeness of their Gettings, that they are supported, and by the Largeness of their Number the whole Country is supported; by their Wages they are able to live plentifully, and it is by their expensive, generous, free way of living, that the Home Consumption is rais'd to such a Bulk, as well of our own, as of foreign Production....

Defoe's reference to the Englishman's 'expensive, generous, free way of living' calls to mind a final aspect of the British domestic market: a consumption pattern favourable to the growth of manufactures. More than any other in Europe, probably, British society was open. Not only was income more evenly distributed than across the Channel, but the barriers to mobility were lower, the definitions of status looser. Nothing is more revealing in this regard than a comparison of contemporary images of society in the different countries of western Europe. For Britain, we have schemes like those of Gregory King or Joseph Massie—congeries of occupational groups ranked according to wealth and so intermingled as to preclude the drawing of horizontal status lines across the whole of the social pyramid. For France, we have a neater tripartite structure: aristocracy, bourgeoisie, *peuple*; within these, to be sure, there are fine distinctions, and it is not always easy to rank people of different occupations or to place borderline groups like artisans and retail shopkeepers; nevertheless, the arrangement is orderly, traditionally logical. For most of west Germany, we have the French system, but more rigid and carefully defined, to the point

[1] [Daniel Defoe], *A Plan of the English Commerce* (Oxford: Blackwell, 1928), pp. 76–7.

where status, even of sub-groups, is often written into law. And east of the Elbe, society was simpler yet: a small landholding aristocracy; the large mass of personally dependent peasants; in between, a thin layer of commercial bourgeois, spiritually and often ethnically alien to the body social within which they lived and moved encapsulated.

So far as the rate of consumption is concerned, the implications of greater equality of income are a matter of some debate.[1] Similarly, mobility is ambiguous in its effects: some people will save to climb; others will consume to announce their arrival. The net result will depend on circumstances.

Quality and direction of consumption, however, are something else again. In non-primitive societies, where skills are fairly advanced and there has been some accumulation of wealth, inequality fosters a taste for extravagant luxuries and services among the few, whereas equality encourages a demand for more sober, solid comforts among the many. Great riches amid a sea of poverty are generally the product of a low capital–labour ratio (or of misinvestment of capital). They give rise to a prodigal expenditure of labour on pleasure and elegance: an over-abundance of domestics—to the point where the mistress of the house spends more time supervising her staff than more modest wives spend doing their own chores; ornamental garments of great price; lavish decoration of residences; the production of exquisitely difficult works of art.

A more even diffusion of wealth, however, is the result of costly labour. This was indeed the case in Britain, where wages—allowing for the uncertainty and partial incomparability of the estimates—ran about twice as high as in France and higher yet than east of the Rhine. In such an economy, production functions are more capital-intensive, while the rich consumer caters less to whim and satisfies himself with a greater abundance of those goods that are available on a smaller scale and in lower quality to his poorer fellows. On the other hand, the relatively high purchasing power of the poorer elements of the population implies a correspondingly greater demand for the things they need

[1] The traditional assumption is that inequality does increase the savings ratio. But there is some question whether this is justified for a pre-industrial society, especially one in which a small privileged group commands the levers of power and can draw a kind of tributary income from the rest of the nation. It seems quite probable, for example, that the court aristocracy of eighteenth-century France lived beyond its income, consuming freely in the knowledge that there would always be ways of obtaining more from the crown. Cf. Milton Friedman, *A Theory of the Consumption Function* (Princeton, 1957), pp. 235 f., who argues that inequality of 'permanent' (as against measured) income *per se* does not affect the consumption–savings ratio; that it is uncertainty about future income that promotes savings, against a rainy day as it were.

and can afford—the cheaper, plainer articles most susceptible of mass production.[1]

Mobility in such a society is a force for standardization. For mobility implies emulation, and emulation promotes the diffusion of patterns of expenditure throughout the population. Where there is no movement between status groups, clear, inviolate distinctions of dress and way of life mark the gradations of hierarchy. Where there begins to be movement, as in the late Middle Ages, sumptuary laws are often needed to keep people in their place. And where mobility has become so commonplace as to seem to many a virtue, discriminatory controls over expenditure are unenforceable.

In England, sumptuary laws were dead letters by the end of the sixteenth century; they were repealed by James I in 1604. Over the next two centuries, the trend toward homogeneity of expenditure—the effacement of vertical regional differences as well as horizontal social distinctions—continued. Contemporaries complained of the luxury of the lower classes, who dressed so as to be indistinguishable from their betters. This was an exaggeration; social lament as a literary genre is invariably hyperbolic. Besides, much of the elegance of the populace was meretricious, the result of an active trade in second-hand clothes. Even so, the very demand for cast-offs was evidence of the absence or decay of customary distinctions: the poor man could and did wear the same kind of coat as the rich. Similarly, contemporaries complained of the farmer's imitation of city ways, his abandonment of the rustic simplicity of yore. Again an exaggeration—yet the truth was that in no economy was the countryside so closely integrated into the commercial circuit; nowhere were the local pockets of self-sufficiency so broken down.

All of this was part of a general process of urbanization, itself a reflection of advanced commercialization and industrialization. London alone was a monster: Defoe estimated in 1725 that it contained a million and a half inhabitants, almost a quarter of the people in the

[1] On the implication of inequality of income, or, more precisely, inequality of consumption, for the nature and composition of industrial output, see the suggestive article of W. Paul Strassman, 'Economic Growth and Income Distribution', *Quarterly J. of Economics*, LXX (1956), 425–40; also S. Kuznets, 'Economic Growth and Income Inequality', *Amer. Econ. Rev.* XLV (1955), 1–28, which is more concerned with the reverse relation.

The best single index to relative factor costs and the pattern of consumption is the extent and character of domestic service: the twentieth century, sometimes called the era of the common man, is also the age of the disappearing maid. And while the British merchant of the eighteenth century had less cause for frustration than his present-day descendants, he had his servant problem. Cf. Defoe, *Everybody's Business Is Nobody's Business*, in *The Novels and Miscellaneous Works of Daniel Defoe* ('Bohm's Standard Library'; 7 vols.; London, 1889), II, 499–500. See also J. Jean Hecht, *The Domestic Servant Class in Eighteenth-century England* (London, 1956), esp. chs. I and VI.

kingdom. This figure is testimony, not to Defoe's accuracy, but to the impression the 'great wen' made on contemporaries; yet even conservative estimates put the population of the metropolitan area at about half that number. In the provinces, the cities and towns developed steadily after the Civil War; among the most rapidly expanding were unincorporated 'villages' like Manchester, which had perhaps 12,500 inhabitants in 1717 and 20,000 by 1758. An estimate of 15 per cent of the population in cities of 5000 and over by mid-century and 25 per cent by 1800 is probably close to the truth. By contrast, the French figure on the eve of the Revolution was something over 10 per cent; and Germany was even more rural.

But it was not only that England had more people living in cities than any other European country except perhaps Holland;[1] it was the character of British urban life that made the pattern of settlement particularly significant. On the Continent, many of the cities were essentially administrative, judicial, ecclesiastical in function. Their populations consisted essentially of bureaucrats, professionals, soldiers, and the shopkeepers, artisans, and domestics to serve them. The city was not so much a node of economic activity, trading manufactures and mercantile services for the products of the countryside, as a political and cultural centre drawing tax revenues and rents from the rural population in return for government and by traditional right. Madrid is the classic example of this kind of agglomeration; but Paris was much like this, and perhaps a majority of the larger French provincial cities—including places like Arras, Douai, Caen, Versailles, Nancy, Tours, Poitiers, Aix, and Toulouse—were little else. In Germany, of course, the very fragmentation of political power was an incitement to the multiplication of semi-rural capitals, each with its court, bureaucracy, and garrison.

By contrast, the relatively smaller size of Britain's political apparatus and its concentration in London left the older provincial centres to somnolence and decay. Nothing is more striking about the map of Britain in the eighteenth century than the modernity of the urban pattern. The medieval county seats—Lancaster, York, Chester, Stafford—were overshadowed by younger places like Liverpool, Manchester, Leeds, and Birmingham, and there was already a substantial shift of population in favour of the North and Midlands. Much of the increase, moreover, did not take place within the cities proper, but took the form of a thickening of the countryside. Numerous overgrown industrial villages sprang up—concentrations of hundreds of spinners and

[1] And Holland's urban population was declining sharply, both relatively and absolutely. Cf. William Petersen, *Planned Migration: the Social Determinants of the Dutch-Canadian Movement* [University of California Publications in Sociology and Social Institutions, vol. II] (Berkeley and Los Angeles, 1955), p. 20.

weavers in the manufacturing districts of Lancashire and Yorkshire, similar in many ways to the earlier rural agglomerations of East Anglia.

The pattern throughout was one of close contact and frequent exchange between city and land. Trade and shops went to the customers: the late A. P. Wadsworth noted the numerous advertisements of cottages-to-let for tradesmen in the villages around Manchester, reflecting on both sides the keen response to economic opportunity.[1] In spite of the sparseness of the data, it seems clear that British commerce of the eighteenth century was, by comparison with that of the Continent, impressively energetic, pushful, and open to innovation. Part of the explanation is institutional: British shopkeepers were relatively free of customary or legal restrictions on the objects or character of their activity. They could sell what and where they would; and could and did compete freely on the basis of price, advertising, and credit. If most shopkeepers continued to haggle, many followed the lead of the Quakers in selling at fixed, marked prices. In so far as such methods prevailed, they conduced to a more efficient allocation of economic resources and lower costs of distribution.

In sum, the home market for manufactures was growing, thanks to improving communication, increase in population, high and rising average income, a buying pattern favourable to solid, standardized, moderately priced products, and unhampered commercial enterprise. How much it grew, however, one cannot say precisely; we have no statistics on domestic consumption.

We are better informed about foreign trade, if only because most of the commodities that came in or went out of the country had to pass under the eyes of the customs officers. Admittedly, the trade statistics are incomplete, inaccurate, and biased by the use of fixed values in a world of fluctuating prices. But they do furnish an order of comparison, showing for example a three- or fourfold gain in British exports (including re-exports) in the century from 1660 to 1760.

We have seen that the growth of Britain's sales abroad, as at home, reflected in large part her natural endowment; to this should be added some institutional and historical advantages. She had a strong maritime tradition, and, unlike most of her continental rivals, did not divert her energies into the maintenance of costly armies and territorial aggrandizement. Rather she concentrated her efforts on securing trading privileges and a colonial empire, in large part at the expense of her leading continental rivals, France and Holland. This kind of thing cost less than European territory and in the long run paid better. No state was more responsive to the desires of its mercantile classes; no country more alert

[1] A. P. Wadsworth and Julia de L. Mann, *The Cotton Trade and Industrial Lancashire, 1600–1790* (Manchester, 1931), p. 276, n. 2.

to the commercial implications of war. Mr Ramsey perceptively notes the role of London in promoting this harmony of trade and diplomacy, contrasting in this regard the isolation of Bordeaux, Marseilles, and Nantes from Paris and Versailles.[1]

At the same time, Britain developed a large, aggressive merchant marine and the financial institutions to sustain it. Of all the Continental countries, only Holland again could rival her in this regard, and the comparative advantage of Holland lay in trade, not industry. Between Dutch mercantile power and Britain's combination of mercantile and industrial strength, the issue was never in doubt; the greatest asset of a port is a productive hinterland.

In the long run, this was Britain's forte: the ability to manufacture cheaply precisely those articles for which foreign demand was most elastic. The most promising markets for Britain in the seventeenth and eighteenth centuries lay not in Europe, whose own industries were growing and whose mercantilist rulers were increasingly hostile to the importation of manufactures, but rather overseas: in the New World, Africa, the Orient. These areas were very different in needs and tastes. The tribesmen of Africa and the plantation hand of the Antilles wanted thin, cool fabrics, bright colours, flashy metal—light woollens, the cotton-linen checks of Manchester, the cheap stampings of Birmingham. The requirements of the Indian or Chinese peasant were similar (excluding most cotton goods) though more sober. The New England farmer or Philadelphia merchant, confronted by a harsher, more variable climate and more sophisticated technologically, bought heavier cloth and sturdier hardware. For all, however, there was one common denominator, and that a negative one: they were not especially interested in costly, highly finished luxuries.

The effect of increased export, then, was to reinforce the pressures toward standardization as against differentiation, quantity as against quality. The sacrifice of quality to quantity was an old story in English manufacture. By this I do not mean adulteration or the sale of inferior goods as first quality—this was an international evil, as the iteration of government and guild regulations on the Continent evidences. Rather I mean the adoption of new methods of production that save costs at the expense of solidity or appearance; the use of coal in place of wood in glass-making or brewing is the best example.[2]

[1] G. D. Ramsay, *English Overseas Trade during the Centuries of Emergence* (London, 1957), pp. 247 f.

[2] J. U. Nef has argued in a number of works that the adoption of mineral fuel itself gave strong impetus to the production of 'quantity and utility rather than quality and elegance'. See, *inter alia*, his *Cultural Foundations of Industrial Civilization* (Cambridge, 1958), pp. 52–3. Yet it is clear that the readiness to accept coal was itself indicative of a deeper rationality; such nations as France, confronted with the same choice, obdurately

This readiness to abandon old ways for new, to place profit above craft pride and even the appearances of pride, implies a certain separation of the producer from production, an orientation to the market instead of to the shop. To some extent it reflected the early domination of British manufacture by mercantile interests and the reduction of the rural artisan to a mere employee of the putter-out. Clearly, however, this is not enough to account for the phenomenon; in the wool industry, for example, the most enterprising centre of manufacture was in Yorkshire, a stronghold of the small independent clothier; and in metallurgy, glass-making, brewing, and chemicals—the industries most affected by the introduction of mineral fuel—the organization of production had nothing to do with putting-out.

Instead, this cost-mindedness must be seen as part of a larger rationality, itself in some measure the result of material circumstances—above all, the greater cohesiveness of the British market and the effectiveness of competitive pressures—but also as an ideological force of its own, whose sources still remain to be explored. In no country in the eighteenth century, with the possible exception of Holland, was society so sophisticated commercially. Nowhere was the response to profit and loss so rapid; nowhere did entrepreneurial decisions less reflect non-rational considerations of prestige and habit. We shall have occasion to consider this again when we speak of investment and the supply of capital for industrialization. At the moment, my only concern is to explain where market pressures were pushing the producers and why the producers responded.

How much of the increase in demand and the trend toward mass production of cheaper articles is to be attributed to the expansion of home as against foreign markets is probably impossible to say. We have only the grossest, global estimates of the proportion of domestic to overseas sales, and these presumably comprise everything, including agricultural products. What interests us here, however, is the demand for manufactures, and only certain manufactures at that. One may perhaps attempt this kind of comparison for the wool industry: at the end of the seventeenth century English exports of wool cloth probably accounted for upwards of 30 per cent of the output of the industry; by 1740, the proportion had apparently risen, possibly to over half, and in 1771–2, something under a half.[1] In this important branch, then, the

rejected coal—even where there were strong pecuniary incentives to switch over to the cheaper fuel.

[1] These figures are based on Phyllis Deane, 'The Output of the British Woollen Industry in the Eighteenth Century', *J. Econ. Hist.* XVII (1957), 209–10, 211–13, 215–16, 220. The article itself makes clear the limitations of these figures, which are essentially informed inferences from informed guesses.

major impetus seems to have come from the export trade, and the most active exporting area in the industry, Yorkshire, was also the most rapidly growing centre of manufacture. It has indeed been argued that not only was the minimum critical market required to induce a technological breakthrough too big for any one country to provide, but that only a large fraction of the growing world demand could supply the necessary push; and that it was that peculiar combination of economic and political circumstances that permitted Britain to win for herself in the eighteenth century so large a share of the trade in manufactures that accounts for the successful leap to the 'higher' mode of production.[1]

Yet the answer is not so simple. Such figures as we have on British exports (overwhelmingly manufactures) show a distinct levelling off in the third quarter of the century. The volume of woollen shipments falls from the late fifties; cottons falter in the late sixties and seventies; the break comes later in iron and steel—in the late sixties—but it is sharp and the drop persists until the nineties.[2] David Eversley argues cogently against the easy acceptance of exports as the leading sector of the economy in process of revolution: noting the weight and relative stability of home demand, he reasons that only the existence of this kind of dependable market justified and permitted the accumulation of capital in manufacture.[3] On the other hand (as in many historical questions, one can fairly shuttle back and forth between pros and cons), this very variability of exports was surely a stimulant to industrial change and growth. It is not only that the marginal increment of sales often spells the difference between profit and loss; the bursts of overseas demand placed abrupt and severe burdens on the productive system, pushed enterprise into a position of rapidly increasing costs, and enhanced the incentive for technological change. Certainly, from the late eighteenth century on, the waves of investment seem to follow on increases in sales abroad.[4]

In any event, this rising demand contained the seeds of difficulty. Every mode of industrial organization has, built into it, opportunities for conflict between employer and employed. These are particularly

[1] Cf. Kenneth Berrill, 'International Trade and the Rate of Economic Growth', *Econ. Hist. Rev.* 2nd ser. XII (1960), 351–9; also P.A., 'The Origins of the Industrial Revolution' [summary of a symposium], *Past and Present*, no. 17 (1960), pp. 71–81.

[2] Deane and Cole, *British Economic Growth*, pp. 46, 59.

[3] D. E. C. Eversley, 'In Pudding Time: the Early Stages of Industrialization in England, 1730–1780' [Faculty of Commerce and Social Science, University of Birmingham, Discussion Papers, Series D, 'Economic and Social History', no. 1 (May 1962)].

[4] Cf. François Crouzet, 'La formation du capital en Grande-Bretagne pendant la Révolution Industrielle' (paper presented to the World Congress of Economic Historians, Aix-en-Provence, 28 August to 4 September 1962; to be published).

serious in putting-out because the system furnishes the arms as well as the causes of hostility: the worker has custody of the materials of the employer and transforms them in his own good time, in his own home, free of supervision. The only resource of the merchant is his limited control over the income of his employees: if he pays them little enough, they are compelled to work for fear of hunger; and if he abates their pay for any departure from standards of quality, they are compelled to maintain a minimum level of performance. To be sure, the exercise of such constraints is contingent on the establishment of some kind of monopsonistic bond between employer and worker; otherwise the employer can do no more than accept the prevailing market price for labour. That such a nexus did in fact often exist—because of actual monopsony in some areas, or personal ties, or debt—and that it led to abuses, seems incontrovertible.[1] There is a substantial body of folklore built around the figure of the grasping clothier and his even greedier minion, Jimmy Squeezum.

On the other hand, it is equally clear that these controls were at best spotty and limited in effect; that the worker early learned to eke out his income by setting aside for his own use or for resale some of the raw materials furnished by the merchant. Such embezzlement was usually effected at the expense of the finished product: the yarn was sized to give it false weight; the cloth was stretched up to and beyond the point of transparency. Nor was there any feeling of moral compunction about such abstraction; it was looked upon as a normal perquisite of the trade, more than justified by the exploitation of the manufacturer.

The employer's control over labour was strongest in a declining market. At such times, the menace of unemployment hung heavy over the domestic workers, and indeed, from the manufacturer's point of view, one of the greatest advantages of the putting-out system was the ease of laying off labour; overhead costs were minimal. (Later on, when the alternative of concentrated factory production became available, many an entrepreneur, in the continental countries especially, delayed shifting over because of the flexibility of the older arrangements.) In the eighteenth century, however, the British putter-out was confronted with a secularly expanding market, which sapped industrial discipline while aggravating the conflicts endemic in the system. Thus the worker's predilection for embezzlement, sharpened in depression by the desire to compensate for increased abatements and lack of work, was nowise dulled in prosperity; on the contrary, the reward for theft was greater.

[1] For one example of the role of debt in holding a worker to his employer, cf. T. S. Ashton, 'The Domestic System in the Early Lancashire Tool Trade', *Econ. Hist. Rev.* I (1926), 136.

What is more, though the system was flexible downwards, expansion of output was difficult. Up to a point, rural manufacture expanded easily by opening new areas—moving from the environs of the manufacturing towns into nearby valleys, invading less accessible mountain regions, spreading like a liquid seeking its level, in this case the lowest possible wage level. It was in this way that the woollen industry filled the dales of Wiltshire and Somerset and came to thrive all along the Welsh marches by the end of the sixteenth century; on the Continent, the growing woollen *fabriques* of Verviers and Monschau were seeking their weavers in the Limburg by the mid-eighteenth century, while the cotton manufacture of Normandy, after covering the Pays de Caux, was spilling over into Picardy.

But in eighteenth-century Britain, the possibilities of geographical expansion had been largely exhausted. The most accessible areas had been explored and drawn into the system. The worsted weavers of the West Riding were buying yarn in the northern dales and as far afield as East Anglia. In Lancashire, by mid-century, weavers were walking miles to collect the weft needed to keep their looms busy the rest of the day and buying the spinsters with ribbons and other vanities. Much of the difficulty was due to the difference in labour requirements for spinning and weaving: it took at least five wheels to supply one loom, a proportion ordinarily at variance with the composition of the population. So long as it was merely a question of finding rural spinsters—whose husbands worked in the fields—to furnish yarn to urban weavers, there was no problem. But once weaving spread to the countryside and the men gave up cultivation for industry, the imbalance was bound to become an obstacle to expansion. There is evidence that some spinners had begun to specialize in particular types of yarn by the middle of the eighteenth century, that a division of labour had come about, in parts of Lancashire at least, in response to the pressure of demand. But this was hardly enough, given the state of technology, and the price of yarn rose sharply from the late seventeenth to the mid-eighteenth century.

Essentially the increase was due to the ever-wider dispersion of the labour force, for nominal spinning wages changed little. The cost of transport was high to begin with; even more serious, in a world of poor communications, the price of moving goods is not a smooth function of distance; costs jump sharply each time one has to cross a natural barrier or bridge gaps in the network of roads and waterways. Barring sharp increases in the price of the finished product, the expanding manufacturer was thus caught in a cost cage and compelled to seek higher output from within his zone of operations.

In the long run, to be sure, he could expect immigration and natural increase to augment his labour force. Thus there was considerable movement of population in spite of restrictions due to the laws of settlement; Lancashire in particular was a kind of internal frontier, attracting thousands from the adjacent counties as well as from Ireland and Scotland well before the coming of machinery and the factory. And industrial activity, by providing new resources, made possible extensive division of the land, encouraged early marriage, and gave rise to densities of settlement that would otherwise have been inconceivable. Professor Habakkuk and others have called attention to the attraction of industry for overpopulated areas;[1] but here, as so often in history, the process is one of reciprocal reinforcement: rural industry frequently laid the basis of what was eventually to become overpopulation.

Yet migration and natural increase are slow-acting palliatives. In the short run, the manufacturer who wanted to increase output had to get more work out of the labour already engaged. Here, however, he again ran into the internal contradictions of the system. He had no way of compelling his workers to do a given number of hours of labour; the domestic weaver or craftsman was master of his time, starting and stopping when he desired. And while the employer could raise the piece rates with a view to encouraging diligence, he usually found that this actually reduced output. The worker, who had a fairly rigid conception of what he felt to be a decent standard of living, preferred leisure to income after a certain point; and the higher his wages, the less he had to do to reach that point. In moments of affluence, the peasant lived for the day; gave no thought to the morrow; spent much of his meagre pittance in the local inn or alehouse; caroused the Saturday of pay, the sabbath Sunday, and 'Holy Monday' as well; dragged himself reluctantly back to work Tuesday, warmed to the task Wednesday, and laboured furiously Thursday and Friday to finish in time for another long weekend.[2]

[1] Cf. H. J. Habakkuk, 'Family Structure and Economic Change in Nineteenth-Century Europe', *J. Econ. Hist.* xv (1955), 1–12; Joan Thirsk, 'Industries in the Countryside', in F. J. Fisher, ed., *Essays in the Economic and Social History of Tudor and Stuart England in Honour of R. H. Tawney* (Cambridge, 1961), pp. 70–88.

[2] Adam Smith (*Wealth of Nations*, Book I, ch. VIII) perceptively noted the connection between intense application and prolonged relaxation, and argued that the former gave rise to the latter. Cf. T. S. Ashton, *An Economic History of England: the Eighteenth Century* (London, 1955), p. 205. This interpretation seems to put the cart before the horse. It was because the worker preferred this kind of leisure and could achieve it by working at full speed for two or three days that he adopted this sporadic pattern; not because he enjoyed working himself to the limit for a few days and needed a long weekend to rest. The latter position is equivalent to arguing that students rest the first three months of the term because of the heavy 'cramming' they do for final examinations.

Thus precisely at those times when profit opportunities were greatest, the manufacturer found himself frustrated by this unreasonable inversion of the laws of sensible economic behaviour: the supply of labour decreased as the price rose. Nor was the other tack more effective. Outright wage cuts were not feasible in the face of increasing demand, for there was a limit to the employer's hold over his workers. More common were surreptitious increases in the worker's task: he was given longer warps or less credit for waste; or procedures of measuring and weighing were altered in the employer's favour. This kind of cleverness, however, brought with it its own penalties. The resentful workers were incited thereby to embezzle the more, and frictions built into the system were correspondingly aggravated. The eighteenth century saw a persistent effort to halt the theft of materials by making embezzlement a criminal offence, providing employers and law officers with special rights of search and seizure, placing the burden of proof on any person holding materials he could not account for, and repeatedly increasing the penalties for violation. These last included corporal punishment, for fines were of no effect on penniless spinners and weavers. The very iteration of these acts is the best evidence of their ineffectiveness; by the last quarter of the century the black market in wool and yarn had become an organized business and many a cotton manufacturer was said to have begun his career by buying materials from this source.[1] Similarly the laws to compel workers to finish their tasks promptly and to fulfil their obligations to one employer before hiring out to another—a problem that apparently grew with the demand for labour—were little more than admissions of difficulty and expressions of intent. The discipline of the industrial system was breaking down.

The shift in attitude toward the labouring poor in the late seventeenth and early eighteenth centuries reflects in part the employer's frustration and vexation. Where once poverty had been looked on as an unavoidable evil and the poor man as an object of pity and a responsibility to his neighbour, now poverty was a sin and the poor man a victim of his own iniquity. Defoe is only the clearest and most effective spokesman for this viewpoint, which castigated the worker for the sloth that made him waste his time in idleness and low diversion, and the vice that led him to squander his scanty resources on alcohol and debauchery. This virtuous indignation seems to have softened from the middle of the century; at least writers on economic matters were beginning to argue that labour was not incorrigibly lazy and would in fact respond to higher wages. Mr Coats has suggested that this shift owed much to

[1] Travis, *Notes...of Todmorden and District* (1896), p. 56, cited by Wadsworth and Mann, *Cotton Trade*, p. 399.

the introduction of machinery and the promise of a definitive solution to the problem.[1] Perhaps; in the meantime, the businessman continued sceptical, and in places like Manchester people were still told in 1769 that the 'best friend' of the manufacturer was high provisions.[2] One can understand why the thoughts of employers turned to workshops where the men would be brought together to labour under watchful overseers, and to machines that would solve the shortage of manpower while curbing the insolence and dishonesty of the men.

Yet if the presence of this growing need for a change in the mode of production clarifies the demand side of technological innovation, it will not suffice to explain the supply side: the conditions that made possible the devising of new methods and their adoption by industry. One thing seems clear: if Britain was the country that felt most keenly the inadequacy of the prevailing system, she was not the only one. The major continental centres were also disturbed by shortages of labour and the abuses of domestic manufacture. As noted above, the weavers and merchant-manufacturers of Normandy and Verviers, of the Rhineland and Saxony, were obliged to find their yarn over an ever wider radius, often in the face of laws in the country of origin forbidding its export to competitors. Nor was this the first time in history that demand had pressed hard on the capacity of craft and domestic manufacture: in medieval Italy and Flanders analogous difficulties arose without calling forth an industrial revolution.

The problem may be broken down into two aspects: the conditions governing the invention of labour-saving devices; and those determining the adoption of these devices and their diffusion in industry.

On the first, it would seem clear, though by no means easy to demonstrate, that there existed in Britain in the eighteenth century a higher level of technical skill and a greater interest in machines and 'gymcracks' than in any of the other countries of Europe. This should not be confused with scientific knowledge; in spite of some efforts to tie the Industrial Revolution to the Scientific Revolution of the sixteenth and seventeenth centuries, the link would seem to have been an extremely diffuse one: both reflected a heightened interest in natural and material phenomena and a more systematic application of empirical searching. Indeed, if anything, the growth of scientific knowledge owed much to the concerns and achievements of technology; there was far less flow of

[1] A. W. Coats, 'Changing Attitudes to Labour in the Mid-Eighteenth Century', *Econ. Hist. Rev.* 2nd ser. XI (1958), 46–8.

[2] Arthur Young's famous testimony, from his *Six Months Tour Through the North of England* (4 vols.; London, 1770), III, 248–9. Cf. Edgar S. Furniss, *The Position of the Laborer in a System of Nationalism* (New Haven, 1920), pp. 98–105.

ideas or methods the other way; and this was to continue to be the case well into the nineteenth century.[1]

All of which makes the question of British mechanical skill the more mysterious. The testimony of contemporary observers on this point is mixed: some found the British creative as well as highly gifted craftsmen; others looked upon them as simply clever imitators; there is no evidence before the great innovations of the eighteenth century of any exceptional reservoir of talent in this sphere. To be sure, there were the millwrights, clock-makers, joiners, and other craftsmen whose experience in construction and contriving trained them in effect to be the mechanics of a new age. But England was not the only country with such artisans, and nowhere else do we find this harvest of inventions.

Yet if there is no positive evidence of a superior level of technical skill in Britain, there is a strong indirect argument for this assumption: even after the introduction of the textile machines (and the new metallurgical and chemical techniques, as well), the continental countries were not prepared to imitate them. The most effective of the early copies were almost all the work of British emigrant mechanics, and it was a matter of decades before the rest of Europe freed itself from dependence on British skills. Nor was this long apprenticeship drawn out simply because of a desire to employ more productive workers. The English artisans who came to the Continent were costly, homesick, insubordinate. Their employers could hardly wait to be rid of them.

Why the British developed these skills earlier and faster than others is another matter. Was it because corporate controls of production and apprenticeship had largely broken down by the end of the seventeenth century, whereas the continued influence of guild organization and the active supervision of mercantilistic governments on the Continent tended to fix techniques in a mould and stifle imagination?[2] Is the *Encyclopédie*, with its careful descriptions of the proper way to do things, a symbol of this rigidity? Or was it because the avenues of social advancement were different in Britain than in the aristocratic monarchies of the Continent, that talent was readier to go into business,

[1] This was true even of the steam-engine, which is often put forward as the prime example of science-spawned innovation. See below, p. 333.

[2] See Gabriel Jars's comparison of Sheffield, where industry was still fettered by the guild system in 1764–5 (though the growth of certain enterprises had burst these bonds), and Birmingham, where any man could engage in any business and at most 20 per cent of the workers had properly served their apprenticeships. 'The multiplicity of trades has given rise to emulation such that each manufacturer is ceaselessly occupied in inventing new means of cutting down labour costs and thereby increasing his profits. This has been pushed to such an extent that it seems unthinkable that ironmongery can be produced anywhere so cheaply as in Birmingham.' Chevalier, 'La mission de Gabriel Jars', *Trans. Newcomen Soc.* XXVI (1947/8 and 1948/9), 63.

projecting, and invention than in more traditionalistic societies? One is struck by the middle-class origins of most of the creators of the first textile machines. John Kay was the son of a 'substantial yeoman'; Lewis Paul, the son of a physician. John Wyatt's background is vague, but he had attended grammar school and was presumably from the kind of family that felt schooling was desirable. Samuel Crompton's father was a farmer who produced cloth on the side and was apparently comfortably situated. Edmund Cartwright was the son of a gentleman and a graduate of Oxford. It was not discreditable in the eighteenth century for children of good families to be apprenticed out to weavers or joiners.[1] Manual labour and dexterity were not stigmata of the *peuple*, as opposed to the *bourgeoisie*.

A further consideration suggests itself. Was it not only that the English atmosphere was more favourable to change, but also that special experience in certain areas provided unique facilities for training? What, for example, was the role of the Newcomen engine in shaping English metallurgy and machine construction? Or does the explanation lie simply in the *greater* need for innovation on the island (a matter of degree, to be sure, but questions of degree can often be decisive): need for labour-saving devices in a textile manufacture whose products lent themselves to mass production; for effective pumping equipment in mines; for ways to make use of mineral fuel in a country with the largest appetite for iron in the world?

The fresh and important researches of A. E. Musson and Eric Robinson offer an impressive picture of the energy with which Lancashire mobilized and trained technological skill in the second half of the eighteenth century—importing craftsmen from as far away as London and Scotland and capitalizing on its own strong traditions of skilled labour to turn joiners into millwrights and turners, smiths into foundry-

[1] Thus Peter Ewart, son of a Scots clergyman, one of whose brothers became minister to the Prussian court, another a physician, a third, partner to John Gladstone in the Liverpool trade: because of his talent for mechanical matters, he was apprenticed as a millwright to John Rennie. W. C. Henry, 'A Biographical Notice of the Late Peter Ewart, Esq.', *Memoirs of the Literary and Philosophical Society of Manchester*, 2nd ser. VII (1846). Or James Watt, father of the famous inventor: son of a mathematics teacher who was an Elder of the presbytery and Kirk Treasurer at Cartsdyke (Scotland), he was apprenticed to a carpenter-shipwright. His brother was trained in mathematics and surveying. S. Smiles, *Lives of Boulton and Watt* (London, 1865), pp. 81–3. Or Charles Tennant, son of a farmer and 'factor to the Earl of Glencairn', who was apprenticed to a weaver. E. W. D. Tennant, 'The Early History of the St Rollox Chemical Works', *Chemistry and Industry*, 1 November 1947, p. 667. Similarly, there was no derogation in marrying a craftsman. See the pedigree of the Pilkington family in the eighteenth and early nineteenth centuries. T. C. Barker, *Pilkington Brothers and the Glass Industry* (London, 1960), pp. 20–30.

men, clock-makers into tool and die cutters.[1] Even more striking is the theoretical knowledge of these men. They were not, on the whole, the unlettered tinkerers of historical mythology. Even the ordinary millwright, as Fairbairn notes, was usually 'a fair arithmetician, knew something of geometry, levelling, and mensuration, and in some cases possessed a very competent knowledge of practical mathematics. He could calculate the velocities, strength, and power of machines: could draw in plan and section....'[2] Much of these 'superior attainments and intellectual power' reflected the abundant facilities for technical education in 'villages' like Manchester during this period, ranging from Dissenters' academies and learned societies to local and visiting lecturers, 'mathematical and commercial' private schools with evening classes, and a wide circulation of practical manuals, periodicals, and encyclopaedias.

Whatever the reasons for British precocity in this domain, the results are clear; and equally clear is the relative ease with which inventors found financing for their projects and the rapidity with which the products of their ingenuity found favour with the manufacturing community—if anything, too much favour, for many of the earlier inventors spent more time enforcing their patent rights than earning them.[3] Some have accounted for this swift diffusion of change by the relatively greater accumulation of capital in Britain than anywhere else in Europe except Holland (which was kind enough to send some of its surplus funds to England, rather than invest them in its own industry). They argue that the greater supply of capital was reflected in lower interest rates, which tended to decline in the course of the eighteenth century, and that this in turn made change that much less costly and, *pari passu*, that much more profitable and attractive.[4]

[1] A. E. Musson and Eric Robinson, 'The Origins of Engineering in Lancashire', *J. Econ. Hist.* xx (1960); 'Science and Industry in the Later Eighteenth Century', *Econ. Hist. Rev.* 2nd ser. XII (1960). Also G. H. Tupling, 'The Early Metal Trades and the Beginnings of Engineering in Lancashire', *Trans. Lancashire and Cheshire Antiquarian Soc.* LXI (1949), 25 f.

[2] Wm. Fairbairn, *Treatise on Mills and Millwork* (2nd ed.; 2 vols.; London, 1864), I, vi.

[3] A number of writers have laid stress on the incentive effect of patent legislation. I am inclined to doubt its significance. This kind of protection was not new; the basis of the system was laid by the Statute of Monopolies of 1624. In our period, the cost and difficulty of obtaining a patent was rising steadily. Cf. Witt Bowden, *Industrial Society in England Towards the End of the Eighteenth Century* (New York), 1925, pp 26–30. At the same time, there was good reason to doubt the efficacy of patents against determined competitors, as numerous inventors learned to their sorrow, and many an entrepreneur placed his reliance on secrecy, rather than the law.

[4] This was the position of Prof. T. S. Ashton in his *Industrial Revolution, 1760–1830* ('Home University Library', London, New York, and Toronto, 1949), pp. 9–11,

The argument is persuasive, but the historical facts tend to modify it at a number of points, diminish its import at others. On the one hand, it is most unlikely that differences in the rate of interest of the order of two, three, even half-a-dozen points are a decisive consideration where the mechanical advantage of innovation is as great as it was for the early textile machines. One can understand that the timing of canal and road construction, or similar costly projects of slow gestation, was affected by shifts in the rate of interest, in part because the very possibility of flotation was frequently dependent on an easy money market. But for the prospective textile entrepreneur, the problem was not whether his profits would cover 6 per cent or 12 per cent on borrowed capital, but whether he could raise the capital at all.

In this regard, the cotton manufacturer of the eighteenth century was favoured by the very newness of the Industrial Revolution. The early machines, complicated though they were to contemporaries, were nevertheless modest, rudimentary, wooden contrivances, which could be built for surprisingly small sums. A forty-spindle jenny cost perhaps £6 in 1792; scribbling and carding machines cost £1 for each inch of roller width; a slubbing billy with thirty spindles cost £10. 10*s.*[1] And these were new. Similar equipment was frequently advertised in used condition at much lower prices. The only really costly items of fixed investment in this period were buildings and power, but here the historian must remember that the large, many-storeyed mill that awed contemporaries was the exception. Most so-called factories were no more than glorified workshops: a dozen workers or less; one or two jennies, perhaps, or mules; and a carding machine to prepare the rovings. These early devices were powered by the men and women who worked them.[2] Attics and cottages were reconverted for the

90–1, but he has since modified it considerably, emphasizing, not cost of capital but its availability. The rate of return on government securities was important, he argues, because of the 5 per cent ceiling on the rate of interest: when the funds fell and the return (including the prospect of capital gains) rose, capital would shift in that direction, diminishing the supply to industry and trade. *An Economic History of England: the Eighteenth Century*, pp. 26–9. There is an excellent discussion in L. S. Pressnell, 'The Rate of Interest in the Eighteenth Century', in Pressnell (ed.), *Studies in the Industrial Revolution* (London, 1960), pp. 190–7.

[1] W. B. Crump (ed.), *The Leeds Woollen Industry, 1780–1820* (Leeds: The Thoresby Society, 1931), pp. 212–13, 293; also Herbert Heaton, 'Benjamin Gott and the Industrial Revolution in Yorkshire', *Econ. Hist. Rev.* III (1931), 52. For purposes of comparison, a cotton weaver earned perhaps 7*s.* 6*d.* a week in 1770, a hand spinner between 2*s.* and 3*s.* Thus the 40-spindle jenny cost about two weeks wages of the forty women it replaced. Wage figures from Wadsworth and Mann, *Cotton Trade*, pp. 402–3. A traditional hand-loom cost more than a jenny; anywhere from £7 to £10.

[2] The first application of water power to the mule was apparently in 1790 at the New Lanark mills. The increased drive made possible 'double mules' of 400 spindles;

purpose; later on a steam-engine might be added to this kind of improvised structure. Moreover, there were premises to rent—here we have another example of the responsiveness of English capital to economic opportunity. Not only were complete buildings offered to prospective tenants, but larger mills were subdivided and let in small units. So that an industrialist could in fact start with a minimal outlay—renting his plant, borrowing for equipment and raw materials, even raising funds for payment of wages by contracting in advance for the finished product. Some no doubt began with nothing more than the capital accumulated by petty local trading in yarn and cloth; others, as noted above, apparently built their fortunes in the black market for embezzled materials.

On the other hand, a good many of the early mill owners were men of substance—merchants whose experience in selling finished commodities had alerted them to the possibilities of large-scale, mechanized production; putters-out, who had had direct experience in manufacture; even independent small producers with enough set by to change their methods and expand. And here, of course, the direct accumulation of capital within the textile manufacture itself and the trades allied with it was a major factor in the rapid adoption of technological innovation—as it was in industries like iron and chemicals. We are now come full circle: the inventions came in part because the growth and prosperity of the industry made them imperative; and the growth and prosperity of the industry helped make their early and widespread utilization possible.

All of which serves to emphasize an important caution: it was not capital by itself that made possible Britain's swift advance. Money alone could have done nothing; indeed, in this regard, the entrepreneurs of the Continent, who could often count on direct subsidies or monopoly privileges from the state, were better off than their British counterparts. What distinguished the British economy, as we have already had several occasions to remark, was an exceptional sensitivity and responsiveness to pecuniary opportunity. This was a people fascinated by wealth and commerce, collectively and individually.

Why this was so is a question worthy of investigation. Certainly the

thus capital bred capital. George W. Daniels, *The Early English Cotton Industry* (Manchester, 1920), p. 125. It is not clear when the steam-engine was first so used—perhaps in the late 1780's, certainly in the early 1790's. A large proportion of these early engines were employed, not to drive the machinery directly, but to raise water upon a wheel; some of them, indeed, were Savery-type steam pumps (see below, p. 331), which were preferred to more efficient machines because of their lower initial cost. An engine delivering 2–4 h.p. could be had new for between £150 and £200. A larger Boulton and Watt rotative engine (15–20 h.p.) cost four or five times as much. Steam came earlier to frame spinning: Arkwright's atmospheric engine at Shudehill (Lancs.), to raise water for a wheel, was installed in 1783. A. E. Musson and E. Robinson, 'The Early Growth of Steam Power', *Econ. Hist. Rev.* 2nd ser. XI (1959), 418–39.

phenomenon was closely related, as both cause and effect, to the already noted openness of society; and this was linked in turn to the peculiar position and character of the aristocracy.

Britain had no nobility in the sense of the other European countries. She had a peerage, composed of a small number of titled persons, whose essential and almost unique perquisite was the possibility of sitting in the House of Lords. Their children were commoners, who often received, to be sure, courtesy titles in token of their high birth, but were no different in civil status from other Britons. Even the peers had only the most modest privileges: trial by their fellow noblemen in criminal proceedings, for example, or the right of direct access to the sovereign. They did not enjoy fiscal immunities.

Below the nobility stood the gentry or so-called squirearchy, an amorphous group, without legal definition or status, that had no equivalent on the Continent. Its edges were blurred, its ranks loosely and heterogeneously constituted. Some gentry were of noble antecedents; others had made their fortunes in trade or the professions or government service and had purchased estates as much for their social prestige as for their income; others were scions of old country families; still others were farmers or yeomen grown wealthy. They had two things in common: land ownership and a way of life that was a vestige of medieval seignorialism. These were the local notabilities—lords of the manor, justices of the peace, county sheriffs. With the peers, they were the true rulers of provincial Britain.

Both nobility and gentry generally practised primogeniture: the oldest male child inherited both title (where pertinent) and land. This had two large consequences: it increased the economic burden of the head of the family; and it compelled most of the children to earn their living, in whole or part.

Thus it was no easy task to preserve and if possible increase the family estate for transmission to one's heir while finding places for younger sons and dowries for daughters. Daniel Finch, Earl of Nottingham, put the point well in a letter of 1695 to his executors; he favoured primogenitary strict settlement, he wrote,

> not so much out of a vain affectation of continuing a great estate in my family, as because [my son] will thereby be under a necessity of observing some good economy that he may be able to provide for his younger children, and consequently will not run into that foolish or extravagant way of living which debauches and corrupts the manners of many families, as well as ruins their fortunes. . . .[1]

[1] Quoted by H. J. Habakkuk, 'Daniel Finch, 2nd Earl of Nottingham: His House and Estate', in J. H. Plumb (ed.), *Studies in Social History: A Tribute to G. M. Trevelyan* (London, New York, Toronto, 1955), p. 156.

To be sure, British society had provided careers for cadets of good family: remunerative offices in government; Church livings; commissions in the armed forces; a growing number of potentially lucrative situations in the colonies (not really important until the second half of the eighteenth century). Yet excessive and otiose as many of these places appeared to contemporary reformers, they were not enough to satisfy the demand, as the competition for patronage testifies, and they had to be shared with such other groups as the legal profession and the mercantile interest. Sinecures and offices came high, and it was a rare father who could place more than two or three sons well. To quote Nottingham again: 'no estate can provide so fully for younger children, but that they must in great degree help themselves'. The fourth and fifth sons, of gentry and even noble families, would have to be apprenticed to trade—not the trade of the shop, to be sure, but the international commerce that was at once the pride of the English economy and the seed-bed of new houses.[1]

Admittedly, there were rarely as many as four or five sons that survived to manhood, and the flow of gentle talent into business was presumably small.[2] Certainly it was less important in the eighteenth century than it had been in earlier periods, partly perhaps because the royal house brought with it from Hanover strong German prejudices against this kind of mobility, partly because the proliferation of office and the extension of British dominion were opening up alternative opportunities that were more attractive and preferentially accessible.[3] Yet it was not so much the substantive contribution to enterprise that counted, as the symbolism of the example, the sanction that this legiti-

[1] Much depended, however, on necessity and opportunity. In Scotland, the gentry were poor and had few claims to preferment. The Established Church was closed to them; the prospect of fighting the battles of England, unattractive; foreign commerce and colonial enterprise offered few employments until the century was well advanced. The more intellectual could prepare for the Bar, but these were necessarily few. Many sought their livelihood, therefore, as shopkeepers, alias 'merchants'. As Henry G. Graham puts it in his classic study, '...in those days, a gentleman's son felt it as natural to fall into trade as for a rich tradesman to rise out of it'. *The Social Life of Scotland in the Eighteenth Century* (4th ed.; London, 1950), p. 33.

[2] We really do not know how important, absolutely or relatively, was the participation of these younger sons of noble or gentry families in business. A systematic survey of the entrepreneurs of the Industrial Revolution would be immensely valuable, though difficulties of definition (what is active participation?) would confuse the issue, especially as regards the aristocratic contribution. In the meantime, we are reduced to discrete impressions. Cf. Walter E. Minchinton, 'The Merchants in England in the Eighteenth Century', *Explorations in Entrepreneurial History* [henceforth *Explorations*], x (1957), 62.

[3] For the earlier period, see Lawrence Stone, 'The Nobility in Business', *ibid.* pp. 54–61.

mate participation, however small, conferred on trade as a respectable activity and on pecuniary rationality as a way of life.

In the meantime, the head of the family had to build the patrimony and make it work for him and his children. Not all landowners did well in the contest for fortune or even tried, but at their best they formed a class of 'spirited proprietors' that warmed the hearts of improvers like Arthur Young. Noble or gentle, they lived on their estates (and not at the court), rode their lands and noted their yield, sought improvements to enhance traditional revenues, conceived new ways to produce income. They rarely cultivated or operated directly—though one can cite contrary instances like Thomas Fitzmaurice, brother of the Earl of Shelburne, who, among other enterprises, bleached and sold the linens woven by his Irish tenants.[1] (Even if they had made their fortune in trade and continued in the firm, they inevitably gave less time to mercantile concerns.) Rather they leased their land to tenants—peasants, commercial farmers, or industrial contractors. When they financed business ventures, it was as creditors more than as partners; or they bought shares in joint-stock companies and trusts. Their interests were generally handled by stewards, agents, and solicitors, and this exposed them to abuses of confidence. Yet that was all the more reason to supervise their affairs closely, and many of them made the decisions that are the hallmark of active entrepreneurship. A significant few opened mines, built iron works and mills, dug canals, developed ports, and leased their urban properties for building. What is more, they anticipated demand, undertook investments on speculation, advertised if necessary for tenants, and stood ready to operate their installations through agents or partners if no lessees were forthcoming.

They also enclosed the land, concentrated their holdings, introduced or found tenants who would introduce better crop rotations and techniques of cultivation, helped spread new ideas about the country. This is not the place to discuss the so-called 'agricultural revolution' of the eighteenth century, or to assess the benefits or injustices that accompanied it. My aim here is to underline the generality of this spirit of innovation and its effects; and also to recall the well-known fact that this was a society that interposed relatively few institutional barriers to a fundamental change of this kind. The Tudor monarchy may have been concerned about enclosures; the parliamentary regime of the eighteenth century was less paternalistic. For good or evil, Britain's countryside was being kneaded like dough; and the improving landlords were a powerful leaven.

[1] A. H. Dodd, *The Industrial Revolution in North Wales* (Cardiff, 1933), pp. 32–3.

It is probable that such industrial ventures as people 'of name' undertook were on the whole less remunerative than competitive efforts by 'professional' businessmen; or for that matter, that the great estates could not bear comparison with the lands of the small proprietors in their neighbourhood.[1] It is also true that the nobility and gentry tended over time to change from active entrepreneurs into rentiers; this was the experience, for example, of areas like Wales and Northumberland, where large coal and ore deposits had initially provided a favourable base for industrial activity by landowners. Yet the significance of these efforts lay in the efforts themselves, not in their return. Once again it lay in the legitimacy conferred on innovation and the pursuit of wealth as a way of life.

A comparison will illuminate the argument. Thus far I have treated this permeation of country life by the spirit of enterprise and calculation as a consequence of social structure and the system of inheritance. There was undoubtedly more to it. In part, it reflected the quickening pace of the economy: new men kept moving up and the older families had to move faster to hold their own. In part, it was probably a response to new opportunities, in particular, to the increasing demand for resources that lay in the hands of the landed proprietors. Yet this response was neither necessary nor inevitable. It would have been just as easy for the nobility and gentry to turn their backs on their new rivals and break off competition by defining the means of their ascension as intrinsically ignoble. This is what the aristocracy of Europe had done in the Renaissance, when it had developed the very idea of the gentleman as a weapon against the pretensions of the bourgeoisie.[2] And this was to be the reaction of much of Europe's aristocracy in the nineteenth century in the face of industrial revolution and a corresponding shift in the balance of political power. The British nobility and gentry chose to meet the newcomers on middle ground: they affirmed their distinction of blood or breeding; but they buttressed it with an active and productive cultivation of gain.

This momentous decision was self-reinforcing. The concern of the British gentleman for the accretion of his fortune made him a participant in society rather than a parasite upon it—whatever judgement one may pass on the character of this participation. Business interests promoted a degree of intercourse between people of different stations and walks of life that had no parallel on the Continent. 'We used to sit down to dinner,' wrote Lord Hervey in 1731, 'a little snug party of

[1] Cf. Adam Smith, *Wealth of Nations*, Book III, ch. II; Arthur Young, *Travels during the Years 1787, 1788 and 1789*, I, 90, 99, 108, and especially 198.

[2] See the article by Arthur Livingston on 'Gentleman, Theory of the' in the *Encyclopaedia of the Social Sciences* and the references given there.

about thirty odd, up to the chin in beef, venison, geese, turkeys, etc.; and generally over the chin in claret, strong beer and punch. We had Lords Spiritual and Temporal, besides commoners, parsons and free-holders innumerable.'[1] Compare Arthur Young's reflections on a visit with the Duc de la Rochefoucauld:[2]

At an English nobleman's, there would have been three or four farmers asked to meet me, who would have dined with the family amongst the ladies of the first rank. I do not exaggerate, when I say, that I have had this at least an hundred times in the first houses of our islands. It is however, a thing that in the present state of manners in France, would not be met with from Calais to Bayonne, except by chance in the house of some great lord that had been much in England, and then not unless it was asked for.

Or, to return to Britain, consider the friendship of Robert Hewer, successful supercargo and trader turned landowner and lord of the manor of Manadon (near Plymouth), with the Duke and Duchess of Bedford: he spent weeks as their guest at Woburn Abbey, where he was 'of all their parties of pleasure!'; and the visit was repeated several times.[3] Such a continuing relationship is more significant of true society than a dozen marriages between noble blood and bourgeois gold *pour redorer le blason*.[4]

Below the level of the gentry, there was no barrier between land and trade—not even a permeable membrane. Given the rural character of most industry and the intermittent claims of agriculture, many cultivators were at the same time manufacturers or middlemen or both. This was true not only of textiles, but of branches like metallurgy where one might think the nature of the manufacturing process would have imposed a more rigorous division of labour: witness Isaac Wilkinson, Aaron Walker, Jedediah Strutt and others. Note that where similar conditions prevailed on the Continent, one found the same combination of activities: the land brought forth industrial enterprise and enterprisers. Yet again, what sets Britain apart is a question of degree. Nowhere else, as we have seen, was the countryside so infused with manufacture; nowhere else, the pressures and incentives to change greater, the force of tradition weaker. It was all of a piece: improving

[1] A. Goodwin (ed.), *The European Nobility in the Eighteenth Century* (London, 1953), p. 4.

[2] Young, *Travels*, I, 207.

[3] Conrad Gill, *Merchants and Mariners of the Eighteenth Century* (London, 1961), p. 138.

[4] There was much of that too—on the Continent as well as Britain. But inter-class alliances are to be found in all but rigid caste societies. The real test is not the union; it is what follows: how many great families in such circumstances are willing to know their new relations after the wedding?

landlords, enclosures, commercial farming, village shops, putting-out, mines and forges, the active mortgage market—all combined to break the shackles of place and habit, assimilate country and city, and promote a far wider recruitment of talent than would have otherwise occurred. In a society of which four of five people lived on the land, this was a powerful stimulus to overall development.

By the same token, the flow of entrepreneurship within business was freer, the allocation of resources more responsive than in other economies. Where the traditional sacrosanctity of occupational exclusiveness continued to prevail across the Channel, enforced sometimes by law but in any case by habit and moral prohibitions, the British cobbler would not stick to his last nor the merchant to his trade. It was not merely a kind of negative phenomenon—that is, the absence of confining regulations or opprobrious strictures; rather it was a positive drive, an ambitious versatility that was always alive to the main chance. One cannot but be impressed by a man like this Thomas Griggs, grocer and clothier of Essex in the mid-eighteenth century, who invested and speculated in real estate, fattened cattle for market, malted barley, lent money on pawn.[1] Or like Thomas Fox, Quaker clothier of Wellington, who was moved by hard times in the wool trade to look into the possibilities of lead, calamine, or copper mining—or all three.[2]

One could extend the list considerably, but one final example will have to suffice: Samuel Garbett of Birmingham, originally brassworker, then merchant and chemist, partner in spinning, chemical (Birmingham and Prestonpans, near Edinburgh), iron-smelting (Carron works, Scotland), and flour-milling (Albion Mills, London) enterprises, and shareholder in the Cornish Metal Co. (copper mines). To appreciate the force of this drive for wealth, one must remember that these men were risking their fortunes at each throw of the entrepreneurial dice. With rare exceptions, there was no haven of limited liability. Garbett went bankrupt in 1772 because of the failure of one of his partners.

Similarly, the structure of the firm was more open and rational in Britain than in the continental countries. Everywhere, the fundamental business unit was the individual proprietorship or the family partnership, but where, in a country like France, the family firm was almost always closed to outsiders, British entrepreneurs were far more willing to enter into association with friends or friends of friends. Indeed, this seems to have been the preferred way of raising capital to expand or of attracting and attaching special skills to the enterprise. In textiles, a capitalist like George Philips would seek out and take as partner an

[1] K. H. Burley, 'An Essex Clothier of the Eighteenth Century', *Econ. Hist. Rev.* 2nd ser. XI (1958), 289–301.

[2] Herbert Fox, *Quaker Homespun* (London, 1938), pp. 46 f.

experienced manager like George Lee, late employee of Peter Drinkwater; or an already hardened flax spinner like John Marshall, faced with a crisis in the trade, could throw out his partners ('As they could neither of them be of any further use, I released them from the firm and took the whole upon myself'); and when, shortly after, though mortgaged to the hilt, he determined to expand, would bring in new ones for much larger amounts.[1] In brewing, where the need for capital was so great and urgent 'that it could not be produced fast enough from the profits of the firms', 'established concerns welcomed into their partnerships bankers and merchants, who of necessity brought in the social and political consequences of vast wealth made in other fields'.[2] In machine building, it was probably skill more than capital that was the scarce factor, though it took thousands of pounds to turn a repair shop into an engineering plant. Boulton and Watt are perhaps the best model of this alliance of money and talent, but one could cite many similar associations, with varying division of contributions and responsibilities.[3] The pattern was probably most widespread in metallurgy; there the requirements of both capital and talent were heavy, and the partnership was the normal business form.[4] Even where a firm was essentially a family affair—the Crawshay smelting mill at Cyfarthfa or ironmongery in London, for example—outsiders were brought in as needed; bought out later if advisable; and new partners found. Professor Ashton has pointed out the importance in this connection of interfamilial associations of Dissenters: the bonds of a common, persecuted religion proved almost as effective a business tie as blood itself.[5]

The cohesiveness and mutual support of the nonconformist business community was only one element in their commercial success. They

[1] These were the Benyons, woollen merchants of Shrewsbury. W. Gordon Rimmer, *Marshall's of Leeds, Flax-Spinners 1788–1886* (Cambridge, 1960), pp. 40–4.

[2] Peter Mathias, 'The Entrepreneur in Brewing, 1700–1830', *Explorations*, x (1957), 73–6.

[3] Thus Bateman and Sherratt of Salford: Fenton, Murray and Wood of Leeds; Hazeldine, Rastrick and Co. of Bridgnorth (Salop); and somewhat later, Maudslay, Son and Field of London; Nasmyth, Gaskell and Co. of Manchester; Sharp, Roberts and Co. of the same city; *et al.*

[4] To be sure, some of the facilities available in textiles were present in metallurgy as well. Thus landowners, desirous of increasing their incomes, were often ready to let mineral rights on favourable terms and otherwise promote enterprise on their estates; sometimes, as at Cyfarthfa and Dowlais, leases fixed at absurdly low rates ran for a century. Also, it was often possible to rent plants already built for a moderate sum. Nevertheless, it took a thousand pounds or more to launch even a modest furnace or forge.

[5] Ashton, *Iron and Steel in the Industrial Revolution* (2nd ed.; Manchester, 1951), ch. IX: 'The Ironmasters'.

suffered numerous disabilities because of their religion, and business was in many ways the most convenient outlet for their energy and ambition; their faith itself, with its stress on diligence, thrift, and rationality as a way of life, was often a competitive advantage. It is surely no coincidence that they were most numerous in the North and the Midlands, the centres of most rapid industrial development. On the other hand, Britain was not alone in having Calvinists, who played the role of an entrepreneurial leaven throughout Europe. What distinguished Britain was the extent to which her religious nonconformists conformed to a wider social pattern; the entrepreneurial differences were differences of degree, and not of kind.

The same observations are relevant to the oft-cited thesis that price inflation brought bigger profits, and that bigger profits made possible industrial change.[1] Even if it could be shown that profits did increase over the century and that it was higher prices that were responsible—and the usual demonstration proves nothing of the sort—the fact remains that Britain was not the only nation to have price inflation in this period; that the best enterprises on the Continent made just as high rates of profit and relied more, if anything, on self-financing.[2] The point again is not so much the rate of return as the manner of its use: where British firms ploughed profits back into the business, their competitors abroad too often transferred them from trade to more honorific callings, or held them as a reserve in the form of land, mortgage loans, and similar non-industrial placements.

Finally, a word should be said about the role of banks and bank credit. In no country in Europe in the eighteenth century was the financial structure so advanced and the public so habituated to paper

[1] The *locus classicus* is Earl J. Hamilton, 'Profit Inflation and the Industrial Revolution', *Quart. J. Econ.* LVI (1941–42), 257–70. See also his earlier article, 'American Treasure and the Rise of Capitalism, 1500–1700', *Economica*, IX (1929), 338–57, and his reply to the criticisms of John U. Nef, 'Prices and Progress', *J. Econ. Hist.* XII (1952), 325–49.

[2] See the excellent analysis of the problem in David Felix, 'Profit Inflation and Industrial Growth: the Historic Record and Contemporary Analogies', *Quart. J. Econ.* LXX (1956), 441–63. One should note that most of the price increase in the second half of the eighteenth century occurred in the 1790's. Felix argues that such expansion of profits as did take place was the result of greater productivity rather than a combination of price inflation and wage lag. This is clearly so: the very industries that were making the most rapid technological advances were the ones whose prices were falling and the nominal wages of whose workers (or real wages, for that matter) were rising through most of this period (1760–1830). The mule spinners were a privileged group. What is more, there is considerable direct evidence that rates of profit in these industries were not increasing over the long run, but rather reached a peak with the introduction of the critical mechanical innovations (Schumpeter's entrepreneurial profits) and then declined as new firms were attracted into the field.

instruments as in Britain.[1] Nominally, the credit offered by the growing multitude of private banks was for short periods—up to ninety days—to cover commercial transactions; in fact, much of it was in the form of revolving and open credits, or even of standing overdrafts, which served as quasi-capital.[2] What is more, the development of a national network of discount and payment enabled the capital-hungry industrial areas to draw for this purpose on the capital-rich agricultural districts. The system was just developing in the last quarter of the eighteenth century. By the 1820's and 1830's, however, when the problem of disposing of the products of British factories had become more difficult than that of financing technological change, bank credit was a pillar of the industrial edifice.

Yet while these institutional developments clearly promoted an easier, more abundant flow of resources from land to industry, the nature and direction of the balance of payments between these two sectors are less obvious. It is a commonplace of economic literature that one of the major aspects or criteria of development is the shift of resources from agriculture to manufacturing; and that a condition of rapid development is an increase of productivity in husbandry that will generate the savings to finance industrial expansion. The best example of such a sequence is Japan, where output per head in agriculture almost doubled in the space of a generation (1878/82–1903/7) at little expense of capital; and where, especially in the early years, the land tax drained a substantial fraction of farm income for investment in development.[3] The British case, however, differs sharply. For one thing, gains in farm productivity were assuredly far lower. The statistics available are in no way comparable to the Japanese, but, such as they are, they have led one authority to speculate that 'output per head in agriculture increased by about 25 per cent in the eighteenth century, and that the whole of this advance was achieved before 1750'.[4] The same source

[1] Note, in this regard, the experience of Robert Owen, who found in 1797 that the toll collectors of the Glasgow–New Lanark turnpike preferred the notes of the local banks to gold coin. *The Life of Robert Owen by Himself*, ed. M. Beer (New York, 1920), p. 71.

[2] Not to speak of accommodation paper, which was a means of obtaining credit, if only for short periods, with or without the co-operation of the banks. With the connivance of banks or discount houses, accommodation paper could be the basis for generous medium- and long-term credit. On all this, see W. T. C. King, *History of the London Discount Market* (London, 1936); L. S. Pressnell, *Country Banking in the Industrial Revolution* (New York, 1956); D. S. Landes, *Bankers and Pashas: International Finance and Economic Imperialism in Egypt* (London, 1958).

[3] Kazushi Ohkawa and Henry Rosovsky, 'The Role of Agriculture in Modern Japanese Economic Development', *Econ. Devel. and Cult. Change*, IX, no. 1, part II (October 1960), 43–67; also G. Ranis, 'The Financing of Japanese Economic Development', *Econ. Hist. Rev.* 2nd ser. XI (1959), 440–54.

[4] Deane and Cole, *British Economic Growth*, p. 75.

suggests that the real output of the farm sectors rose about 43 per cent in the course of the century, 24 per cent during the critical decades from 1760 to 1800;[1] by contrast, Japanese agricultural product about doubled from the late 1870's to the early 1900's.

Moreover, the increase in British farm output was due in large measure to enclosures and the improvements they made possible: concentration of holdings, elimination of fallow, more productive choice and rotation of crops, selective breeding of livestock, better drainage and fertilization, more intensive cultivation. It is still a matter of dispute how rapidly these new techniques spread or how quickly they followed on enclosure itself. What is clear is that both the division of the land and the subsequent improvements in its use cost money: for legal expenses, roads, ditches, and fences, to begin with; and eventually for buildings, equipment, drains, and materials. Unfortunately, we have no figures of the area affected, but such partial statistics as are available—enclosure of commons and waste, for example, by parliamentary act—suggest that from 1760 to 1815 Britain enclosed millions of acres at an initial cost of redistribution of upwards of £1 per acre, and at an eventual cost of anywhere from £5 to £25 per acre, depending on the original condition of the soil and the nature of its use.[2] Such investments paid, as the higher yields and rents on consolidated land show. But it may well be that in the early decades of heavy enclosure, that is, the very years that also saw the birth of modern industry, British husbandry was taking as much capital as it was giving; while in the period from 1790 to 1814, when food prices rose to record levels, the net flow of resources was probably toward the land. The great contribution of agriculture to industrialization came after 1815, when both enclosure and the breaking of marginal soil slowed and proprietors and tenants reaped the fruits of earlier efforts. Yet even then, these returns depended on protection against foreign corn and were therefore not a net addition to the savings generated by the economy. Rather, they were bought at the price of a certain misallocation of resources, and abundant and responsive though they may have been, they were probably less than what the land would have provided under more competitive conditions. Still, it was thanks to enclosures and what is sometimes called the 'Agricultural Revolution' that Britain fended off as well as she did Ricardo's 'stationary state'—that end of growth and accumulation wherein the pressure of population on the supply of food has so raised the cost of subsistence and hence wages, that manufacturers can no longer make a profit and the wealth of the nation flows as rent to the owners of the land.

[1] Deane and Cole, *op. cit.* p. 78.

[2] On the cost of enclosure, cf. Great Britain, Board of Agriculture, *General Report on Enclosures* (London, 1808), p. 97. On subsequent expenses of improvement, Albert Pell, 'The Making of the Land in England: A Retrospect', *Journal of the Royal Agricultural Society of England*, 2nd ser. XXIII (1887), 355–74.

To sum up: it was in large measure the pressure of demand on the mode of production that called forth the new techniques in Britain, and the abundant, responsive supply of the factors that made possible their rapid exploitation and diffusion. The point will bear stressing, the more so as economists, particularly theorists, are inclined to concentrate almost exclusively on the supply side. The student of economic development, impressed on the one hand by the high cost of industrialization, on the other by the low level of savings in underdeveloped countries, has devoted most of his attention to the problem of capital formation: on ways to raise the rate of net investment from, say, 5 per cent to 12 or more; and on devices to prevent increased income from dissipating itself on increased consumption.[1] His approach is essentially analogous to that of an economic historian like Hamilton, with his thesis of industrial revolution born of and fed by profit inflation. And indeed, many a planner has seriously contemplated the deliberate use of inflation, which tends to shift resources from consumers to savers, to promote industrialization.

Yet however justified this concern with saving and capital may be in this age of costly equipment and facilities and abysmally poor would-be industrial economies, it is less relevant to the British experience. For one thing, eighteenth-century Britain enjoyed far more wealth and income per head than the backward nations of today; it started, in other words, from a higher base. For another, the capital requirements of these early innovations were, as we have seen, small—within the reach of a single person or family. Moreover, they were concentrated at first in a small sector of the economy, and their effects—and thus their appetite for capital—spread from this narrow base by a process of derived demand that fed on earlier success.[2] In Britain, starting from scratch, the investment of one period was built on the productivity gains of the one before, plus the injection of new labour from population increase and immigration. It was the flow of capital, more than the stock, that counted in the last analysis; so much for the Marxist preoccupation with primitive accumulation.

[1] One could cite numerous examples. For some idea of the wide spectrum of thought along this one line, cf. W. A. Lewis, *The Theory of Economic Growth* (London, 1955), pp. 201 f.; W. W. Rostow, 'The Take-off into Self-Sustained Growth', *Econ. J.* LXVI (1956), 25–48; and a review by O. Ehrlich of Gerald M. Meier and Robert Baldwin, *Economic Development: Theory, History, Policy* (New York, 1957), in *J. Econ. Hist.* XVIII (1958), 74.

[2] So much for hypotheses of balanced growth; the historical experience, under conditions of relatively unplanned development, followed other lines. See below, pp. 314 ff., 321, 338. Also John Hughes, 'Foreign Trade and Balanced Growth: the Historical Framework', *Amer. Econ. Rev.* XLIX, no. 2 (May 1959), 330–37; and Goran Ohlin, 'Balanced Economic Growth in History', *ibid.* pp. 338–53.

WHO UNBOUND PROMETHEUS? SCIENCE AND TECHNICAL CHANGE, 1600–1800[1]

by Peter Mathias

I

An economic historian is interested in science not for its own sake (which for an historian of science is doubtless the only academically respectable way of looking at it) but for his own utilitarian purposes. He asks the questions: how was science related to technology at this time? how far did scientific change influence the process of technological change? to what extent was the Industrial Revolution associated with scientific advance? Taking the very long view from medieval times to the present day is to see a dramatic change in these relationships. Broadly we may postulate the earlier position as a context where empirical discoveries and the development of industrial processes in such industries as metals, textiles, brewing, dyeing took place and advanced without being directly consequential upon knowledge of fundamental scientific relationships in the materials concerned. The chemistry of what happened inside a blast furnace was not known until the mid-decades of the nineteenth century. The secrets of fermentation were first revealed by Pasteur. There might be close links between science and technology in other ways, but this was nonetheless a world very different from our own where industrial advance becomes more directly consequential upon the advancing frontier of scientific and technological knowledge, with a developing institutional relationship between science and industry to consolidate the connection.

For the pivotal period of the seventeenth and eighteenth

[1] An early version of this article appeared in the *Yorkshire Bulletin of Economic and Social Research*, XXI (1) (1969).

centuries, however, which saw dramatic advances in both scientific knowledge and industrial techniques, varying answers have been offered to these questions by economic historians and scholars generalising about the relationships from the side of science. Professor A. R. Hall summed up for the earlier period 1660–1760 'we have not much reason to believe that in the early stages, at any rate, learning or literacy had anything to do with it [technological change]; on the contrary, it seems likely that virtually all the techniques of civilisation up to a couple of hundred years ago were the work of men as uneducated as they were anonymous.'[1] Sir Eric Ashby concludes for the period 1760–1860: 'There were a few 'cultivators of science' (as they were called) engaged in research, but their work was not regarded as having much bearing on education and still less on technology. There was practically no exchange of ideas between the scientists and the designers of industrial processes.'[2] Professor Landes is equally firm in this opinion to as late as 1850.[3] A. P. Usher is in the same tradition.[4]

Equally forthright assertions crowd the other side of the stage. 'The stream of English scientific thought', wrote Professor Ashton, 'was one of the main tributaries of the Industrial Revolution ... The names of engineers, iron-masters, industrial chemists, and instrument makers on the list of Fellows of the Royal Society show how close were the relations between science and practice at this time.'[5] Professor Rostow, considering the whole sweep of economic change in Western Europe, gives the two essential features of post-medieval Europe as 'the discovery and re-discovery of regions beyond Western Europe, and the initially slow but then accelerating development of modern scientific knowledge and attitudes'.[6] When considering the essential propensities for

1 A. R. Hall, *The Historical Relations of Science and Technology* (inaugural lecture) (London, 1963); J. D. Bernal, *Science in History* (London, 1954), pp. 345–6, 352, 354–5, 365–6, 370, argues in a similar vein.

2 Sir E. Ashby, *Technology and the Academics* (London, 1958), pp. 50–1.

3 D. Landes in *Cambridge Economic History of Europe*, vol. VI, pt. 1 (Cambridge, 1965), pp. 333, 343, 550–1; also *The Unbound Prometheus* (Cambridge, 1969), pp. 104, 113–14, 323.

4 A. P. Usher, *A History of Mechanical Invention* (Boston, 1954 ed.) contains very little reference to the role of science in this period.

5 T. S. Ashton, *The Industrial Revolution* (London, 1948), pp. 15–16.

6 W. W. Rostow, *The Stages of Economic Growth* (Cambridge, 1960), p. 31.

Peter Mathias

economic growth (relationships that he does not specifically limit in time or place) the first two on his list are: 'the propensity to develop fundamental science and to apply science to economic ends'.[1] For the English case Musson and Robinson have recently sought to demonstrate how extensive the linkages were between innovation and science, between scientists and entrepreneurs.[2] They see this co-operation assisting England to 'retain that scientific lead over the continent upon which she established her industrial supremacy'.[3] The Lunar Society, now documented at great length, has been called 'a pilot project or advance guard of the Industrial Revolution' on the argument that 'strong currents of scientific research underlie critical parts of this movement'.[4]

Many more such summary assertions could be deployed on either side. It seems likely that, as historians explore more systematically and in more local detail the development of different branches of the chemical industry and other industrial processes involving chemistry (following up the seminal work on the Chemical Revolution by A. and A. N. Clow, published in 1952); as they find out more about the various local societies of gentlemen meeting in small towns up and down the country in the eighteenth century on the lines of the Lunar Society of Birmingham, the balance will tip heavily towards the positive equation. This theme is captured in the remark: 'science is the mother of invention; finance is its father'.[5]

The question, therefore, invites discussion. The arguments, however, should be prefaced with one or two comments. Without the assumption that a simple, linear, cause-and-effect

[1] W. W. Rostow, *The Process of Economic Growth* (New York, 1952), p. 23.

[2] A. E. Musson and E. Robinson, 'Science and Industry in the late 18th Century', *Economic History Review*, XIII (1960), 222–4; *Science and Technology in the Industrial Revolution* (Manchester, 1969).

[3] E. Robinson, 'The Lunar Society and the Improvement of Scientific Instruments II', *Annals of Science*, XIII (1957).

[4] R. E. Schofield, *The Lunar Society of Birmingham* (Oxford, 1963), pp. 437, 410. The argument is summed up on pp. 436–40. See also the special issue of the *University of Birmingham Historical Journal*, XI, no. 1 (1967), devoted to the Lunar Society, particularly the articles by E. Robinson, M. J. Wise and R. E. Schofield; E. Robinson, 'The Lunar Society Its Membership and Organisation', *Transactions of the Newcomen Society*, XXXV (1962–3).

[5] T. H. Marshall, *James Watt* (London, 1925), p. 84.

relationship exists between phenomena like scientific knowledge and innovations in technique, multi-dimensional historical developments such as the Renaissance or the French Revolution or the Fall of the Roman Empire or the Industrial Revolution, cease to be analysable in terms of single-cause, single-variable phenomena. In the last analysis, quantification of contributory causes of them is impossible, given the intractable nature of the evidence and the subtlety of the inter-relationships, direct and indirect, involved. Therefore, no intellectually satisfying proof becomes possible that one answer is demonstrably 'correct' in a scientifically provable way. Quantification does not offer any obvious solution either. One might hope that, taking a defined population of innovations, it would be possible to determine the percentage which depended upon scientific knowledge, or to allocate degrees of such dependence upon some quantified scale. But establishing the criteria of such a scale would be subjective enough, while yet greater discretion would remain in allocating most innovations to the different boxes. Moreover innovations form a most heterogeneous collection, differing very greatly in relative importance. Bringing qualitative considerations into the argument would imply further discretionary allocation of innovations into a scale of importance so that the degree of dependence of innovations upon scientific knowledge could be construed against some norm of economic significance. Were the scientifically-orientated innovations in the 'population' more, or less, important than their arithmetical proportion suggests?

The question of the *strategic* importance of innovations raises a further issue. For example, despite the percentage of total technical change subject to the linkage with science being small, a strategic blockage on a narrow front at the frontier of technical possibilities might hold up innovation in a wider span behind it. One strategic science-linked innovation could make possible a large number of empirically-based innovations which were, to a degree, dependent upon that initial advance, and vice versa. Moreover, it is impossible to demonstrate the potential quantitative importance of this by being able to indicate what would have happened if an absolute blockage at the frontier had occurred without substitute arrangements bypassing the obstruction. Perhaps detailed analysis can be applied in the micro-study of particular innovations (carefully chosen), but it is difficult to see how a quantified assessment can be made for the wide sweep of innovations under discussion here. History is a depressingly inexact science as economists – let alone natural scientists – discover to their frustration.

Conclusions in this field are also much influenced by methodological or definitional problems. Controversies on such general themes characteristically sink under the weight of semantic disagreement and pleas for more systematic research. What do we include in (or exclude from) the concept 'innovation'? Were the activities of these seventeenth- and eighteenth-century people, properly speaking 'scientific'? Was it *real* science, identified by some later, designated, objective norm – in the 'Baconian' mechanistic tradition – or was it bogus, mistaken, irrational – and following a magical, alchemical or Hermetical tradition?[1] How much, for example, can one claim for Jethro Tull, eagerly pursuing 'scientific' technique in agriculture on the assumption that air was the greatest of all manures and that the fertility of soil consequently varied in direct correlation with the amount of ploughing and pulverising that it received, to the exclusion of all else. Bogus science, quasi science, mistaken science, amateur science which was so very prominent in the seventeenth and eighteenth centuries, particularly in the field of chemistry (where the direct linkage between science and industry are probably most diffused) does raise interesting issues. Does one judge these practitioners by their intentions, their motivations or by their results, however mistaken their assumptions, looked at *ex post facto* with hindsight.

Arguments about distinctions between 'pure' and 'applied' science relate to these controversies, for the seventeenth and eighteenth centuries no less than the nineteenth and twentieth.[2]

This paper will first explore the positive case and then

[1] P. Rattansi, 'The Social Interpretation of Science in the Seventeenth Century', above, pp. 1–32.

[2] R. K. Merton, 'Science, Technology and Society in Seventeenth Century England', *Osiris*, IV (1938); A. R. Hall, 'Merton revisited: Science and Society in the Seventeenth Century', *History of Science*, II (1963); C. C. Gillespie, 'The Natural History of Industry', *Isis*, XLVIII (1957).

consider its qualifications.[1] The key question to be answered is not what examples can be found of links between science and industry in the period but rather how important relative to other sources of impetus was scientific knowledge to industrial progress? Can it be judged 'an engine of growth' for innovation, or a pre-condition? In short, how extensive were the linkages, how strategic, how direct?

II

If economic history is written from the evidence of intention, of aspiration and endeavour, rather than the evidence of results (which is often less accessible) then these connections appear very intimate indeed. In the first place, a very large number of persons – scientists, industrialists, publicists, and government servants – said loudly in the seventeenth century and have gone on saying ever since then, even more loudly, that the linkage was important and ought to be encouraged. For most of the 'professional' scientists of the Restoration the improvement of techniques in the material world, science in the service of a technological Utopia, was a subordinate quest, a relatively low priority. But, even so, many such as Robert Boyle were active on both sides of the watershed between searching for knowledge and applying knowledge to practice, and certainly acknowledged that *one* of the roles of science was to help where it could. Boyle's *Usefulness of Natural Philosophy* (1664) was a systematic survey of the methods then used in industry and of the ways in which science was improving them and would continue to do so. 'These [mechanical] arts', he wrote, 'ought to be looked upon as really belonging to the

1 An equivalent, and connected debate, which will not be considered here, is in progress over these relationships, and that of religion in the seventeenth century. See H. F. Kearney, *Origins of the Scientific Revolution* (London, 1964); C. Hill, *The Intellectual Origins of the English Revolution* (Oxford, 1965); C. Hill, H. F. Kearney, and T. K. Rabb in *Past and Present*, XXVIII, 81; XXIX, 88; XXXI, 104, 111; XXXII, 110. The thesis was formulated by R. K. Merton, 'Science, Society and Technology in Seventeenth Century England', *Osiris* (1938). See also S. F. Mason, 'The Scientific Revolution and the Protestant Revolution', *Annals of Science*, IX (1953); D. S. Kemsley, 'Religious Influences in the Rise of Modern Science', *Annals of Science*, XXIV (1968). This, in turn, is an extension of the much older debate about the links between protestantism and capitalism, from Max Weber.

history of nature in its full and due extent.'[1] 'There is much real benefit to be learned [from mathematical or philosophical inquiries?]', wrote Dr J. Wilkins in 1648, 'particularly for such gentlemen as employ their estates in those chargeable adventures of Drayning, Mines, Cole-pits, etc. ... And also for such *common artificers* as are well skilled in the *practise* of the arts.'[2] Boyle was himself active particularly in investigating the techniques of mining, assaying and agriculture. In evidence of intention, if not of result, John Richardson changed the title of his book on *Philosophical Principles of the Art of Brewing*, much taken up by the largest brewers in London, to *Philosophical Principles of the Science of Brewing*.[3] R. Shannon's more empirically titled work *Practical Treatise on Brewing* was primarily a plea that brewers and distillers should profit from contact with 'men of reflection acquainted with first principles who have more methodically considered the subject'. 'Chemistry', he remarked, 'is as much the basis of arts and manufactures, as mathematics is the fundamental principle of mechanics.'[4] They were echoing a traditional sentiment which continued to reverberate until scientific discoveries with major implications for technology in the industry really were made by Pasteur and others in the mid-nineteenth century.

Two eminent Victorians out of many, may be quoted to show the canon during the nineteenth century. Charles Babbage, writing *On the Economy of Machinery and Manufactures* (1835) concluded: 'it is impossible not to perceive that the arts and manufactures of the country are intimately connected with the progress of the severer sciences; and that, as we advance in the career of improvement, every step requires, for its success, that this connexion should be rendered more intimate'.[5] Dr Lyon Playfair, the forward-looking Scot who helped to organise the Great Exhibition of 1851 wrote, with justly famous perception: 'Raw material, formerly our capital

[1] Cf. A. R. Hall, *Ballistics in the Seventeenth Century* (Cambridge, 1952), p. 3; also G. N. Clark, *Science and Social Welfare in the Age of Newton* (Oxford, 1937), p. 14.

[2] J. Wilkins, *Mathematical Magick* ... (London, 1649), p. vi.

[3] J. Richardson, *Philosophical Principles of the Art of Brewing* (Hull, 1788); *Philosophical Principles of the Science of Brewing* (Hull, 1798).

[4] R. Shannon, *Practical Treatise on Brewing* (1804), pp. 48–9.

[5] C. Babbage, *On the Economy of Manufactures* (London, 1835), p. 379, para. 453.

advantage over other nations, is gradually being equalised in price, and made available to all by improvements in locomotion, and Industry must in future be supported, not by a competition of local advantages, but by a competition of intellects.'[1] The assertion of the linkage has formed a continuum; and still does.

Apart from such aspirations look also at what endeavours actually took place. The state actively sought to press scientists into utilitarian endeavour. A long list of instances can be drawn up. Typical examples are ballistics and navigation (improvements in cartography, scientific instruments, astronomy, mathematical tables, accurate time-keeping lay behind this). Much medical experimentation went on sponsored by the Admiralty, facing particular problems of maintaining efficiency in fleets, long on foreign station, from scurvy and other diseases. Standardisation in production, in dockyards, of interchangeable parts, exact measurement techniques, were much encouraged. Industrial and scientific skills likely to be useful in war received particular attention. More widely, national rivalries became important in the seventeenth century for stimulating inventions in many industries where there was most technical progress – export industries, sugar refining, distilling, glass blowing, silk, tobacco, book printing, paper making and others.[2]

The Royal Society in England, of 1662, as the French *Académie* of 1666, personified such state patronage (although in England with virtually no public resources) for utilitarian ends, an intention explicitly stated in its first charter. The draft preamble of the statutes of the Royal Society ran: 'The business of the Royal Society is: to improve the knowledge of natural things, and all useful arts, Manufactures, Mechanic practices, Engynes and Inventions by experiment.'[3] Nothing could be more explicit. Its first historian stressed this need to focus the work of scientists upon technology; in the words

1 Lyon Playfair, *Lectures on the Results of the Great Exhibition of 1851* (London, 1852). Royal Commissions joined the chorus in 1864 with the publication of the Taunton Commission report on technical and scientific education. Mark Pattison made the same plea in *Suggestions on Academical Organisation* ... (Edinburgh, 1868).

2 Clark, *Science and Social Welfare*, pp. 51ff.

3 See also M. Ornstein, *The role of Scientific Societies in the 17th Century* (Chicago, 1928), pp. 108–9. For details of the utilitarian aims of the French *Académie des Sciences* see ch. 5.

of Thomas Sprat in 1667 its work was intended 'for the use of cities and not for the retirements of Schools'.[1] Pepys urged its members to 'principally aim at such experiments or observations as might prove of great and immediate use', and had the record searched for helps to navigation. The King, petitioned by 'projectors' with secret weapons to save an industry or confound the French, referred such proposals to the Society for vetting and report. Members divided themselves into special committees for this purpose. The *Philosophical Transactions* in the seventeenth century exemplify the common concern; experiments and reports intended to have practical applications, to agriculture as well as industry, had as much space or more devoted to them as any other. This surely, is the breeding ground for innovation. The spark then jumps from the metropolitan scene of the Royal Society in its early days to the many provincial societies linking amateur scientists with gentlemen-manufacturers in the Lunar Society of Birmingham and very many others of lesser renown. Relatively obscure towns like Spalding, Northampton, Peterborough and Maidstone for example, boasted such gatherings. Almost thirty are known to have existed.[2]

William Shipley called the Northampton Philosophical Society specifically a 'Royal Society in Miniature' – 'a Society of Gentlemen that are much addicted to all manner of natural knowledge'.[3] Most of these local societies had the specific aim of popularising science and using scientific knowledge for practical ends in the improvement of practical skills in industry and agriculture – as with the national institutions of the Society of Arts (1754) founded by William Shipley and the Royal Institution (1799), founded by Count Rumford, both of whom were passionate advocates of the application of science.

1 T. Sprat, *History of the Royal Society* (London, 1667). See also J. G. Crowther, *The Social Relations of Science* (1967 ed.), pp. 274–87; C. R. Weld, *A History of the Royal Society* . . . (London, 1848), vol. I, pp. 146ff; vol. IV, section V; M. Purver, *The Royal Society: Concept and Creation* (London, 1967).

2 D. McKie, 'Scientific Societies to the end of the 18th Century' in A. Ferguson ed. *Natural Philosophy Through the 18th Century* (1948); E. Robinson, 'The Derby Philosophical Society', *Annals of Science*, IX (1953); R. E. Schofield, *The Lunar Society* (Oxford, 1963); D. Hudson and K. W. Luckhurst, *The Royal Society of Arts* (London, 1954).

3 D. G. C. Allan, *William Shipley* (London, 1968), p. 169.

Who unbound Prometheus?

Next look at a growing list of examples of innovations which sprang, or appeared to spring, from this fertile soil of scientific discourse and social nexus between the men of science and industry. Steam-power above all; but also the adolescent chemical industry with chlorine-bleaching, sulphuric-acid production, soda making, coal distillation.[1] James Watt, Dr John Roebuck, Josiah Wedgwood, Lord Dundonald, George and Charles Macintosh are the most well-known individuals who personify these connections. The *extent* of interest in 'amateur science' coupled with the extent of endeavour in relating science to industry is remarkable, and in this England is certainly unique in Europe. The important research of Musson and Robinson has placed all economic historians in their debt by revealing how extensive these interests were – almost, one might say, a 'sub-culture' of interest in science, faith in the possibilities of applying science, and enthusiastic advocacy.

In fact, mathematics may well have played a wider role in these relationships than science until the end of the eighteenth century. Navigation techniques and improvements at sea (not alone sponsored by the navy), land surveying techniques for estates, accountancy for business, assaying, architectural drawing, spectacle making are examples of practical skills that gained and were seen to gain, from mathematical knowledge. The Nonconformist and Quaker groups gave a prominent place to modern studies, particularly mathematics, that had a greater presence in new educational movements than science.[2] The observations of a distinguished user of the new mathematical knowledge for practical purposes underline this truth:

> We are sure of finding a Ship's place at Sea to a Degree and a half and generally to less than half a Degree, [wrote Captain James Cook]. Such are the improvements Navigation has received from the Astronomers of the Age by the Valuable Table they have communicated to the Publick under the direction of the Board of Longitude ... [By] these Tables the Calculations are rendered short beyond conception and easy to the

[1] A long bibliography is contained in A. E. Musson and E. Robinson, *Science and Technology in the Industrial Revolution* (Manchester, 1969).

[2] S. Pollard, *The Genesis of Modern Management* (London, 1965), ch. 4; J. D. Bernal, *Science in History* (London, 1954) p. 346; N. Hans, *New Trends in Education in the Eighteenth Century* (London, 1951). A Mathematical Society was established in Spitalfields in 1717, and another at Manchester in 1718 (T. Kelly, *G. Birkbeck – Pioneer of Adult Education* (Liverpool, 1957) p. 66).

meanest capacity and can never be enough recommended to the Attention of all Sea Officers ... Much credit is also due to the Mathematical Instrument Makers for the improvements and accuracy with which their instruments are made for without good instruments the Tables would lose part of their use.[1]

The utility of such mathematical expertise, coupled with precision measurement by new instruments, for a trading, industrial, seafaring nation was sufficient for it to become institutionalised in schools on a fairly wide scale in eighteenth-century England. Rather than enlarge the catalogue of instances, however, let us now move to looking at some of the problems – acknowledging that a long list of such individual instances exists. It is the nature of the connections between science and technical change, no less than the extent of the association between them which is in question.

III

The first complication is, perhaps, not a fundamental one within the European scene, although it raises important questions when relating science to innovation within a single country; or perhaps even more fundamentally when one compares scientific knowledge and its relation to technique (the general level of diffused technique rather than individual instances of 'best practice' technology) in Europe and beyond – say in China. The point is, simply, that we are much concerned with differences between national performances in industrial growth and innovation, in striving to explain the fact that the British economy advanced more extensively than others in this way, and became *relatively* so much more forward in adopting new techniques and developing new industries in 1750–1850 than other economies. This is particularly true of the general level of technique, productivity and output characterising growth industries (textiles, metal production, metal-using techniques, machine tools, machine making, particularly power engineering, chemicals, pottery, glass).

Scientific knowledge does not show, at all, the same concentration within Britain, particularly in the case of chemistry

1 *Journals of Capt. James Cook (1768–1779)* ed. A. Grenfell Price (New York, Heritage Press, n.d.) pp. 112–13 (for 14 January 1773).

where the linkage between scientific knowledge and industrial innovation was probably most intimate. The advance of scientific knowledge was a European phenomenon. There was, in France, much greater state patronage for science through the *Académie des Sciences*, by military sponsorship, and direct industrial sponsorship, as with the research department attached to the Sèvres porcelain factory working on glazes, enamels and paints. Provincial academies also flourished in the main regional cities.[1] In the *Description des Arts et Métiers* of Réamur 1761 one had a more elaborate schema published than any known in Britain. On balance, more systematic work was carried out in technology by scientists in France than on this side of the Channel. Countries innocent of industrialisation (but with pressing military needs) also established equivalent academies, with state patronage for the useful arts – especially arts useful for military success – supported by much private interest. Sweden, Russia, Prussia and Italy are examples.[2] A Royal Irish Academy also flourished. The 'Dublin Society for Improving Husbandry, Manufactures and other Useful Arts' was the first of the 'popularising' associations, established in 1731. The Welsh Society of Cymmrodorion followed in 1751. One of the earlier of the agricultural improvement societies was that of Brecknockshire, founded in 1753.[3] Evidence of motivation the institutionalising of practical science in these societies clearly was; but it may well have come, in such instances, where the need was greatest, rather than where the links were closest. It should also be said that the English societies flourished with very tiny material resources indeed, being amateur and self-financing. The very small cash premiums or medals they offered as inducements to inventors cannot be seen as 'research and development' costs in the modern sense of capital investment in innovation. The fact that endeavour was stimulated by the chance of winning a medal offered by such a private society or

[1] F. A. Yates, *The French Academies of the 16th Century* (London, 1911); H. Brown, *Scientific Organisations in 17th Century France* (New York, 1934).

[2] R. Hahn, 'The application of Science to Society: the Societies of Arts', *Studies on Voltaire and the Eighteenth Century*, XXIV–XXVII (Geneva, 1963), 829–36. This article lists a dozen such societies in different countries.

[3] H. F. Berry, *A History of the Royal Dublin Society* (London, 1915); Allan, *William Shipley*, p. 61.

appearing in its transactions, says much for the prestige attached to science and to the quest for 'improvement' in practical matters. But, clearly these investments and endeavours could be made on a scale more extensive absolutely than England (in the case of France) and on a scale relatively greater (judged against the resources of the country) without much of a 'spin off' giving a boost to industrial growth.

The French record of scientific growth and invention in the eighteenth century was a formidable one.[1] Berthollet first revealed to the world the bleaching possibilities of chlorine, first isolated as a gas in 1774 by a Swedish chemist Scheele, which was followed by energetic efforts to promote its manufacture in France. A similar sequence followed with Leblanc making soda from salt and sulphuric acid.[2] Very sophisticated work was done in the production of dyestuffs in France; with varnishes, enamels and many other techniques and materials. Yet the difference in the rate of industrial growth based on these advances in chemistry between France and Britain in the period 1780 and 1850 was remarkable. Almost all the theoretical work on structures, stresses and the mechanics of design in civil engineering was French. This did not appear to have much relationship to the speed of development, or even innovations in these fields, as far as economic progress was concerned. The same was true of power engineering and hydrodynamics.[3] The record of development and implementation was also significantly different from the record of invention.

The wider question, not to be pursued here, is even more interesting. The sophisticated scientific mechanical knowledge in China produced even less impetus to the general levels of industrial technique representative of that vast region, or to industrial growth. It remained more sealed up in a small

1 See, in illustration, S. T. McCloy, *French Inventions of the Eighteenth Century* (Lexington, Ky, 1952) and *Government Assistance in Eighteenth Century France* (Durham, N.C. 1946).

2 C. C. Gillespie, 'The Discovery of the Leblanc Process', *Isis*, XLVIII (1957).

3 D. S. L. Cardwell, 'Power Technologies and the Advancement of Science, 1700–1825', *Technology and Culture* (1965) vol. VI; 'Some Factors in the Early Development of the Concept of Power, Work and Energy', *British Journal for the History of Science*, III (1966–7); R. Hahn, *L'hydrodynamique au XVIIIe siècle* (Paris, 1965); D. Landes, *Cambridge Economic History of Europe* (Cambridge, 1965), vol. VI, p. 333.

enclave of scholars, civil servants, and isolated groups under noble and royal patronage than in – say – St Petersburg. By itself, therefore, it becomes difficult to argue that a flow of new scientific knowledge and applied science is a key variable; it may be a pre-condition for advance, but it does not necessarily give the operational impetus.

Secondly, the problem of time lags between knowledge and action raises awkward problems for the 'positive' one-to-one equation in its simple form. The economic historian is more interested in innovation and the diffusion of innovations than in invention for its own sake. Putting inventions to productive use involves all the costs and problems of translation from laboratory technique into industrial production, from the largely non-commercial context of the pursuit of knowledge to profitability as a condition of existence. One is not even very interested in isolated examples of new industrial techniques but rather in their diffusion to the point when innovations begin to affect general levels of output, costs, productivity in an industry; when their adoption is on a sufficient scale to affect the performance of the industry significantly. To mention a few of these astonishing time lags. The screw-cutting lathe, foundation of the precision engineering skills which made an efficient machine-making industry possible, was clearly documented by Leonardo da Vinci in the *Note-Books*, laid out again in the section on watch-making tools in the *Description des Arts et Métiers* in mid-eighteenth-century France and developed, spontaneously, again by Maudslay, to become – from that innovation – the basis of a progeny of machine tools. Sir T. Lombe's silk throwing machine, which was used for the first time in a factory in England in 1709, had been used and known in Italy since 1607 – with an accurate engraving in a book on the open shelves of the Bodleian Library by 1620.[1] The same is true of gearing and the design of gear wheels, bridge design, pumps, Archimedean screws, the 'pound' lock, mass production needle-grinding machines and a host of others (all to be found in Leonardo's work).[2] In certain respects, steam power is another example. The pound lock – being the basic technology of a dry dock – was known in

[1] Quoted Clark, *Science and Social Welfare.*
[2] I. B. Hart, *The World of Leonardo da Vinci* (London, 1961).

Peter Mathias

Dutch shipyards in the fifteenth century, perhaps much earlier, and appeared in England in the sixteenth century. But a still-water canal system of which this is the only important piece of technology was an eighteenth-century phenomenon in Britain.[1] Equally dramatic time lags exist in the opposite direction – between empirical improvements in technique and the beginning of scientific interest in explaining them.[2]

Bound up with this problem of time lags between knowledge and invention, invention and adoption, adoption and diffusion are correlated phenomena such as simultaneous inventions (developed spontaneously and independently in different places at about the same time), re-inventions of lost techniques, 'alternative' inventions coming very close together in time for providing different ways of getting the same thing done.[3]

The 'profile' of technical change usually shows an evolutionary curve as well as revolutionary discontinuities. The interstices between the discrete advances made by identifiable individuals are filled by 'continuum' improvements made on the job, by countless improvements without known, or identifiable and published authorship. Collectively the latter may yield a cumulative advance in productivity greater than the identifiable discrete innovations. This has been likened to biological change, improvement and survival by the techniques most efficiently and economically adapted to their function – a kind of technological Darwinism.[4] The burden of all this is, of course, that invention waits upon economic opportunity before it can come to fruition in innovation and the diffusion of new techniques. The determinants of timing are usually set – in the long run – by non-technical criteria. These determinants may be economic criteria of different sorts – the widening of the market giving inducement for larger production

1 Surveying techniques were certainly advanced enough in the late sixteenth century to facilitate canal cutting. The New River project bringing water from Hertfordshire to North London involved very sophisticated routing and exactness in calculating levels.

2 A. R. Hall and T. S. Kuhn in M. Clagett, *Critical Problems in the History of Science* (Madison, Wisc., 1959), pp. 16–17.

3 R. K. Merton, 'Singletons and Multiples in Scientific Discovery', *Proceedings of the American Philosophical Society*, CV (1961); W. R. Maclaurin, 'The Sequence from Invention to Innovation . . . ', *Quarterly Journal of Economics*, LXVII (1953).

4 S. C. Gilfillan, *The Sociology of Invention* (Chicago, 1935); 'Invention as a Factor in Economic History', *Journal of Economic History*, Supplement (1945).

and hence new methods, greater facility in the supply of capital, a change in factor prices[1] (for example, labour becoming relatively more expensive or intractable, raising the incentives to cut labour costs). Boom conditions, creating bottle necks in supply, higher profits, and greater incentives to expand may create the operational incentives. In a dynamic sequence, when an economy is on the move with innovations flowing, a depression may equally induce further innovation by creating pressures to cut costs. The process of innovation itself creates a dis-equilibrium in various ways – that dis-equilibrium, to be resolved, creating the need for further changes, which become self-reinforcing. These may indeed be technical in nature, but they are need-creating in the way they operate. The causal arrows flow from industrial demand towards the absorption of new knowledge. The timing is set from within the industrial rhythm and the economic context, rather than given to it exogenously by new acquisitions of knowledge.[2] There are other determinants – social, political and legal – affecting the condition of risk. 'Entrepreneurship' may also prove to be greater than the sum of these other criteria.

These sorts of motivations tend to be the operational criteria in this period, I believe, determining which bits of scientific knowledge were taken up, developed, applied, and which lay unused; which inventions remained known, but sterile, and which quickly became adopted, perhaps outside the country which gave them birth. Clearly, this is the style of explanation

1 'Every price change, by creating cost difficulties in certain fields and opportunities for profit making in others, provides a double stimulus to invention.' (A. Plant in *Economica* (1934), 38, quoted Clark, *Science and Social Welfare.*)

2 J. Schmookler, *Invention and Economic Growth* (Cambridge, Mass. 1966); 'Economic Sources of Inventive Activity', *Journal of Economic History*, XXII (1962); W. F. Osburn and D. Thomas, 'Are Inventions Inevitable?', *Political Science Quarterly*, XXXVII (1922); R. C. Epstein, 'Industrial Inventions: Heroic or Systematic?' *Quarterly Journal of Economics*, XL (1926); R. K. Merton, 'Fluctuations in the Rate of Industrial Invention', *Quarterly Journal of Economics*, XLIX (1934–5).

The very large literature about technical change and innovation now developing is seeking to establish criteria for measuring and evaluating this phenomenon in economic theory. Almost all of it relates to twentieth-century examples and assumptions, particularly that concerned with research costs and applied science. Conclusions are therefore not *ipso facto* applicable to innovation as a phenomenon in the seventeenth and eighteenth centuries.

behind the very rapid adoption of 'chemical' bleaching in the cotton industry in Britain more than in France: the enormous expansion in the output of cloth made a more rapid means of bleaching imperative. Great stress has recently been placed by historians of science on the ways in which empirical processes and skilled artisan technology in the mechanical arts stimulated scientific advance in these centuries.[1]

This does not, however, necessarily subvert the core of the 'positive' equation. It can be argued that applied science needs to build up a capital 'stock' or a 'library' of knowledge, so to speak, which is thereupon available for industrialists to draw upon in innovation – across national frontiers, no doubt, for science now enjoyed, in printing, a very effective means of diffusing knowledge in the seventeenth century and after; and with a timing doubtless profoundly influenced by conditions and incentives within industry. But, without that capital stock produced by the advance of scientific knowledge, runs the argument, a limit would have been placed upon the range of advance.

This begs a question as to how far the impetus deriving from industry was able to produce the conditions – of innovation as well as other things – needed to sustain its own progress. To what extent was the flow of innovations produced from within the empirical world of industry and not given to it from an 'exogenous' world of science, advancing under its own complex of stimuli? That in turn begs the question as to how far industrial demand, with the needs of the 'empirical' world, was itself the stimulus for creating new scientific knowledge in this period, of what 'feeds-back' or 'feeds-forward' there were?

Judging the effectiveness of the contributions of science by results, *ex post facto*, rather than by endeavour, is to greatly reduce their importance. Little of the mass of experimentation

[1] E.g. P. Rossi, *Francis Bacon* (London, 1968), p. 9; E. J. Dijksterhuis, *The Mechanisation of the World Picture* (Oxford, 1961), pp. 243-4; T. S. Kuhn, 'Energy Conservation ... ' in Clagett, *Critical Problems in the History of Science;* A. R. Hall, *From Galileo to Newton, 1630-1720* (London, 1963), pp. 329-43; A. R. Hall, *The Scientific Revolution* (London, 2nd ed. 1962), pp. 221, 225, 236; J. D. Bernal, *Science in History* (London, 1954), pp. 345-6, 371. At its extreme, this case becomes the doctrinaire Marxist position that advances in scientific knowledge were determined purely by the bourgeoisie's commercial and industrial needs. See B. Hessen, *Science at the Cross-Roads* and the debate given in condensed form in G. Basalla (ed.), *The Rise of Modern Science* (Boston, 1968).

in agricultural projects of the Royal Society in its early years, for example, seems to have had much direct effect upon improving the efficiency of farming.[1] Most of the more direct links between advancing knowledge of chemistry and the expanding chemical industry came only at the end of the eighteenth century. This disconnects the timing of much of the new knowledge, particularly in chemistry, from the initial phases of industrial growth. The great advances in mechanics in the seventeenth century – then one of the most advanced of the sciences – had given birth to very sophisticated theoretical schema about ballistics, which do not appear to have significantly affected the processes of innovation in making metals or working metals, of gunfounding or of gunnery, again judging by results, until after the Crimean War. The 'science' remained almost purely abstract. Each cannon cast and bored remained slightly different from every other; each shot and each charge of powder were equally 'unique', which kept techniques of gunnery strictly empirical. A precision engineering industry to produce the guns, and a precision chemical industry to produce the propellants, were required before this theoretical knowledge could become operational.[2]

The empirical stimulus creating response within the immediate context of production accounts for a very high proportion of the advance in productivity, even in those industries most exposed to the impact of science, and a great determinant of timing and the rate of diffusion of new techniques. Great areas of advance were relatively untouched by scientific knowledge, judging by result rather than by intention or endeavour, until the nineteenth century: agriculture, canals, machine-making, the mechanisation of cloth-making (as distinct from bleaching and dyeing), iron- and steel-making. Taking the occupational census of 1851, a very small percentage indeed of the labour-force was engaged in trades where the linkages were – superficially at any rate – high, as in chemicals.

IV

Steam power is important enough to merit separate attention.

[1] See below, pp. 73–6.

[2] A. R. Hall, *Ballistics in the Seventeenth Century* (Cambridge, 1952).

Peter Mathias

Here was the greatest gift from science to industry, it has often been claimed, born exactly from the world of the Royal Society, of noblemen's laboratories, from an international competition amongst scientists and their leisured patrons in the seventeenth century. Watt, in his generation, carried on this precise linkage between scientific knowledge and commercial application in the series of formal experiments to analyse the properties of steam and the conductivity of metals which lay behind his own inventions of the separate condenser and steam power proper (as distinct from 'atmospheric' power). This can be called the classical example of science in alliance with practice. But other factors also conditioned the rates of advance of efficiency in steam power and the timing of stages of growth of this innovation. Thomas Newcomen and Savery were not so directly within this educated scientific tradition. And historians of science continue to push back the genealogy of the basic scientific awareness of steam power – knowledge as distinct from laboratory experiments that worked.[1] The jump to the world of Thomas Newcomen, who had no personal contacts with the leading scientists of the day,[2] fashioning an effective commercial device, meant problems of manufacture, of standards of accuracy in metalworking, that alone made effective use possible on a commercial basis. This, it can be argued, more than anything else set the limits of efficiency. And this, as a blacksmith, was Newcomen's world, not that of the Royal Society. Once again, the context within which Watt's inventions had to become operational meant that the accuracy of working metal, of fitting a piston to a cylinder throughout its length, of getting steam-proof valves and joints set the limits to the rise in the degree of efficiency which potentially resulted from the new inventions. These efficiencies came from the empirical world of John Wilkinson and Matthew Boulton, with rising standards of its own, a world increasingly working to rule, but still mainly innocent of formal scientific thought.

1 A. R. Hall has gone so far as to state: 'No scientific revolution was needed to bring the steam engine into existence. What Newcomen did could have been done by Hero of Alexandria seventeen hundred years before, who understood all the essential principles' (*From Galileo to Newton, 1630–1720* (London, 1963), p. 333).

2 D. S. L. Cardwell, *Steampower in the Eighteenth Century* (London, 1963), p. 18.

Who unbound Prometheus?

And, subsequent to Watt, most of the pioneering of 'high-pressure steam', the adaptation of steam to traction, to small-bench engines, to ships and the continuum of improvement to the Watt-style engine itself, belonged, for the most part, to the empirical world of the obscure colliery engineers, the captains of Cornish mines, the brilliant mechanics such as Murdock. Some of them were trained in the best precision workshops of the country, such as Maudslay's, but remained nevertheless innocent of scientific fundamentals and were not seeking to create their improvements in the light of awareness of such fundamentals. Yet the cumulative total effects of 'continuum' innovation, effected on the job, at the work bench, bit by bit, were profound. Taking steam-power efficiency again as one example, the first Newcomen engine had a duty[1] of *circa* 4.5 m., it has been calculated.[2] This had been raised to 12.5 m. by the time of Smeaton's improvements in 1770. The initial Watt separate condenser engine raised this duty to *c.* 22 m. By 1792 it had been raised by continuing improvement to over 30 m. The best recorded Watt-type engine in 1811, working in Cornwall, had a duty of 22.3 m. In 1842–3, under continuous gradual improvement, the best duty was recorded at 100 m. Average duty rates recorded on Cornish engines quadrupled between 1811 and 1850 as a result of this continuum-type improvement.

V

Although this essay is primarily concerned with sources of innovation in industrial techniques, improvements in agriculture are relevant to the issue, both because agriculture potentially stood to gain from the application of science, comparably to industry (and has gained as dramatically from its connection with science as has industry in the twentieth century), and because contemporaries certainly gave agricultural improvement at least as high a priority as industrial

[1] The number of pounds of water raised one foot by the consumption of 1 bushel of coal.

[2] D. B. Barton, *The Cornish Beam Engine* (Truro, 1966), pp. 28, 32, 58. These figures are suspect, but there is no reason to suppose that the degree of suspicion advanced with time – rather the reverse. Other factors went into these duty counts as well as the intrinsic technical potentiality of the engines.

Peter Mathias

advance. Agricultural improvement also had a more general appeal to the upper and middle-classes of English society than any other branch of production, if only because larger and more influential social groups were concerned with the land. Agricultural innovations and scientific experiments in agriculture featured in virtually all the scientific and philosophical societies mentioned above; while other societies were specifically agricultural in their terms of reference. The Georgical Committee of the Royal Society was established in 1665 (as one of eight sectoral groups). Many experiments in husbandry found a place in the *Philosophical Transactions* over the years; and eleven reports are known to have been made about agricultural practices, produced from national enquiries.[1] A 'Society of Improvers in the Knowledge of Agriculture in Scotland' was founded in 1723 (forty years before that in Brecknockshire) and many local agricultural societies followed, particularly in the 1790s.[2]

Scientists concerned themselves with agricultural experiments, advocating the experimental method and lauding the claims of science in farming no less than in industry. Francis Bacon's *Sylva Sylvarum* (1651) included a comparative study of different modes of fertilising. In 1671, Boyle urged farmers to experiment. 'Chymical experiments ...' he wrote, 'may probably afford useful directions to the Husbandman towards the melioration of his land, both for Corn, Trees, Grass and consequently Cattell.'[3] John Evelyn's various works contained a curious – if typical – mixture of magic and shrewd commonsense, in circumstances where virtually anything organic, and much inorganic, could be thrown onto the land with advantage.[4] Hale's *Vegetable Statics* (1727) continued the line, for the first time challenging the view that plants were composed simply of water. Francis Home, both a doctor and subsequently a professor in Edinburgh, also deliberately set

1 R. L. Lennard, 'Agriculture under Charles II', *Economic History Review* (1932).

2 These include the Canterbury Agricultural Society, Odiham Society, London Veterinary College, Bath and West and Southern Counties, Norfolk Agricultural Society.

3 R. Boyle, *Some Considerations Touching the Usefulness of ... Natural Philosophy* (Oxford, 1671).

4 J. Evelyn, *Sylva* (London, 1644); *Kalendarium Hortense* (London, 1664); *A Philosophical discourse of earth* ... (London, 1676).

out to apply science to agriculture to see 'how far Chymistry could go in settling the Principles of Agriculture'.[1] Lavoisier set up agricultural experiments and ran a model farm.[2]

Thus, taking the evidence of intention and endeavour, agricultural innovation shares with industry a common link with science in the seventeenth and eighteenth centuries. The problems about concluding from these aspirations and associations that applied science was the prime source of such innovation in agriculture as actually occurred are even greater, if anything, than with industry. There, at least, in the specialised sector of the chemical industry and immediately related fields there is solid evidence for the connection in a direct way (though casual links could flow in both directions). Much theorising was plainly mistaken, based on quite irrational premises. The innovations which characterised progressive farming did not owe much, if anything, to such science; while only the very exceptional farmer or landowner was directly influenced by the scientists. Russell, as a leading present-day agricultural scientist is profoundly sceptical of the relevance of chemistry to agricultural advance before the generation of Leibig, Davy's *Elements of Agricultural Chemistry* (1813) and the Rothamsted experiments of Lawes and Gilbert in the 1840s.[3]

But scepticism about the significance of the direct application of formal scientific knowledge to agrarian improvement in these two centuries does not end the story of the relevance of this evidence. To anticipate a tentative conclusion drawn below, one can certainly take this large body of data as strong evidence of *motivation* for agrarian advance. Coupled with false premises about chemical reactions were urgent pleas for experimentation, shrewd observation and recording, the comparative method, seeking alternative ways of doing things which could be measured and tested to see if they were superior to the old. This was a programme for rejecting traditional methods justifiable only because things had always

1 F. Home, *Principles of Agriculture and Vegetation* (Edinburgh, 1757).

2 E. J. Russell, *A History of Agricultural Science in Great Britain, 1620–1954* (London, 1966), p. 53.

3 E. J. Russell, *A History of Agricultural Science*, pp. 25, 37, 46. See also G. E. Fussell, 'Science and Practice in Eighteenth Century British Agriculture', *Agricultural History*, XLIII (1969). But see the comments of D. J. Brandenburg.

been done in that way (even though such customs, hallowed by the passing of time, often did embody strict rationality, even if unselfconscious and inarticulate on the lips of their practitioners). Scientific procedures and attitudes encouraged by the scientists may have been more influential than the scientific knowledge they dispensed. 'I should not, therefore, proceed a single step', wrote Francis Home, 'without facts and experiments.'[1]

The publicity given to new methods, new crops, rotations and implements by these same groups may also have increased the pace of diffusion of innovations in agriculture. Certainly the flood of writings, at all different levels, is evidence of an intellectual world where progress was written into the assumptions of the age.

VI

The institutional development of science creates certain problems, because the patterns of development in science do not always fit sequences of innovation and development in industry. The foundation of the Royal Society and the Lunar Society are always quoted as evidence of the developing nexus between science and industry. But one must then face the issue of the decline in the utilitarian orientation towards applied science of the Royal Society after 1670, its decline during the first half of the eighteenth century and in parts of the nineteenth century, and its great weaknesses compared with the equivalent academy in France. The Lunar Society also withered into a state of collapse after the seventeen-nineties. In some fields also it seems perfectly possible – even in that most applied of sciences, medicine – for the accumulated advances of several generations, well institutionalised, not to result in any major impact upon national demographic trends in death-rates or disease-rates. Medicine, we are told, did not begin to have such a major impact – that is, upon a statistically significant proportion of the population – until the second half of the nineteenth century, apart, perhaps, from the effects of inoculation and vaccination on smallpox. And this was in a

1 F. Home, *Principles of Agriculture and Vegetation* (Edinburgh, 1757). He urged the Edinburgh Society to raise 'a spirit of experimental farming over the country'.

field where there was very great interest, considerable advances in scientific knowledge, a greater flow of money, and probably more scientifically trained persons than in all other branches of science put together.[1]

The problem of numbers is also relevant. Persons professionally trained in medicine, in chemistry, the 'scientific' members of the Royal Society (as distinct from the much larger number of 'gentlemen' innocent of professional commitment) remained the merest handful. The average number of elections to the Royal Society in the early eighteenth century was about ten. The average number elected in each year to the College of Physicians was five or six before 1700, supplemented only marginally with those with degrees from foreign universities. These numbers did rise after 1700 but remained very tiny. This may be taken as an index of the professionalisation of science to the extent that this term possessed its modern connotation in the eighteenth century. Beyond this there were, of course, much larger numbers of amateurs and people in business, such as opticians and distillers, who practised empirical science in a commercial way. As Professor Hall has said:

> the impact of any question of abstract science upon a human brain was exceptionally infrequent – it could only happen to, say, one individual in a hundred thousand. But, in the history of technology the situation is very different; the proportion of human beings who could have been very well acquainted with handmills and ploughs and textile- and horsegear has always been very large indeed, until the last century or so in the West. Very few of these ever effected the slightest variation in any technique; but the potentiality for effecting a variation was virtually universal. It is only when the use of relatively uncommon machines or techniques is introduced that the potentiality for innovation becomes restricted.[2]

In England no great expansion of institutions in 'professional' science developed in this period outside the Royal Society, either within the universities or outside them, apart from the amateur groups and the popularisers. Science did not become an established part of the educational system, either within the traditional institutional hierarchy or beyond it in special organisations of its own. The Mechanics' Institutes subsequently

[1] This conclusion may need to be modified in the light of research being currently undertaken by Dr E. Sigsworth of York. See below, pp. 97–110.

[2] A. R. Hall, *The Historical Relations of Science and Technology* (London, 1963).

Peter Mathias

became the only widespread national movement to get institutionalised in this way in the first half of the nineteenth century, and they trained the aspiring literate artisan, not the research chemist. The greatest contrast existed between the English experience and French and German developments in the *École Polytechnique* and the *Technische Hochschule*. The School of Mines and the Royal College of Chemistry in London were the two institutions that challenged this generalisation. They remained very small, isolated geographically, socially and technically from having any significant impact upon mining or industry in Britain as a whole during the first half of the nineteenth century.

VII

In conclusion: it was the same Western European society which saw both great advances in science and in technological change in the great sweep of time and region across the fifteenth to the nineteenth centuries. It would be carrying nihilism to the point of dogma to write this off as a mere accident, even though the case of China suggests that it is perfectly possible for sophisticated scientific and technological knowledge in some fields to produce a very small impetus towards lifting general levels of industrial technique. The simplest assumptions of causation flowing directly and in one direction need to be questioned; the presumption that connections between science and industry were direct, unitary, simple. Negatively it can be argued that the many other conditioning factors in technical change were collectively of much greater importance during the first century of industrialisation and that, in the immediate context of manufacture, formal scientific knowledge was much less strategic in determining commercial success than some modern studies have suggested. In longer perspective we may see that the main impetus from formal applied science to innovation came after 1850 on an ever widening front, but in a context which was highly favourable for many other reasons. That this was the real pivot in the connections between science and industry was shown by default, to a large extent, in the case of Britain lagging most in exactly those fields of innovation where the connection was becoming most intimate.

But much depends upon whether we are looking at the immediate context of innovation or at the general nature of the society, and its intellectual parameters, within which industrial advances were burgeoning. 'We have to see', as Sir George Clark concluded, 'not a gradual and general mutual approach of these elements in society, but the joining of contact, first at isolated points, then at more points; finally almost everywhere.'[1] Until the end of the eighteenth century – that is, until long after systematic, cumulative change on a scale quite uncharacteristic of medieval technical change was under way – that inter-penetration was confined to fairly small areas, even if some of them were strategic.

It should also be acknowledged that scientific attitudes were much more widespread and diffused than scientific knowledge. Attitudes of challenging traditional intellectual authority, deciding lines of development by observation, testing, experimentation and adopting – indeed, actively stimulating the development of – scientific devices such as the thermometer and hydrometer, which enabled industrialists to reduce their empirical practices to rule wherever possible, were certainly being strengthened.[2] The quest for more exact measurement and research for the means to fulfil it was certainly characteristic of these linkages, even where the object was not to subvert empirical techniques, of which the chemistry remained unknown, but to standardise best practice within them. Scientific devices and techniques were thus often used to buttress empirical techniques rather than to challenge them. In this sense, the developing Baconian tradition of the experimental sciences, the tradition of research based upon systematic experimentation (as in late-eighteenth-century chemistry) had closer links with the process of innovation than did advances in cosmology, mechanics or physics in the seventeenth century. And in such linkages science probably learned as much from technology as technology from science until the nineteenth century: scientists were much concerned with trying to answer questions suggested from industrial

1 Clark, *Science and Social Welfare*, p. 22.

2 The brewing and distilling industries, with the excise authorities anxious to have more precise calculations in gauging for taxation, offer a good example of such a sequence. See P. Mathias, *The Brewing Industry in England, 1700–1830* (Cambridge, 1959), pp. 63–78.

Peter Mathias

techniques. 'Technological progress implied the idea of intellectual progress, just as chance discoveries implied the possibility of systematic ones.'[1]

We may conclude that together both science and technology give evidence of a society increasingly curious, increasingly questing, increasingly on the move, on the make, having a go, increasingly seeking to experiment, wanting to improve. This may be the prime significance of the new popularisers of science and technology, the encyclopaedias, the institutions like the Society of Arts, the Royal Institution, the Lunar Society and the various local philosophical and scientific societies, the new educational movements, the intriguing links between radical non-conformist scientific and business groups in the eighteenth century or between Puritans and the founders of the Royal Society in the seventeenth century. So much of the significance, that is to say, impinges at a more diffused level, affecting motivations, values, assumptions, the mode of approach to problem-solving, the intellectual milieu, rather than a direct transference of knowledge. In this sense, of course, the conclusion is banal, that the advances in science and in technical change should both be seen as characteristics of that society, not one being simply consequential upon the other.

[1] A. R. Hall, *The Scientific Revolution* (London, 1962), p. 369.

SCIENCE AND INDUSTRY IN THE LATE EIGHTEENTH CENTURY [1]

by A. E. MUSSON AND E. ROBINSON

It is a deeply entrenched tradition that the technical achievements of the early Industrial Revolution were the products of uneducated empiricism.[2] It is gradually becoming evident, however, that the Industrial Revolution was not unrelated to the Scientific Revolution, which became so marked in Restoration England, with the founding of the Royal Society and the great discoveries of Newton and Boyle, and which was developed in the later eighteenth century by such as Priestley, Black, Cavendish, and Herschel.[3] James Keir, F.R.S., the famous industrial chemist, declared in his *Dictionary of Chemistry* (1789):

> 'The diffusion of a general knowledge, and of a taste for science, over all classes of men, in every nation of Europe, or of European origin, seems to be the characteristic feature of the present age'.[4]

And the *Philosophical Magazine* referred in 1798 to the 'well known' fact that 'the Arts and Manufactures of Great Britain .... have been much improved by the great Progress that has lately been made in the various Branches of the Philosophical Sciences.'[5]

Professor Ashton has already suggested that collaboration between men of science and of industry, in societies such as the Literary and Philosophical Society in Manchester and the Lunar Society in Birmingham, was an important factor in the Industrial Revolution.[6] It is clear that outstanding figures like Boulton, Watt, and Wedgwood were men of considerable intellectual and scientific as well as technical and commercial ability. Other important industrialists such as John Kennedy of Manchester, the Strutts of Derby, and Thomas Bentley of Liverpool were also members of their local philosophical societies and were not strangers to liberal conversation and intellectual pursuits.[7] In recent years historians of education have done much to illuminate this question and their findings have suggested that a knowledge of science was more widely

[1] Based on a paper read on 3 April 1960 at the annual conference of the Economic History Society.

[2] This view has been reiterated recently in R. Mousnier, *Progrès Scientifique et Technique au XVIIIe Siècle* (Paris, 1957), a work based almost entirely on secondary sources, with little new evidence. For a very critical review of this book, see *Annales Historiques de la Révolution Française*, Oct.–Dec. 1959, pp. 376–82.

[3] See, for example, G. N. Clark, *Science and Social Welfare in the Age of Newton* (1938); A. Wolf, *A History of Science, Technology and Philosophy in the Eighteenth Century* (1938); A. Ferguson (ed.), *Natural Philosophy through the 18th Century* (commemoration number of the *Philosophical Magazine*, 1948); J. D. Bernal, *Science in History* (1954); A. R. Hall, *The Scientific Revolution, 1500–1800* (1954); A. and N. L. Clow, *The Chemical Revolution* (1952).

[4] *Op. cit.* Preface, p. iii.

[5] *Op. cit.* I (1798), Preface.

[6] T. S. Ashton, *The Industrial Revolution, 1760–1830* (1948), pp. 16 and 20–21.

[7] See R. Angus Smith, *A Centenary of Science in Manchester* (1883); R. B. [Richard Bentley], *Thomas Bentley, 1730–1780* (Guildford, 1927); and Eric Robinson, 'The Derby Philosophical Society', *Annals of Science*, IX (1953), 359–367.

diffused through industrial society than has hitherto been suspected.[1] In this paper we shall make a detailed examination of the relationship between science and industry in Manchester – centre of the most rapid and important technical advances in the late eighteenth century – and then make a briefer general survey of the situation in other industrial centres.

I

When the Scientific Revolution began in the seventeenth century it was principally a metropolitan movement centred on the Royal Society. By the end of the eighteenth century it was largely a provincial movement, healthily expressed through the many philosophical societies in Norwich, Northampton, Exeter, Bristol, Bath, Plymouth, Birmingham, Derby, Manchester and many other places. The preface to the first volume of the Manchester Literary and Philosophical Society's *Memoirs* pointed out the utilitarian as well as the scientific value of these learned societies. Those founded in different parts of Europe in the seventeenth and eighteenth centuries had 'been not only the means of diffusing knowledge more extensively, but have contributed to produce a greater number of important discoveries, than have been effected in any other equal space of time'. Their journals had publicized 'many very valuable discoveries, or improvements in arts, and much useful information in the various branches of science'. But although several French provincial societies had been instituted, 'in England, they have almost been confined to the Capital'. It was therefore urged that 'the promotion of arts and sciences' would be 'more widely extended by the forming of Societies, with similar views, in the principal towns of this kingdom'.[2]

The Manchester Society was founded in 1781, growing out of informal meetings, which had been held for some years previously, of 'a few Gentlemen, inhabitants of the town, who were inspired with a taste for Literature and Philosophy', and who had 'formed themselves into a kind of weekly club, for the purpose of conversing on subjects of that nature'.[3] These meetings had first been held in the house of Dr Thomas Percival,[4] the leading figure in the Society's foundation and later its president for many years. Its subjects of conversation were to include 'Natural Philosophy, Theoretical and Experimental Chemistry, Polite Literature, Civil Law, General Politics, Commerce, and the Arts' [manufactures].[5] The medical profession was strongly represented in its membership and a few clergy also participated, but 'the great majority of the members were either engaged or interested in the extension of Science and Art to manufacturing purposes'.[6]

[1] E.g. N. Hans, *New Trends in Eighteenth Century Education* (1951); W. H. G. Armytage, *Civic Universities* (1955); T. Kelly, *George Birkbeck* (Liverpool, 1957); B. Simon, *Studies in the History of Education 1780–1870* (1960).

[2] *Memoirs of the Manchester Literary and Philosophical Society*, I (1785), v–vi. Erasmus Darwin's *Address to the First Meeting of the Literary Society at Derby, July 18, 1784* (Derby Public Library) makes the same kind of case.

[3] *Manchester Lit. & Phil. Memoirs*, I (1785), vii. There is no modern history of the Society, but see Angus Smith, *op. cit.* and F. Nicholson, 'The Literary and Philosophical Society, 1781–1851,' *Manchester Lit. and Phil. Memoirs*, LXVIII (1923–24), 97–148. The Society was strongly influenced by Warrington Academy, several of its leading members, such as Thomas Percival and Thomas Barnes, being Warrington *alumni*. See H. McLachlan, *Warrington Academy: Its History and Influence* (Manchester, 1943), pp. 114–28.

[4] E. Percival, *Memoirs of the Life and Writings of Thomas Percival, M. D.* (1807), p. lxvii.

[5] *Manchester Lit. and Phil. Memoirs*, I (1785), xii–xiii.

[6] A. A. Mumford, *The Manchester Grammar School 1515–1915* (1919), p. 218. This statement is borne out by the membership lists in the *Memoirs*.

The attempt to build up a provincial culture based on the study of modern science was undertaken in a spirit of both utility and idealism. Thomas Henry pointed out in Manchester that 'several branches of Natural Philosophy seem peculiarly adapted to fill up the vacant hours in which the tradesman can withdraw from his employments. A general knowledge of all will tend to open and enlarge his understanding .... While the study of some, *in particular* [such as mechanics, hydrostatics, hydraulics, and chemistry], may .... supply him with a kind of information which he may turn to good account ....' [1] In a similar mood the Rev. Thomas Barnes expressed the hope that 'the happy art might be learned, of CONNECTING TOGETHER, LIBERAL SCIENCE and COMMERCIAL INDUSTRY'.[2] It is clear, however, from remarks by both Henry and Barnes that there was considerable antipathy among tradesmen in Manchester to liberal education, but this antipathy was being combatted. It is true that Thomas Henry was an apothecary and that Barnes was a religious minister, but a large number of their fellow members were men of business. John Barrow, John Drinkwater, the Philips, George Walker and John Wilson as well as several others were engaged in trade.

The dividing line between professional man and manufacturer was not so marked then as it is today. Thomas Henry, F.R.S., an apothecary, developed into a manufacturing chemist, and though he trained his sons as doctors brought them into business alongside him. James Dinwiddie, A.M., lecturer in natural philosophy, appears to have entered the calico-printing business.[3] Thomas Cooper, who failed to become a F.R.S. though his certificate was signed by Priestley and Watt, contributed his chemical knowledge to the bleaching firm of Baker and Cooper. Charles Taylor, the Manchester dyer, became Secretary to the Society for the Encouragement of Arts, Commerce and Manufactures.[4] Men of chemical knowledge quickly put it to practical test in the bleaching and dyeing trades.[5] The result was that in the Manchester Literary and Philosophical Society there was from the beginning an easy exchange of opinion, and in its rooms a manufacturer could consult with half-a-dozen Fellows of the Royal Society among the Manchester members, without taking the Corresponding Members into account. In the very first published list of members the names of Charles White, Thomas Henry, Thomas Butterworth Bayley, and Thomas Percival appear, all of whom were Fellows of the Royal Society,[6] and there were more to come in a few years.

As a result of the papers read to the Manchester Literary and Philosophical Society by Thomas Henry and the Rev. Thomas Barnes, it was decided to establish a 'College of Arts and Sciences' in Manchester in 1783, apparently the

[1] Thomas Henry, 'On the Advantages of Literature and Philosophy in General, and especially on the consistency of Literary and Philosophical with Commercial Pursuits', read October 3, 1781, *Manchester Lit. and Phil. Memoirs*, I (1785), 7–29.

[2] Thomas Barnes, 'A Plan for the Improvement and Extension of Liberal Education in Manchester', read April 9, 1783, *Manchester Lit. and Phil. Memoirs*, II (1785), 16–29. See also his paper, 'On the Affinity subsisting between the Arts, with a Plan for promoting and extending Manufactures by Encouraging those Arts, on which Manufactures principally depend', read January 9, 1782, *Manchester Lit. and Phil. Memoirs*, I (1785), 72-89.

[3] *Manchester Mercury*, 22 November and 6 December 1796. (Advt. re bankruptcy of Messrs. Dinwiddie and Bewicke.)

[4] Sir Henry Trueman Wood, *A History of the Royal Society of Arts* (1913), 334–35.

[5] We have found so much evidence of this that we propose to publish shortly another paper on applied chemistry in the Lancashire cotton industry during this period. Thomas Henry, F.R.S., was especially important in this field.

[6] *Manchester Lit. & Phil. Memoirs*, I (1783), xvi–xvii.

first institution of its kind in the country.[1] Barnes was the chief architect of this scheme. He stressed the great need for an institution of higher education, after elementary tuition at private or grammar schools. Here young tradesmen would be able to study '*Natural Philosophy*, the *Belles Lettres*, and *Mathematics*; together with some attention to *History*, *Law*, *Commerce*, and *Ethics*'. The course on Natural Philosophy 'should pay a very particular attention to CHEMISTRY and MECHANICS because of their intimate connection with our manufactures'.[2] This scheme was supported by the Literary and Philosophical Society.[3] It was 'principally accommodated to young men designed for a respectable line of *trade*', but would also provide a general education for those entering the professions. Lectures would be given in the evenings, 'so as not to interfere with the regular hours of business'. Those on natural philosophy would be illustrated by experiments on fire-engines, air-pumps, microscopes, etc., while the chemical lectures would have 'a reference to the arts of *Dyeing*, *Bleaching*, &c. which, depending upon chemical principles, might probably, by the knowledge of those principles, be very greatly extended and improved'.

The 'College of Arts and Sciences' was instituted at Manchester on 6 June 1783.[4] Its president, Thomas Percival, and eight governors were all members of the Literary and Philosophical Society, and so were all the lecturers: Henry Clarke on Practical Mathematics, Natural and Experimental Philosophy, and Geography; Thomas Henry on Chemistry with reference to Arts and Manufactures; George Bew on the Theory and History of the Fine Arts; and the Rev. Thomas Barnes on the Origins, History, and Progress of Arts, Manufactures and Commerce, Commercial Law, etc. But the College – started with such high hopes – had a brief existence, lasting only until 1787–8.[5] The most successful of the science lectures, 'delivered .... to numerous and attentive audiences', were those of Thomas Henry on chemistry, which included 'a course on the Arts of Bleaching, Dyeing, and Calico-Printing; and to render this course more extensively useful, the terms of access to it were made easy to the superior class of operative artisans'.[6] John Banks also lectured at the college on natural philosophy.[7] Charles White, F.R.S., and his son Thomas White, M.D., delivered lectures on anatomy.[8] The causes of the College's demise are not clear, but are said to have included 'a superstitious dread of the tendency of science to unfit young men for the ordinary details of business'.[9]

1 In addition to the papers of Henry and Barnes, previously cited, see the first report of the College, 'College of Arts and Sciences, Instituted at Manchester, June VI, MDCCLXXXIII' (Manchester Central Reference Library, S. & A. 94/42. Reprinted in *Manchester Lit. & Phil. Memoirs*, II (1785), 42–46); *Manchester Mercury*, 29 July, 30 Sept. 1783; E. Percival, *op.cit.* pp. lxx–lxxi; W. C. Henry, 'A Tribute to the Memory of the Late President [Thomas Henry] of the Literary and Philosophical Society of Manchester', *Manchester Lit. & Phil. Memoirs*, 2nd ser. III (1819), 224–6. There are brief accounts of the College in R. Angus Smith, *op. cit.* pp. 104 ff.; J. Thompson, *The Owens College* (1886), pp. 7–10; W. Bowden, *Industrial Society in England towards the end of the Eighteenth Century* (New York, 1925), pp. 45–6; H. McLachlan, *op. cit.* p. 123.

2 *Manchester Lit. & Phil. Memoirs*, II (1785), 22.

3 *Ibid.* 29–41.

4 *Manchester Mercury*, 29 July 1783, and the College's first annual report, cited above.

5 *Manchester Mercury*, 28 August, 25 September, 2 October 1787. James Watt junior attended lectures by Charles White as late as December 1788. J. Watt, jun., to J. Watt, 7 December 1788. Doldowlod MS. Cf. Percival, *op. cit.* p. lxxi.

6 W. C. Henry, *op. cit.* p. 226. Thomas Henry continued his lectures 'long after' the College's collapse.

7 See below, pp. 232–3.

8 *Manchester Mercury*, 10 and 31 October 1786. See also 'A Discourse delivered at the Commencement of the Manchester Academy, September XIV, MDCCLXXXVI. By Thomas Barnes, D.D.', p. 25.

9 W. C. Henry, *op. cit.* p. 225.

The College of Arts and Sciences was a non-denominational institution which has sometimes been confused with the Manchester Academy, or 'New College', forerunner of the modern Manchester College, Oxford, founded by local Dissenters on 22 February and opened on 14 September 1786.[1] The Manchester Academy was formed in order to save something from the wreckage of the Warrington Academy.[2] Hackney College, London, received the scientific equipment from Warrington, while the library, of about 3,000 volumes, was inherited by Manchester Academy. The most important figures in the promotion of the Manchester Academy were Thomas Percival, president of the Literary and Philosophical Society, Dr Thomas Barnes, co-secretary of that society, and the Rev. Ralph Harrison, another member. These three, however, involved the Society too far in the projected academy, especially by their suggestion in the trustees' first report that the Literary and Philosophical Society supported the foundation of the new college. On 15 March 1786, in a statement signed by Thomas Barnes and Thomas Henry as secretaries, the Society, *as such*, declared its independence of the Manchester Academy on the grounds that it had always maintained its independence of any religious opinion or sect.[3] Nevertheless individual members of the Literary and Philosophical Society continued to be prominent in the Manchester Academy. Barnes and Harrison were appointed professors of theological and classical studies and were to be responsible for choosing a professor of mathematics and natural philosophy. Most of the officers and committee were members of the Literary and Philosophical Society. Many of the founder members were also former pupils of Warrington Academy or other dissenting academies.

The main purpose of the Academy was to train candidates for the Dissenting ministry, but it would also prepare students who wished to enter the medical and legal professions and give a general education to those 'designed for Civil and Commercial life'.[4] In regard to the third category, Barnes proceeded to 'imagine .... a system of education, for a commercial man, which shall contain all the parts of science proper for him to know, as much as possible in a practical form'. The first tutor in mathematics and natural philosophy at the Academy was Thomas Davies, appointed 1787, followed in 1789 by Francis Nicholls. He was succeeded in 1793 by the famous scientist, John

[1] Examples of this confusion are in Hans, *op. cit.* p. 159 and E. Halevy, *History of the English People in the Nineteenth Century* (revd. ed. 1949), I, 561.

[2] See the trustees' report on its institution, 22 Feb. 1786 (Manchester Central Reference Library, S. & A. 94/48), which is reprinted, together with later reports, in a booklet devoted mainly to two sermons on the Academy by Dr Thomas Barnes and the Rev. Ralph Harrison, ministers at Cross Street Chapel, Manchester. (Volume entitled *Manchester Academy* in Manchester Cen. Ref. Lib., ref. no. 378.42/H2.) These sermons were 'A Sermon Preached at the Dissenting Chapel in Cross Street, Manchester, March XXVI, MDCCLXXXVI, on the Occasion of the Establishment of an Academy in that Town. By Ralph Harrison', and 'A Discourse delivered at the Commencement of the Manchester Academy, September XIV, MDCCLXXXVI. By Thomas Barnes, D.D.' The latter was reprinted in 1879 under the ambiguous title *Free Teaching and Free Learning in 1786* (Manchester, 1879.) See also *Manchester Mercury*, 21 March, 18 July, 12 Sept. 1786, 3 April 1787, 16 Dec. 1788, 6 July 1790; *The Monthly Magazine*, IV (July–Dec. 1797), 105–7, containing a report of 'the Academical Institution, or New College, at Manchester', reprinted in J. Harland, *Collectanea Relating to Manchester and Its Neighbourhood* (Manchester, 1867); E. Percival, *op. cit.* pp. lxxvii–lxxxi; F. Nicholson, *op. cit.* p. 140; R. Wade, *The Rise of Nonconformity in Manchester* (Manchester, 1880), pp. 148–9; McLachlan, *op. cit.* pp. 119–20, 128–37, and *English Education under the Test Acts* (Manchester, 1931), pp. 255–62. There is a good modern history of the Academy by V. D. Davis, *A History of Manchester College* (1932), but this shows little appreciation of the Academy's early significance as a training ground for industry and trade.

[3] *Manchester Mercury*, 18 July 1786; Nicholson, *op. cit.* p. 140; McLachlan, *op. cit.* pp. 119–20.

[4] Barnes's 'Discourse'.

Dalton, who had previously been running a school at Kendal.[1] In 1794, when Dalton became a member of the Literary and Philosophical Society, he is said to have had twenty-four pupils for mathematics, mechanics, geometry, algebra, bookkeeping, natural philosophy, and chemistry.[2] He left the Academy in 1800, to engage in private teaching. In that year he was elected secretary of the Literary and Philosophical Society, by which time he had begun to publish important scientific papers.

When Barnes resigned as principal of the Academy in 1798, he was succeeded by George Walker, F.R.S.,[3] who had been a student at Kendal Academy under Caleb Rotheram,[4] then at Edinburgh and Glasgow Universities. He had been a tutor in mathematics and natural philosophy at Warrington Academy in 1772–4, after which he became minister of High Pavement Chapel, Nottingham. A distinguished mathematician as well as theologian, he was elected president of the Manchester Literary and Philosophical Society on Percival's death in 1804.

Up to Barnes's retirement in 1798, the total number of students who had entered the Academy was 137,[5] of whom 89 were destined for commerce and industry. Even before Barnes's resignation, however, the Academy was in difficulties, which finally resulted in its transference to York in 1804.

II

Some impression of the intellectual life available to a young businessman in Manchester in the late eighteenth century can be seen from the correspondence of James Watt junior, though it must be admitted that he was a rather exceptional individual. He was apprenticed to the firm of Maxwell and Taylor, fustian manufacturers and dyers; the Taylor was Charles Taylor to whom we have already referred.[6] The introduction to this firm in 1788 seems to have been made in the first place by Thomas Henry who had been corresponding with Watt about bleaching. In a short while the young Watt struck up friendships with Thomas Henry's sons, with Joseph Priestley junior, who was also working in Manchester, with John Ferriar, an assistant of Thomas Percival, with Thomas Cooper, the bleacher, and with the merchants, Thomas and Richard Walker. By November 1788 he was attending meetings of the Literary and Philosophical Society, and read extracts from his translation of a German treatise by Professor Winterl[7] even before he was made a member, in February 1789.[8] He was also encouraged by James Keir in his translation

1 On Dalton see W. C. Henry, *Memoirs of the Life and Scientific Researches of John Dalton* (1854); R. Angus Smith, 'Memoir of John Dalton, and History of the Atomic Theory', *Manchester Lit. & Phil. Memoirs*, 2nd ser. XIII (1856); H. Lonsdale, *John Dalton* (1874); H. E. Roscoe, *John Dalton* (1901); J. P. Millington, *John Dalton* (1906); H. McLachlan, 'John Dalton and Manchester (1793–1844)', *Manchester Lit. & Phil. Memoirs*, LXXXVI (1943–45), 165–77; F. Greenaway, 'The Biographical Approach to John Dalton', *ibid.* (1958–59), 1–98.

2 R. Angus Smith, 'Memoir of John Dalton', p. 18.

3 Davis, *op. cit.* pp. 67–9.

4 On Kendal Academy and the Rotherams, see below, p. 231.

5 *Roll of Students entered at the Manchester Academy* (Manchester, 1868); Percival, *op. cit.* p. lxxxi; McLachlan, *Warrington Academy*, pp. 136–7. Davis, *op. cit.* p. 62, followed by McLachlan, 'John Dalton and Manchester', p. 171, gives 135, but appears to have overlooked two students at the end of the official *Roll*.

6 See above, p. 224.

7 James Watt, junior (Manchester) to James Watt (Birmingham), 27 November 1788 and 7 December 1788, Doldowlod MS. (Dol.)

8 James Watt, junior (Manchester) to James Watt (Birmingham), 24 February 1789, Dol.

of a work by Christophe Meiners, Professor of Philosophy at Göttingen,[1] and he was visited by several foreign scientists while in Manchester. At this time he helped Keir with an article on bleaching for his *Dictionary of Chemistry* (1789), to which his master, Charles Taylor, also contributed articles on cotton dyeing and calico printing.[2]

In 1789 Watt junior was appointed joint secretary of the Literary and Philosophical Society with John Ferriar. Part of the society's programme was to establish relationships with foreign scientists and scientific societies. Presumably with this aim in view, Watt wrote a letter of congratulation to the famous German scientist and editor, Lorenz Crell, and at the same time sent him copies of two papers he had published in the *Memoirs* of his own society. Crell accepted the compliments and the papers for publication in his *Annalen* and asked for a regular scientific correspondence to be maintained between himself and the Manchester society.[3] Watt was eventually instrumental in securing Crell's election as a Corresponding Member of the Society.

Apart from establishing connexions with foreign scientists, young Watt did something to bring foreign publications on science and engineering to the knowledge of his Manchester friends. He wrote to his father in 1789, saying: 'I am glad to hear that Messr^s Lavoisier, Bertholet & Morveau intend translating Crells Annals it will be doing an essential service to Chemistry & will, I have no doubt, amply repay their trouble. Mr [Charles] Tayler & Mr [Thomas] Henry wish to subscribe for one copy each & Mr [Thomas] Cooper for two, one for himself and one for his partner, Mr[Joseph] Baker. These are all our chemical men here, Dr Percival not professing any partiality to that science ....'[4] On another occasion, George Lee, who had been manager of Drinkwater's mill and later became a partner in the cotton-spinning firm of Philips and Lee, wrote to James Watt junior in Paris:

> 'if in the course of yr travels you meet with any good or curious Books upon Mechanics and it does not incommode you materially I wish you would purchase them upon my Account'.[5]

In the same letter he delivered a message from Peter Ewart, the Manchester millwright, who wanted to buy a copy of '*Architecture Hydraulique par Monsr Porny*'.[6] Even before he left Manchester, Watt had been buying German scientific books for his father. Watt's connexions with Manchester ended for a time in February 1792, when he left for Paris, together with Thomas Cooper, as a delegate of the Manchester Constitutional Society to the Club des Jacobins.

It must not be thought that young men like Watt and Joseph Priestley junior were the only men in Manchester who had contacts with French and German scientists. Charles Taylor described to James Watt, senior, a day he spent in Berthollet's company,[7] and when he applied for the post of secretary to the Society for the Encouragement of Arts, Commerce and Manufactures in 1799 he stated to the committee that he was known to almost all the chemists in

[1] James Keir (Smethwick) to James Watt, junior (Manchester), 20 November 1789, A.O.L.B.
[2] *Ibid.* 10 February 1790, A.O.L.B.
[3] Lorenz Crell (Helmstedt) to James Watt, junior (Manchester), 26 January 1790, Dol.
[4] James Watt, junior (Manchester) to James Watt (Birmingham), 1 February 1789, Dol. Taylor, Henry, Cooper, and Baker played leading rôles in the application of chemistry to chlorine bleaching, dyeing, and calico printing in Lancashire.
[5] George Lee (Manchester) to James Watt, junior (Paris), 18 April 1792, M.N., Birmingham Reference Library (B.R.L.).
[6] On Ewart and Lee, see below, pp. 229–30
[7] Charles Taylor (Manchester) to James Watt (Birmingham), 21 June 1789, Dol.

Europe.[1] Thomas Henry, in a paper on dyeing, refers to similar publications by foreign chemists such as d'Apligny, Macquer, Hellot, Berthollet, and many others.[2] In 1776 he translated some of Lavoisier's works, *Essays, Physical and Chemical*, and in 1783 a further selection of the same writer's *Essays on the effects produced by various processes on atmospheric air*. Adam Walker, the Manchester schoolmaster and lecturer, had several friends among French scientists, and Henry Clarke was of international repute.[3] In 1794 John Dalton at the Manchester Academy is said to have been teaching science to his pupils from Lavoisier's *Elements of Chemistry* and the similar work of Chaptal.[4] The *Memoirs* of the Literary and Philosophical Society teem with references to the works of foreign chemists, while the Manchester *Memoirs* were themselves translated into French and German.

It may be felt, however, that our argument has not yet been sufficiently concerned with practical men of business of the sort who must have formed the majority in Manchester's industrial life at this time. Let us examine three such men. The obituary of Thomas C. Hewes (1768–1832), 'a machine maker and general millwright', who settled in Manchester in 1792, describes him as 'a man of science' with 'eminent talents in the various branches of the mechanical art'. He is said to have had 'a valuable knowledge of mechanics in general' and 'a particular acquaintance with hydrostatics' and to have 'studied with most satisfactory results the effects of currents of wind as a means of power'. Hewes planned and erected mills, made mill-gearing and spinning machines, and constructed water-wheels all over the country. He was famous for his improvements to water-wheel machinery, especially for the introduction of the suspension wheel.[5] Such a man evidently had a sound knowledge of the mechanical principles underlying the machines he built.

A rather more celebrated millwright than Hewes was Peter Ewart (1767–1842), a pupil of John Rennie and later an employee of Boulton and Watt, who established himself in Manchester in the 1790's first of all as an engineer and then as a cotton manufacturer. He was specially interested in theoretical mechanics and he was 'sufficiently conversant with mathematics to read with facility most works, which were then the standard authorities in mechanics'. He was 'profoundly learned in all that has reference to steam', both in theory and in practice. He had a close friendship of nearly half a century with John Dalton and published a paper in the *Annals of Philosophy*, vol. VI, on Dalton's 'Theory of Chemical Composition'. He was also a friend of Thomas Henry and one of the executors of his will. He was elected a member of the Manchester Literary and Philosophical Society in 1798 and was a vice-president from 1812 to 1835 when he left the Town. He read several papers to the Society, chiefly on mechanical theory.[6]

George A. Lee (1761–1826), of the great Manchester cotton-spinning firm of Philips and Lee, is said to have been 'imbued with a love of the sciences, and

[1] Sir H. T. Wood, *A History of the Royal Society of Arts* (1913), p. 335.

[2] T. Henry, 'Considerations relative to the Nature of Wool, Silk, and Cotton, as Objects of the Art of Dyeing', *Manchester Lit. & Phil. Memoirs*, III (1790). A translation of the French works mentioned was advertised in the *Manchester Mercury*, 30 June 1789.

[3] For Walker and Clarke, see below, pp. 231–2, 237.

[4] R. Angus Smith, 'Memoir of John Dalton', p. 18. Dr William Henry, of Manchester, also recommended these and other French works in his *Epitome of Chemistry* (1801).

[5] *Manchester Guardian*, 11 February 1832. Hewes had one of the biggest engineering works in the country in 1824, when he gave evidence before the Select Committee on Artizans and Machinery (*Fourth Report*, pp. 340–50). See J. H. Clapham, *An Economic History of Modern Britain* (1939), I, 355. His name is misspelt 'Herves'.

[6] Obituary notice in *Manchester Lit. & Phil. Memoirs*, Second Series, VII (1846).

was .... remarkable for the extent and precision of his acquirements in them'. It was for this reason that he was so perceptive of 'the advantages to be derived from applying to useful purposes the great inventions that distinguished the era in which he lived'.[1] The spinning machinery in his mills, constructed under his supervision, was remarkable for its 'correct and excellent workmanship'. He was on friendly terms with James Watt and Matthew Boulton and also with Peter Ewart, and was 'fully sensible of the advantages of the steam engine'. He soon made himself 'master of the abstrusest parts of its theory', with the result that his firm's engines were 'the finest specimens of perfect mechanism'. He was the first to improve on the fire-proof mills of his friend William Strutt, by the employment of cast-iron beams,[2] and he was among the first to employ steam for warming mills. He was friendly with Boulton and Watt's foreman, William Murdoch, whose invention of coal-gas lighting he immediately utilized, Philips and Lee's mill being the first (1805) in Manchester to do so;[3] and he gave Dr William Henry assistance in his chemical researches to improve the process.[4]

These men – Hewes, Ewart, and Lee – were among the leading engineers and cotton spinners in Manchester. But even a common engine erector like Isaac Perrins was not an absolutely unlettered man, as is evident from the style in which Boulton and Watt wrote to him: 'From long experience & knowledge of Steam Engines we doubt not but you are fully competent to discriminate between a variation of form & a variation of principle in the construction of an Engine ....'[5] Even the ordinary millwright in the eighteenth century, so William Fairbairn tells us, was generally 'a fair arithmetician, knew something of geometry, levelling, and mensuration, and in some cases possessed a very competent knowledge of practical mathematics. He could calculate the velocities, strength, and power, of machines: could draw in plan and section ....'[6] Indeed, Fairbairn states that millwrights were commonly regarded as men of 'superior attainments and intellectual power': they were by no means unlettered rude mechanics.

III

One is naturally prompted to ask how these men acquired such scientific and technical knowledge. They were undoubtedly influenced to some extent by the itinerant lecturers in natural philosophy who regularly visited the principal

[1] Obituary notice in *Manchester Guardian*, 12 August 1826 and *Gentleman's Magazine*, n.s. XCVI, pt. ii (July-December 1826), 281–2. Robert Owen described Lee as 'one of the most scientific men of his day'; he had been 'highly educated', and was 'a finished mathematician'. He 'possessed great talent as a scientific machinist and engineer'. He had been manager of Drinkwater's mill, the first Manchester mill (1789) to instal one of Watt's rotary engines. R. Owen, *The Life of Robert Owen. Written by Himself* (1857), I, 26–8, 32. See above, p. 228, for his request to James Watt junior in Paris to acquire for him 'any good or curious Books upon Mechanics'.

[2] The Salford mill of Philips and Lee (1799–1801) was one of the earliest iron-framed buildings. See T. C. Bannister, 'The First Iron-framed Buildings', *Architectural Review*, Vol. 108, April 1950, 231–46; H. R. Johnson and A. W. Skempton, 'William Strutt's Cotton Mills 1793–1812', *Newcomen Soc. Trans.* Vol. 30 (1956), 179–205.

[3] *Philosophical Transactions*, February 1808.

[4] W. Buckley and A. McCulloch, 'A Record of Some Early Experiments on the Carbonisation of Oil and Coal, by Dr. W. Henry, F.R.S. (1808–1824)', *Manchester Lit. & Phil. Memoirs*, LXXIV (1929–30), 69.

[5] Boulton and Watt to Isaac Perrins, 7 January 1795 (should be 1796), Boulton and Watt Letter Book (Office), September 1795 to August 1796, Birmingham Reference Library.

[6] W. Fairbairn, *Treatise on Mills and Millwork* (1861–3), I, v–vi.

towns in the later eighteenth century.[1] In Manchester, for example, men like Caleb Rotheram, master of Kendal Dissenting Academy, and his son, John Rotheram, James Ferguson, James Arden, Adam Walker and his son, of the same name, John Warltire, Henry Moyes, John Banks, Thomas Garnett, Gustavus Katterfelto, and many others gave lectures on science and mechanics, besides those lecturers we have already mentioned in connexion with the College of Arts and Sciences and the Manchester Academy.[2] Most of these lecturers were itinerant. Some of them were nationally famous in their day; others were rather more local figures, but their importance in the general pattern of scientific education has not been generally recognized. Several of them took with them on their tours quite large collections of working models and scientific apparatus. In addition, they usually published their lectures in book form, or if they did not publish the complete lectures, they printed and sold a synopsis of them. Some of the earliest lectures in Manchester on 'experimental philosophy', those delivered in 1743 by Caleb Rotheram, of Kendal Academy, and his son John, are still preserved in manuscript in Chetham's Library.[3] It is worth recalling that one of the greatest engineers of the age, John Wilkinson, was educated at Kendal Academy under Caleb Rotheram, and so were George Walker, principal of the Manchester Academy from 1798 and president of the Manchester Literary and Philosophical Society, 1804–07, and John Banks, one of Manchester's lecturers on science at the College of Arts and Sciences.

One of the most famous itinerant lecturers of the century, Adam Walker (1731–1821),[4] had close connexions with Manchester, where he established a school in January 1762.[5] He is said to have 'conceived a system of education more adapted to a Town of Trade than the Monkish system still continued in our Public Schools'.[6] His ideas were widely approved and 'many of the first people in that town' are said to have benefited from his tuition during the five years (1762–66) which he spent in Manchester. He gave several public lectures there on subjects of natural philosophy, especially on astronomy. Then in 1766 he purchased 'the celebrated Philosophic Apparatus' of another lecturer, a Mr Griffith,[7] and thereafter proceeded to travel throughout the north of England, southern Scotland, and Ireland, lecturing on 'Mechanics, Hydro-

1 The influence of these men has been seriously underestimated by economic historians, e.g. W. H. B. Court, *The Rise of the Midland Industries, 1600–1838* (1938), p. 225: 'A sign of this side of the town's [Birmingham's] life was the occasional giving of scientific lectures by visitors, whose qualifications, however, are not always apparent .... The circle of people affected by such lectures would be exceedingly small, the effects on invention probably negligible ....' The best general account of the lecturers is given by N. Hans, *op. cit.* There are also important articles on individual lecturers such as: Douglas McKie, 'Mr. Warltire, a good chymist', *Endeavour*, X (1951), and J. A. Harrison, 'Blind Henry Moyes, "An Excellent Lecturer in Philosophy",' *Annals of Science*, XIII (1957), 109–125. See also D. Layton, 'Newton's Effect on English Thought', *History Today*, VII (1957), 388–95.

2 We have compiled a long list (omitted here for reasons of space) from the *Manchester Mercury*, which shows that almost every year for the last forty years of the eighteenth century Manchester was visited by one and sometimes by several of these lecturers. Each usually gave a course, occasionally two courses, of lectures, numbering sometimes as many as thirty, spread over several weeks or even months.

3 Chetham's Library, Mun. A. 2.68, and Mun. A. 2.81.

4 For accounts of Walker's life see *European Magazine, and London Review*, XXI (Jan.-June 1792), 411–3; *Gentleman's Magazine*, XCI, pt.i (Jan.-June 1821), 182–3; *D.N.B.*; Bowden, *Industrial Society in England*, pp. 20–21; Hans, *op. cit.* pp. 146–8.

5 *Manchester Mercury*, 26 January 1762.

6 *European Magazine*, XXI, 411. The subjects taught in his school included English, writing, accounting, mathematics, book-keeping, drawing, and geography (*ibid.* 412).

7 *Manchester Mercury*, 21 January 1766.

statics, Pneumatics, Chemistry; on Optics and Astronomy; and on Magnetism, Electricity, and the general Properties of Matter'.[1] After several years' residence firstly in Ireland and then in York, he established himself as a lecturer in London. He was a friend of Joseph Priestley, Henry Moyes, and other scientists. But he was not merely an academic lecturer, for he had a long string of practical inventions to his credit.[2] He was sufficiently well versed in engineering to be consulted by the Richmond Water Company when they were considering whether to instal a Boulton and Watt engine in 1780,[3] and three years earlier Boulton had even thought it worth while to sign an article of agreement with Walker for an option on one of Walker's inventions.[4] Unfortunately the invention is not described, but we know that Walker invented a steam engine, a drawing of which appears in his published lectures and another in the Boulton and Watt Collection at the Birmingham Reference Library.[5] When he lectured in Birmingham in 1781 he dined with Boulton, who expected that he would 'set people a talking about Engines'.[6] Both Adam Walker and his son lectured frequently in Manchester, where they were clearly held in high esteem. Walker senior published his lectures, which had gone into eight editions by 1792.[7]

John Banks, another Manchester lecturer in natural philosophy, stated in the preface to his *Treatise on Mills* (1795) that this work was based largely on the public lectures he had been delivering for twenty years. The book did not attempt to teach the mechanic how to perform the detailed engineering work, but examined the general principles of mechanics and hydrostatics. In justifying the publication of his book, Banks referred to the condition of mechanical knowledge in the country at that time:

> 'It is true, that we have in the kingdom many intelligent engineers, and excellent mechanics; and there are others who can execute better than they can design, otherwise there would not have been so much money expended in attempting what men of science know to be impossible.'

The *Treatise on Mills*, together with his earlier *Epitome of a course of lectures* (1775) and his book, *On the Power of Machines* (Kendal, 1803), which describes many experiments on steam engines and on the strength of materials, did much to inform those who could 'execute better than they can design' by discussing engineering problems in a realistic manner.

There are several references in the *Manchester Mercury* to Banks's lectures in the last quarter of the eighteenth century, and we know also that he lectured in Birmingham, Doncaster and Kendal. By 1785 his lectures on natural philosophy in Manchester were being given under the auspices of the College of Arts and Sciences, and when he lectured in Doncaster he advertised himself as 'Lecturer in Philosophy to the College of Arts and Science in Manchester'.[8] John Kennedy, the great Manchester cotton spinner, refers to Banks in his 'Early Recollections', where he tells how, at the age of fourteen, he (Kennedy)

1 *European Magazine, loc. cit.*

2 *Ibid.* and *Gentleman's Magazine, loc. cit.*

3 M. Boulton to J. Watt, 26 December 1780, B.R.L.

4 M. Boulton to J. Watt, 26 May 1777, B.R.L.

5 Parcel E, Boulton & Watt Collection, B.R.L.

6 M. Boulton to J. Watt, 28 July 1781, Matthew Boulton Letter Book, 1780–83, A.O.L.B.

7 *European Magazine, loc. cit.* The second edition of his *Analysis of a course of lectures*, published at the request of his students in Manchester, appeared in 1770. The date of the first edition is not known. Hans, *op. cit.* p. 147.

8 *Yorkshire Journal and General Weekly Advertiser*, 10 and 24 March 1787. We are grateful to Mr J. A. Harrison for this reference.

was apprenticed with Adam Murray to the firm of Cannan and Smith, machine makers, at Chowbent in Lancashire, and was brought down from Scotland by James Smith, one of his masters. On the journey, Smith had some calls to make:

> 'amongst others, on Mr. Banks, a lecturer on mechanics and natural philosophy, whom he advised to come to Chowbent, where he would strive .... to get him a sufficient audience to attend his lectures. I remember Adam Murray and I had a ticket between us, each paying half-a-crown, and we attended alternately. This was to us a wonderful insight into the laws of nature'.[1]

The price, five shillings, for the course of lectures was evidently not prohibitive even for two apprentices, so that the lecturers at least to some extent anticipated the work of the Mechanics' Institutes. Leaders of science and industry often recommended lecturers to each other so that an audience was guaranteed at the next port of call.[2]

Banks's *Treatise on Mills* was subscribed to by a most impressive array of scientists, engineers and professional men.[3] Such subscription lists for the works of itinerant lecturers show the wide interest in lectures on mechanics and science. A similar impression is formed from the study of the membership lists of the philosophical societies of the day – from the great societies of Birmingham and Manchester to those in market towns. The subscribers to the books and the members of the philosophical societies were largely the same people. Quite often the lecturers were themselves members of scientific societies.

The itinerant lecturers were varied in their educational backgrounds and social antecedents. Many of them, such as the two Rotherams, Warltire and Moyes, had degrees from Scots universities where scientific research was most alive at this period, and they may also have attended one of the Dissenting academies; others, like Adam Walker and James Ferguson, are said to have been almost entirely self-educated. Nearly all of them had strong practical interests and often had one or two minor inventions to their credit; some had watch-making or mathematical instrument making as their principal occupation. Peter Clare, the Manchester clock-maker, was one of the latter type. He lectured in Manchester from about 1772. In 1778 he proposed to establish 'a Philosophical School in which young Persons will be instructed in the general principles of Mechanics, Hydrostatics, Pneumatics, Electricity, Chemistry, &c ....',[4] but he also continued to give public lectures, especially on electricity. Clare was a very close friend of John Dalton, in fact his closest companion and secretary in later years, and was an executor of Dalton's will.[5] He also

[1] J. Kennedy, *Miscellaneous Papers* (Manchester, 1849), p. 14.

[2] Joseph Priestley (Calne) to Matthew Boulton (Birmingham), 28 September 1776, A.O.L.B., introducing John Warltire; John Roebuck (Bo'ness) to Matthew Boulton (Birmingham), 30 October 1777, Dol., recommending Henry Moyes; and Thomas Walker (Manchester) to James Watt, junior (Birmingham), 8 May 1796, Box 3. W, B.R.L., recommending Dr Garnett.

[3] The list of subscribers shows the wide variety of people who were interested in buying such a book. The names included Richard Bradley, engineer, Manchester; David Broad, architect, Manchester; Peter Clare, watchmaker, schoolmaster and lecturer, Manchester; John Dalton; William Hirst, engineer, Gomersal; William Kirk, watchmaker, Manchester; John Knowles, teacher of mathematics, Liverpool; Robert Owen; Joseph Priestley; John Raistrick, Low-moor furnace, Leeds; John Rennie; Thomas Rider, ironfounder, Salford; John Samuels, engineer, Manchester; Michael Satterthwaite, Ulverston, and Benjamin Satterthwaite, Lancaster; William Sharrett [Sherratt], engineer, Manchester; John Southern, Birmingham; Thomas and Richard Walker, merchants, Manchester; and John Wilkinson.

[4] *Manchester Mercury*, 6 January 1778.

[5] H. McLachlan, 'John Dalton and Manchester', p. 177; R. Angus Smith, 'Memoir of John Dalton', p. 277.

became one of the secretaries of the Literary and Philosophical Society.

Another Manchester clock- and watch-maker, as well as optician, was John Imison.[1] In 1785 he published *The School of Arts*; *or, An Introduction to Useful Knowledge*,[2] which dealt with such 'Branches of Practical Science' as mechanics, electricity, optics, astronomy, instrument-making, clock- and watch-making, and other crafts. The first section was separately published in 1787, as *A Treatise on the Mechanical Powers*,[3] to which Imison had added items on mill-work, including the 'Construction of Water Mills', which he hoped would prove useful to 'those for whose information they are principally intended, viz. working mechanics'. This revised section on mechanical powers was included in a second edition of *The School of Arts*, published about 1794,[4] which contained additional sections on pneumatics, hydrostatics, and hydraulics, including pumps and steam engines. Imison also published other works, including *Elements of Science and Art*.[5] The instrument-makers and clock-makers like Imison and Clare thus played a very important rôle in diffusing scientific knowledge, by making apparatus for lecturers, sometimes by demonstrating the experiments or lecturing themselves, by running schools, and by publishing books on mechanical and scientific subjects.[6]

The subscription fee for a course of lectures might be anything from a few shillings to a guinea, according to the duration of the course and the popularity of the lecturer. No doubt professional men and some of the more prosperous manufacturers tended to make up the bulk of the audiences, but, as we have seen, craftsmen like John Kennedy and Adam Murray also attended, before they had risen into the manufacturers' ranks. Practical men must have been attracted to the lectures by the continual attempts to adapt them to local needs:

> 'This Course is confessedly more adapted to every Branch of Business than any ever read in this Town ....'[7]
>
> 'If a sufficient Number wish to attend, a Lecture will be delivered on the application of Water to Bucket Wheels, accompanied with many Experiments ....'[8]
>
> '[Chemistry's] Application to the common Purposes of Life, as well as to the useful Arts of Bleaching, Dyeing, Agriculture, &c.'[9]

A good deal of the apparatus used also showed a sense of practicality in the lecturers. A list advertised by Mr Pitt in 1779, for example, included optical instruments, cranes, pile-drivers, sawing engines, pumps, steam engines, and many other items.[10] These machines could be seen by the public, free of charge. Most itinerant lecturers had a similar collection. Adam Walker, for example, had a number of working models of different steam engines, including a model of a Boulton and Watt engine.[11]

[1] Hans, *op. cit.* p. 151.

[2] There is a copy in the Manchester Central Ref. Lib. The work had been published in parts at an earlier date, under the title of *An Introduction to Useful Knowledge*. See *Manchester Mercury*, 30 Sept. 1783.

[3] Copy in the British Museum.

[4] Copy in Chetham's Library, Manchester. A second edition of *A Treatise on Mechanical Powers* was also published in 1794 (copy in Manchester Cen. Ref. Lib.).

[5] Revised edition (2 vols. 1803) in the British Museum.

[6] Hans, *op. cit.* pp. 150–2. Clock-making, as we have shown elsewhere, was important in the development of mechanical engineering. See A. E. Musson and E. Robinson, 'The Origins of Engineering in Lancashire', *Journal of Economic History*, June 1960, pp. 209–33.

[7] *Manchester Mercury*, 26 November 1771. Adam Walker's lectures.

[8] *Ibid.* 26 January 1790. John Banks's lectures.

[9] *Ibid.* 26 January 1796. Dr Garnett's lectures.

[10] *Ibid.* 16 March 1779.

[11] Matthew Boulton to James Watt, 30 July 1781, Box 20, B.R.L.

No one has yet studied in detail the wealth of books published in the eighteenth century on science and technology. The *Manchester Mercury*, in the second half of the century, printed many advertisements of such books as Benjamin Martin's *General Magazine of Arts and Sciences*;[1] *The New and Compleat Dictionary of Arts and Sciences*, by a Society of Gentlemen;[2] *The Circle of Sciences*, sold by A. Clark, bookseller, Manchester;[3] Charles Hutton's *Treatise on Mensuration*, 'adapted particularly to the Uses of Schools, Mathematicians and Mechanics';[4] William Nicholson's *Introduction to Natural Philosophy*;[5] B. W. Emerson's *The Principles of Mechanics*, described as being 'extremely useful to all sorts of Artificers, particularly to Architects, Engineers, Shipwrights, Mill-wrights, Watch-makers, &c or any that Work in a Mechanical Way';[6] A. Fletcher's, *The Universal Measurer and Mechanic*, 'a Work equally useful to the Gentleman, Tradesman and Mechanic';[7] William Casson's proposed book on fire engines which was to cover every side of the mechanical operation of steam engines;[8] John Imison's books which we have mentioned above;[9] Adam Walker's lectures;[10] William Henry's *A General View of the Nature and Objects of Chemistry And of its Application to Arts and Manufactures*,[11] and many others. Some provided elementary mathematical knowledge, such as J. Bonnycastle's *Scholar's Guide to Arithmetic* and his *Introduction to Mensuration and Practical Geometry*, which was adapted 'as much as possible to practical Uses .... Workmen and Artificers of all Denominations will find it a ready Assistant'.[12] To these must be added the many encyclopaedias and dictionaries which discussed machines and mechanical processes, often illustrating them with drawings. Books and dictionaries were sometimes published in parts so that they were within range of the artisan's pocket. Some were printed locally; others were printed in London for well-known publishers of scientific books such as Joseph Johnson, who was the publisher for at least three famous Manchester scientists at this period – Thomas Percival, Thomas Henry and John Dalton. In addition to scientific books, there was a great increase in scientific periodicals from 1780 to 1800.[13]

Those who could not afford to buy these books and periodicals could read many of them in Chetham's Library, Manchester, which had been founded in 1653, the first free public library in the country. John Dalton found this a great asset: 'There is in this town', he wrote in 1794, 'a large library, furnished with the best books in every art, science and language, which is open to all, gratis ....'[14] Examination of the library accessions book and of its first printed

[1] *Manchester Mercury*, 7 January 1755. For Martin, see Hans, *op. cit.* pp. 141, 145, 152, 206. Many of these books were advertised in the printed catalogue of the first large-scale booksellers of cheap books, *A Catalogue of Books, for the year 1804–5; Comprising upwards of 850,000 Volumes, in all Languages and Classes of Learning .... by Lackington, Allen, & Co. Temple of the Muses, Finsbury Square, London.*

[2] *Ibid.* 21 October 1755.

[3] *Ibid.* 30 December 1755.

[4] *Ibid.* 21 May 1771.

[5] *Ibid.* 3 December 1782.

[6] *Ibid.* 7 December 1773.

[7] *Ibid.* 7 December 1773.

[8] *Ibid.* 18 January 1785.

[9] See p. 234.

[10] *Manchester Mercury*, 23 July 1799.

[11] *Ibid.* 10 December 1799

[12] *Ibid.* 29 January 1782.

[13] Douglas McKie, 'The Scientific Periodical from 1665 to 1798', *Philosophical Magazine Commemoration Number*, 1948.

[14] Letter of Dalton dated 20 February 1794, quoted in Nicholson, 'The Literary and Philosophical Society, 1781–1851', p. 114.

catalogues[1] reveals that a considerable number of scientific works, both British and foreign, were acquired in the later eighteenth century. The Literary and Philosophical Society also began to build up its own library, which was particularly strong in works on natural philosophy. In addition, there was a subscription library established in the Exchange in 1757, which acquired many scientific works.[2] Another subscription library was founded in 1792, by the 'Manchester Reading Society', the first name on whose membership list was that of Joseph Priestley junior, who made a presentation of 85 volumes, including some of his father's works, which apparently formed the library's original nucleus.[3]

When all has been said, however, about philosophical societies, itinerant lecturers, and scientific books and periodicals, the strongest educational influence in the lives of many Manchester men must have been their schooling. There were at this period several schools and academies in and about Manchester which taught the principles of general science and provided a sound basis in mathematics. The most important of all, the Warrington Academy (1757–1786), has been described elsewhere.[4] Joseph Priestley, of course, was the most eminent of the Academy's tutors (1761–7), and though he taught languages and 'belles lettres' there he had an influence on scientific studies. An indication of the content of science teaching at Warrington is given by the *Institutes of Natural Philosophy, Theoretical and Experimental* (1785), by William Enfield, LL.D., tutor there from 1770 to 1786, who wrote this work on account of 'the difficulty which I met with, in providing my Classes with a [suitable] Text-book in Natural Philosophy'. It deals with the properties of matter, mechanics, hydrostatics, pneumatics, optics, astronomy, magnetism, and electricity. It is mathematical in emphasis, requiring 'a previous knowledge of the elements of Geometry, Trigonometry, the Conic Sections, and Algebra', but many parts of it are illustrated with experiments. The Academy's scientific apparatus, which was disposed of to Hackney College on the Academy's demise,[5] had been acquired from John Holt, one-time tutor there in natural philosophy, who had obtained it from Caleb Rotheram, of Kendal Academy, who, in turn, had purchased it from John Horsley (1685–1732), M.A., F.R.S., dissenting minister and lecturer in natural philosophy in Morpeth and Newcastle in the early eighteenth century.[6] Scientific apparatus was commonly handed on in this way. So, too, was scientific knowledge. Thus Enfield states: 'I have made use of the works of NEWTON, *Keil*, *Whiston*, *Gravesande*, *Cotes*, *Smith*, *Helsham*, *Rowning*, and lastly *Rutherforth*'.[7] Undoubtedly the origins of the developments we have been describing go back to the Scientific Revolution of

[1] *Bibliotheca Chethamensis*, I & II (Manchester, 1791) and III (Manchester, 1826). See also Mumford, *op. cit.* pp. 222–3.

[2] J. Aston, *A Picture of Manchester* (1816), p. 175, gives 1757, but the correct date may be 1765. See *A Classed Catalogue of the Manchester Subscription Library, instituted 1765* (Manchester, 1831), and W. E. A. Axon, *Handbook of the Public Libraries of Manchester and Salford* (Manchester, 1877), p. 178.

[3] Axon, *op. cit.* p. 61. *A Catalogue of the Manchester Subscription Library .... instituted Aug. 29th, 1792* (Manchester, 1834).

[4] See H. McLachlan, *Warrington Academy: its History and Influence* (Manchester, 1943); and William Turner, *The Warrington Academy* (reprinted from the *Monthly Repository*, 1813–1815), edited with an introduction by G. Carter (Warrington, 1957). See above pp. 223 *n.* 3 and 226–7 for its influence upon the Manchester Literary and Philosophical Society, the College of Arts and Sciences, and the Manchester Academy.

[5] See above, p. 226.

[6] See *D.N.B.* and McLachlan, *English Education under the Test Acts*, pp. 190, 228.

[7] For the works of these earlier natural philosophers and mathematicians, see *D.N.B.*

the late seventeenth century, and the dissenting academies played an important rôle in preserving and disseminating scientific knowledge.

Manchester was also fortunate in possessing a Grammar School which preserved in the eighteenth century a strong scientific and mathematical tradition.[1] The High Master from 1764 to 1807 was Charles Lawson, a keen mathematician, who promoted the study of his subject in the school and recruited the services of two former pupils of the school, Henry Clarke and Jeremiah Ainsworth. Several eminent scientists were educated at Manchester Grammar School at this time,[2] and it is also interesting to observe that the largest proportion of pupils came from among the artisans, shopkeepers and minor occupations.[3]

Henry Clarke was an important figure in local scientific education. After leaving Manchester Grammar School in 1756, he taught for some years in Leeds and then travelled on the continent, visiting several important mathematicians, among them Lorgna of Verona, whose works he later translated. He established his 'Commercial and Mathematical School', or 'Salford Academy', in Chapel Street, Salford, as early as 1765. In 1772 he began a night-school in Salford in a variety of subjects, including mathematics and natural philosophy, in which subjects he was appointed to lecture at the College of Arts and Sciences.[4] His school at Salford was still being advertised in December 1789.[5] In 1792 he moved to Liverpool and there established the 'Literary, Commercial and Mathematical School', Mount Pleasant (Martindale Hill).[6] But on 25 March 1794 he removed his school back to Manchester.[7] He particularly stressed in his advertisement that he prepared youths for 'the Scientific Parts of Mechanic Professions in general'. About 1797 he moved to Bristol,[8] and in 1802 he was appointed Professor of History, Geography and Experimental Philosophy at the Royal Military College, Marlow. His publications on mathematics were many and varied, and he was altogether a most distinguished scholar and teacher.

We have already referred to the schools of Peter Clare and Adam Walker.[9] These were not the only schools in Manchester with a mathematical and scientific basis. Jeremiah Ainsworth (1743–84), educated at Manchester Grammar School, ran a Mathematical School in Long Millgate for many years and took some of the grammar school boys for mathematical lessons.[10] Holt and Son's 'Mercantile and Mathematical School', Oldham Street, was described in 1788 as having been going for several years, and catered for both day and evening students.[11] In the same year R. Hartley's 'Commercial and Mathematical School', Princes Street, was advertised.[12] In 1796 John Fern opened a 'Mathematical Academy' at 26 Spring Gardens.[13] We can conclude therefore

[1] A. A. Mumford, *The Manchester Grammar School 1515–1915* (1919), p. 193; Hans, *op. cit.* pp. 39–41.

[2] Hans, *op. cit.* p. 40.

[3] Mumford, *op. cit.* p. 170.

[4] See above p. 225.

[5] *Manchester Mercury*, 29 December 1789.

[6] Hans, *op. cit.* p. 96.

[7] *Manchester Mercury*, 25 February 1794.

[8] Rev. George Heath, *The History, Antiquities, Survey and Description of the City and Suburbs of Bristol* (Bristol, 1797), p. 81.

[9] See above, p. 233 and p. 231.

[10] Mumford, *op. cit.* p. 197; Hans, *op. cit.* p. 111.

[11] *Manchester Mercury*, 1 January 1788.

[12] *Ibid.* 18 March 1788.

[13] *Ibid.* 29 March 1796.

that there were numerous schools in Manchester directed towards satisfying the demand for scientific, mathematical and commercial studies.

IV

It would appear, from our investigations in Manchester, that there was a closer association between science and industry in the early Industrial Revolution than has hitherto been thought. Further research is needed on other towns, but there seems to be sufficient *prima facie* evidence for asserting that scientific and technical education was fairly widespread in this period. Most of the books and periodicals to which we have referred had a national market. The itinerant lecturers in natural philosophy visited most of the chief towns. Some of them, such as Adam Walker, had schools in London.[1] The metropolis, of course, had its Royal Society, the Society for the Encouragement of Arts, Manufactures and Commerce (founded 1754), and the Royal Institution (founded 1799): it was still the acknowledged centre of scientific thought and education. But some of the larger provincial cities were challenging its predominance. The Manchester Literary and Philosophical Society was paralleled in Birmingham by the famous Lunar Society. In Birmingham there was also, as Professor Ashton has pointed out, 'much coming and going between the laboratory and the workshop, and men like James Watt, Josiah Wedgwood, William Reynolds, and James Keir were at home in the one as in the other.'[2] They doubtless benefited a good deal from their regular intercourse with such scientists as Erasmus Darwin, Richard Lovell Edgeworth, and Joseph Priestley.[3] 'Never before in history was there such an advantageous syncretism of pure science and advancing industry as in the Lunar Society.'[4]

This orientation of science towards industry in the deliberations of the Lunar Society has been examined more closely in recent years,[5] and though it is clear that the industrial emphasis did not debar the 'Lunatics' from engaging in pure research in electricity and the collection and identification of new gases, at a time when their potential applications to industry were not known, yet a good deal of the correspondence between friends in the society has a very practical flavour. Keir, Watt and Black shared an interest in a patent for the manufacture of alkali, and the course of their interests is described very fully in correspondence between Watt and Black;[6] Keir and Boulton wished to exploit commercially an alloy, discovered by Keir, which had almost the same composition as Muntz's metal;[7] Wedgwood supplied chemical utensils for Priestley's experiments, while Priestley analysed minerals of possible utility in pottery;[8] Small, Boulton, Keir and Watt, with the advice of John Whitehurst, F.R.S., the Derby watchmaker, were involved in a plan to mass-produce clocks, while the tools of the clock- and instrument-making trade assisted Boulton and Watt

[1] For other London schools in which science was taught, see Hans, *op. cit. passim.*

[2] Ashton, *op. cit.* p. 16.

[3] *Ibid.* p. 21. See also P. Mantoux, *The Industrial Revolution in the Eighteenth Century* (1928), pp. 387–8. The stimulus and benefits were mutual. See E. Robinson, 'The Lunar Society and the Improvement of Scientific Instruments', *Annals of Science*, XII (1956), 296–304, XIII (1957), 1–8.

[4] A. and N. L. Clow, *The Chemical Revolution* (1952), p. 614.

[5] R. E. Schofield, 'Industrial Orientation of Science in the Lunar Society of Birmingham', *Isis*, 48 (1957), 408–415.

[6] In the Watt papers at Doldowlod.

[7] H. W. Dickinson, *Matthew Boulton* (Cambridge, 1937), pp. 101–3.

[8] E. Robinson, 'The Lunar Society and the Improvement of Scientific Instruments', *Annals of Science*, XIII (1957), 1–2.

in their development of engineering technique.[1] Even before Watt came to Birmingham, the members of the Lunar Society were interested in the design of steam engines. Boulton had invented a steam-engine of his own which he discussed with Erasmus Darwin and Benjamin Franklin. Boulton's 'toy' business also caused him to be interested in metallurgy. In these pursuits he was aided by William Withering, who translated Bergmann's *Outlines of Mineralogy* (Birmingham, 1783), and by James Keir. Similar examples of collaboration in the industrial applications of science could be multiplied.

Besides those men of industry and science who are known to have been members of the Lunar Society, there were others who long maintained close relationships with members of the Society. Josiah Wedgwood is the first in importance among some very important men. He was a F.R.S. and a member of the Derby Philosophical Society, as well as of numerous other organizations, including Franklin's 'Club of Thirteen'.[2] The interplay between industry and science in Wedgwood's life is best seen in his scheme to combine with other Staffordshire potters to set up a research establishment.[3] He also encouraged scientific interests in his own district, by patronizing the lectures of John Warltire and engaging him to instruct his own children and those of Erasmus Darwin in chemistry.[4] Wedgwood's own laboratory assistant was Alexander Chisholm, a chemist of no mean ability.

Like Wedgwood, John Roebuck and Samuel Garbett cannot be recognized as full members of the Lunar Society, though they were certainly on very close terms with Boulton, Watt and others. The refining laboratory and the vitriol works operated in Birmingham by Garbett and Roebuck are *loci classici* of the union between chemistry and industry in the development of eighteenth century Birmingham. Among the occasional visitors to the Lunar Society were John Smeaton, the engineer, Henry Moyes, the itinerant lecturer, Pieter Camper, the Dutch scientist, and R.E. Raspe, geologist, assayer, and man of science.[5] There must have been many others whose visits cannot be substantiated by written evidence, since the Lunar Society kept no records, but from those who are known to have attended meetings we can compile a most impressive record of the relationships between scientists and industrialists. Indeed many of the industrialists have some claim to scientific prominence in their own right. We should not overlook the fact, moreover, that the 'Lunatics' spread their influence by their membership of other societies. All of them were Fellows of the Royal Society and many were members of the Society for the Encouragement of Arts, Manufactures and Commerce.[6]

In Newcastle-upon-Tyne, in this same period, we come across Henry Moyes, Gustavus Katterfelto, and others lecturing in natural philosophy, and in 1793 the Newcastle Literary and Philosophical Society was founded, following the

[1] *Ibid.* XII (1956), 296–304.

[2] E. Robinson, 'R. E. Raspe, Franklin's Club of Thirteen, and the Lunar Society', *Annals of Science*, XI (June 1955).

[3] R. E. Schofield, 'Josiah Wedgwood and a Proposed Eighteenth-century Industrial Research Organization', *Isis*, 47 (March 1956).

[4] D. McKie, 'Mr Warltire, a good chymist', *Endeavour*, X (January 1951).

[5] R. E. Schofield, in his valuable paper, 'Membership of the Lunar Society of Birmingham', *Annals of Science*, XII (June 1956), denies that Raspe and Pieter Camper ever attended any meeting of the Lunar Society, but their visits can be substantiated from James Watt's pocket diaries at Doldowlod (*e.g.* under 13 June 1779: 'Sunday at Soho with the Lunar Society. Mr Raspe was there & Mr Edgeworth'). They were not, however, regular attenders.

[6] R. E. Schofield, 'The Society of Arts and the Lunar Society of Birmingham', *Journal of the Royal Society of Arts*, June and August 1959.

Manchester and Birmingham precedents.[1] Like the Manchester Society, that in Newcastle was broad in its scope, with moral and literary as well as scientific purposes, but from the first it showed a marked interest in the application of natural philosophy to industrial processes, particularly to the local industries of coal, lead, and iron mining, metal, chemical, and glass manufactures.[2] The papers on coal-mining were, as might be expected, most numerous of all. Indeed, the Newcastle Literary and Philosophical Society was the cradle of the North of England Institute of Mining Engineers, eventually established in 1852.

Regular courses of scientific lectures were instituted by the Newcastle Society in 1803, when the Rev. William Turner was appointed to 'a permanent Lectureship in the several branches of Natural and Experimental Philosophy, Chemistry, etc.',[3] a post which he held until 1833. 'Without mechanics and chemistry', it was pointed out, 'how wretched would be our manufactures, and how worthless the amount of our mineral possessions! Exactly in proportion to the state of these sciences is the condition of the one and the value of the other.' Turner was provided with a lecture room and a 'philosophical apparatus' formerly belonging to Dr Thomas Garnett (recently deceased) of the Royal Institution. He greatly stressed in his introductory discourse 'the application of Philosophy to the arts' [manufactures], especially to local industries, and in formulating his courses he was careful 'to apply the [scientific] principles laid down to the explanation of the machines, etc. most useful in this district'.[4]

The Newcastle Literary and Philosophical Society was in touch with the Lunar Society through Richard Chambers's correspondence with Matthew Boulton. The plans for the society were submitted to Boulton in 1793 and he was also asked to honour them with his 'Observations upon any Subject connected with Coal or the Coal Trade – particularly if you could furnish us with an account of the Wedgeburgh mines . . . .' [5] As early as 1784, Chambers had introduced Henry Cort to Boulton. In April 1795 Chambers wrote on behalf of the Literary and Philosophical Society to Boulton requesting him to distribute some enclosed letters about the coal trade 'among the most intelligent Gentlemen concerned in Coal Works in your Neighbourhood'.[6] It is clear that the philosophical societies could be a means of diffusing applied science throughout the country.

Philosophical societies in smaller towns had not such large intellectual and financial resources as those of Manchester, Birmingham, and Newcastle, but were similar in their characteristics. Derby, for example, had a philosophical society begun in 1783, though formally inaugurated in 1784.[7] The moving spirit in it was Erasmus Darwin, who had previously had a hand in founding the Lichfield Philosophical Society and the Lunar Society. One of the first letters relating to the Derby Philosophical Society is an invitation from Darwin to

[1] R. S. Watson, *The History of the Literary and Philosophical Society of Newcastle-upon-Tyne* (1897). The references to Moyes and Katterfelto are on pp. 203–6. A Unitarian minister, the Rev. William Turner, who was educated at Warrington Academy, played a leading rôle, like the Rev. Thomas Barnes in Manchester. See above, p. 236, for John Horsley's lectures in natural philosophy in Morpeth and Newcastle in the early eighteenth century.

[2] See especially Turner's paper on the objects of the proposed society (*ibid.* pp. 35–40) and the brief survey of papers read at the monthly meetings (*ibid.* chap. vii).

[3] *Ibid.* pp. 208–16. Previously, it was pointed out, scientific education could be acquired only 'by expensive residence in a distant capital', or by 'the precarious arrival of occasional lecturers' (*ibid.* p. 217).

[4] *Ibid.* p. 223.

[5] R. Chambers to M. Boulton, 6 March 1793, A.O.L.B.

[6] R. Chambers to M. Boulton, 30 April 1795, A.O.L.B.

[7] E. Robinson, 'The Derby Philosophical Society', *Annals of Science*, IX (1953), 359–67.

the 'Lunatics' to hold joint meetings.[1] The members of the society were varied. They included local worthies like Richard Leaper, C.S. Hope and John Crompton, all of whom became mayors of Derby; the local aristocracy such as Lord Belper and Sir Brooke Boothby; Josiah Wedgwood and his inventor-cousin, Ralph Wedgwood; a number of local medical men; and last, but not least, local industrialists, such as William Strutt, William Duesbury and Dr Robert Bage of Elford.

William Strutt, son of Jedediah Strutt, was a co-founder with Erasmus Darwin of the Derby Society and on Darwin's death in 1802 succeeded him as President of the Society.[2] Through Darwin, Strutt became a close friend of R. L. Edgworth, another member of the Lunar Society; he also knew Peter Ewart, with whom he discussed his scientific writings, and Ewart's friend, George Lee, while Thomas Hewes built two of the wheels for the Strutts' West Mill.[3] Strutt is typical of the second generation of educated industrialists:

> 'Although it was Strutt's mechanical genius that secured his election as Fellow of the Royal Society in June, 1817, when he was 60, he maintained throughout life an unflagging interest in many branches of pure science. His reading was wide – it included the works of Newton, Euler, Blair, Priestley, Landen and Vince ....' [4]

He was no doubt assisted in obtaining this wide range of reading by the fine collection of books bought by the Derby Philosophical Society which the members of the Society were entitled to borrow.

William Duesbury, another of the industrialist members of the Derby Society, was well known as a porcelain manufacturer of the period, and Robert Bage, the paper manufacturer and friend of William Hutton, also belonged to the Society. The Derby Philosophical Society was probably typical of the societies in Northampton, Exeter, Leicester and other middle-sized towns.

The frequency with which itinerant lecturers visited Manchester in the last decades of the eighteenth century might be thought to be exceptional, but a study of local newspapers in other industrial cities shows that they were frequent visitors to all of them. In Bristol, for example, all the well-known lecturers gave courses at one time or another. Advertisements appeared in the Bristol newspapers for courses by James Arden, James Ferguson, John Warltire, Henry Moyes, Adam Walker junior, and Henry Clarke, while a local personality, Benjamin Donne, gave several courses each year from 1765 to 1798. On several occasions during this period there were two lecturers operating simultaneously within the town, while Donne was often holding courses in the same week both at Bristol and Bath. Bristol was fortunate moreover in numbering among its citizens Dr Thomas Beddoes,[5] formerly Reader in Chemistry at Oxford, and founder of the Bristol Pneumatic Institute, where Humphrey Davy began his scientific career. Beddoes was a close friend of James Watt, Erasmus Darwin and other members of the Lunar Society, and he was related by marriage to R. L. Edgeworth. The Wedgwoods knew him and were in the habit of staying in Clifton. He assisted the Oxford engineer and aeronaut, James Sadler, in his experiments upon steam engines in 1792–3.[6] While Warltire was lecturing in Bristol, his host was Joseph Fry, who corresponded

[1] E. Darwin to M. Boulton, 17 January 1784, A.O.L.B.

[2] R. S. Fitton and A. P. Wadsworth, *The Strutts and the Arkwrights, 1758–1830* (Manchester, 1958), p. 175.

[3] *Ibid.* pp. 175–7, 207 *n.*, 215 *n.* For Ewart, Lee, and Hewes, see above, pp. 229–30.

[4] *Ibid.* pp. 170–1.

[5] E. Robinson, 'Thomas Beddoes, M.D., and the Reform of Science Teaching in Oxford', *Annals of Science*, XI (June 1955).

[6] A. Raistrick, *Dynasty of Ironfounders. The Darbys of Coalbrookdale* (1953), pp. 158–9.

with James Watt about Warltire's attempts to obtain a soft metal from manganese.[1]

The diaries of an eighteenth century Sheffield cutler, Thomas Asline Ward,[2] show how significant a part the itinerant lecturers played in the cultural life of Sheffield industrialists, and Mr J. E. Harrison has collected several references to their activities in Doncaster, which Banks, Warltire, Moyes and others visited.[3] The itinerant lecturers operated with the same frequency in Birmingham, though no thorough survey of their visits there has yet been undertaken. References to them occur quite frequently in Matthew Boulton's correspondence. We have already mentioned Boulton's contacts with Adam Walker; there are also letters from Joseph Priestley to him recommending Warltire [4] and from John Roebuck asking him to collect subscribers for Moyes.[5] In turn, Boulton recommended Warltire to Erasmus Darwin in Derby and Josiah Wedgwood in Etruria. That such lectures were considered to be an essential part of the education of every young businessman can be seen from the concern of John Gilbert, the Duke of Bridgwater's steward, that his son should accompany Boulton to a course of lectures by James Arden in Birmingham, while the young man was Boulton's guest.[6] Boulton himself arranged for his own son, M. R. Boulton, to attend Dr Beddoes's and Dr Hornsby's lectures in Oxford.[7] The taste remained with him in adulthood, since M. R. Boulton's pocket diary refers to his attendance at John Stancliffe's lectures on galvanism.[8] *Aris's Birmingham Gazette* shows that James Ferguson, James Booth, Mr Burton, Mr Long, and many others lectured in Birmingham besides those we have already mentioned. As in Manchester, their advertisements often stress the practical application of the lectures to local industries. Thus John Warltire's lectures on chemistry would give Birmingham manufacturers

> 'an Opportunity of informing themselves of the particular Properties of the various substances imployed in the different Processes they are necessarily conducting every Day; and consequently, a Means of improving them to *their own Advantage*'.[9]

The association between science and industry was particularly strong in Scotland, where there occurred in this period a brilliant flowering of scientific thought and technology.[10] In Edinburgh and Glasgow universities, great scientists such as William Cullen, Joseph Black, James Hutton, and Francis Home made important contributions to industrial technology. Their students included later famous industrialists such as John Roebuck, James Keir, and Charles Macintosh, and they were in close contact with scientists and manufacturers in Birmingham, Manchester, and other English industrial centres. The association between Black, Roebuck, and Watt in the development of the steam engine is famous, but there was important scientific-industrial collabo-

[1] Joseph Fry to James Watt, 1st Month 16, 1783, Dol.

[2] *Peeps into the Past: Extracts from the Diaries of Thomas Asline Ward* (ed. A. B. Bell, Sheffield, 1909).

[3] J. A. Harrison, 'Blind Henry Moyes, "An Excellent Lecturer in Philosophy",' *Annals of Science*, XIII (June 1957).

[4] Joseph Priestley to Matthew Boulton, 28 September 1776, A.O.L.B.

[5] John Roebuck to Matthew Boulton, 3 October 1777, Dol.

[6] He thought it 'of the Greatest Consequence' that his son should attend these lectures. J. Gilbert (Worsley) to M. Boulton (Birmingham), 7 Sept. 1765, A.O.L.B. *Aris's Birmingham Gazette*, 2 and 9 September 1765, contains advertisements of Arden's lectures.

[7] A. Noble Brown (Balliol College, Oxford) to M. Boulton (Soho), 25 February 1791, A.O.L.B.

[8] M. R. Boulton's Pocket Diary, 4 and 5 March, 5 April, 1801.

[9] John Warltire's advertisement in *Aris's Birmingham Gazette*, 11 October 1779.

[10] See A. & N. L. Clow, *op. cit.* especially chap. XXV, for a detailed account.

ration in other fields, such as bleaching and the production of synthetic soda. Edinburgh and Glasgow also had philosophical and chemical societies and academies in which science was taught outside the universities. There was, for example, the Andersonian Institution (established by Dr John Anderson, professor of natural philosophy in Glasgow University), forerunner of the Royal Technical College in Glasgow. Dr Thomas Garnett, who lectured there on chemistry and physics, we have already mentioned in our list of itinerant lecturers in natural philosophy; he eventually moved to the Royal Institution in London. George Birkbeck also came from the Andersonian Institution to London, where he took the lead in establishing Mechanics' Institutes and was associated with the founding of University College.[1]

Ireland also participated in these developments. A Dublin Philosophical Society had been founded as early as 1683 and the (Royal) Dublin Society had existed continuously since 1731. A Medico-Philosophical Society was formed in 1756 and the Irish Academy in 1783 (chartered 1786), the first volume of its *Transactions* appearing in 1787. Towards the end of the eighteenth century the Dublin Society felt 'that the time had come when the formation of schools of science, in which qualified professors might lecture .... in accordance with the example and precedent set in such matters in England and Scotland, more likely to further the purposes for which the Society had originally been founded'.[2] Hence William Higgins (1763–1825) was appointed professor of chemistry and mineralogy to the Society in 1795,[3] equipped with a laboratory, and instructed 'to make such experiments on dyeing materials and other articles, wherein chymistry may assist the arts'.[4] That practical chemical instruction was given in his laboratory is evident from advertisements in *Saunders' News Letter* and the *Hibernian Journal* of 1797 and later years. Systematic courses of lectures in chemistry and natural philosophy were instituted in 1800 [5] and soon became an important feature of the Society's work in Dublin and also in the provinces, to which they were ultimately extended. In 1795 Dr Walter Wade was appointed professor and lecturer in botany, and in 1800 James Lynch, optician, of Capel St., Dublin, was appointed to a similar post in hydraulics, mechanics, experimental philosophy, etc.[6] The most eminent scientist in the Society, however, was Dr Richard Kirwan, F.R.S., president of the Royal Irish Academy, 1799–1812, one of the most notable chemists in Europe at this time.[7] Many of the books and papers published by Kirwan and Higgins were of considerable importance to industrial chemistry, especially to the Irish linen industry.

Thus it appears that there was a widespread interest in applied science in the late eighteenth century. Many scientists or 'natural philosophers' had industrial interests, and many industrialists were interested in science. It may be objected, however, that these interests were too empirical and too fragmen-

[1] T. Kelly, *George Birkbeck* (Liverpool, 1957).

[2] H. F. Berry, *A History of the Royal Society of Dublin* (1915), p. 159. That some of the itinerant natural philosophers visited Ireland is shown by the example of Adam Walker. See above, pp. 231–2.

[3] He was also chemist to the Irish Linen Board. See *D.N.B.*; T. S. Wheeler, *Endeavour*, XI (1952), 47–52; J. R. Partington, *A Short History of Chemistry* (1951), pp. 272–6. He was the nephew of another well-known Irish scientist, Bryan Higgins (*c.* 1737–1818), M.D. of Leyden University, who in 1774 opened a school of practical chemistry in Greek Street, Soho, London, and who was one of the first to take out a patent for the manufacture of synthetic soda. See Sullivan, *op. cit.* 483–7; Partington, *op. cit.* pp. 269–72; Clow, *op. cit.* pp. 97–9; *D.N.B.*

[4] Berry, *op. cit.* p. 355.

[5] See W. Higgins, *A Syllabus of a Course of Chemistry for the year 1802* (Dublin, 1801).

[6] Berry, *op. cit.* pp. 159–60.

[7] See D.N.B.

tary to be called 'scientific'.[1] This criticism can be met by considering what science was at this period. It is true that many of the scientific theories of the day – the whole complex of phlogiston theory, the theories of chemical combinations, of dyeing, bleaching, etc. – were later to be proved erroneous, but they were useful concepts at the time: modern science has evolved through a series of discarded hypotheses. The scientific theories of the late eighteenth century ought to be criticized only in the light of the knowledge then available. At that time, as a contemporary, William Nicholson, pointed out, scientists were 'not sufficiently advanced in any part of the knowledge of nature, to reason with safety, without constant recurrence to the test of experiment'.[2] This widespread adoption of what has been called 'the scientific method' – the numerous attempts 'to improve natural knowledge by experiment', the original aim of the Royal Society – was of considerable importance to industry, for much of the experimenting was of a practical, even utilitarian, kind, in which scientists and industrialists had mutual interests.

A modern chemist may easily take an unhistorical view of the primitive attempts made by such men as Watt, Keir, and Henry to analyse chemical processes too complex for the chemical knowledge of the period. But their efforts were undoubtedly scientific in the sense of being based on rationally ordered experiments. Watt and Keir, for example, certainly thought in terms of controlled experiments in a laboratory when they were considering the manufacture of alkali or the chlorine process of bleaching. Similarly Thomas Cooper asked James Watt to send him a digester so that he could carry out his experiments in laboratory conditions, and Thomas Henry also experimented very thoroughly on dyeing and bleaching.

No doubt the mass of manufacturers in the late eighteenth century had little scientific knowledge, but a substantial minority, including some of the most significant figures, were passionately interested in it, for utilitarian as well as intellectual reasons, and there seems little doubt that such knowledge contributed to the industrial changes of that period. We do not wish to push this thesis too far – to underestimate the contributions of unscientific though intelligent practical craftsmen – but the evidence appears to necessitate some modification of the traditional view of the Industrial Revolution.

University of Manchester
City of Coventry Training College

[1] Such a criticism was made at the Economic History Conference by Dr D. W. F. Hardie.
[2] W. Nicholson, *The First Principles of Chemistry* (1790), pp. 413–4.

Raphael Samuel

WORKSHOP OF THE WORLD: Steam Power and Hand Technology in mid-Victorian Britain

There is no doubt whatever that the people of England work harder, mentally and physically, than the people of any other country on the face of the earth. Whether we take the town or the country population, the same plodding industry is apparent, and the respites enjoyed by either in the shape of holidays are few and far between . . . We have no desire in stating the fact that Englishmen work harder than their neighbours to make them discontented with their lot; far from it. It is due to their untiring industry and to the natural advantages of the country of which we are so justly proud, that England holds the foremost rank among the nations of the earth.

Mechanics Magazine, III, 280, 27 April 1860.

The genius of Great Britain is mechanism; the master-spirit the civil engineer; her tendencies, to relieve labour from its drudgery, and delegate to iron, and steam, and water, the real weight and burden of toil.

Illustrated Exhibitor, 7 June 1851.

Nowhere do we find a more shameful squandering of human labour-power for the most despicable purposes than in England, the land of machinery.

Marx, *Capital*, Dona Torr ed., I, 391.

The Messrs. Whitehead send their Buenos Aires wool to be picked in the prisons of Manchester, that species of raw material being so coarse and dirty that it is difficult to find free labourers to meddle with it. A great deal of the ordinary picking is, however, done by the women in their cottages in the neighbourhood.

Morning Chronicle, 'Labour and the Poor, Manufacturing Districts XIII: Saddleworth', 29 November 1849.

Repavement of the Strand, from *The Illustrated London News,* 26th April, 1851.
The Mansell Collection, London.

I. Introduction. The Machinery Question

Whatever their disagreements about the origins of the industrial revolution, economic historians are in little doubt about its effects. Steam power and machinery transformed the labour process, and acted on society as an independent or quasi-independent force, demonic or beneficient according to the point of view, but in any event inescapable. Commodities were cheapened and new markets opened up for them; labour was made enormously more productive at the same time as the physical burden of toil was eased; mechanical ingenuity took the place of handicraft skill. David Landes' summary in *The Unbound Prometheus* is both influential and representative:[1]

> In the eighteenth century, a series of inventions transformed the manufacture of cotton in England and gave rise to a new mode of production – the factory system. During these years, other branches of industry effected comparable advances, and all these together, mutually reinforcing one another, made possible further gains on an ever-widening front. The abundance and variety of these innovations almost defy compilation, but they may be subsumed under three principles: the substitution of machines – rapid, regular, precise, tireless – for human skill and effort; the substitution of inanimate for animate sources of power . . . thereby opening to man a new and almost unlimited supply of energy; the use of new and far more abundant raw materials, in particular, the substitution of mineral for vegetable or animal substances.

This account has the merit of symmetry, but the notion of substitution is problematic, since in many cases there are no real equivalents to compare. The fireman raising steam in an engine cab, or the boilermaker flanging plates in a furnace, were engaged in wholly new occupations which had no real analogy in previous times. So too, if one thinks of the operations they were called upon to perform, rather than the nature of the finished product, were the mill-hands of Lancashire and the West Riding. And if one looks at technology from the point of view of labour rather than that of capital, it is a cruel caricature to represent machinery as dispensing with toil. High-pressure engines had their counterpart in high-pressure work, endless chain mechanisms in non-stop jobs. And quite apart from the demands which machinery itself imposed there was a huge army of labour engaged in supplying it with raw materials, from the slave labourers on the cotton plantations of the United States to the tinners and copper miners of Cornwall. The industrial revolution, so far from abridging human labour, created a whole new world of labour-intensive jobs: railway navvying is a prime example, but one could consider too the puddlers and shinglers in the rolling mills, turning pig-iron into bars, the alkali workers stirring vats of caustic soda, and a whole spectrum of occupations in what the Factory legislation of the 1890s was belatedly to recognise as 'dangerous' trades. Working pace was transformed in old industries as well as new, with slow and cumbersome methods of production giving way, under the pressure of competition, to overwork and sweating.

Nor is it possible to equate the new mode of production with the factory system. Capitalist enterprise took quite different forms in, for instance, cabinet-making and the clothing trades, where rising demand was met by a proliferation of small producers. In agriculture and the fisheries it depended upon an increase in numbers rather than the concentration of production under one roof. In metalwork and engineering – at least until the 1880s – it was the workshop rather than the factory which prevailed, in boot and shoemaking, cottage industry. The distributive trades rested on the broad shoulders of carmen and dockers, the electric telegraph on the juvenile runner's nimble feet. Capitalist growth was rooted in a sub-soil of small-scale enterprise. It depended not on one technology but on many, and made use, too, of a promiscuous variety of profit-making devices, from the adulteration of soot[2] (in which there was an international trade with the West Indies, as well as a local one with farmers for manure[3]) to the artificial colouring of smoked haddocks.[4] Bread was dosed with liberal sprinklings of alum to disguise inferior wheats; low-grade cloths were camouflaged with 'size'.[5] In domestic housebuilding scamped workmanship kept the speculative builder afloat, while in the East End furniture trade orange boxes provided the raw materials for piano stools and Louis Quatorze cabinets. The 'Golden Dustman', immortalized by Charles Dickens in *Our Mutal Friend*, is as representative a figure of mid-Victorian capitalism as the Bradford millionaires pilloried by John Ruskin for their taste. So too – from the same novel – are the Veneerings, whose provincial counterparts rose to affluence by cotton 'corners' on Liverpool or Manchester Exchange. One thousand needlewomen made the fortunes of Nicoll, the Regent Street sweater,[6] while the railway speculations of the 1840s rested on the muscle power of three hundred thousand navvies.

Economic historians have had remarkably little to say about either labour process or the relationship of technology to work. They are much more concerned with business cycles, and measuring rates of growth. Commercial achievement excites them, and whole histories will be written to celebrate the achievements of individual firms. Railways are discussed as a source of investment, and their comparative contribution to economic growth is a subject of hot debate; nothing at all is said about how the rolling stock was made or the engine cabins staffed or merchandise unloaded. Bricks, too, are treated as an index of investment, without so much as a word being said of the primitive conditions in which they were made, or the ferocious toil imposed on the men, women and children who made them.[7] Production is seen at second or third remove, in terms of inventory cycles and aggregate profitability: we do not learn how the furnaces were de-clinkered, or the iron steam ships coaled.

Except in the 'heroic' age of invention, economic historians have very little to say about machinery. They may tell us what it did for production, but not what it meant for the producers, and their preoccupation in recent years with 'take-off' – 'that decisive interval in the history of a society when growth becomes its normal condition' – means that they give far more attention to the progress of mechanisation, and the constellation of circumstances favouring it, than to measuring its human costs. The plight of the hand-loom weavers in the 1830s is admitted, even insisted upon, but since they are regarded as a solitary and to some extent exceptional case, they do not seriously obstruct the linear march of improvement, and once they have been disposed of the historian passes quickly to the problems of a 'mature' economy, and the triumphs of Free Trade.

For labour historians, the machinery question attracts attention chiefly in the 1820s and 1830s, when Cartwright's loom was throwing thousands out of work, and when the rival merits of an agrarian and an industrial society ('past and present') were being vigorously canvassed on all sides. The scenario is arresting, with midnight raiding parties, rickyard incendiaries and factories besieged. But the drama is short-lived, and once the protagonists have performed their parts they are quickly shuffled off-stage. Opposition to machinery is assigned to the pre-history of socialism, when it was 'utopian' rather than 'scientific'; and the machine-breakers, despite Eric Hobsbawn's pioneering attempt to interpret their action in the light of modern collective bargaining (machine-breaking as a form of strike)[8] take their place in the gallery of 'primitive', pre-industrial rebels, along with such other early 19th century martyrs to oppression as Jeremiah Brandreth and Dic Penderyn. Luddism appears as a doomed, if heroic, resistance to the ineluctable forces of change – a fight against the inevitable – the Swing Riots of 1830 as 'the last labourers' revolt'. Yet in industry after industry the machinery question was still being fought out in mid-Victorian times, and there was a whole spectrum of occupations where mechanisation was still being resisted, or its scope drastically curtailed, in the 1890s: the last great machinery strike in the boot and shoe trade did not take place until 1895; while as late as 1898 a steam saw mill was blown up in the Forest of Dean.[9] There were also striking regional variations in the application of invention and progress of the machine, and in some cases at least the strength or otherwise of the workers' opposition seems to have been the deciding factor. In carpet weaving, for instance, the 'extra speeded' Moxon (an improved power loom of the 1870s) was kept out of Kidderminster entirely, where the weavers' organisation was strong, but installed with apparent ease in Rochdale, Halifax and Durham, the northern centres of the trade.[10] In printing, the Hattersley, an early mechanical typesetter, was widely employed by provincial newspapers (the first was installed in the offices of the *Bradford Times* in 1868) but the London Society of Compositors was successful in keeping it at bay.[11] Similarly in boot and shoe making, the 'stabbing machine' – an application of the sewing machine to waxed threads – was excluded from Northampton, the metropolis of the wholesale trade, after three general strikes against it, fought between 1857 and

1859; but it was widely employed at Leicester, Norwich and Bristol.[12] In metal-working, the treadle-worked 'Oliver', a semi-mechanical stamp which had been common in Staffordshire for 'generations', was still apparently unknown in Manchester in 1865, and when in that year a local manufacturer attempted to introduce it, the nut and bolt makers (or at any rate the anonymous correspondent who wrote on their behalf) threatened to kill him.[13]

Even when machinery was eventually installed, the struggle to control it remained unresolved, and one of the most common complaints of employers in the late 19th century was that tools were not run at their proper speeds, but were being sabotaged by worker lethargy or resistance. In a cotton mill every spindle was potentially a battleground as mules increased in size: in an ironworks every attempted economy in fuel or alteration to the 'heat'. Often the machine proved disappointing to its patentees and promoters, either for want of precision, or because of the recalcitrance of the raw material, or because of the irreplaceability of handicraft skill. Patent could follow patent without anything like profitability being achieved, and the employer's dream of a 'self-acting' mechanism – equal to the best hand labour, but driven by itself – remained elusive. Mechanisation, in short, was a process rather than an event. It did not begin with the great inventions of the 18th and early 19th centuries; nor did it end with their application. The process itself was neither linear nor smooth but, on the contrary, discontinuous and subject to a whole complex of competing claims, pulling in opposite directions. For the most part it advanced by small increments rather than by leaps, and forward movements were often followed by retreat, as workers reasserted their claims. In the study of which this article forms a part, I want to argue that the machinery question, so far from being settled by the defeat of the Luddites, is in some sense coterminous with capitalism itself; that resistance to machinery, though often opaque and only intermittently recorded in the documents, was an endemic feature of 19th century industrial life. I also want to look at the repercussions of machinery on skill, and at the ways in which the labour process was reconstructed both from above and below, under the impact of technical change. Finally I want to look at machinery in relationship to the 'reserve army of labour' and the demographic changes of the early and middle years of the nineteenth century, and to consider the relationship of factory industry to capitalism in the countryside, domestic outwork and the workshop trades.

Readers of *Capital* will know that such a discussion inevitably bears on Marx's 'stages' of capitalist development. In chapters XIII to XVI of *Capital* Vol. I he proposes three great epochs of capitalist development, which are both chronologically and analytically distinct.

1. The handicraft stage, or that of petty commodity production – the chrysalis from which later capitalism grew.
2. Capitalist 'manufacture' – the concentration of artisan and handicraft production under the control of a single capitalist, and the systematic extension of the division of labour.
3. 'Modern' industry – the epoch inaugurated by the coming of machine tools and the factory system.

In Marx's discussion each of these epochs appears to supersede its predecessor and in the case of 'manufacture' and 'modern industry' at least a clear chronology is suggested, the first being assigned to the period from the middle of the sixteenth to about the middle of the eighteenth century, the second to the age of invention. But as Marx's lengthy chapter on 'modern industry' unfolds – it takes up fully 150 pages of the book – it becomes clear that modern industry incorporates older systems of production rather than superseding them, and that it is in fact a mixed development, in which 'modern' domestic industry and 'modern' manufacture play no less distinctive a part than the machine-based factories.[14] Here, as elsewhere in *Capital,* there are plainly shifts of emphasis in Marx's discussion, and one way of elucidating them – as well as of determining their theoretical status – would be to consider the historical phenomena to which they were addressed. The discussion of such questions has in recent years been left to the philosophers and the economists, each of them concerned, in their own way, with the theoretical consistency of Marx's texts rather than the industrial reality which he was attempting to dissect. The historian may be ill-equipped to undertake a work of epistemological clarification, or to explore the more problematical reaches of the law of value. But that does not or should not mean that he or she has no contribution to make to theoretical discussion. The territory of *Capital* Vol. I is, after all, a historian's territory, one whose landmarks are in many cases familiar, and whose signposts the historian will sometimes be better placed than an economist or a philosopher to read.

Another theoretical question which this discussion poses is the relationship of ideology and class struggle. There is no doubt that so far as public agitation is concerned historians have been right to focus upon the 1820s and 1830s as the crucial period for the debate on machinery. All the major questions still seemed open, and the debate was uninhibitedly pursued, with Macaulay and the political economists upholding the 'march of improvement', Cobbett as its most eloquent critic, and Carlyle as the prophet of doom. Even after the defeat of the Luddites the inevitability of mechanisation was by no means accepted. In the 1830s there were widespread working class demands for a tax on machinery, and a vigorous agitation for short-time which was quite openly directed against factory 'slavery'. The economy plunged wildly from hectic bouts of activity to devastating slumps, and in the depths of the 1842 depression even the Prime Minister seemed uncertain as to what the outcome would be.[15] The labour movements of this time openly regarded machinery as an enemy force, one which they held responsible for the huge increases in the reserve army of labour and which, if left to continue unchecked, would make every class of worker redundant. Alternatives to the factory were eagerly canvassed and whenever the opportunity arose they were put into practical effect – spade husbandry and smallholdings on the land, producer co-operatives in the towns. As a public issue the machinery question was settled by the defeat of the Owenite trade unions in the 1830s, and the failure of the producer co-operatives to establish themselves; by the growing confidence of the industrial bourgeoisie; and by the retreat from socialist theory which marked the Chartist years (a subject to be discussed by Gareth Stedman Jones in a future issue). Once the opposition to machinery had been abandoned there was no turning back, and in mid-Victorian times resistance was characteristically fragmented. Trade unions re-wrote their rule books to expunge the anti-machinery clauses; trade union leaders adopted the language of 'improvement' as though it were their own. But at the level of the shop floor the machinery question was not settled at all, either in theory or in practice, and the battle of ideas continued to be waged around the workshop stove long after it had disappeared from the public stage.

Finally there is the whole question of whether technology should be seen as a cause or an effect, an outside neutral force, or what the sociologists call a 'dependent variable'. It is one of the great unresolved questions about the industrial revolution itself, and it is certainly one which should be posed of the mid-Victorian economy, when Britain was the workshop of the world. Marx's texts are ambiguous on this point. In Chapter XV of *Capital* Vol. I he begins with a purely technical explanation for the rise of modern industry, identifying it with the replacement of hand tools by machine-based technologies. The machine is more than just a giant tool. It is a process, a demonic power, a whole productive system – 'Modern Industry'. Machinery takes over from man the role of tool-bearer, manipulating numbers of tools simultaneously, synchronising separate detail processes as one. Instead of labour power, it is the instruments of labour which dominate production. 'It is now no longer the labourer that employs the means of production, but the means of production that employ the labourer'. But as Marx's lengthy discussion of 'modern industry' unfolds, machinery seems less and less essential to its hegemony, which is no less clearly established, and no less exploitative, in manufacture and domestic outwork than it is in factory industry. Mechanisation comes to appear more a result of modern industry than a cause – the capitalist's way of escaping from worker resistance.[16] Earlier, in a letter to Annenkov (28 December 1846) Marx distinguished between two great epochs in the progress of machinery:[17]

> One can say that up to the year 1825 – the period of the first general crisis – the general demands of consumption increased more rapidly than production, and the development of machinery was a necessary consequence of the needs of the market. Since 1825, the invention and application of machinery, has been simply the result of the war between workers and employers . . . This is only true of England. As for the European nations, they were driven to adopt machinery owing to English competition both in their home markets and on the world market. Finally, in North America the introduction of machinery was due both to competition with other countries and to lack of hands, that is, to the disproportion between the population of North America and its industrial needs.

Marx did not go any further in exploring the first of his two epochs of machinery and they remain unexplored by historians to this day. It is possible that research will show that the rise of machinery in eighteenth century England was much more closely related to class struggle than Marx believed, that, for instance, the migration of the silk trade from Spitalfields to Essex was intimately bound up with the strength of the weavers' combinations, and that similar factors were involved in the migration of the woollen trade from the West Country and Norwich to Yorkshire. Conversely it is by no means clear that mechanisation in mid-Victorian times can be explained by class relationships alone. All these questions remain properly open. The general bias of these chapters is to reduce the part assigned to machinery in the making of modern industry, and to look for wider changes *both* in technology *and* in the capitalist organization of work; it is also to reject the idea of mechanisation as a self-generating process, and to look instead at the complexity of competing claims which 19th-century capitalism faced, and with which the workers' movements of the time had to contend.

So far as economic history is concerned one of the main general questions raised by these articles concerns the relationship of mechanisation to overall capitalist growth. It should become clear as the argument proceeds that the two are in no sense one and the same. Not only is the tempo of change different in different trades, but its character is polyglot. Increased investment was by no means synonymous with the growth of large capitalist firms or the installation of elaborate plant. In some trades – classically in the building industry – it was accompanied by a proliferation of small producers. The response to market competition was also exceedingly various, and impossible to account for simply in terms of economic rationality. The conservatism of the Sheffield employers, which in the 1870s and 1880s exposed them to heavy competition from Germany and America, cannot be separated from the extraordinary power of the workers' trade societies in mid-Victorian times, and the very special claim of their skills.

Another question in economic history which these articles address – more particularly the third, which will be about 'sweating' – is that of the standard of life. This is usually discussed statistically in terms of household budgets, or else of take-home pay. 'Optimists' – i.e. 19th century capitalism's academic partisans – point to the progress of the nation, measured the statistical indices of consumption; 'pessimists' point to the vast numbers living at or below subsistence, or to the sanitary condition of the towns. Neither however address themselves centrally to the question of work, or to the momentous transformation which 19th century production imposed on working pace. Yet it is surely impossible to consider the standard of life apart from the conditions under which it was earned. The modest prosperities of mid-Victorian Lancashire were purchased at the cost of putting whole families to work in the mill; those of Victorian railwaymen – as Frank McKenna pointed out in *History Workshop Journal* I – at the price of working hours which turned day into night, and working shifts which could extend to 24 hours at a stretch. High wages in mid-Victorian England were quite often a species of death money, for those who could not hope to live long at their trades; or else a compensation for the fact that they had to work like horses: the Sheffield grinders and the Staffordshire iron puddlers might come into the first class, railway navvies into the second, while the coal miner, with more and more dust in his lungs as deeper shafts were sunk, and more explosives used, could qualify for either. Low wages, as Henry Mayhew pointed out, were in the London trades inseparably associated with over-work and sweating. It is hardly in question that the nineteenth century saw an enormous deterioration in working conditions, yet the matter has inspired little research. Historians are willing to delve into the remotest crevices of the stock market, in searching for the origins of railway investment or the effects of overseas trade, but so far as I know there is not a single modern article attempting to compute the comparative mortality of the trades, or to reconstitute the aetiology of industrial disease, despite the mass of evidence bequeathed to us by the more radically-minded 19th century doctors.

The materials for an inquiry into 19th century work are inexhaustible. Quite apart from the ordinary printed and manuscript sources available to historians, there is a huge technical literature, very little of which is used, and it is hoped that this article may indicate something of its potentiality. There are literally hundreds of treatises devoted to the 'rudiments' of individual trades – over 100 titles in Weale's series alone*. Then there are the compendious dictionaries of

* The Patent Office Library is a useful resort for British Museum readers who find a title they are searching for has been consigned to that elastic category 'Destroyed by Bombing in the War'.

manufactures, such as those of Tomlinson and Ure, which are often very well illustrated with prints; the industrial surveys undertaken, from time to time, by provincial newspapers (examples can be seen in the Goldsmith's collection of economic literature in the University of London library, Senate House); and the guides to industrial 'curiosities', a genre of investigative writing, developed around the time of the Great Exhibition, which in subsequent years kept many a Grub Street hack employed, as well as providing Sunday School superintendents with useful prize books: *The Busy Hives Around Us* (1860?) is a representative title. Trade newspapers, ranging from the *Horticultural Times* to the *Iron and Coal Trades Review* have been a rewarding primary source. They are often more informative about industrial conditions, though written from the point of view of capitalists, than labour movement journals. Periodical publications such as the *Journal of the Society of Arts* act as a sensitive register to some of the 'intermediate' technical changes; so too, for heavy industry, do the 'transactions' and 'proceedings' of the various societies of civil and mechanical engineers. As well as reprinting papers these latter journals also contain the discussions which followed them, so that you can sometimes hear the Victorian engineer thinking aloud about his work. Finally there are the Patent journals which, together with the Board of Trade papers at the Public Record Office, make it possible to track the course of speculative invention, and the euphoric hopes invested in self-acting mechanisms of all kinds.

The historian can only be grateful for such a wealth of documentation. Yet at the same time he or she should be on guard against its bias. The technical treatises, for instance, rarely deal with the industrial and commercial setting, so that it comes as something of a shock to turn from, say, James Facey's *Elementary Decoration* (1882) with its detailed specifications about the nature of the painter's craft, to the realities of 'scamping' in Robert Tressell's *Ragged Trousered Philanthropists*. And beyond an occasional reference to 'the operator' they may tell one nothing about labour at all. Tilt-hammers rise and fall, with never a gang of men to position the metal; convex parts are jointed as by an invisible hand. The trade journals, too, are unsatisfactory as a guide to work. They persistently exaggerate the importance of invention, so that even in the most resolutely handicraft sectors of production it often seems – on the evidence of single instances – that mechanisation is about to take off. The trade reports from Sheffield in *The Ironmonger*, for instance, are filled with trials of machinery in the 1860s and 1870s yet the Sheffield trades remained overwhelmingly handicraft right down to 1914. The *Boot and Shoe Trade Journal* is even more misleading: to read it in the 1870s is to have the impression of an industry on the very edge of automation rather than one in which domestic outwork was still, in the leading centre of manufacture, a major part of the production process.

The Factory Inspectors' reports, another major source for any inquiry into nineteenth century work, also need to be treated with care. Their evidence is strongest on the textile mills, the first great object of factory legislation and, until the Extension Act of 1867, its only subject. But even here it is limited and partial. The Inspectorate's investigations were largely confined to the objects specified by the Factory Acts, working hours and industrial safety. The historian concerned with job control and machinery, or with the class struggle at the point of production, will have to look elsewhere. Another limiting bias is the restriction of the inspectors' activities to the sphere of women's and children's employment. As a result they have little to say about industries, such as shipbuilding and engineering, where adult male labour predominated; while artisan trades are in many cases neglected, and the building industry ignored. Backyard industries and domestic outwork are also excluded from notice, because the small scale of the production unit disqualified them from protective legislation. It is not until the sweating inquiries of the 1880s that they were exposed to investigation, and then as a special problem rather than as an integral part of 19th century production. The map suggested by the factory inspectors' reports is thus cumulatively misleading; the lateral spread of 'modern' (factory) industry is exaggerated, while its artisan components are ignored, and comparatively little notice is taken of production in the workshop and the home.

To correct these biases the historian can make use of other more fugitive sources. Strikes throw up a great deal of incidental information about working arrangements, and show how custom and practice was shaped in the crucible of struggle; so do the arbitration proceedings which in the 1860s and 1870s were sometimes reprinted verbatim. County Court cases tell one a great deal which could never be learned elsewhere, prosecutions can often illumine the flash-points of resentment which at other times are hidden. Medical evidence can take one closer to the realities of industrial life than the Factory Inspectors' reports, as in Dr. Greenhow's 1861 survey of industrial lung diseases, or Dr. Edward Smith's report on the sanitary circumstances of tailors, printers and dressmakers. A rare though valuable source of evidence is that of workers writing about their own industries at the time, as in the reports by English artisans on the Paris Exhibitions of 1878 and 1889, or the accounts (mainly from the industrial crafts) collected by F. W. Galton in *Workers on their Industries* (1896). Branch reports in trade union journals can tell one a good deal about bargaining conditions, while the proceedings of annual or quarterly trade union meetings often contain extensive reference to the ways the machinery question was discussed. Above all there is the priceless evidence of autobiography, which makes it possible to translate technical process in terms of the individual's experience of work. Enough of them survive for the 19th century – or can be reconstituted through oral history – to provide the historian with a benchmark against which every other class of evidence can be measured.

These articles are based on an unwritten chapter of a half-finished book. The material had been lying fallow for some years and was brought to life as a result of discussions in the History Workshop collective. We were looking for a piece which would address itself to some area that was central in history teaching. 'The Machinery Question in 19th Century England' – the subject of my unwritten chapter – seemed to fit the bill, and was adopted in place of other alternatives on which I had been working. The piece was started in April 1976, but as a result of critical discussion in the collective, each of the sub-sections began to grow into chapters, and the piece is now far too long to be published in the journal, even in serial form. We have finally taken out three closely-related studies. The first, published in this issue of *History Workshop Journal,* is about the balance of steam power and hand technology in the 1850s and 1860s. It offers an epitome, brief, but in intention comprehensive, of the situation in the main areas of industrial activity, and discusses some of the obstacles to mechanisation. Its basic argument is that labour power was much more important than capital equipment in making Britain, at mid-century, 'the workshop of the world'. A second article, to be published in a future issue of *History Workshop*, is called 'Industrial design and the rise of alternative technologies'. It is about the consumer goods industries of mid-Victorian Britain and the ways in which mass production methods were pioneered. It also attempts to relate work to Victorian industrial design, and to show how the simplification of labour process was accompanied by an increasing elaboration of the product, as manufacturers competed with one another for the market. A third article, called 'Sweating', is primarily about factory industry, though it also discusses the domestic outworkers to whom the writings on 'sweating' usually refer. This argues that the capitalist achievement of the nineteeth century was accompanied by a radical deterioration in working conditions, which can be measured by the incidence of industrial disease, the rise in workplace temperatures, and the intensification of work loads. It looks at some of the hidden components of industrial activity, such as industrial haulage and cleaning, the stoking and charging of furnaces, the sorting and packing of goods, and argues that mechanisation in one department of production was usually accompanied by an increase of drudgery in others.

Glass blowers' furnace *Sir Benjamin Stone Collection, Birmingham Reference Library.*

II. Labour Power

Capitalism in the nineteenth century grew in various ways. Mechanisation in one department of production was often complemented by an increase of sweating in others; the growth of large firms by a proliferation of small producing units; the concentration of production in factories by the spread of outwork in the home. Sugar was refined in factories, like Messrs. Tate and Lyle's at Silvertown;[18] but sweets were manufactured for the million in back street kitchens and courts, as also were such popular children's purchases as ginger beer, sarsaparilla and ice-cream (among the manufacturers, in 1890s East London, were out-of-work dockers, victimized by the employers as a result of their activities in the Dock Strike).[19] Timber was sawn at the saw mills, where steam-driven machinery was, by the 1850s, very general; but it was shaped at the carpenter's bench, on the cabinet-maker's trestles, and at the cooper's coke-fired cresset. In ironmaking, the giant furnaces of the Black Country existed cheek-by-jowl with thousands of back-yard smithies, complementary in their action, yet radically distinct. The same was true of steelmaking and the cutlery trades in Sheffield, where 30 or 40 rolling mills supplied the working material of some 60 handicraft trades in which production was organized by outworking journeymen-masters.[20] Textiles were mechanised and accounted for far more steam power than any other trade – as can be seen from the figures in Table 1 – but the clothing trades, which increased by leaps and bounds in the 1840s and 1850s, depended on the poor needlewoman's fingers.

TABLE I
Steam Power in 1870
(Source: P.P. 1871 LXII (440), Return of Factories and Other Manufacturing Establishments)

	Number of Works	Steam H.P.
Textile Trades		
Cotton factories	2,371	280,602
Woollen factories	1,550	45,148
Worsted factories	599	47,140
Silk factories	692	7,465
Lace factories	223	998
Flax factories	155	8,670
Hosiery factories	126	777
Bleaching, dyeing, calico printing	341	14,789
Other textile factories	369	9,159
Clothing Trades		
Millinery, mantle, stay, corset and dressmaking	9,625	110
Tailoring	5,918	79
Ironmaking		
Blast furnaces	511	164,551
Iron mills	201	36,970
Foundries	1,310	20,022
Metal Working Trades		
Nails and rivets	1,424	3,925
Cutlery	379	447
Files, saws and tools	738	3,634
Locks	111	378
Brass finishing	256	1,786
Electro-plate	102	873
Buttons	254	353
Iron ship-building	48	3,536
Manufacture of machinery	1,762	36,473
Leather Trades		
Tanners and curriers	569	2,054
Saddlery	1,144	82
Boots and shoes	8,865	420
Gloving	68	8
Mineral Working Trades		
Potteries	517	4,679
Bricks and tiles	1,630	7,614
Other earthenware	446	2,797
Glassmaking	213	3,944
Glass cutting	81	110
Building Trades		
Builders	2,890	3,391
Marble and stone masons	1,213	1,372
Carpenters, joiners, etc.	5,518	3,016
Woodworking Trades		
Cabinet and furniture makers	1,821	718
Boatbuilding	167	114
Coopers	712	383
Food Manufactures		
Bakehouses, biscuits, confectioners	4,381	724
Breweries	860	4,466
Sugar refineries	33	2,356

The most complete triumph of the machine was in the cotton trade of industrial Lancashire. Elsewhere its progress was more halting, and there were major sectors of the economy – as Table 1 suggests – where down to the 1870s steam-power had made very little impression at all. Often its effects were secondary, applying only to the preparatory process of manufacture – or to the finishing – while leaving the main body of the work untouched: the case, for example, with firebricks.[21] In other instances it served to make handicraft labour more productive without impairing its skill – as in the example of glass cutting, where steam power turned the grinding wheels, previously worked by a man or boy assistant, but the delicate work of grinding, smoothing and polishing remained in the hands of the craftsman who had traditionally performed it.[22] In yet other instances steam power and machinery were chemical rather than mechanical in their action, and fuel-saving rather than labour-saving in effect. This was the case, in ironmaking, with Neilson's Hot Blast, which cut down coal consumption by about a half,[23] and in glassmaking Siemens' tank furnace. Even when machinery was extensively applied it by no means necessarily reduced workers to the status of mere hands; often its role was ancillary rather than commanding, and it may be useful to suggest a broad line of distinction between the textile industries on the one hand where, by mid-Victorian times, repetition work largely prevailed, and metalwork and engineering on the other, where the production process was discontinuous, and depended on craftsmanly skill. Mechanisation and steam power, in short, were by no means inseparably linked, and a vast amount of 19th century work was affected by them only at second or third remove.

Even in textiles the progress of mechanisation was uneven. It was faster in spinning than in weaving, in worsteds than in woollens, in cotton than in linen or silk. Silk was mainly a hand-loom trade at the time of the Cobden Treaty of 1860;[24] so were many branches of West Riding woollens,[25] while in the linen industry the power loom had only recently made its first appearance. The Nottingham lace trade was a factory trade almost from the start (though much of the machinery was hand rather than steam-powered).[26] But in hosiery, the other great East Midlands staple, the manufacture, so late as 1861, was 'largely . . . domestic'[27] and though the steam-powered rotary made swift progress in the following two decades, the finishing branches of the trade – seaming and stitching – remained almost entirely in the hands of out-workers.[28] In cotton spinning itself, the original site of the 'industrial revolution', hand-mule spinners held their own in the cotton mills right down to the 1870s, where they were employed on the finer counts, and maintained their own trade union organization. Before the Cotton Famine of the 1860s 'only the more venturesome' manufacturers would trust their fortunes to the self-actor mule entirely, and down to the 1880s the self-actor and the hand-mule, instead of competing with one another, were allotted separate departments of the work.[29]

Many trades in mid-Victorian times remained divided between machinery and handicraft sections. In armaments, for instance, the Enfield Rifle was manufactured by machinery on a system of interchangeable parts. But ammunition continued to be made by hand – at Woolwich Arsenal, when Andrew Wynter visited it in 1860, some 800 children 'aged from 8 to 12' were employed at packing cartridges[30] – while in the 'Gun' quarter of Birmingham the small arms trade was diffused among a multitude of small producers, with lock, stock and barrel as separate handicraft trades.[31] In newspaper printing, press work was transformed by the steam-powered cylinder machines, installed at *The Times* in 1814, but typesetting down to the 1860s, and in London until the introduction of Linotype in the 1890s, remained a work of the individual hand compositor, piecing together letters from a case.[32] In brewing, the twelve big London firms were among the largest manufacturing establishments in the kingdom.[33] They had been among the very first to order James Watt's steam-engine, when it was still in its patent stage,[34] and they conducted operations on a factory scale. But in the 1850s there were still thousands of handicraft brewers – no fewer than 1,800 in Birmingham alone.[35] Steam-powered flour mills were built on the waterfront to process imported wheat, but in the 1850s and 1860s by far the greater part of flour was ground by stone – in some places still in hill-top windmills.[36]

Dual technologies also existed together – in large establishments – under the same roof, as in the example already given of hand mules and self-actors in different departments of cotton spinning, or that of hand looms and power looms in silk mills and woollens.[37] At the Atlas Works, Sheffield, armour plates were tempered in the furnace, but dragged about the factory floor by teams of men in chains.[38] Another striking example from Sheffield, also illustrated in an accompanying engraving, is that of steelmaking, where the ingots were melted in the furnaces, but the crucibles which held them were hand-made on the same premises, and the ganister or clay from which they were made was prepared for manufacture by puddling it underfoot. Pot-making is a well remembered Sheffield occupation, of which we are fortunate to have an excellent autobiographical account,[39] and the process of treading it underfoot – very much akin to the wine-growers trampling of grapes – remained a skilled occupation down to at least 1914. Here is an account of it incorporated in a technical treatise of the time:[40]

Treading the clay,
The Mansell Collection, London.

When the materials are to be mixed by treading, they are spread out on a concrete floor and are sprinkled with water. The mass is turned over repeatedly with spades and, when it becomes too pasty to be worked in this way, it is again spread out and is trodden by men with bare feet, who squeeze the clay between their toes, and so mix it thoroughly. Each portion of the clay has to be squeezed between the toes, compressed and then pressed on to the previously worked paste. The treader stands in the middle and, working his toes, goes over the whole surface of the clay in a spiral direction, always working towards the edge. Having reached the edge, he turns round and walks in the opposite direction until he arrives at the starting point. Some treaders prefer to walk in straight lines instead of in a spiral direction. The trodden mass is then made up into balls of 40 to 45lbs. weight, and is afterwards beaten into a dense mass. In some works it is 'pugged' after being trodden.

In transport, too, dual technologies existed side by side. Iron steam-ships shared the coasts with thousands of collier brigs and sailing drifters and their cargo was taken ashore by dumb barges and dockers walking planks.[41] Railway engines revolutionised passenger transport, but road haulage remained the province of the one or two horse cart. Goods carried by rail increased from 60 million tons in 1851 to 410 million tons in 1900, while those carried by carts, waggons and vans increased in the same period from 106 million to 671 million tons.[42] Over the same period the number of railwaymen increased from 29,000 to 318,000, while those in road transport from 139,000 to 565,000.[43] The greater part of the mineral traffic was carried by canal, long after the coming of the railways.[44] At Messrs. Chance Brothers, Smethwick, the largest glass factory in the world, 'not a few' of the barges bringing iron, coals and glass-making materials were drawn by human beings: a visitor in December 1850 noticed, in a very short journey, 'no less than five barges, to each of which a man and an ass were yoked together', (the factory was engaged at the time in making the Crystal Palace).[45]

'Improvement' in 19th century industry – as the above paragraphs may suggest – by no means worked in a single direction, and in the following pages I want briefly to indicate, industry by industry, some of its characteristic paths.*

A. *Mining and Quarrying*

In coal mining, steam power transformed the scale of operations, while leaving the technology of hewing unchanged. Steam-driven fans were applied to ventilation, and allowed working places to proliferate, instead of being tied to the foot of the shaft. Steam-driven pumps were applied to underground drainage, and allowed the mining engineers to explore new and deeper levels, more especially in the second half of the century, as the shallower seams showed signs of exhaustion. But there was a total absence of mechanisation at the point of production, where the coal was still excavated by shovel and pick – 'tools of the most primitive description, requiring the utmost amount of bodily exertion to render effective'.[46] Mechanical coal-cutters were frequently patented and in times of strike high hopes were entertained by employers of the 'revolution' they might affect;[47] but in 1901, forty years after the first wave of patents, only 1½% of total output could be attributed to them[48] – a percentage which had still only risen to 8½% in 1913.[49] Output was increased not by mechanisation but by recruiting extra men. More and more hewers were needed as workings were extended both laterally and in depth. The numbers of hauliers (mainly boys) also increased: there was more coal for them to handle, and longer galleries to travel. Longer galleries also meant more roofs to prop, more roads to keep up, more rails to be laid down, while the increased use of blasting meant more hand-bored holes. The nineteenth century saw the creation of whole new classes of underground worker – 'stonemen' or rippers who had the job of extending the levels, timbermen to do the propping,[50] shot-firers to bore the holes. The mining labour force, which had stood at little over 200,000 in 1841 rose to 1,202,000 by 1911.[51] Animal power, too, was brought to production's aid, with the introduction of pit ponies for underground haulage: there were an estimated 11,000 of them in 1851, 25,000 in 1881, 70,000 by 1911.[52]

* Detailed discussion of textiles and the clothing trades is held over for a subsequent article.

Stone quarrying, like mining, was a sweat and muscle job, and little that took place in the nineteenth century impaired its labour-intensive character. Technological change characteristically took the form of 'improved' hand tools rather than of steam-powered machinery, as with the introduction of the 'patent axe' ('an instrument composed of thin slips of steel tightly bound together') which first found its way into Scotland about 1818 and by the end of the century was in use at granite-working centres all over the country;[53] or the spread of the 'feather and tare' method of stone-splitting, which replaced the old method of wedge and groove, but still required the quarryman's whole strength (the 'tares' had to be repeatedly hit with a sledge-hammer until the rock eventually broke).[54] Gunpowder was increasingly used in quarrying for the rougher class of stone, but boring holes for them was hardly less taxing than working with crowbar and pick: in Clydach Vale a foot an hour was considered a workmanlike progress ('You kept on turning and twisting the bar . . . like buttermaking in a churn').[55] The most widespread measures of 'improvement' were in haulage. Cranes were introduced into the Aberdeenshire quarries as early as 1835,[56] and by the 1900s there were few large quarries without at least the hand-operated version, while in some the stone was carried away on the overhead rails known (after the acrobat of that name) as 'Blondins'.[57] But in the smaller quarries – still by far and away the most common – the stone was dragged about by main force with men or horses to provide the traction. Fred Bower, who worked there briefly in the 1880s, describes the laborious procedure used on the Isle of Purbeck: the quarrier, having cut his stone, loaded it on to a 'bogie' or trolley, to which he harnessed himself 'just like a horse' and dragged it to the foot of the incline, from where it was drawn up to the top by a donkey- or mule-driven windlass.[58]

Clay-getting, though providing two major industrial raw materials, for pottery and bricks, was less affected by nineteenth century 'improvement' than any other branch of extractive industry, with the possible exception of salt. In the china clay industry of Cornwall, which shipped vast quantities of working materials to the Potteries of North Staffs., there were no mechanical aids and only the most primitive plant. The clay was puddled (i.e. sifted, washed and homogenized) in open-air pans, 'forty feet in diameter, and from six to ten feet deep';[59] it was dried by the wind and sun, with reed thatches to cover it when it rained. Pumping engines (bought second-hand from bankrupt tin and copper mines) were an innovation of the 1860s and 1870s; kilns came even later. The clay diggers worked with a heavy pick (the heaviest pick in Cornwall, according to an authoritative account published in 1875), and a long, square-mouthed shovel.[60] In the settling pans they stirred the mass of slime with a 'dubber' and brought it to consistency by trampling. The boy-runners, who had the job of haulage, carried the blocks of clay on boards, with a stiff leather shield 'to keep the wet clay from reaching one's chest'.[61] 'Balmaidens' (the women workers who cleaned the final impurities from the clay) used a small iron scraper 'resembling a . . . Dutch hoe'. In the 1860s they were expected to clean two or three tons a day for the princely sum of 1s.

In the Devon ball-clay industry, another major source of raw materials for North Staffs., the clay was cut dry in open-cast or underground pits, and the intermediate processes of puddling dispensed with, but the technology was no more sophisticated than that in china clay. When it came to carrying the clay down the Stover Canal to Teignmouth, men were used instead of horses for haulage, dragging the barges by rope.[62] Loading the ships at Teignmouth was also primitive, the clay being hoisted aboard by a simple expedient known as 'jumping': 'A rope was suspended from a pulley-block attached to the ship's rigging. One end was attached to the basket while the lumper grasped the other ... end and jumped from the ship's deck into the barge below with the effect that the laden basket was raised by his own weight'.

B. *Agriculture and Market Gardening*

In agriculture, cheap labour rather than invention was the fulcrum of economic growth, and the changes inaugurated by the agricultural revolution were accompanied by a prodigious increase in the work force, as well as by an intensification of their toil. Heavier crop yields needed more hands to harvest them, new methods of cultivation demanded a more intensive preparation of the soil – more draining and ditching, deeper ploughing, heavier and more frequent dressings of manure. Turnips in particular, the pivot of the new rotation of crops, demanded an almost gardenly care. The soil had to be in such a state of pulverisation 'as to fall from the plough like meal'[63] with ploughings, harrowings, rollings, pickings and liberal dressings of manure before it was ready for the seed, and once the crop had begun to appear the earth had to be continually prodded and poked; sometimes there would be three different hoeings.[64] The extension of arable farming into upland, heath and fen – another feature of the 'New Husbandry' – also made heavy demands on labour, with more claying or marling to bind the sandy soil, more manuring to bring it into a state of cultivation, more stone-picking and weeding to keep it clean.[65] This was the origin of the gang system in Norfolk and the Lincolnshire Wolds. In harvesting, the most labour-intensive sector of farm work, the main innovation in the early and middle years of the nineteenth century, the replacement of the sickle by the scythe, took place within a hand technology.[66] Although reaping machines had been in use since the 1820s, it was not until the last quarter of the nineteenth century that they began to make serious inroads into the work, and not until the coming of the reaper-binder, at the very end of the century, that farmers could dispense with the seasonal need for extra hands.

Market garden crops were even more labour-intensive than wheat. Wage rates, even by the miserly standards of the British farmer, were exceptionally low, and much of the labour was performed by the grower and his own family: even so labour costs per acre were calculated at anything up to £11 per acre in the 1870s, compared with between £1 and £2 per acre on an arable farm.[67] Much of the ground was dug by spade husbandry – i.e. trenched rather than ploughed;[68] many of the seeds were dropped by hand rather than sown by horse-drawn drill; potatoes were individually planted.[69] Heavy and frequent dressings of manure were administered to offset the intense exploitation of the soil. In rhubarb farming, wrote Shaw in 1879, describing the practice in the Surrey market gardens, 'rank litter' was used, but because of the softness of the newly-dug soil it had to be carried by basket, some people being employed to fill them, others to carry them on their heads, 'and a few more to empty their contents over the crowns of the plants'.[70] Earthing-up was by hand-held trowels: it had to be done two or three times in the case of celery 'to suit the growth of the plants'.[71] Other crops, too, needed an individual treatment: cauliflowers had to be transplanted, onions and cucumbers tied, rhubarb pruned of leaves 'sometimes thrice in the season'.[72] Weeding – generally a job for women and children – was done with a sharp-bladed knife rather than a hoe, in order to get close to the crops, and baskets rather than wheelbarrows were used to carry the weeds away.[73] Much of the harvesting was done on all fours.

A great deal of market garden work was done by contract. The grower, who bore many resemblances to a peasant proprietor (and some to a kulak), worked by rule of thumb. For much of the year he might employ no regular labour at all except for that of his family. But he would engage poor people in the locality – women and children especially – to take on particular tasks, and for the larger jobs he would engage with labour contractors or itinerant labourers, drawn from further afield.[74] They would be employed not only for picking and pulling but also for other jobs, such as hoeing and weeding. Patrick MacGill was one of those who engaged at it, working with a gang of men and women for a potato farmer in Buteshire, Scotland. The men would dig potatoes from the ground with a short three-pronged 'grapis', a kind of fork. The women followed behind, crawling on their hands and knees, and dragging two baskets a-piece into which they lifted the potatoes thrown out by the men:[75]

> The job, bad enough for men, was killing for women. All day long, on their hands and knees, they dragged through the slush and rubble of the field. The baskets which they hauled after them were cased in clay to the depth of several inches, and sometimes when emptied of potatoes a basket weighed over two stone. The strain on the women's arms must have been terrible .... Pools of water gathered in the hollows of the dress that covered the calves of their legs. Sometimes they rose and shook the water from their clothes, then went down on their knees again. The Glasgow women sang an obscene song, 'just by way o' passing the time', one of them explained .... Two little ruts ... lay behind the women in the black earth. These were made by their knees.

c. *Food Industries*

Food processing in mid-Victorian England was perhaps less subject to technological improvement than any other branch of production. Vegetables were dressed for the market by hand. Earth-stained crops, such as celery and radishes, were washed and bunched by women and children, working for market gardeners in their sheds.[76] At Covent Garden, peas were podded in the market itself and sold 'ready for the saucepan' according to their respective size (the work was chiefly in the hands of old women, working at the rate of 1s. or 18d. a day and recruited by salesmen from the local workhouse).[77] Pickling too, though increasingly a factory trade, was mostly done by hand. The vegetables had to be soaked in brine, diced or sliced to size, liberally sprinkled with vinegar and then 'artistically arranged' in jars.[79] It was cold winter work (the vegetables had to be kept in the cold for fear that they would rot) and chapped hands and cuts are remembered occupational hazards.[79] Onions were particularly labour-intensive as they had to be individually peeled. 'Consequently in a bottle of pickles every onion is always visible from the outside though perhaps a cabbage may fill up the middle – an onion is never allowed to enjoy oblivion.' There was more machinery in jam-making, where by the 1880s steam-jacketed boilers reduced the fruit to pulp, but the preparatory stages were performed by hand. The fruits were sorted out into their respective qualities and the damaged parts cut away, oranges peeled, lemons squeezed, soft fruit separated from brambles. Extra labour was taken on at the jam factories during the height of the fruit-picking season, and sacked when it was over.[80] In the meat trade, cattle continued to be brought live to market and to be sold by dealers 'on the hoof' – i.e. for the purchaser to kill. The great majority were taken by family butchers, who did the slaughtering on their own premises[81] (it took four men to kill a bullock, fewer for pigs or sheep).[82] Not until the 1880s did anything like a wholesale trade develop on any scale, though in the 'bad meat' trade – whose physiognomy can be studied in the cases which came before the police courts – it seems that there was a class of travelling dealers who supplied the 'slink' butchers with their wares.[83] Cheese, one of the very few proteins to form a regular part of the labourer's diet (even workhouse inmates had a ration) was made by a long and protracted process in the farmhouse. Stilton required up to half-a-year for ripening; Cheshire and Cheddar cheeses had to be turned daily for three months. Cheese-making involved a great deal of heavy lifting and carrying, and for the dairymaids who performed the work – or the farmer's wife and daughters – the process was intensely laborious, as well as long drawn-out: 'in many dairies . . . not only the hand, but also the whole arm is immersed in nearly every operation'. Most farms seem not to have made use of either chemical or mechanical aids. The heat of the milk was tested by finger-tips rather than thermometers; curds were broken by hand or knife. A semi-mechanical separator was patented in 1834, but farmers were very slow to take it up, even when it proved successful, and down to the 1880s it seems to have been very little used in Somerset and Cheshire, two of the leading cheese-making counties. 'The women do all the work', wrote Acland in his report on Somerset, the home of Cheddar cheese:[84]

> It is true the men see the cows milked at a very early hour in the summer, and have some trouble with them in the winter, but the real hard labour falls on the women; and very active and industrious they are; but it is a sad sight to see a man standing by doing nothing, while his wife or daughter is turning many times a day a weight above half a cwt . . . Cheese-turners used in the Midlands are not in use in Somerset . . . the farmers who have heard of them say they would occupy too much space.

In the fish trade, too, a purely hand technology prevailed. In the middle years of the nineteenth century more and more boats were engaged at it, especially in the East Coast herring trade, and there was an enormous increase in the volume of the catch. New docks and harbours were laid down to cope with the increased traffic, while the coming of the railways opened up the inland markets. There was a vast increase, too, in fish manufacturing (over half a million barrels of herring a year were being cured in the 1860s), but working practices remained primitive. The Scotch herring girls, who followed the herring down the coast, from North-east Scotland to Lowestoft and Yarmouth, worked ankle-deep in the fish, equipped with nothing more than a short-bladed knife. (The barrels themselves were hand-made by journeymen coopers who migrated to the herring ports for the season.) Smoked haddocks – 'the especial dainty of the breakfast table' – were cured over peat and sawdust fires. (Coke was substituted in the back slums of Limehouse and Camberwell, where fish curing remained a cottage industry down to the 1900s.) Bloaters, the favourite fish of the London working class, were prepared in even simpler conditions, since they did not need so much fire. Whelks and winkles were boiled and salted in the backyards of the dealers, and made more attractive by polishing the shells (one of Mayhew's costermonger informants employed children at ½d. an hour to do the works). The potted shrimp trade, which provided the working class of Lancashire and Yorkshire with a tea-time relish, was hardly more advanced: for the most part it was conducted in the outhouses of Marsh Side, Southport.[85]

In baking the only mechanical triumph which 19th century capitalism could record was the mass production of biscuits. Milling remained dispersed among thousands of local millers, and depended for motive power on wind and water rather than on steam: not until the spread of roller milling in the 1880s, and the massive importation of foreign wheats, did modern capitalist industry take their place. Bread-making itself was notoriously backward. Marx in *Capital* scathingly described its technology as 'pre-Christian', and Seymour Tremenheere, in his report of 1862 on the journeymen bakers of the metropolis, was hardly more complimentary. 'There is probably no trade supplying a vast and constant demand which has so completely remained in its primitive condition . . . as the baking trade'. Numerous projects for improvement were canvassed, but except in Birmingham, where steam bakeries seem to have been established in the 1860s, results were disappointing: Dr. Daugleish's much trumpeted 'aerated' bread – which used carbonic acid instead of yeast – proved both expensive to manufacture and tasteless; Mr. Stevens' 'Patent Dough-Making Machine', though it impressed Seymour Tremenheere, seems to have been ignored. Rising demand was met not by mechanisation, but by a proliferation of 'under-cutting' bakers, and a marked increase in over-work and sweating.

The London bakehouse of the 1850s and 1860s was as often as not the underground appurtenance of a shop, a single cellar which served simultaneously as storeroom, manufactory and oven, as well as providing the journeymen with planks to take their naps on, and sacks to serve as pillows, while they waited for the dough to rise. The only machinery in use was the rotary action of the workers' hands and fists. Blending was in many cases performed by manual labour entirely, the different flours being emptied into a bin or trough and 'thoroughly mixed' by hand or spade. 'Making the ferment' involved mashing boiled potatoes, and then mixing them with yeast, flour and water as a thin, runny mass in a trough. 'Putting in the sponge', the next stage of the process, involved 'very hard work' in a swill. 'Making the dough' which followed was by far the most laborious part of the work.[86]

> The process is usually carried on in some dark corner of a cellar, by a man, stripped naked down to the waist, and painfully engaged in extricating his fingers from a gluey mass into which he furiously plunges alternately his clenched fists.

The bakehouse visited by Andrew Ure was little different, it seems, from those described by Eddie Dare (see below p. 138).

In the sweet trade, as in bread, increasing demand was met by a proliferation of small handicraft producers. In some cases they worked out of their own kitchens – like William Luby, the Ancoats sugar-boiler, whose early experiences are recorded in *Useful Toil*; others used the back of a shop. Apart from access to a fire, very little was needed in the way of plant. Syrups and creams were made with a spatula, boiled sweets with a ladle and pan, lozenges with a rolling pin and cutters. Technology was not necessarily more sophisticated in the little factories which enterprising capitalists began to put up in the 1850s and 1860s. Messrs. Wotherspoon, for instance, the Glasgow confectioners, who produced 'really fabulous' quantities of sweets – their trade list included no fewer than fifty-two different classes of lozenges – patented several machines for superseding hand-cutting, but they were all found impracticable 'in consequence of the difficulty of keeping the cutters clean', and when J. D. Burn visited their works in 1858 they had reverted to cutting by hand. Messrs. Schoolings of Bethnal Green were equipped, according to an admiring observer in 1866, with 'the most perfect' steam appliances in the trade. But 'Surprise Nuts', one of their most popular selling lines, were all individually filled by hand, and the nuts themselves hollowed out by girls with pen-knives and rosecutters. Here is an 1858 account of comfit-making at an Edinburgh sweet factory:[87]

> A large copper pan, probably 3 feet in diameter, containing the seeds, is suspended by a double sling from the roof over a low fire or stove. A quantity of sugar is dissolved in water and kept standing in a vessel by the workman's side. He then pours over the seeds a spoonful of this dissolved sugar, and, by a process of shaking and mixing in the hot pan, every seed is made to take on an individual layer. The heat and . . . friction soon cause the water to evaporate and the sugar to dry on the hot seed. Another dose of the liquid is then given, another process of shaking and mixing, and thus the manufacture proceeds, until at length the seeds have attained to an enormous bulk, and then the heat is slightly diminished, and the friction increased for the finishing coat. The whole process . . . requires great skill in the manipulation, and it also requires the most severe and continuous muscular labour. We know, indeed, of no other kind of labour that requires more. Not a muscle or joint of the whole body remains inactive. It has quite a marvellous affect in taking down superfluous fat. It is well known that a stout man, taken perhaps from lozenge-making, and put to work this hot pan, becomes in six weeks converted into a living skeleton.

D. *Building and Construction*

The building industry, like mining and agriculture, was labour-intensive and increased output was achieved by putting on extra men: the work force recorded in the census was 376,000 in 1841, by 1901 it had risen to 1,216,000. Building and construction was one of the fastest-growing sectors of the economy in mid-Victorian times, and accounted for between 20 and 30 per cent of gross domestic fixed capital formation, rather more than twice the amount attributed to cotton.[88] But the scale of enterprise was characteristically small, and investment, whether by master-builders or sub-contractors, went on labour and materials, not on plant. The main thrust of technical innovation, such as it was, came in the direction of labour-saving materials rather than of mechanical devices. In the 1850s and 1860s their influence was comparatively slight. The painter still mixed his own colours; bricklayers still cut and shaped their own bricks (so late as 1874 it was considered a more important part of their work than setting); carpenters and joiners worked, very often, to their own designs.[89]

Road-making, one of the earliest 'improvements' associated with the industrial revolution, was transformed by the coming of the turnpikes, and the new surfaces of Telford and Macadam. Large numbers found employment at it, both paid and unpaid: among them, in 1832, were some 32,000 parish paupers.[90] But though road surfaces were revolutionised, the technology of labour was unchanged. Stones were broken up by sledge-hammer, a weary labour at the roadside or in the parish pits; and the new surfaces were trodden down by foot and pitched and packed by hand (the great turnpike roads had to be very closely wedged, 'so as to form a regular convexity' between the centre and the channels).[91] The coming of asphalt seems to have made the work, if anything, more severe. Smoking loads of it were shot from wheelbarrows, raked to and fro to eliminate the harder knobs, and then pressed with red-hot irons, hissing and scorching as the surface set.[92] Many English workmen refused to be employed at the work (despite comparatively favourable wage-rates) on grounds of the intolerable heat; and the London companies who pioneered the process used French and Italian workers to take on some of the more fiery work.[93] Steam-rollers, first manufactured and marketed in England by Aveling and Porter in 1867,[94] were slow to be adopted, and it seems that stone-crushers, too, were little used. In Chamberlain's Birmingham, a go-ahead city when it came to 'improvement', they were grudgingly accepted by municipal officials: the Corporation Surveyor, in a report of 1874, complained that with Blake's Patent Stone-Crusher, one of the most advanced models on the market, some 16% of the stone was lost in dust.[95]

Paving stones – such as those which figure so largely in our cover illustration – were entirely hand-made, and notwithstanding the vast increase in their manufacture, little seems to have been done to alter the character of the work. At Mountsorrel, Leicestershire, one of the great centres of the industry, a variety of hammers were used: the 'burster', 'an immense tool weighing 30 lbs', which was used for breaking up the large blocks and irregular lumps, the 'knob hammer' for knocking off the larger knobs, and the 'squaring' hammer, 'universally employed' for squaring setts. In Aberdeenshire too, the other great centre of granite paving, sett-making remained a work of the hammer down to the end of the 19th century. 'The dressing of granite by mechanical means is a problem that has engaged the attention of engineers for many years', wrote Powis Bale in 1884, 'and is still, practically speaking, unfulfilled'.[96]

Mid-Victorian town 'improvement' owed more to artisans than to civil engineering, and the monumental art forms favoured at this time gave a wide field for the exercise of their skills. Banks competed for grandeur in columns of Bath and Portland stone, Doric, Ionic or Corinthian, according to the architect's whim. Churches – a major growth industry down to the 1880s, and by far the wealthiest of the architect's clients – blossomed out in revived Gothic; warehouses and offices went baronial, with turrets, gables and keeps. Stonemasons were very largely responsible for these ornate facades though bricklayers played more part in the revived Queen Anne style of the 1870s. Stonemasons also serviced the mid-Victorian boom in town halls and exchanges, the rage for civic statuary and clock towers, and the proliferation of museums, galleries and halls.[97] All of their work was hand-tooled, and it was not until the very end of the century that stone-working machinery began to make serious inroads on their craft. Most of the decorative work was done by banker masons in the yards, but the stone carvers worked on site, as did the 'roughers out' – the walling masons who shaped the stones for fitting.[98]

Railway building was very largely pick-and-shovel work, notwithstanding the achievements of the civil engineers. Contractors were niggardly in their fixed investment, lavish in their use of hands, and at the height of the 'mania' of 1845-7 were thought to have had some 300,000 men in their employment. Bricks were manufactured for the viaducts and embankments on site. Millions of tons of earth, sand and gravel were shifted by the spade – twenty tons a day was the normal navvy stint in the 1850s.[99] In tunnelling, the rock was shifted with gunpowder, but bored by handworked drills; heavy clays, such as those encountered by the London and Birmingham railway at Primrose Hill, were broken up by hatchet and cross-cut saw.[100] Hoisting devices were primitive, with nothing more elaborate in the way of handling equipment than cumbersome horse-drawn gins. On deep cuttings, like the Tring works of the London and

Birmingham line, where some 1,400,000 yards of chalk had to be removed before the track could be laid, tipping was conducted on the principle of the see-saw, with a man and horse as weights. There is a vivid description of the work in F. S. Williams *Our Iron Roads* (1852):[101]

> Runs, as they are called, are also made by laying planks up the sides of the cutting, on which barrows may be wheeled. The running is performed by stout young men, round the waist of each of whom is a strong belt, fastened to which is a rope running up the side of the cutting, and turning on a wheel at the top, whilst to the other extremity a horse is attached. The barrow being laden . . . a signal is given to the driver, who leads the horse quickly out a given distance into the field, and thus the man is drawn up the acclivity; the contents of the barrow are emptied, and the horse being led back the rope is slackened, and the man runs down the plank again, drawing the empty barrow after him.

In domestic housebuilding activity expanded with little more in the way of mechanical aids than the wheelbarrow (even this could be controversial – Manchester building workers came out on strike against it in 1864).[102] There were no pile drivers to dig the footings, or, in many cases, hoists to raise the parts: 'it was quite customary to see labourers staggering up a ladder, carrying a window-head or a sill balanced on their shoulders, using both hands to grasp the sides . . . of the ladders'.[103] All of the work was done by hand tools. House carpenters would have an elaborate set of them, which it could take them an hour or two to grind.[104] Others were simpler: the slater worked with an all-purpose hatchet, pointed at one end, called a 'zax',[105] the gas-fitter with soldering-irons and a blow-pipe,[106] the bell-hanger with a pair of combination pliers which in the 1880s could be bought for as little as 2s. 4d.[107]

Despite the existence of a few large contractors, such as Thomas Cubitt in London, and Johnson of Manchester (the man whose home was blown up by the local brickmakers in 1870), domestic housebuilding fell very largely within the province of the small master economy, with jobbing men acting as sub-contractors for portions of the work, or venturing into business on their own. Often the master would be an enterprising carpenter with nothing to lose by failure except a return to the journeyman's ranks, or a man out of work for whom starting up as a 'capitalist' was an alternative to unemployment.[108] 'A £5 note and plenty of cheek' was said in the 1880s to be enough for a man to set up on his own. The plumbing and the slating could be farmed out, the scaffolding hired, the raw materials obtained on tick (the initial load of bricks had to be paid for on the nail, but the cash for that could be borrowed).[109] Rowland Kenney has left a good example of a job like this which he took on, together with a mate, in the 1890s.[110]

> Fred learned of a possible job at Standish, a few miles the other side of Wigan. A small local contractor had been offered the job of building a couple of cottages . . . a considerable cut above the small places in which most of the local miners lived. The 'contractor' was a joiner who had the sense to see that there was going to be some building development in the neighbourhood, and that he would do better contracting than following his trade as a journeyman joiner; but he had no knowledge of bricklaying . . . It suited him better to sub-let the bricklaying part of the job, and here was where I could perhaps be useful. I was 'quick' at figures and Fred was not; so the suggestion was that he should contract to do the brickwork at so much a square yard. I should be labourer, and also do the measuring up and generally act as clerk of the works. As occasion served I was gradually to take over trowel. We had visions of becoming masters instead of men, and tramping and roughing it seemed likely to be things of the past for me. The scheme did not work out well at all. Whether we had taken the job too cheaply I do not know – the amount Fred had asked per square yard struck me at the time as remarkably small . . . he had been very reassuring when I questioned about it . . . but the job never paid even full wages.

E. *Building Materials*

In the building materials industry, as in building work, increased activity was not matched by technical change. Of the major building materials only wood was seriously affected by machinery, and its influence – for reasons which will be indicated later – was definitely limited. Window-glass was hand-blown at the furnaces, window fittings and door furniture (in ways which will be described in a subsequent article) were hand-hammered at the forge. Builder's lime, the chief ingredient of stuccos and mortar, was reduced to powder at hand-charged furnaces; builder's sand was dug by spade.[111] Cement technology was a bit more advanced, with clays being washed in a mill, but most of the work seems to have involved heavy digging, lifting and carrying, in dismal conditions (at Harwich in 1849 some 500 people were apparently engaged at dredging the harbour to get stones for a local cement works).[112]

Except for marble (extensively used in the manufacture of chimney pieces)[113] almost every class of builder's stone was hand-dressed. Aberdeen granite, a popular architect's embellishment for the more imposing class of shop-front, was both shaped and polished by hand, with emery paper to produce the final gloss when the surface had been smoothed. The hearthstone raised at Godstone, Surrey, was hewn into shape by a peculiar double-headed axe.[114] The window sills and doorsteps manufactured on the Isle of Purbeck, and exported by stone boat to the southern English towns, were shaped by mallet and chisel.[115] Slates, which in the third quarter of the nineteenth century largely displaced tiles as roofing materials, were both hewn and shaped by hand, principally in the slate quarries of Bethesda and Dinorwic. Each slate had to be individually treated, according to the peculiarity of the rock. 'The splitter, the real aristocrat of the quarry craft, had to be able to tell at a glance what size and quality . . . he could coax out of a particular block . . . to glance at a slab and recognize . . . posts, crychs, bends, sparry veins, faults, joints and hardened rock.'[116]

Bricks, like slates, were individually hand-made, though manufactured in prodigious quantities (150 million a year, for instance, on the great Cowley brickfield, which stretched from Southall to Slough).[117] The work, which was sometimes undertaken by family groups or gangs, sometimes sub-contracted, was dirty and laborious, and involved a whole series of exhausting tasks, from the clay digging of the autumn and winter months to the setting of the clamps or kilns (20,000 bricks was an average firing in the 1850s). Stones were picked from the clay by hand, and the puddling or tempering, which reduced the clay to plasticity, depended on bodily weight, the clay being turned over repeatedly by the spade, mixed with water, and then trodden underfoot: the labourers who did this (sometimes horses, sometimes men, sometimes children) were said to develop 'puggers' feet' from the sensitivity with which they could detect the slightest intrusive pebble.[118] Moulders, too, worked in close physical proximity to the clay, pressing their bricks into shape with rapidly working fingers, and taking off surplus pieces with a scraper, or dextrous flicks of the thumb ('wrist-up' or 'lack of joint oil', according to a Chesham brickmaker, was an occupational hazard, 'if you started off too rash').[119] Numbers of different machines were projected to streamline the work – in 1856 over 230 patents were in existence for machine-moulding alone[120] – but down to the 1880s (and in the southern brickfields a good deal longer) most of them came to grief, either because of the obduracy of the raw materials (London clays were particularly unyielding) or because of the inferiority of the finished product. Only the pug-mill, a horse-ginned mixer, was widely adopted, but it was too expensive for the smaller yards, and in the 1870s there were still many districts where it was comparatively rare. 'Pug mills are not general in Essex and Suffolk', Factory Inspector Lakeman reported in 1872, 'the earth is trodden by children who are kept at work tempering a heap of clay from morning to night'.[121]

The ingredients of the brick were also got by hand. In London they were drawn from a wide hinterland, though the clay itself was dug on site. Lime, which the moulder used to bind his clay together, came from places like the Merstham limeworks, Surrey.[122] Sand, sprinkled on the inside of the moulds to prevent the clay from sticking, was dredged from the river Thames at Woolwich, spread out thinly to dry, and repeatedly raked over 'so as to expose every particle in succession to the sun's rays'.[123] Ashes, 'thoroughly incorporated' with the brick earth while the clay was being puddled, both to prevent the warping of the bricks while they were drying, and to assist in their colouring when burnt,[124] came from the great metropolitan dust heaps at places like Paddington Basin. So did the 'breeze' or cinders which the brickmakers used as fuel (the larger suburban makers would order as many as 155,000 chaldrons at a time). Sifting the dust for 'breeze' was a woman's occupation in mid-Victorian London, paid for at a rate of 1s. a day plus a perquisite of cinders. The 'hill women' who performed the work had the job of sorting out the cinders according to their respective size (the larger ones were reserved for the laundries), making separate piles for the different classes of debris (linen rags went to the papermakers, woollen ones to the shoddy mills of the West Riding), and heaping up the ashes and the 'breeze'. They were armed with large iron sieves which they held waist-high and riddled, while the men threw up the stuff.[125]

Painter's colours were produced in even more barbarous conditions – those of the Victorian lead works, the most poisonous, perhaps, of all the 'dangerous' trades.[126] Paint was a standard product used in great quantities. It should have been a prime candidate for self-acting machinery, especially in view of the known dangers of lead-colic and 'drop-hand'. But in fact, machinery was only used for crushing and grinding – in most factories all the other processes were done by hand. The women who did the stacking (men, it seems, could not be found to do the work because it had such a bad name) had to carry heavy pots of lead on their heads, from 30 to 50 lbs in weight, and to climb ladders from 10 to 15 feet high. They did the work, as a Newcastle manufacturer gratefully acknowledged, with 'truly marvellous skill and rapidity'. Lifts were installed in place of ladders at a leading Glasgow works, partly to save the women fatigue, and also because it was 'upon all accounts . . . a very much more respectable way of doing it than going up ladders for women'. But the Newcastle manufacturers refused to follow suit, partly, it seems, on grounds of productivity, and partly on that of cost ('ladders would be, perhaps . . . less expensive . . . because they last a long time, and it is not like machinery that wants to be put in order . . . I think we could fill a stack almost as rapidly by ladders as with the hoist'). The puddling and firing of the lead, to reduce it to powder, also involved a more or less continuous risk of poison. Mechanical stoves had been invented to take on portions of this work, but manufacturers in 1893 were still making little use of them, both on grounds of quality ('there was a prejudice in the trade as to the lumps') and also on grounds of cost. 'No perfectly satisfactory mechanical stove has yet been devised, and their very high price is also a bar to their general introduction'.[127]

F. *Glass and Pottery*

Manufacturing in other classes of mineral work was as labour-intensive as it was in lead or bricks, though demanding varying degrees of athletic strength and skill. In chemicals the crucial operations were performed by men with long iron rods stirring vats, and furnacemen sifting the powder in kilns. In saltmaking the lumpers, who had the making of the finer class of salt, worked as high-speed rakers over boiling pans of brine. In glassmaking everything was made by hand, whether it was destined for the sideboards of the aristocracy, the shelves of the pharmacist, or the counter of a back-street bar. Even in glass bottlemaking, a manufacture for the millions, every item was individually made, the gatherer taking the 'metal' from the furnace at the end of an iron rod, the blower shaping the body of the bottle with his breath, while the maker who finished the bottle off ('the most difficult task') tooled the neck with a light, spring-handled pair of tongs. Glassmaking was a factory industry in which, however, as late as 1895, 'no machinery' had been applied at all, in spite of the vast proliferation of its products. The glassblower's tools were few and simple, indeed the fewer he used the better so far as the quality of the work was concerned: 'the more these shape the article, the more imperfections are likely to be produced by the scratching and rubbing of the irons'. The chief instrument of production was an iron blow-pipe – 4 or 5ft. long, and with a slightly enlarged nose at the end. In bottlemaking the glass was gently rubbed against a small metal plate known as a 'marver' or blown into a mould; in window glass rolling, the sheets were first blown up into cylinders and then flattened on a rolling table, 'a cast-iron slab of sufficient size to accommodate the largest sheet which has to be rolled'.[128]

Sheet or 'cylinder' glass, the great innovation of early Victorian times (it was first introduced to this country by Messrs. Chance Brothers of Smethwick in 1838),[129] cheapened production by making it possible to use inferior sands in the mix, and brought a whole new trade in window glass. But it owed everything to skilled workmanship, nothing at all to plant. On the part of the blower, swinging and rotating a molten mass of glass, it required both delicacy and strength, delicacy to prevent the glass being mis-shapen or marked, as it swelled on the end of his blow pipe, strength to blow with sufficient force to make a six-foot cylinder.[130] The sixteen acres of glass which covered Mr. Paxton's 'Palace of Industry' were blown into shape in this way, and there is a vivid description of it by George Dodd, written after visiting Messrs. Chances' works at Smethwick:[131]

> The workman takes up, on the end of his iron tube, a large mass of glass, and whirls it about in an extraordinary way. He swings it round in a vertical circle, blows into it, swings it again, rolls it, until at length it protrudes beyond the end of the tube, as a cylinder, perhaps four feet long by ten inches in diameter. This cylinder is easily opened at the two ends and detached from the tube. A hot iron wire severs it in a line from end to end; and the cylinder by careful heating and running, opens out into a square sheet, flat, but slightly wrinkled on the surface.

Glass finishing – i.e. ornamentation and polishing – was also a handicraft work, though steam power was applied to the turning wheels. It was a distinct class of labour, carried on for the most part in separate establishments and had more of both drudgery and art than glassblowing. In the best class of ware, such as wine decanters or cut-glass chandeliers, the worker was his own designer, shaping the vessel on the wheel. Accuracy of eye and steadiness of hand were indispensable. 'Those arabesques, stars, running lines, vandykes, and often intricate traces which appear on the most costly glass fittings... are produced... without any previous drawing... the workman, holding the glass in his hands, with no more indication of the pattern than two or three pencilmarks dividing the globe or plate, grinds out the pattern on a wheel of Craig Leith stone, guiding the brittle material by eye and hand with a precision and rapidity perfectly marvellous'. In the cheaper classes of ware, such as hall lamps for boarding houses, or frosted glass for pubs, much of the work was done by women, pressing palmfuls of wire and sand against the globes. Polishers, who made the dull surface bright, rubbed it up with pumice stone, and then gave it a finishing sheen with jeweller's rouge or putty. This was largely women's and children's work, and involved a series of processes that were wet, monotonous, and, in the case of the puttying, poisonous, since the putty was made of lead. (Puttying was especially a feature of high class work, and was used to bring out the fine gloss and white colour 'so much admired in cut glass of the most expensive qualities...').[132]

Despite the absence of mechanisation, glassmaking was subject to a whole series of developments which cheapened production and led to a proliferation of new lines. One important influence was the removal of the Excise Duty in 1845, which led to a great burst of activity – the bottle manufacture is said to have increased fourfold in the following twenty years, while the population of Castleford, a new Yorkshire centre of the industry, increased from 2,000 to 10,000 in the same period.[133] Another was the introduction of the Siemens tank furnace, which allowed for a much more continuous working than the clay pots it displaced, as well as bringing substantial economies in fuel.[134] Then there was the development of new types of commodity, such as the stained glass windows fostered by Ritualism and the Gothic Revival. In the flint glass trade the most striking improvements were in what an 1866 commentator described as its 'artistic' development[135] – the adoption of engraving as a method of ornamentation, the introduction of coloured glass and the use of new decorative processes, such as embossing.[136] None of these changes affected the basic technology of work. The blowpipe, 'invented sometime during the three centuries preceding the Christian Era',[137] remained the fundamental tool in every class of manufacture except for pressed glass.*

The manufacture of china-ware and crockery presents an even more striking instance than glassmaking of mass production on the basis of hand technology. In spite of the early appearance of a factory system (pioneered in the 1760s and 1770s by Josiah Wedgwood), and an intense development of the industry in North Staffordshire, a very simple technology prevailed – 'the same essential

* The persistence of handicraft methods no doubt owed a good deal to the glassworkers' trades unions, which were quite exceptionally strong in all the main centres of the trade – the glass bottlemakers of Castleford no less than the flint glassmakers of Stourbridge. They maintained strict production quotas or 'tantums', controlled worker hiring and firing, and operated their own work sharing systems when employment in the trade was slack. 'The relation of master and man in the blown flint and bottle trades amount to a chronic strike', a Tyneside manufacturer wrote in 1864, complaining of the 'insensate' worker opposition to improvements: 'Large orders have been transferred from the Tyne to Belgium, and the manufacturers here purchase foreign glass, for the production of which they have every appliance at home, except labour at a reasonable cost'.

appliances as were used in Egypt four thousand years ago'. 'There are none of the noisome adjuncts and deafening sounds of the huge cotton or woollen factory, where hundreds of hands are working together, and have to feed the gigantic machinery which dwarfs them all', an American visitor to the Potteries wrote in 1871, '... There are a large number of rooms, and only a few operatives in each. Each potter works independently'. In the 1870s machinery, in the form of steam-powered 'jiggers' and 'jollies', was slowly making its way. But it advanced faster in the out-potteries, such as those of Glasgow and Newcastle-upon-Tyne, than in the hinterland of the industry, at least partly, it seems, because of the strength of the workers' opposition. In 1889 its impact was still sufficiently limited for a Factory Inspector to report that 'by far the larger number of china manufacturers do not employ steam power'.[138]

Pottery work was very physical, and involved, in every department of production, a more or less continuous personal handling of the clay. The thrower shaped his ware on the wheel 'by the exquisitely fine touches of... thumb... finger and palm'. He had to keep his hands constantly wet 'so as to mould the clay with the least possible friction'. So did the handlemakers. 'The operator... gives it the required shape... with no other tools than his wet fingers, a drop of adhesive liquid, and a moist sponge.'* Platemakers worked in a similar way, except that the final shape of their ware depended on flicks of the wrist. Dippers prepared the ware for firing by steeping it in a glaze tub. 'The right hand and arm... are plunged nearly up to the elbow as he passes the piece of ware through the liquid... Then with a rapid circular movement of the hand... a movement that only years of practice can teach... he makes the glaze flow thin and even over the surface, and shakes off what is superfluous'.[140] Motive power in the Potteries, such as it was, also depended on the human body. The thrower's wheel was turned by a 'baller' ('generally a female') whom he employed full-time to keep the wheel on the go, and feed him with small lumps of clay.[141] The turner's wheel was kept moving by a lathe-treader, again a woman, who was paid (in the 1890s) 4d. for every 1s. that he earned.[142] She rested the weight of her body on the left foot while turning the wheel with her right – Dr. Arlidge described her work in 1878 as a sort of 'perpetual jumping on one foot'.[143] Platemakers also had a wheel turner, though in their case boys were employed rather than women or girls.[144] Attempts to introduce steam to these departments of the work were bitterly resisted by the operatives, and also found little favour with the Staffordshire manufacturers, '... With steam it was difficult to regulate the different speeds required', Clara Collet was told in 1893, when enquiring into this particular branch of women's employment.[145]

G. *Leather Trades*

In the leather trades, every process of production, from the preparatory work to the finishing, depended on manual dexterity and strength. The industry employed some quarter of a million people in the 1830s and McCulloch, in his

* The fitting of handles was at one time even more physical, as 'Old Potter' recalls in his account of childhood work in the 1840s: '(Handles) were made by two half moulds made to fit into one another by notches on one side and holes on the other. The piece of clay to form the handle was placed in the bottom half of the mould, then the top half was put on and pressed down by the boy's stomach, with a sort of wriggle'.[139]

Statistical Account of 1837, estimated it as third or fourth in the kingdom, 'inferior only . . . to . . . cotton, wool, and iron'. But then and down to a much later date the peculiar nature of its raw material seemed to make it impervious to the machine. 'I do not think you will ever get machinery into our trade', a clicker told the Royal Commission on Labour in 1892, 'until you can grow all the animals of one size with just the same blemishes'. Tanning (the preparation of leather from raw hides) was a dirty business, and for the yardsmen who had the job of lugging heavy animal carcasses in and out of pits, it was also a wet one, which needed a strong constitution (in Bermondsey, Mayhew tells us, the majority were Irish). The hide took a tremendous beating when it was not soaking in the pits. The flesher and the unhairer slashed away at it with their pokes and knives, the shedman pummelled it with a double-headed stave, while at the end of the process the creases were taken out of the leather by a triangular steel pin, with a labourer's weight behind it. Tanning was a protracted process, though the period varied according to the stoutness of the hide, and the manufacture for which it was destined: in the case of sole leather it could take a year or more to complete. Patent improvements abridged the period of the work ('almost every tanner has some process peculiar to his establishment') but they did not alter its essential character. Even in such a large tannery as the Avonside works of Messrs. Evans, 'the most modern and complete' in the Bristol region, according to the *Boot and Shoe Trade Journal* in 1887, it took more than twelve months for a hide to progress through the successive stages of its treatment. The firm used a Tangye pump for pumping water; there was a machine for grinding bark; and there were three boilers, though more than one was rarely used at a time. But beyond these the sound of machinery was 'scarcely heard'.

Currying – i.e. the softening up of leather for saddlery, coach linings and shoes – was classed as a superior trade, but it required a labourer's strength to perform it, as well as an artisan skill. 'Shaving' for example, 'the most difficult part of the work, which none but the best workmen can perform', involved long heavy strokes with the knife – 'driven with great force' from the top to the bottom of the beam – while the rest of the worker's strength was devoted to keeping the hide in place by the weight and pressure of his knees. As in the case of tanning there was a great deal of pummelling and slashing away with knives, though the more delicate nature of the material required a greater care: 'The unskilful currier is constantly liable to injure the leather by cutting through it, as well as by failing to produce a regular substance'.[146]

In the manufacturing branches of leatherwork, machinery was hardly more in evidence than it was in tanning and currying. In gloving 'getting . . . the stretch of the leather' was the great point, and the finer and more flexible the leather, the more individual care and attention was needed. Yeovil, Torrington, Worcester and Woodstock were the centres of the trade, which largely depended on outwork. The delicate work of cutting out – 'anything but . . . mechanical' – was done on the master's premises, the rest of the work was farmed out. In Worcester, where the famous 'dogskin' glove was made, the fingers and thumbs went to women in the villages, while the back and 'topping' went to outworking sewers in town. Saddlery and harness were hand-stitched down to the 1870s, and workers in the trade enjoyed a large measure of self-government; many worked at home, and those who worked on the master's premises elected their own shop constable to govern workshop affairs, and owned their own tools. In

Walsall, which established itself as the centre of the trade in the third quarter of the century, the sewing machine was introduced to the work in the 1870s, but as late as 1898 W. J. Gordon could describe saddle-making there as 'a genuine trade of the old sort . . . all handwork'.[147] In London the introduction of the sewing machine was bitterly and successfully resisted. 'Through the saddlers' windows', wrote Japp in *Industrial Curiosities* (1880):[148]

> We still see sturdy men slowly stitching away like so many women; and on inquiry we find it asserted that the machine cannot produce sufficiently strong work. Now that waxed-thread machines are common this is a peculiar excuse. All the harness used in the United States is machine-made, even that employed in the artillery, consequently it cannot be said that it will not stand wear and tear. Saddlers and harness-makers are, as a trade, conservative, and obtain very high prices for their present work, and what keeps the machine at arm's length is the fear of strikes and the lowering of prices.

Boot and shoemaking, though much less aristocratic than saddlery (London saddlers liked to refer to themselves as 'gentlemen stitchers') was also, at mid-century, entirely a hand-sewn trade, with vast numbers of men, women and children engaged at it, mostly as domestic outworkers. The industry was revolutionised not by machinery (which did not begin to appear until the 1860s), but by the increasingly elaborate sub-division of labour, and by the rise of wholesale manufacturing. In Northampton, the first great centre of the wholesale trade, the 'factories' of the 1840s and 1850s were little more than putting-out stations – warehouses in which the skins were cut, and the finished parts received on the various stages of their progress through production.[149] Not until the 1870s did anything like a modern factory system begin to appear, and not until the great strike of 1895 was its predominance ratified.

H. *Woodworkers*

In wood, as in leather, the variability of the raw material, and the delicacy, in many cases, of the finished product, made mechanisation problematical, and, as in the case of shoemaking, there was a superabundance of labour ready to take up new openings on the basis of handicraft skill. Economic growth took place almost independently of the machine. Steam-power was applied at the saw mills, turning timber into deals, or slicing them up thinly as veneers. And in the 1860s and 1870s steam joineries began to appear, supplying ready-made mouldings and parts. But woodworkers themselves, with the exception of the sawyers, were only indirectly affected by these changes.

In country districts, remote from the steam-powered saw mills, timber still came to the workshops rough from the woodman's axe; and down to the end of the century the saw-pit remained an inescapable adjunct of the work at the wheelwright's shop or the carpenter and joiner's. At Haddenham, in Buckinghamshire, the village carpenter in the 1880s purchased standing trees, cut them down, and stored them in the yard for twelve months. 'Then the sides of the logs were trimmed with the axe and afterwards rolled on the saw-pit to be sawn by hand into boards or planks'. At Codnor, Derbyshire, in the carpenter's shop where Joseph Severn began his apprenticeship in 1877, there was a circular saw 'ginned by a horse', but all the rest of the work was done by hand. 'The first thing . . . was to learn to plane and saw well . . . to sharpen planes, saws and chisels, and . . . to saw straightly long planks nine or eleven by three into spars and scantling.' The wheelwright's shop at Farnham, Surrey, where George Sturt began work some seven years later, had even less in the way of plant:[150]

The want of machinery was most evident in the daily task of cutting up plank or board for other work, and of planing and mortising afterwards. We had neither band-saw nor circular saw. Most of the felloes were shaped by adze or axe; the pieces for barrow-wheel felloes were clamped to a woodman's bench (they were too short and small for an axe), and sawed out there by a boy with a frame-saw. (I hated the job – it was at once lonely and laborious); the heavy boards were cut out (and edged up) with a hand-saw, being held down on the trestles with your knee (it was no joke to cut a set of one-inch elm boards – for a waggon-bottom – your arm knew about it); but all the timbers for framework of waggon or cart, or harrow or plough or wheelbarrow, were cut out by two men on a saw-pit.

The influence of machinery was greater than this in the towns, but the 'Mechanical Joiner', exhibited at the Crystal Palace in 1851, proved as phantasmagoric in real life as that later creature of the capitalist imagination, hopefully publicised in the *Builders' Trade Circular* for 1862, 'the self-acting Trowel'. Saw mills did not abolish the need for the hand-saw (the Sheffield manufacture of them was never more flourishing than in mid-Victorian times, when the grinders held the masters in their thrall) nor planing-machines the plane. They could supply the rougher classes of deal, such as flooring boards, railway sleepers, and ships' timbers. But their action was too crude for many hard woods and too indiscriminate for the different lengths, breadths and thicknesses required in, say, the making of a writing desk, the framing of a roof, or the fitting of a steam-ship cabin. What such machinery did do was to provide a much broader base for handicraft activity by cheapening the raw materials of the trade. The case of machine-cut veneers, one of the earliest and most brilliant successes of 19th century timber technology, is particularly striking. It robbed the sawyers of what had been by far the most skilled and paying branch of their trade. But at the same time it helped to produce a vast proliferation of activity in furniture making, especially at the 'slop' or lower-priced end of the trade. It encouraged the wholesalers and dealers – the 'slaughterhouse' men who bought the cabinet-makers' work – to extend the influence of novelty and fashion to new, more plebeian markets, and brought highly-polished furniture within the reach of the comparatively impecunious (by the early 1870s, there were numerous complaints in the capitalist press of miners buying pianos). The process was already apparent in Mayhew's time, though it was not perhaps until the closing years of the century that its effects were universally felt. 'I think that machinery has been a benefit to us', a fancy cabinet-maker told Mayhew, 'it increases the material for our work. If there wasn't so much veneering, there wouldn't be so much fancy cabinet-work'.[151]

In the third quarter of the nineteenth century a number of machine tools were introduced into the woodworking trades, but though they speeded up portions of the work, they did not dispense with the need for either hand tools or handicraft skills. The application of steam power to the turner's lathe, for instance, enabled a man to produce double the quantity of work in the same time as when the lathe had been turned by hand, but it was still the turner who guided the lathe rather than vice-versa: 'Steam . . . is only used . . . as the motive power, for a man must still be employed to "turn"'.[152] Similarly in the case of the band-saw, a treadle-worked machine which was very widely adopted in the East End cabinet-making trade, the work of cutting out depended on the versatility of man rather than of the machine which served his purposes – indeed the machine was so constructed that it could be handled almost as dextrously as if it had been a chisel or a plane.

The secular shift from wood to iron represented a more serious threat to the woodworkers' position, but in mid-Victorian times its effects were masked by the growth in the overall volume of work. In factory industry woodworkers found a place as coachmakers and body-builders in the railway works, patternmakers in the foundries, and coopers in the chemical works and breweries.[153] In the shipyards, in spite of the coming of the iron steamship – or because of it – the number of shipwrights recorded in the census increased from 27,805 in 1851 to 70,517 in 1891.[154] In furniture-making the third quarter of the century saw the advent of cheap mass production, though on the basis of hand labour rather than the machine. Windsor chairs, the new staple trade of High Wycombe, were being turned out in the 1860s at the rate of one a minute 'all the year round';[155] Kentish Town pianos serviced the musical needs of both shabby-genteel and the newly-respectable, while in the East End of London there was a whole new industry of one-room cabinet-makers, using ramshackle materials and highly-polished veneers.

One sign of their handicraft status was that so many woodworkers owned their own means of production. The coopers, in Mayhew's time, found all their tools themselves – 'a kit for a general workman being worth £12' – except for the jointers to weld the staves.[156] A London cooper, who started work in 1912, remembers that those who went on tramp to take part in the cider-making in the Vale of Evesham, or the herring harvest at Great Yarmouth, would leave the heavier tools in the shop, and travel with a hammer and driver, 'a knife or two', and an adze.[157] The value of a good cabinet-maker's chest of tools was reckoned at between £10 and £20 in 1853 Edinburgh.[158] In London at the same time it was stated by Mayhew to be about £10, but less in the more specialist branches of the trade, where the tools were lighter: a buhl-cutter's tools ('very small and niggling') cost £5 to assemble, the wood-carver's about the same. The garret-makers in the slop end of the trade would make their kit up second-hand, and Mayhew has a fine account of the makeshift way it would be assembled:[159]

> The tools are generally collected by degrees, and often in the last year of apprenticeship, out of the boy's earnings. They are seldom bought 'first-hand', but at the marine store shops, or at the second-hand furniture broker's, in such places as the New Cut. The purchaser grinds and sharpens them at any friendly workman's, where he can meet with the loan of a grindstone, and puts new handles to them himself, out of pieces of waste wood; 10s., or even 5s., thus invested has started a man with tools; while 20s. has accomplished it in what was considered 'good style'. Old chisels may be bought from 1d. or 1½d. to 2d., to 5d.; planing irons from 1d. or 1½d. to 3d.; hammer heads from 1d. to 3d.; saws, from 1s. to 2s 6d., and rules and other tools equally low.

The carpenter and joiner's tools – inscribed with his name – served him as a very title to employment, when he moved from job to job, and also as his means of production when he was forced to set up on his own. They cost a great deal to assemble, and their loss through theft or fire was a personal tragedy.[160] 'Tool benefit' was one of the chief provisions of the Amalgamated Society of Carpenters and Joiners, as it was of other trade unions in the woodworking trade; and the detection of tool-stealing offences seems to have been a recurring preoccupation of the local branches, to judge by the notices they sent up for printing in the Society's Monthly report.*

In factory industry, too, woodworkers characteristically owned their own means of production. A patternmaker's tools cost (in 1884) between £8 and £16 to assemble, and about 6d. a week to keep in repair.[161] They would be insured

* For reasons of space, examples have been eliminated.

with the trade union and jealously guarded against misuse.[162] The young apprentice was advised to respect them. 'A patternmaker's tools are idols', wrote a Liverpool engineer in 1878:[163] 'and one of the first duties the tyro must learn is, to return all tools to their owners properly sharpened and ready for immediate use, and above all things, he should be careful not to leave them kicking about his bench after having made use of them . . .'.

I. *Metallurgy*

In metallurgy steam power was massively harnessed to the primary processes of production, notably in puddling and rolling; but at the same time new fields were opened up for handicraft skills. In foundry work machine moulding was introduced in the 1850s (during the lock-out of 1852 some engineering employers fondly believed that it would deliver them from their men);[164] but it remained confined to the most inferior branches of the trade, such as the making of cast-iron drain-pipes.[165] Moulders – 'the wildest, the most grimy, the most independent, and, unfortunately, the most drunken and troublesome of any English workmen who have any claim to the title of "skilled"'[166] – were virtually untouched by it and for the most part worked with the very simplest of tools, whilst their brethren, the dressers, smoothed the rough castings with hand-files.[167] The Friendly Society of Iron-moulders was composed uniquely of handicraft workers right down to the 1900s and (as employers complained) maintained a high rate of wages, restrictive shop practices, and unsleeping hostility to the machine-based class of worker.[168]

Engineering, though less severe than foundry work, was a heavier kind of labour than it is to-day, as well as one which demanded much more versatile skills. To begin with it involved working very largely with iron, which was much more difficult to shape than the high speed steel and light alloys of twentieth century engineering. The smith who had the forging of it was engaged in a muscle-hardening work, and one that was no less pivotal to production than it was at a country blacksmith's. Fitting was almost entirely done by hand. It involved the making of parts rather than the mere assembly of them – a protracted process of chipping and smoothing at the bench with no more sophisticated mechanical aid than a sharp-edged chisel and file.[169] In the 1860s production engineering was still in its infancy, and the system of manufacture by interchangeable parts, introduced from the United States at the time of the Crimean War, was for the most part confined to the making of sewing machines and small arms. Engineering firms, large and small, went in for a variety of different products rather than limiting themselves to standardised lines, and much of their work was 'bespoke' – carried out according to customer specifications and requirements.[169a] In the heavier classes of work, steam power was applied to the coarse, preparatory stages of production – rolling, stamping and cutting out – but most of the precision work was still hand-tooled, whether by smiths, fitters or turners. At the St. Pancras Ironworks where Wal Hannington worked as a boy, making spiral staircases, the iron handrails were bent in a rolling mill, but the shaping of the finished curvature depended on hand-hammered work of an exceedingly laborious kind:[170]

> Joe Hands was a working foreman and I worked with him when he was fitting a hand-rail I suffered physically and mentally. I had to hold the rail with both hands across a heavy V-block

fitted to an anvil, whilst he held the form block in one hand against the rail and struck it with a weighty club-hammer to get the required bend. It was a crude, old-fashioned trial-and-error method, requiring much hammering before the correct shape was obtained. Holding the curved rail . . . on the V-block was not as simple as it would appear to be. The correct position was for it to rest evenly on both crowns of the block at the moment the hammer struck it. But if I moved slightly and the rail was a fraction of an inch off one crown the amount of error would be multiplied by the length at which I was holding it and I would get the shock of the hammer blow in my hands and arms. I was always in fear of that happening and when it did I suffered. But my distress never evoked any sympathy from the foreman, on the contrary, he always abused me with the foulest language.

The mid-Victorian engineer was still characteristically a craftsman, an 'artisan' or a 'mechanic' rather than an 'operative' or 'hand'. He was engaged for the most part in speciality work, in which precision was more important than rapidity. He moved freely about the work-place – too freely, according to his critics, who would have preferred to tie him down in one place. He was expected to find his own tools (often he made them), and keep them in good fettle by grinding and sharpening them himself.[171] In a small workshop, the most frequent form of enterprise in the 1850s and 1860s, he would turn his hand to a variety of different tasks, working sometimes at the forge, sometimes at the bench, sometimes on outside jobs.[172] Even in a larger works he would often have to improvise the materials for a job, cutting extra tools and accessories, finding additional jigs, and acting as his own draughtsman and designer.[173] He would also regulate, in some degree, his own work pace (opposition to piecework was a cardinal tenet of the Amalgamated Society of Engineers, defended in some notable strikes). Machine tools were increasingly introduced from the 1840s, but they were very far from automatic, and did not dispense with the need for a keen eye, a steady hand, and an ability to work to fine limits. Fitters, the most numerous class of engineer, were largely unaffected by them.[174] In turners' work, steam power was applied to the lathe, at least in the larger works, and Nasmyth slide-rests attached to them, but it was still the worker who guided the tool rather than the other way round – accuracy and speed depended on personal judgement rather than pre-set chucks or jigs.[175] Not until the technical innovations of the 1890s did the 'machine question' (bitterly fought out in the lock-out of 1896-7) begin to pose a fundamental threat to craftsmanly skills.

Machinery, the chief object of mid-Victorian engineering activity, was built up very largely from hand-made components, all of them requiring individual attention, whether at the foundry where components were cast, at the smithy where they were forged, or at the bench where they were chipped and filed into serviceable parts. It was all precision work, much of it being undertaken to order, and mounted or fitted up on site.[176] A first class reputation was much more important than cheapness, whether for the 'general' engineer or 'jobbing' foundry, servicing local customers, or the manufacturer of specialist machinery exported to the markets of the world. The majority of workers at a machine-builders were artisans, though with a complement of boy labourers and apprentices to take on the more routine drudgeries.[177] Messrs. Platt Brothers, Oldham, the leading manufacturers of textile machinery, were employing some 10,000 workers in the 1890s, and almost every one of them, according to Paul de Rousiers, was skilled:[178]

The work done there is of an extremely delicate and technical kind, and the mounting of the looms in particular requires great care and the most exact finish in all the component parts. Every workman is a specialist. A significant fact in proof of this is that not a single woman is employed by the firm . . . There is no room for occasional hands or casuals, who take up any

> sort of work to-day to drop it for something else tomorrow. Every workman must be a master of his business. Except the porters, who are superseded as far as possible by lifts and locomotives, almost every individual employed by the firm is a skilled workman.

Locomotive engineering, like machine building, was precision work which was simultaneously delicate and exacting. A single locomotive was said in 1852 to contain no fewer than 5,416 separate components, 'and it will readily be conceived that nothing short of the utmost completeness and accuracy, in the finish of these parts, could enable the workmen to combine them in one harmonious and efficient unity... The failure of one screw or bolt, or the bending of one rod, may hereafter involve, not only the costly fabric itself in ruin, but occasion the destruction of property and life to a terrible extent'.[179] Hand tools were an indispensable condition of production. The worker would own a chest of them himself; and he would also have access to the 'shop tools' provided by the management.[180] At the G.W.R. works, Swindon, in 1852, there were two Nasmyth steam hammers to shape the heavier metals, but no fewer than 176 smith's hearths where the smaller components were forged – axles, piston-rods 'and other pieces too numerous to mention'.[181] At Crewe in 1894 – the locomotive engine works of the London and North Western Railway – the smaller parts of the engines were forged at 120 smith's hearths. At the Wolverton works of the London and Birmingham railway there was practically no machinery at all when Hugh Stowell Brown went to work there in the 1840s. All the forging was done by hand ('it was a fine sight to see seven or eight stalwart strikers, at the forging of a crank-axle, plant their huge hammers in rapid succession upon the spot indicated by the smith with a piece of rivet rod iron'). There was no travelling drill ('all keyways had to be first drilled in round holes, and then cut out with a cross-cut chisel, and finished with the file'), and no shaping machine of any account save a machine for cutting nuts.[182]

> Every surface that could not be formed by the planing-machine had to be chipped and filed... Yet there were men who could do wonderfully true work. I have seen a fitter take two rough pieces of wrought iron of more than one pound weight each. I have seen him chip them to a surface almost perfectly smooth, and then with files so perfect the surface that when placed one upon the other the lower piece would hang to the upper by the force of molecular attraction, as if glued to it.

In marine engineering – one of the foundations of Britain's trading supremacy, as well as an important focus for mid-Victorian capital investment – skill was no less at a premium than it was in a railway factory. In a highly competitive market, the reputation of a firm depended not on speed but on quality workmanship and attention to customer needs. Bespoke work was the norm, since every ship had to be individually suited, and it seems to have affected the manufacturers of components in much the same way as the marine engineers themselves. Lockwood and Carlisle of Sheffield, for instance, whose patent piston rings were adopted by some of the leading Clyde shipbuilders, met all their orders in the 1880s with a work force of about a dozen craftsmen: they would work up to 75 hours a week when a new order came in. 'Most of the orders... reflected the views of designers of ships' engines... Each had to be made individually'.[183] As in other classes of heavy engineering, the work was both laborious and skilled. Here is an account of it by one who served his apprenticeship on Clydeside in the 1880s, working for a firm which he describes as 'the greatest marine engine works in the world':[184]

The most powerful engines which have ever been made were then under construction in the works . . . They were enormous things, and they stood fifty-four feet high when they were finished. Machine tools were not nearly so accurate then as they are now, and nearly everything had to be finished by hand. Bolts weighing over a hundredweight were not uncommon, and these were all fitted in place. It was hard work lifting these monsters all day long and driving them in and then driving them back and scraping them until they fitted perfectly. My wrists used to ache at night and my ribs were sore with swinging the big hammer. These hammers were known as 'mondays' and they weighed forty-two pounds. There was a certain amount of skill required to get in the full force of the blow, and unfortunately for me it was discovered that I had this knack. At first I was flattered when the foreman took me away from my work to knock out a bolt with a monday hammer, but it was a dearly bought pleasure and very exhausting.

Steam boilers – the high-pressure engines which gave life to machinery and put railway trains and steamships on the move – were in the 1850s almost entirely hand-made. Some of the work, such as plating, remained beyond machinery's reach because of the intricacy or peculiarity of the parts; other parts resisted it because hand labour was more adaptable and precise.[185] The plates which gave the boilers their jackets were punched out by machinery, but they had to be forged and welded at the hearth and they were then subjected to a tremendous hammering to make them both ductile and strong, to bend them to the right shape and to bring them to the right density. The joints were made steam-tight by a succession of blows with the sledge-hammer. Hydraulic riveting was introduced in the 1850s, but it made slow progress because it was considered much less trustworthy than hand-hammered work.[186] In the G.W.R. factory at Swindon, the premier locomotive plant in the kingdom, the preference for hand-hammered work continued right down to 1914: 'very little is left to the chance work of the machine, which is often faulty and unreliable. Rivets put in by hand are far more trustworthy'.[187]

There was a great deal of hand riveting in iron shipbuilding too, despite the massive concentration of capital and ownership. In 1884, according to Pollock's *Modern Shipbuilding* published in that year, hydraulic riveting was general in interior work, but shell plating – i.e. the ship's water-tight covering – was still riveted by hand. In 1905, according to a later work by the same writer, the situation had not changed: 'Probably a complete solution of the shell-riveting problem is not now far distant, but meantime the binding of the shell to the ship's framework is mostly accomplished by the old and laborious way by the "hand-and-hammer" of the craftsman'. Hand riveting was also widely employed in constructional engineering, and indeed in every class of wrought-iron work where there were metal plates to be joined. The Britannia Tubular Bridge (1850), one of the mechanical wonders of the age at the time of the Great Exhibition, was riveted entirely by hand, with some 2,000,000 holes to be seamed and jointed.[188]*

Ironmaking itself – i.e. the production of pigs and bars from iron ores – was very much a sweat and muscle job, as well as one requiring great vigilance to stave off injuries. Despite the vast mobilization of steam power all of the strategic operations were performed by hand – the preparation of the ores, which were broken up by sledge-hammers; the loading and tapping of the blast furnaces; the puddling of the iron; the shingling and rolling of the bars. In blast furnace work, the first stage of production, steam power was applied to the bellows, magnifying the blast a thousand-fold compared to the old-time

* For reasons of space, a section on boilermaking squads has been cut.

bloomeries. Everything else depended on men, working with heavy weights at almost inhuman temperatures. As Lady Bell wrote of Middlesbrough in *At the Works* (1908): 'From the moment when the ironstone is lifted off the trucks, then dropped into the kilns, afterwards taken to the furnace, and then drawn out of it, it has not been handled by any other means than the arms of powerful men, whose strength and vigilance are constantly strained almost to breaking-point'.[189]

Technology in the rolling mills, the second great department of production, was more sophisticated, with steam-powered hammers to pound the metal and steam-powered rollers to squeeze it into shape. But all the strategic operations were performed by men. Cort's puddling furnace, which revolutionised the work, was chemical rather than mechanical in its action. It cheapened the production process and made possible an enormous increase in output, but it made labour itself a great deal more severe. The puddler, who had the key role in the new process, was given a task that was simultaneously highly skilled and exhausting, turning a viscous mass of liquid into metal. He worked in conditions of tremendous heat, violently agitating the metal as it boiled, rabbling it from side to side in the furnace, and then gathering it at the end of a rod while the molten liquid thickened and the supercharge of carbon burnt out. The men had to relieve each other every few minutes, so great was the exertion and so intense the heat, but even so it was said that every ounce of excess fat was drained from them and that cataracts 'induced by the intensely bright light of the furnaces' were a frequent complaint. Few puddlers, it was said, lived beyond the age of 50. 'Shingling', the next stage of the process, was also very physical, a kind of blacksmith's work in which heavy balls of red-hot metal had to be heaved into place for the steam-hammer (shinglers wore iron masks to protect themselves from the raging heat).[190] Sir George Head, visiting the Low Moor ironworks at Wibsey Slack in 1835, likened the working conditions to those of an inferno:[191]

> Athletic men, bathed in perspiration, naked from the waist upwards, exposed to severe alternations of temperature, some, with long bars, stirring the fused metal through the door of the furnace, whose flaming concavity presented to the view a glowing lake of fire – were working like Cyclopes. By continued and violent applications of strength, visible in writhing changes of attitude and contortions of the body, raking backwards and forwards, and stirring round and about the yielding metal, they contrived to weld together a shapeless mass gradually increasing in size till it became about an hundred pounds weight; this, by a simultaneous effort of two men with massive tongs, was dragged out of the furnace, radiant with white heat . . . Now subjected to the blows of a ponderous hammer, it was wonderful to mark the vigour and dexterity with which the men contrived to heave the mass round and round at every rise of the hammer . . . while the fiery ball was now turned one side, again the other side uppermost, with the same facility apparently to the operators as if it had been a horseshoe.

The rolling mills were also fearsome places to work in, with giant plates to be drawn from the furnaces and positioned for 'Demon Crusher' hammers. At the Atlas Works of John Brown's, Sheffield – illustrated in the engraving reprinted on p. 7 – the men who had the job of rolling armour plate would gather some forty strong at the furnace mouth, dragging the plate by main force to subject it to the rollers: 31 tons was an average weight to drag. The men wore thin steel leggings, aprons of steel, and a thin curtain of steel 'dropping over their faces like a large, long vizor'. All the rest of their bodies were muffled in thick, wet sacking, to keep out the infernal heat. 'In spite, however, of every precaution that the best workmen can employ, they cannot always escape splashes of melted iron.'[192]

The predominance of hand technology was yet more pronounced in the so-called light metal trades, such as the manufacture of domestic hardware. In the third quarter of the 19th century an increasing number of mechanical aids were being brought into use, such as fly wheels for polishing, and treadle lathes for turning parts; but the vast growth of output was mainly due to new alloys, cheaper iron and skilled workmen. In the Wolverhampton tinware trade stamping machinery was very common in the 1860s, though operated by treadle or crank rather than by steam. But workers in the trade depended on a galaxy of small hand tools for the body of the work, with hammer and mallet as the 'old indispensables' for finishing – 'polished hammers on polished anvils' in the case of quality goods, such as the better class of saucepan, 'hammered . . . till they acquire a beautifully smooth and silvery surface'.[193] In the Sheffield cutlery trades steam power was applied to the grinding and polishing wheels, but it provided motive power only, leaving the grinder's work intact, while the actual work of forging remained a purely manual process, with a hearth of fine coke 'forced into intermittent activity by a hand bellows', and frequent beatings with a hammer, to give the blade temper and tact.[194] In the Birmingham brass trades, which in the third quarter of the 19th century emerged as the town's leading employment, steam power was sometimes rented for polishing, but fittings were turned on a treadle-worked lathe with the guiding power as the worker's. ('In the turner's hand the tool must never be allowed to rest, but must have a proper rotation and a correct inclination to avoid furrowing the work'). Burnishing and decoration – a major feature in a trade wedded to ornamentation – was largely a work of hand-files. 'Any man who can use a hammer and file can be a bedstead maker', wrote Edward Peyton in 1866. Brass founding, a mainly Birmingham trade, carried on in ramshackle premises and open-fronted sheds, was largely in the hands of small masters: the 'plant' consisted for the most part of small moulding tubs where fittings and components were cast.[195]

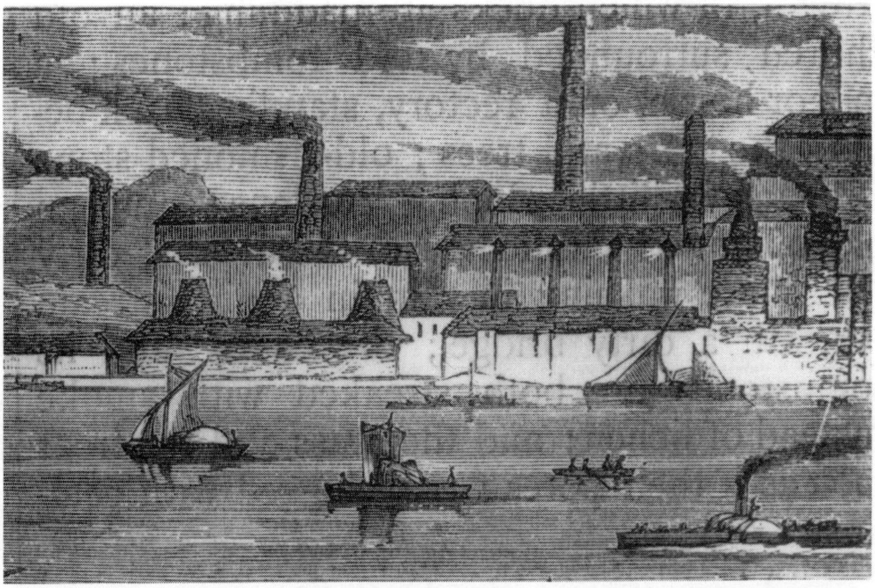

Coal pits and shutes on the Tyne *from an original engraving in the Mansell Collection.*

III. Combined and Uneven Development

The foregoing epitome, though necessarily abbreviated, may be enough to suggest that in speaking of the primacy of labour power one is referring not to single instances, or to curious survivals, but to a dominant pattern of growth. In manufacture, as in agriculture and mineral work, a vast amount of capitalist enterprise was organised on the basis of hand rather than steam-powered technologies. In Marxist terms, the labour process was dependent on the strength, skill, quickness and sureness of touch of the individual worker rather than upon the simultaneous and repetitive operations of the machine. The restraints 'inseparable from human labour power'[196] had not yet been cast aside. On the contrary, a great deal of entrepreneurial ingenuity was employed in turning them to advantage. Commercial progress depended quite largely on the physical adaptability of the worker, whether it involved crawling on all fours to gather the woad harvest,[197] climbing up and down perpendicular ladders (in a Cornish tin mine the ascent would take an hour or more each day),[198] or working, like boilermakers on repair jobs, upside down in tanks.[199] The lungs of the glassblower, working as bellows, or those of the gas-fitter, soldering pipes, were not the least of the forces of production which 19th century capitalism summoned to its aid, nor were there any more important in the clothing trades than the needlewoman's fingers and thumbs. In the Potteries, dinner plates were shaped by dextrous jerks of the flat-presser's wrists, and surfaces varnished with the dipper's bare arms in a glaze tub (in 1861 Dr. Greenhow estimated they were immersed for eight of a twelve hour day). Ironmaking depended on violent muscular exertion, and an ability to withstand white heat, engineering on precision of judgement and touch. In the metal-working trades no action was more highly valued than the ability to deliver well-directed blows with the hammer, while those engaged in press-work were in almost perpetual motion with their arms and wrists: 'practiced workers' in the metal button trade were said to make from 14,000 to 20,000 strokes a day, 'the whole strength of a woman' being needed on the heavier class of press.[200]

Human beings were quite often used instead of horses for haulage, not only on the canals, but also on the brickfields, where the children who acted as 'off-bearers' or 'pushers-out', taking cart-loads of bricks from the moulder's table to the brick-setter's kiln, carried an average weight variously computed at between 12 and 25 tons a day.[201] They were used as lifts in the lead factories, and as shunters in the docks. In the workshop trades, delivery boys were strapped to the heavier loads very much in the manner of Volga boatmen; so were some of the bakers' boys who went on 'rounds'. Men were also used like this in heavy industry, when there were plates to be drawn from the furnaces, or castings from their beds. At J. G. Thomson's, Clydeside (Peter Taylor recalls) there were a hundred and fifty men at the ropes when there was a ship's engine to deliver – they started off at a canter, outside the factory gates.[202] Sturdy legs were essential for mould-runners in the potteries, carrying plates from the flat-pressers' wheels to the drying rooms; they were also much in demand at a glassworks, where a whole army of juvenile runners were kept on their feet all day: in a glass bottle factory of the 1860s it was calculated that they travelled the equivalent of between 13 and 17 miles a day. 'The smaller the articles . . . and the rougher the workmanship, the greater is the number of them turned out . . . and the greater also is the demand on a boy's exertion'.[203]

Human beings were also used as balances and weights, whether to give motion to machinery, like the lathe-treaders, or to act as see-saws when there were heavy loads to hoist. In clay-treading they had to act as mangles, in baking as rotary knives. In the London fur trade grown men were employed in the 1890s to bring seal skins to plasticity by jumping on them. 'It is a curious sight, on entering a room, to see a row of . . . tubs each with its Jack-in-the-box bobbing up and down', wrote one of Booth's investigators. 'Every man is naked except for a vest, and a rough cloth which is tied round his waist and attached to the rim of his barrel. With hands resting on either ledge up and down he treads, and earns 20s to 25s piece-work. Skins cured by this process are said to be softer and silkier'.[204]

In mid-Victorian England there were few parts of the economy which steam power and machinery had left untouched, but fewer still where it ruled unchallenged. At both top and bottom a mainly hand technology prevailed, at top because of the irreplaceability of human skill, at bottom because of the plentiful supply of drudges. High technology industry – what some Marxists call 'machinofacture' – was for the most part confined to the factories, but even here mechanisation was very far from complete. One might refer to the pot-makers at a steel-works, the rag-sorters at a paper-mill or to the bottle hands whom Dodd describes at Day and Martin's, Holborn, meting out blacking from vats.[205] In the Lancashire mills the 'self-acting mule' depended on the nimble fingers of the piecers, while power-loom weavers kept their machines at work by 'shuttle-kissing', threading the weft with their lips.[206] In ironmaking the whole scale of enterprise was transformed by technical innovation, but labour remained absolutely primary at the point of production itself. As the *Morning Chronicle* Commissioner wrote in 1850, after investigating the Monmouthshire iron works: 'Although capital is the motive power, it is upon the rude virtues of the workman that the entire system of manufacture . . . rests'.[207]

Capitalist development in the 19th century depended on numbers as well as strength. In coal mining, the labour force expanded from some 218,000 in 1841 to 1,202,000 in 1911, at the height of the industry's prosperity. In the building trades, over the same period, it more than trebled. British agriculture was prodigal in its use of hands, and the labour force in the 1850s was greater than at any other time before or since. Factories, too, despite the introduction of labour-saving machinery – and in some cases because of it – required a phenomenal number of hands, both to take charge of the preparatory processes, and to make good the deficiencies of the machine. In the Nottingham lace trade of the 1860s there were many more people employed as outworkers, mending and making up, than there were in the factories themselves.[208] In the workshop trades, employers could rely on the services of a whole army of job-hands and strappers, permanently under-employed, who were only taken on when trade was seasonally brisk.

The slow progress of mechanisation in mid-Victorian times had many different causes, but one of them was undoubtedly the relative abundance of labour, both skilled and unskilled. In striking contrast to the earlier years of the industrial revolution, every branch of employment was over-stocked. In agriculture there was a huge labour surplus, men, women and children who never had full employment except in the harvest months. Railway building and construction sites depended upon a great army of free-lance, tramping navvies, who took up employment only for the duration of a job. The reserve army of labour was no less a feature of the workshop trades. The supply of needlewomen was infinitely elastic – the number recorded in the census tripled between 1841 and 1861 – while that of carpenters and joiners, tailors and shoemakers, printers and bookbinders was always far greater than the number of regular berths. 'Tramping artisans' were very much a feature of the labour market in the new industrial crafts, such as boilermaking.[209] In iron shipbuilding, where most employment was on a job-and-finish basis, they constituted the bulk of the labour force: Samuel Kydd in 1858 described the Clydeside shipbuilders, restlessly scouring the riverside for work, as being more like 'wandering tinkers' than regular mechanics.[210] The trade union records of the mid-Victorian ironmoulders show that there were seldom less than 5% of members out-of-work, and often more than 10%.[211] In the Sheffield trades as many as a fifth of union members in the 1860s were 'on the box' (out of work, and supported from union funds), and indeed it was the strain of maintaining them at a fair level of benefit which seems to have been indirectly responsible for the notorious Sheffield 'outrages' (industrial terrorism against members in arrears with subscriptions). Amongst Yorkshire glass bottlemakers the mean level of unemployment seems to have been even higher, to judge by the figures reproduced in Table 2.[212]

This superabundance of labour was a pre-condition of Victorian economic expansion, and it also helped to determine its distinctive physiognomy and style. It encouraged capitalists to engage in capital-saving rather than labour-saving investment, to perpetuate low-intensity technologies, and to rely on workers'

TABLE 2

Unemployment among Yorkshire Glass Bottlemakers, 1867-1897

Year	No. of Men Unemployed	% of Unemployed	Year	No. of Men Unemployed	% of Unemployed
1867	163	31.3	1883	223	24.7
1868	53	11.3	1884	305	28.0
1869	61	12.6	1885	432	35.3
1870	94	17.3	1886	796	63.9
1871	39	9.5	1887	246	25.6
1872	10	4.2	1888	380	32.5
1873	18	5.6	1889	201	16.7
1874	39	7.9	1890	210	16.9
1875	57	8.8	1891	231	15.6
1876	188	22.7	1892	545	32.6
1877	204	24.5	1893	430	26.2
1878	353	40.2	1894	950	49.3
1879	632	70.7	1895	580	31.5
1880	247	32.2	1896	528	30.6
1881	242	36.5	1897	444	25.3
1882	171	25.8			

Source: London School of Economics, Webb T. U., Collection, Glass Bottlemakers of Yorkshire, Quarterly Reports, XXII, p. 76, 12 June 1909.

skills even when there was machinery ready, in principle, to replace them.

In the United States, by contrast – a new world, sparsely populated by immigrants – labour-saving improvements were a very condition of capitalist growth, and self-acting machinery, which in many cases the American invented, and in others they adopted as their own, made much more rapid strides than it did in mid-Victorian England.[213] The reaper-binder was transforming the harvest on the prairies while in England it was still being taken by the sickle or the scythe. In construction work the steam navvy was so extensively employed in America, and from such an early date, that the very word 'navvy' was attached not to the men (as it was in England, and as it is to-day) but to the machine;[214] the steam-navvy was patented in America as early as 1841; it did not appear in England until the 1870s, when the manufacture was taken up by Ruston and Proctor of Lincoln;[215] and its capabilities only began to be widely recognized in the 1890s, when it was employed on such big construction jobs as the Great Central Railway and the Manchester Ship Canal.[216] There was an equally pronounced disparity in lock-making, where, in the 1870s, machine-made, cast-iron locks from America were sweeping the markets of the world, while the Willenhall lock was still hand-made: it was widely attributed to the superabundance of cheap handicraft labour, penniless small masters, 'and . . . the perpetuation of an ancient and wasteful garret system of working'.[217] Much the same was true in nailmaking, where machinery had entirely conquered production in America when the Committee on Machinery reported on it in 1854 (the first American machine had been patented as early as 1810); in England, down to the 1880s, the great bulk of production was in the hands of Black Country out-workers, using hammers or treadle-worked 'Olivers'.[218] In boot and shoemaking the slow adoption in England of the McKay and Blake machines – patented in America during the Civil War – was also related directly to the disparities in labour supply. 'In America weekly wages were quite double those ruling in the English shoe trade. Many machines which were profitable to use when they displaced labour at a shilling an hour were hardly worth investing in when hand labour cost but sixpence. And this was about the relative position in the two countries at the time the revolution of manufacturing methods was in its most active phase.'[219] Similarly the slow progress in England of the Danks' puddling furnace – an American innovation of the 1860s in ironmaking – was attributed in 1898 (thirty years after the first patents had been taken out) to the fact that 'good puddlers can be obtained here at a lower price than is paid for even inferior workmen in the United States'.[220]

The contrasting pace of English and American mechanisation was already beginning to arouse notice at the time of the Great Exhibition.[221] By the 1860s it was a commonplace and Marx makes it the subject of a characteristically caustic comment in *Capital*: 'The Yankees have invented a stone-breaking machine. The English do not make use of it, because the "wretch" who does this work gets paid for such a small portion of his labour, that machinery would increase the cost of production to the capitalist.' Elsewhere in *Capital* he drew attention to 'the invention now-a-days of machines in England that are employed only in North America'[222] – an observation admirably borne out by the subsequent history of automatic looms and ring spinning.[223]

Another reason for the slow progress of mechanisation was the possibility of increasing productivity within a hand technology, either by the introduction of improved tools, or by a more systematic exploitation of labour, or both. Agriculture provides a prime example, with the change from sickle to the scythe, the extension of soil-improving crops and manures, and the mid-Victorian improvements in field drainage. Coal-mining, too, advanced on the basis of improved hand technology. Between 1850 and 1880 output in the industry doubled, and this was due not only to the increase in the number of underground workers, but also to improved haulage methods, harder work, and improvements in the miner's pick, with the substitution of steel for iron.[224] At the same time better transport, both by sea and land, helped to end local monopolies, and brought down prices to the industrial and domestic consumer. Another striking example, to which Eric Hobsbawm drew attention some years ago, is that of gas-making, an industry which down to almost the end of the century was entirely dependent on the physical strength of the stokers. The amount of coal carbonised in the London gasworks rose by some 75% between 1874 and 1888, while the labour force increased by under a third. The employers' gain may be attributed partly to severer labour discipline, partly to the replacement of the one by the three-man scoop (which gave a much greater charge to the firing though leaving its manual character untouched), and partly to the adoption of new chemical processes.[225] In outwork productivity was increased by 'sweating' – screwing down piece-rates so that earnings could only be made up on the basis of unpaid family labour. This was the very principle of expansion in the clothing trade, and it played a large role in such important domestic or semi-domestic industries as nailmaking, chainmaking and the furniture trade. With rates of 2d. a gross, and children to work all hours as 'little human machines'[226] – the basis on which Bryant and May's got their match-boxes made – the incentives to mechanisation must have been low, particularly when the domestic workers had to pay for their own paste, heat and light.

Cheap, labour-saving materials were another alternative to the machine. In tanning, for instance, the cost structure of the industry was transformed by the use of new tanning extracts imported from abroad. By the 1870s they had cut down the average time taken in treating the individual hide from a year to as little as four months, though the individual stages of production remained unchanged.[227] In bookbinding, the substitution of cloth for leather bindings (an innovation in the 1820s), and the introduction of 'case' work – the sub-division of work on covers – made hand labour much more productive while at the same time reducing raw material costs.[228] Mechanical aids, though bitterly contested by the journeymen bookbinders, were comparatively few, at least until the 1850s, and speeded up production only on the simplest classes of work, such as edge-cutting and flattening.[229] But the amount of hand tooling needed on each individual volume was drastically reduced. In the new cloth binderies books were numbered in thousands rather than hundreds – tens of thousands in the case of the new cheap bibles[230] – and they were produced with an open market rather than individual subscribers in view. Employment increased rapidly and the 'mystery and art' of bookbinding was transformed into something approaching a mass production industry. Cheaper iron was the main basis for expansion in the metalworking trades, together with the invention of new alloys, such as Britannia metal and German Silver. In the Wolverhampton holloware trade, one of the chief sources of mid-Victorian kitchenware, the processes of manufacture remained 'essentially the same' in the fifty years from 1810 to 1860, but costs were reduced by the introduction of enamel linings, and progressive falls in the price, or quality, of iron.

A third alternative to mechanisation – and another avenue to more rapid workmanship – was the division of labour and simplification of the individual task. In mid-Victorian times it was just as likely to take place off the master's premises as on them. A prime example is the introduction of 'riveting' in the boot and shoe trade which brought a new and cheaper class of boot on to the market, and revolutionised the wholesale trade. Under the new system of work the soles were nailed to the uppers, instead of being stitched, and the work of 'making', previously performed by one man, was now divided between two – the riveters and the finishers.[231] Riveting was a spectacular commercial success, and Leicester, where the invention was patented in 1861, rapidly established itself as the largest producer of ready-made footwear.[232] Skill was reduced, labour costs fell, and there was a sharp increase in productivity. 'The old crafts would make about three boots or two pairs a day . . . the riveter and finisher can produce ten pairs in the same time'.[233] The new labour, however, was unmistakeably handicraft in character:

> No machinery was used; the soles and uppers were cut by hand, then the upper was moulded round a last, with the edges pulled inwards. A 'nailer' or 'riveter', as he was variously called, would fill his mouth with 'sprigs', and taking them one by one would hammer on the sole and heel. When this had been done, the edges were trimmed with a sharp knife. Finally, the sole and edges were . . . polished with a hot iron and a heelball.[234]

Riveters often worked at home; finishers did so 'almost invariably'; and attempts to mechanise their work and put them into factories were for a long time unsuccessful. Indeed so far from hand-riveting being displaced by machine, it seemed at one time as if the reverse was taking place. 'The English workman prefers to drive his rivet by hand, which he can do more quickly than the machine', wrote the *Shoe and Leather Trades Record* in 1878, '... thus the rivet machine has been entirely superseded in England'.[235]

In the Sheffield trades the division of labour was the primary axis of 19th century growth, and as in the case of boot and shoemaking it was accompanied by a wide extension of outwork. Saw-making was distributed between seven different classes of work, each branching out as a little industry on its own – cutting and paring, hardening and tempering, grinding, glazing, toothing, handlemaking, finishing and polishing.[236] Similarly in the manufacture of table-knives, the work was dispersed into separate, sub-divided processes, with the cutler himself confined to the intermediate and final stages of fitting: 'Forging the blade is now one trade, grinding it another, and forging the scales and springs another ... besides ... several subsidiary branches, as the cutting up of stag and buffalo horns, and the working of hard wood, pearl or tortoiseshell'.[237] The work was passed backwards and forwards between these various specialists – in the best class of work a blade might travel three or four times between the grinders and forgers alone, quite apart from its journeyings to and from the cutler.[238]

The division of labour in needle-making – a manufacture for the million largely based on Redditch in Worcestershire – had more in common with Adam Smith's classic example of pin-making in *The Wealth of Nations*. Here the division of labour was associated with minutely sub-divided repetition work rather than with a proliferation of skills – the needle 'however unconsiderable its size' passing through the hands of 120 different workers. Despite the scale of production, there was little in the way of plant. The needle was cut from the wire with hammer or mallet; the points were ground on a stone; the eyes were pierced with a hand-lever press; the sorting was done by touch. Here is an account of it published in 1861:[239]

> The little girl who performs this office places a rag or dolly upon the forefinger of her right hand, and with the left presses the needle against it; the points stick into the soft cotton, and are thus easily withdrawn and laid in the contrary direction. Little children 'rag' with inconceivable rapidity, and with equal speed the process of sorting, according to lengths, is performed, the human hand appreciating even the sixteenth of an inch in length, and separating the different sizes with a kind of instinct with which the reasoning power seems to have nothing to do.

Technical Difficulties

Another obstacle to mechanisation was the gap between expectation and performance. In many cases the machines failed to perform the 'self-acting' miracles promised in the patents, and either needed a great deal of skilled attendance, or failed to execute their appointed tasks. Even if brought 'nearly ... to perfection[240] by its inventor, a machine would often prove difficult to operate. Unexpected snags would be encountered, unintended effects would appear, and it was possible for patent to follow patent without anything like continuous flow production being achieved. Wright's pin machine of 1824, which, according to its promoters, 'during a single revolution ... produced a perfect pin',[241] turned out to be so far from perfect that forty years later, despite thousands of pounds spent on costly experiments, the 'nobbing' or heading of the pin had still very often to be done by hand (in Gloucester this was a cottage industry, though the body of the pin was made in factories).[242] Wall's 1880 machine for manufacturing cheap pottery failed more quickly, though causing a brief sensation among the operatives. '... There was one defect in nearly all the ware independent of the want of polishing; air cracks almost invariably made their appearance in the backs of the ware after firing.'[243] The steam-powered 'Jolly', which had caused such a panic in the Potteries thirty-five years earlier (the Potters' Union set up an Emigration Society, and planted a colony in America, as a way of escaping it)[244] failed 'partly, it is supposed, through the desire of the employers not to come into conflict with the men', but chiefly 'owing to some defects in . . . construction'.[245] (Later it resurfaced, and by the 1890s was in general use.)[246]

Often the action of machinery was too crude and indiscriminate for the tasks it was appointed to perform. Power-looms, for instance, were too rough in their action for the more delicate processes of weaving, or the finer classes of thread. In hosiery, intricate patterns defeated the steam-powered rotary frame, and 'fancy' hosiery (a rapidly-expanding branch of the trade) remained the province of the hand-frame knitters long after the plainer classes of ware had gone to the machine. Steam threshers, though successful with wheat, were much too violent in their action for barley, which continued to be hand-flailed down to the end of the nineteenth century.[247] Mechanical potato-diggers were too clumsy and impetuous in their grasp: J. R. Wallace's model was a great success when it was put on trial at the Glasgow show of 1875, but trials under more difficult conditions showed that the potatoes were so badly damaged that they could not be stored in pits. Milking machines, despite numerous patents and improvements, were also much too rough: the siphon type injured the cow's udders and were troublesome to fit; the suction type was abandoned because it was liable to force out blood.[248] In manufacturing industry the difficulty often lay with the raw materials which were too delicate, or too variable, for machinery's harsh beat. No two skins were ever precisely the same in the leatherworking trades,[249] no two grains in furniture.[249] In the Birmingham trade of pearl-button making, where union members were fined £5 if they worked with steam,[251] the 'brittle nature' of the raw material did not allow for a faster rate of working than a foot-lathe would accomplish.[252] In cutlery, the difficulty lay with the steel: '. . . Cutlery . . . is not an article that lends itself to alteration', a Sheffield manufacturer said in 1903, 'High-class steel is so hard that attempts to manipulate it by machinery break the tools, and heating it to make it work easily would destroy its temper'.[253]

Even when such difficulties were overcome, machinery was rarely self-acting, but required skilled hands to guide and to complete its work. Nor, for the most part, could it be adapted to the finer and more delicate classes of work, where quality counted for more than yield. Machinery was thus used to do some sorts of things and not others, and many manufacturers settled for a mixed development, in which machinery was installed for the coarse, preparatory, stages of production, while the shaping and finishing was done by hand. This was notably the case in the Potteries, where pug-mills were used for mixing the flints and clay, but all of the actual making was done by hand. It was also the compromise reached in the Sheffield trade of saw-making, where the teeth were punched out by machinery, but bent into shape by hand, while the blade itself was tempered on the anvil by repeated hammer blows. 'The sharpening and setting of a saw requires considerable skill of hand and accuracy of eye; for if any one of the teeth projects either edgewise or sidewise beyond the true line, it renders the sawing harsh'.[254]

Profitability and costs

Economic considerations also limited machinery's scope. In some cases the gains in productivity were comparatively modest, either because the intricacy of the process slowed down working pace (the case in full-fashioned hosiery, and in certain branches of lace-making, where the machine was subject to frequent stoppages),[255] or – in the simpler classes of work – because hand labour could be worked at an almost machine-like pace. In the linen industry of the 1850s the slow tempo of mechanisation in weaving as compared with that in spinning seems to have been due to this cause. A power-loom could produce four times the amount of a hand-loom weaver in any given day, but in spinning the differential was 320:1. 'The enormous saving of expense . . . made the transition from . . . hand-spinning . . .an unavoidable necessity . . . The saving by power-looms not being nearly so great, the transition state will . . . be prolonged'.[256] Then again, the saving in labour might be counterbalanced by a greater wastage in raw materials (machinery was seldom employed where they were costly) or a need for higher quality inputs. In the galvanised iron trade, for instance, the new machinery of the 1850s and 1860s required a more costly material because the metal could not be 'humoured' as it could when it was shaped by hand;[257] while in cork-cutting it was thought likely to fail because of lack of discrimination: 'a skilful cork-cutter will so manage his material as to lead to a minimum of waste'.[258] A further disincentive was the possibility of breakdown, which could be expensive to repair, as well as bringing production to a halt.[259]

In agriculture mechanisation brought no overall gain in output. Steam cultivation might be more effective than horse tillage, and reaper-binders than harvesting by hand, but they did not increase the yield of a cornfield. The saving was principally on wages, but since these were screwed down to the barest minimum, farmers had no great incentive to invest in plant. 'In other arts machinery not only renders operations formerly done by hand more effective, but multiplies results to an almost infinite extent, while from the very nature of agriculture such an effect is not to be looked for'.[260] In metallurgy the gain in yield was offset by the loss in quality. Machine-riveting, for instance, was credited with performing the work at ten times the rate of hand-riveting, but even Fairbairn, a great advocate of hydraulic riveting, had to admit that the machine-made joints were not as strong as those which had been hammered by hand, 'though the faults would not appear in a comparatively small experiment'.[261] In the making of ship's chains even speed was not on machinery's side. In 1860 the welding of links was still being done at small forge fires with a top tool and hammer to fasten the stay. A recent attempt to abridge the work by introducing mechanical compressors had been abandoned on grounds of both quality and speed. 'However completely this and other machines may do the work, hand labour does the work quicker and better, almost beyond comparison.'[262]

Another set of difficulties presented themselves on the question of fixed and variable costs. A machine had to be kept running all the time to justify the expense of outlay, but even so the rise in output might not be sufficient to offset the running charges, depreciation costs, and investment. For the small producer, engaged in a week-by-week struggle for survival, the cost of installing machinery made it, quite simply, unthinkable – 'economy in time, energy and manpower was . . . less important than economy in cash outlay'.[263] But even for a large and well-established firm it could be daunting. Mechanisation involved not merely the substitution of a machine for a tool, but an entire revolution in the production process. It involved exchanging well-established and familiar routines for new and untried methods, either with a brand-new work force, which would have to be trained to perform new tasks, or with an old one determined to protect their jobs (workers at Waterlow's, the Finsbury papermakers, held a party when the 'striker', a new-fangled American machine of 1874, broke down, taking the day off to celebrate and going to Epping for a swim.)[264] In the workshop trades, with their cramped, overcrowded premises, want of space made it difficult to accommodate more than a modicum of plant.[265] Even in a large firm want of space could present an intractable problem. A potteries, for instance, was a rambling maze of small work-rooms, and machinery could only be installed if the entire premises were rebuilt.[266] Lay-out could be difficult elsewhere. At the Saltley gasworks, Birmingham, in 1895, two of the three retort-houses, with 960 mouthpieces to feed, depended on manual labour entirely because: 'These houses being too narrow for . . . machinery of any kind, the retorts are still drawn and charged by hand, and the coal is delivered . . . and the coke removed . . . by barrows.'[267] Extra fuel costs also militated against the introduction of machinery, especially in districts which were remote from the coalfields. A steam saw consumed a ton of coal and a quart of machine oil a day[268] – ginned by a horse it would be as cheap as the local grass or hay. All the big concentrations of mid-Victorian steam-power were to be found in close proximity to the coalfields, while in southern England, where coal prices were about double those in the manufacturing districts of the north, it was comparatively sparse. Nevertheless fuel costs do not seem to have acted as an independent force: the production census of 1871 shows that the variations *within* the coal districts were almost as steep as those which separated the coal districts from manufacturing industries further south. In Staffordshire, for instance, 66,425 horse-power were concentrated in the blast furnaces and iron mills, 2.09 h.p. per worker, while in the whole of the Staffordshire Potteries there was only 3,101 h.p. (0.09 h.p. per worker). Birmingham, despite the proximity of the Staffordshire coal-field, had only 11,272 h.p.; Oldham, in the cotton district of East Lancashire, had 31,025 h.p.[269]

Market Uncertainties

Another obstacle to mechanisation was the irregular nature of demand, and its often limited character. Steam power and machinery were only profitable if they were geared to large-scale production. But in the workshop trades short production runs were endemic, and output fluctuated sharply not only with the trade cycle, but also from season to season and in many cases from week to week. 'Little makers' like the Willenhall locksmiths, the Sheffield cutlers, or the cabinet-makers of Bethnal Green, could only afford to make up goods in small quantities at a time, because they had to meet wages and costs out of weekly earnings. Warehousemen and buyers-up, for their part, were niggardly in their purchases, and preferred, as a matter of policy, to trade from week to week rather than to run the risk of carrying unsold stock on their hands. Consumer demand also tended to favour limited production runs, alternating between periods of heavy pressure, when there was a helter-skelter rush of work (as in the 'bull' weeks immediately preceding Christmas) and others when trade was dead.[270] In conditions like these it was easier, when faced with a rush of orders, to take on extra hands, or sub-contract the work, than to install expensive machinery and plant: less risky in the long run, and in the short run at least a great deal more profitable.

The position was not necessarily different in heavy industry, despite the vast scale of many works. Tinplates – the most recent historian of the South Wales industry tells us – 'were not manufactured ahead of demand but were rolled to order'.[271] According to Menelaus, the manager of the Dowlais Works, this was also very frequently the case in heavy iron. 'When rolled iron is wanted either in large masses, or of difficult sections and . . . lengths', he told the South Wales Institute of Engineers in 1860, 'the quantities generally are so small that even if you have suitable machinery, before you get properly to work . . . the order is finished'.[272] In shipbuilding and engineering, a great deal of work was done to order rather than for stock, while the willingness of British engineering firms to make large numbers of products in small quantities – and to fit them up, if necessary, on site – was the very basis of the world-wide reputation for excellence they enjoyed.[273] Ransome's of Ipswich, the East Anglian manufacturers of agricultural machinery, maintained some 700 models in their catalogues, and any of these might be modified on the job at the pattern stage.[274] Much the same was true of many of the machine-tool makers discussed by Roderick Floud in his recent book.[275]

Limited production runs were also a feature of the hardware trades, partly because of market uncertainties, but also because demand was highly specific, and new lines were being continually pioneered to meet particular customer requirements. No fewer than 300 different classes of nail were being manufactured in the 1860s, for instance, with at least ten different sizes to each sort, so that in all upwards of 3,000 different kinds of nail were on the market.[276] 'Machine-made nails of certain kinds have largely invaded the market, but there seems no likelihood of machine nails superseding the general varieties now made by hand', runs a trade report from the Black Country in August 1878, 'for, considering the demand, it would not pay anyone to make these particular classes of nails by machinery'. The Willenhall lockmakers were even more individualistic: at Messrs. Carpenter's, the oldest works in the town, there were not less than 500 distinct patterns 'and when it is remembered that each pattern is made in various sizes ranging inch by inch from five to twelve inches an idea may be formed of the bewildering array'.[277]

'Specialty' was also rife in the manufacture of workmen's tools. In 1850s Birmingham, for instance, 45 different kinds of axe were being made – 14 of them specifically directed at the American market; there were upwards of 70 different types of hoe,[278] and no fewer than 500 varieties of hammer: 'not only is each adapted to one particular purpose, but several varieties often serve exclusively for the different operations in one and the same process'.[279] The larger firms in toolmaking added complications of their own. Wilkinson, the Grimesthorpe saw-makers, with a world-wide market for their hand-made saws, were making them to some 200 different patterns in 1880, and constantly adding improvements to their design;[280] Messrs. Elwell and Co. at Wednesbury Forge showed some 1200 different patterns in their catalogue 'including mattocks, plantation hoes, pickaxes, axes, adzes, spades, shovels, and forks'.[281]

Short production runs were also encouraged by the continuing vitality of local markets. Baskets, for instance, took on quite different shapes and sizes according to the district in which they were made, and the specific purposes to which they were directed. In Furness they had to serve as corves for the local coal and iron pits, in Nottingham as skips for the dyers and bleachers, in Southport as hampers for the shrimps.[282] Much the same was true of other commodities which were locally manufactured to meet specific local needs, coopers' ware,[283] for instance, fancy boxes and containers,[284] hay wagons, carts and barrows.

Consumer Preference

In many cases hand labour retained favour against machinery because of better performance. In paper-staining, for instance – i.e. the manufacture of wallpapers – block-printers held their own against machine-made competitors because the colours (which they mixed themselves)[285] were less liable to fade. 'Cheap machine-made papers quickly lose their colour', an architect advised potential clients in 1872, 'and as the process of manufacture does not permit of their being properly "set", the tints rub off and the patterns fade'.[286]

In edge-tool making hand-made planes were preferred on grounds of their greater accuracy, and the superior smoothness of their cut. 'Machinery has been employed for the manufacture of hand-planes, but with little success', wrote a commentator in 1866. 'The irregularity in the thickness of the irons requires the exercise of considerable skill in "bedding" them upon the wood, and hand-labour is indispensable to effect this . . . properly. Unless this point is attended to, the tool jumps over the wood it has to plane, and makes it rougher instead of smoother.'[287] Hand-cut files held their own notwithstanding numerous attempts by Sheffield manufacturers to market a cheap, machine-made alternative.[288] 'The wonderful sympathy of the eye and hand of the file-cutter are as yet totally unapproached by any machinery', wrote *The Ironmonger* in August 1878, 'The machine will certainly cut the teeth in the forged shape, but it does not give them that peculiar feather which is the essential feature of the file-cutter's craft'.[289]

In consumer durables there was a well-established preference for hand-made goods, not only because of their better quality, but also because of their superior finish, and of those additional embellishments which manufacturers were apt to call 'artistic'. The preference for hand-made goods was particularly marked in mid-Victorian times, when the bourgeoisie, in one or other of its manifestations, not only dominated public taste, but very often constituted the bulk of effective demand. A very common commercial practice of the 1870s and 1880s was for manufacturers and merchants to send machine-made goods to the Colonies and America, while marketing higher-class, hand-made goods at home.[290] In domestic hardware and the cutlery trades, the preference for hand-hammered goods was virtually unchallenged, and continued until the end of the century. When, for instance, about 1900, the Bilston manufacturers began to go over to machine-stamped frying pans, they took good care to do so in disguise. 'The knowing housewife still looks for the marks of the hammer, so they are added afterwards, just as a little sand is added to sponges.'[291] Hand-made locks held their own in the home market, not only because of their cheapness, but also because they were more cumbersome than their streamlined American competitors and might therefore be more formidable to burglars. They also possessed that priceless quality of individuality which bourgeois householders have always sought after so eagerly, and held in such esteem. 'American locks are, it must be granted, marvels of cheapness', *The Ironmonger* wrote disparagingly in October 1877,

> and the expenditure of manual labour in their production is almost *nil*... But in thus practically dispensing with manual and patient labour; in thus turning out a gross of locks faster than English makers turn out a dozen... certain features of the article – which used to be thought indispensable – have necessarily to be sacrificed. The ordinary American cast door-lock has... no wards, the only security being in the 'fancy' or shape of the keyhole. A few strokes with a file will of course modify this... and then all "security" is destroyed... Machine-made keys, however ingenious... do not, and cannot in the nature of things sufficiently "differ".[292]

Industrial design – which will be discussed in a second article – served to reinforce the role of speciality, and the extension of fashion to a new and cheaper range of commodities brought a feverish search for new styles. In boot and shoemaking patterns changed regularly two or three times a year, quite apart from being subject to short-lived crazes such as glazed heels. Wallpaper manufacturers were continually bringing out new lines, with dazzling new tints and a bewildering variety of motifs. Carriage-makers, too, were subject to continual changes of taste, and this seems to have been one of the reasons why, even in the largest works, so little machinery was used. 'As a class of manufacturers, we labour under many disadvantages in adopting machinery', a Bath capitalist told the Institute of Carriage Manufacturers in 1890, '... Owing to the numerous sizes and great variety of design, even the very largest establishments can only place a limited number of any one carriage in hand at one time.[293] In Birmingham, with its multiplicity of small producers, marginal differentiation of the product was a very principle of growth, and responsiveness to changes in public taste a condition of business survival. Patterns were constantly going out of date, and manufacturers continually experimenting with new lines. '... in Birmingham the different varieties and sizes and patterns of article are so numerous, that the adjustment of the steam-engine to do the work would be almost impracticable, and unprofitable if practicable', wrote Charles Knight in 1846, spelling out some of the technological consequences. 'The adjustments required by the ever-varying tastes and wants of the age can be effected only by men's fingers: the steam-engine being appealed to for that kind of service which may be common to all the works required'.[294]

Combined and uneven development

Steam power and hand technology may represent different principles of industrial organisation, and to the historian they may well appear as belonging to different epochs, the one innovatory, the other 'traditional' and unchanging in its ways. But from the point of view of 19th century capitalist development they were two sides of the same coin, and it is fitting that the Great Exhibition of 1851 – 'the authentic voice of British capitalism in the hour of its greatest triumph'[295] – should have given symbolic representation to them both. 'Steam power', an admiring commentator noted, 'wholly turned the mahogany which runs round the galleries of the Crystal Palace'.[296] But the 300,000 panes of glass which covered it were blown by hand,[297] and so was the Crystal Fountain which formed the centre-piece of the transept, 'glittering in all the colours of the rainbow'.[298] The promoters were intoxicated with the idea of 'self-acting machinery', and the technological miracles it might perform. But they devoted a great deal of their space to – among other things – needlework; and in demonstrating the competitive capabilities of British industry they were heavily dependent on artisan skills. Most of the manufactures on display were handicraft products, and even in the Machinery Court many of the exhibits were assembled from hand-made components. 'Few objects' excited more attention among foreigners than the displays of Sheffield cutlery and edge-tools (the Sheffield Court was one of the most extensive in the building),[299] while domestic visitors, it seems, were no less enraptured by the impenetrable locks, 'myrio-permutation' keys and incombustible safes of Messrs. Chubb, Bramah and Mordan. Superimposed on the idea of mechanical progress there was also a nascent commercial aesthetic which the Exhibition's promoters rather grandly labelled 'the marriage of industry and art'. In subsequent years it was to make 'taste' a very principle of production and the marginal differentiation of products a primary axis of growth.

The balance of advantage between steam power and hand technology was, in mid-Victorian times, very far from settled, and many manufacturers though experimental in marketing new products and multiplying novelties of design, remained wedded to conservative production routines. Human beings, the main alternative to machinery, were, from a commercial point of view, often a much more attractive proposition. They were a great deal cheaper to install than a power house, and much more adaptable in their action than a self-acting stamp or press. When they broke down, the master did not have to pay for repairs; when they made a mistake, he could fine them; when there was no work for them to do he could give them the sack. Skills too were cheaper than machinery to come by. A steam sawing machine, in 1850, cost £700 to install;[300] a pair of travelling sawyers could be hired to do a job for five shillings, while a circular saw – such as the one used at Joseph Severn's shop in Codnor – could be ginned by a horse for free. Machinery was thus often adopted as a last resort, when every alternative means of extracting surplus value had failed to yield an adequate return, and it is no accident that manufacturers – like the Sheffield file makers of 1866 – so often turned their eyes to it when they were faced with demands for higher wages.

The orthodox account of the industrial revolution concentrates on the rise of steam power and machinery, and the spread of the factory system. It has much less to say about alternative forms of capitalist enterprise (such as those to be found in mining and quarrying), about the rise of sweating, or the spread of back-yard industries and trades. Nor does it tell us much about the repercussions of technology on work. Landes' picture has the compelling power of paradigm, with mechanisation on an 'ever-widening front' and steam power – 'rapid, regular, precise' – effortlessly performing labour's tasks. But if one looks at the economy as a whole rather than at its most novel and striking features, a less orderly canvas might be drawn – one bearing more resemblance to a Bruegel or even a Hieronymus Bosch than to the geometrical regularities of a modern abstract. The industrial landscape would be seen to be full of diggings and pits as well as of tall factory chimneys. Smithies would sprout in the shadows of the furnaces, sweatshops in those of the looms. Agricultural labourers might take up the foreground, armed with sickle or scythe, while behind them troops of women and children would be bent double over the ripening crops in the field, pulling charlock, hoeing nettles, or cleaning the furrows of stones. In the middle distance there might be navvies digging sewers and paviours laying flags. On the building sites there would be a bustle of man-powered activity, with house-painters on ladders, and slaters nailing roofs. Carters would be loading and unloading horses, market women carrying baskets of produce on their heads; dockers balancing weights. The factories would be hot and steamy, with men stripped to the singlet, and juvenile runners in bare feet. At the lead works women would be carrying pots of poisonous metal on their heads, in the bleachers' shed they would be stitching yards of chlorined cloth, at a shoddy mill sorting rags. Instead of calling his picture 'machinery' the artist might prefer to name it 'toil'.

Skill was as important as toil (the two often went hand in hand) and in mid-Victorian times it was plentifully available. The domestic housebuilder could draw on a vast substratum of carpentering skills: so could such booming industries as Kentish Town pianos and High Wycombe chairs. The new iron shipyards were quickly filled with artisans and mechanics drawn from a dozen different trades; by the 1870s they were already a very cockpit of sectarian craft rivalries. Engineering employers recruited their labour from those who had served their apprenticeships in the 'country' branches of the trade, with wheelwrights, blacksmiths, and in the small town foundries; and it was a matter of real anxiety in the industry when, towards the end of the century, this source of recruitment began to dry up. 'There is no evidence that labour supply impeded any of the machine tool firms', writes Roderick Floud in his recent book, 'Even as early as the 1830s, Nasmyth was able to break a strike in his works, aimed at forcing him to employ only men who had served an apprenticeship, by importing sixty-four Scottish mechanics and he remarked that "we might easily have obtained three times the number . . ." No other machine tool maker appears to have had difficulties in securing labour, or, indeed, . . . in dismissing it when times were bad, in the confident expectation that the men could be re-employed if trade improved'.[301]

It was not only in craft industry that capitalism drew on reservoirs of skill, but in every branch of economic activity where a mainly hand technology prevailed. Tunnel bricklayers on the railway works were in their own way as skilled as stonemasons; so were the coal heavers and timber porters in the docks, the carters and wagoners in road haulage, the ploughmen and rick-builders on the farms, the shot-firers and hewers in the pits. As well as the 'aristocracy of labour', on whom British historians have lavished such continuous attention, there was also a whole army of rural mechanics and small town artisans, like the 'ragged trousered philanthropists' of Mugsborough, who still await their chronicler. So do the poor artisans of Shoreditch and Bethnal Green, in East London, who at the time of the 1891 census constituted no less than 60 per cent of the local working population.

Nineteenth century capitalism created many more skills than it destroyed, though they were different in kind from those of the all-round craftsmen, and subject to a wholly new level of exploitation. The change from sail to steam in shipping led to the rise of a whole number of new industrial crafts, as well as providing a wider arena for the exercise of old ones. The same may be said of the shift from wood to iron in vehicle building, and of horse to steam in transport. In the woodworking trades a comparatively small amount of machinery supported a vast proliferation of handicraft activities, while in metallurgy the cheapening of manufacturing raw materials led to a multiplication of journeymen-masters. The mid-Victorian engineer was a tool-bearer rather than a machine minder; the boilermaker was an artisan rather than a factory hand. In coal mining activity increased by the recruitment of a vast new class of workers who were neither exactly labourers, nor yet artisans, but who very soon laid claim to hereditary craft skills. Much the same was true of workers in the tinplate mills and ironworks. The number of craftsmen in the building trade increased by leaps and bounds, though the rise of new specialities led to a narrowing of all-round skills.

In juxtaposing hand and steam-powered technologies one is speaking of a *combined* as well as of an *uneven* development. In mid-Victorian times, as earlier in the 19th century, they represented *concurrent* phases of capitalist growth, feeding on one another's achievements, endorsing one another's effects. Both were exposed to the same market forces; both depended for their progress upon the mobilisation of wage labour on a hitherto unprecedented scale, and both were equally subject to the new work discipline, though it affected them in different ways. The industrial revolution rested on a broad handicraft basis, which was at once a condition of its development and a restraint on its further growth. In mid-Victorian times – as I shall attempt to show in a second article – the handicraft sector of the economy was quite as dynamic as high technology industry, and just as much subject to technical development and change. It was indeed in the first rather than the second that mass production methods in many cases were pioneered; that new classes of commodity were created; and that modern capitalist methods of exploitation – both of producers and consumers – were most clearly prefigured and explored.

GUIDE TO ABBREVIATIONS AND REFERENCES

MSS – manuscripts; RC – Royal Commision; S.C. – Select Committee; M.B. – Minute Book; M.O. – Medical Officer.
Webb Coll. – London School of Economics, Webb collection of trade union journals and MSS.
Booth MSS. – London School of Economics, Charles Booth's MSS. for *London Life and Labour*.
Howell Coll. – Bishopsgate Institute, George Howell trade union collection.
Goldsmith Coll. – University of London, Goldsmith's collection of economic literature.
P.R.O. – Public Record Office.
P.P. – Parliamentary Papers.
Iron & Steel J. – Journal of the Iron and Steel Institute.
Journ. Soc. Arts – Journal of the Royal Society of Arts.
Royal Ag. Soc. J. – Journal of the Royal Agricultural Society of England.
Stat. Soc. J. – Journal of the Royal Statistical Society.
Newcomen Soc. – Transactions of the Newcomen Society.
Econ. H. R. – Economic History Review.
Bevan, *Manufacturing Industries* – Ed. George Bevan, *British Manufacturing Industries*, 14 unnumbered volumes, London 1876.
Ure, *Dictionary* – Andrew Ure, *Dictionary of Arts, Manufactures and Mines*, 3 vols., London 1860 ed.
Timmins – ed. Samuel Timmins, *The Industrial Resources of Birmingham and the Midland Hardware District*, London 1866
Marx, *Capital* – Karl Marx, *Capital*, Vol. I, Dona Torr ed. London 1949.
Rousiers, *Labour* – Paul de Rousiers, *The Labour Question in Britain*, London 1896.
Cox, *British Industries* – ed. Harold Cox, *British Industries under Free Trade*, London 1903.
Clapham, *Economic History* – J. H. Clapham, *An Economic History of Modern Britain*, 3 vols., Cambridge 1926-1938.
Mitchell and Deane – B. R. Mitchell and Phyllis Deane, *Abstract of British Historical Statistics*, Cambridge 1971.
Labouring Men – E. J. Hobsbawm, *Labouring Men, Studies in the History of Labour*, London 1964.
Workshop Trades – ed. R. Samuel, *The Workshop Trades*, (forthcoming).

For reasons of space, the issue editors have been obliged to omit some footnotes and conflate others. The *Morning Chronicle* references, for which date only is given, are to the series 'Labour and the Poor', covering 'Metropolitan Districts' 'Rural Districts' 'Manufacturing Districts' 'Birmingham' 'Liverpool' and 'The Mining and Manufacturing Districts of South Wales'.

[1] David Landes, *The Unbound Prometheus*, Cambridge 1969, p. 41.

[2] For an autobiographical account by one who traded in adulterated soot, George Elson, *The Last of the Climbing Boys*, London 1900, pp. 78-81. For complaints of the master chimney sweeps about 'pirate' soot merchants, Booth MSS, B 160, fol. 16. For a soot merchant in Poplar, *ibid*, B 120, fol. 80. For the soot trade in Westminster, Westminster archives, SKT. Martin's in the Fields Paving Commissioners M.B., 24 April 1844. For fluctuations in the price of soot, J. Thomson and Adolphe Smith, *Street Life in London*, London 1877.

[3] *Liverpool Mercury*, 10 Oct. 1845, p. 400. I am grateful to Tim Mason for this reference.

[4] Henry Mayhew, *London Labour and the London Poor*, London 1861.

[5] For the effect of heavy sizing on the health of those who had to work with it in the mills, P.P. 1872 (203) LIV, P.P. 1884 (c.3861) LXXII. For a moving autobiographical account, Alice Foley, *A Bolton Childhood*, Manchester 1973, p. 51.

[6] Goldsmith's Coll., Henry Mayhew, *Labour and the Poor*, London 1851, p. 40. Nicoll was under-sheriff of the City of London. He and his brother were said to be worth £80,000 a-piece.

[7] H. A. Shannon, 'Bricks, a trade index', *Economica* 1934.

[8] Hobsbawm, 'The Machine Breakers', in *Labouring Men*.

[9] P.R.O., For. 3/559,3/263. The explosion, like others in the district, seems to have been the work of Walter Virgo and the Blakeney gang.

[10] Howell Coll., *The Moxon Loom Arbitration Proceedings*, Kidderminster 1879. Cf. also J. N. Bartlett, 'The Mechanization of the Kidderminster Carpet Industry', *Business History*, 1967, IX.

[11] J. Reynolds, *The Letterpress Printers of Bradford*, Bradford 1971, pp. 5-6; Sid Wills, 'The Compositor's Frame' in *The Workshop Trades*.

[12] John Ball, 'Account of the Northamptonshire Boot and Shoe Makers Strike in 1857/8/9', *Trade Societies and Strikes*, London 1860. There are very full reports of these strikes in the *Northampton Mercury* and the *Northamptonshire Free Press*.

[13] 'Murderous Threats to Workmen in the Screw Bolt Trade', *The Ironmonger*, 30 Dec. 1865, VII 180-1.

[14] *Capital* XV 464-96; XVI 519.

[15] Peel to Croker, 27 July 1842, in ed. C. S. Parker, *Sir Robert Peel . . . Private Papers*, London 1899, II 529. Cf. also G. Kitson Clark, 'Hunger and Politics in 1842', *Journal of Modern History*, 1955.

[16] *Capital* XI 297, cf. XV 422.

[17] Karl Marx and Friedrich Engels, *Selected Correspondence*, London 1935, pp. 10-11.

[18] There is an account of the Tate and Lyle works in W. Glenny Crory, *East London Industries*, London 1876.

[19] Booth MSS. A.24 part A, fols. 17-23 for two usefully detailed biographies; *ibid* B.10, fol. 125, B.16, fols. 43, 91, B.44, fols. 26, 41; B.45, fols. 60, 147 for some other back-street manufacturers.

[20] For the rolling mills, P.P. 1876 (c.1443-1) XXX, Rep . . . Fact. & Workshops Act, QQ. 12058-61; for a list of the 55 trades 'in union', Frank Hill, 'Combinations in Sheffield', *Trade Societies and Strikes*, London 1860, pp. 564-5.

[21] P.P. 1876 XXX, Rep . . . Fact. & Workshops Act, QQ. 5529, 5644; Webb Coll., Sect. A, Vol. X, fol. 386; A. B. Searle, *Refractory Materials*, London 1917, p. 144.

[22] *Morning Chronicle*, 23 Dec. 1850, Birmingham.

[23] Ellis A. Davidson, *Our Houses*, London 1869, p. 96; Prof. Barff, 'Glass and Silicates', pp. 76-7, in Bevan, *Manufacturing Industries*.

[24] Mathew Blair, 'The Silk Trade' in Cox, *British Industries*, p. 65. P.P. 1861 (161) XVI, 3rd Rep, M.O. Privy Council, p. 159.

[25] Clapham, *Economic History*, II 83; H. Heaton, *The Yorkshire Woollen and Worsted Industries*, Oxford 1965 ed. p. 358. For a wages agitation of the Pudsey handloom weavers, *Beehive*, 11 Nov. 1871, and for an attractive autobiographical account from Colne Valley in the 1870s, Ben Turner, *About Myself*, London 1930, pp. 34-35.

[26] Sir R. Lloyd Patterson, 'The Linen Industry' in Cox, *British Industries*, p. 58. For some strike movements among the hand-loom linen weavers of Barnsley, *Beehive*, 10 Feb.; 10, 24, 31 March; 10, 17, 24 Nov.; 8 Dec. 1866.

[27] P.P. 1861 (161) XVI, 3rd Rep. Med. Off. Privy C., p. 169. For the slow spread of steam in the industry, W. Felkin 'The Lace and Hosiery Trades of Nottingham', *Stat. Soc. J.* XXIX, 1866; Roy Church, 'Technological Change and the Hosiery Board of Conciliation . . . 1860-1884' *Yorkshire Bull. of Econ. and Soc. Research*, May 1963, XV/I, p. 53.

[28] Dodie Weppler, 'Seamers and Stitchers', unpublished paper given at Ruskin History Workshop 8, 'Women in History', May 1972.

[29] The Bolton hand-mule spinners numbered 1,326 in their Association in 1874, *Capital and Labour*, 16 Sept. 1874. For their amalgamation with the self-actor minders, Bolton, Spinners' Hall MSS., Hand Mule Spinners Association Minutes, 21 Feb. 1880. For an employer's view of the hand-mule spinners, 'The Bolton Strike' in *Capital and Labour*, 12 Sept. 1877, IV 466. Clapham, *Economic History*, II 81; for a more detailed discussion of hand-mule spinning inside the factories, Harold Catling, *The Spinning Mule*, Newton Abbot 1970, pp. 46, 51, 63, 119, 115-16, 118-19, 123. Readers might be interested in Marx's discussion of the survival of hand-mule spinning: *Capital*, pp. 376-7.

[30] Andrew Wynter, *Curiosities of Civilization*, London 1860, pp. 262-3.

[31] J. D. Goodman, 'On the Progress of the Small Arms Manufacture', *Stat. Soc. J.* XXVIII, 1865, pp. 495-8.

[32] Sid Wills, 'The Compositor's Frame' in *The Workshop Trades*.

[33] For a good account of one of them, George Dodd, *Days at Factories*, London 1843, pp. 17-40.

[34] John Lord, *Capital and Steam Power, 1750-1800*, London 1923; for a very helpful general discussion of the diffusion of steam power, A. E. Musson, 'Industrial Motive Power in the United Kingdom, 1800-1870', *Econ. H. R.*, Aug. 1976, 2nd Ser. XXIX/3.

[35] Clapham, *Economic History* II, 123. In the year ending 31 March, 1873, 30,671 brewers took out licences in England: T. A. Pooley, 'Brewing', p. 202, in Bevan, *Manufacturing Industries*. London brewers recruited many of their workers from the small country brewers: Booth MSS., B122, fols. 31, 81.

[36] Nineteenth century windmills seems to have been as much subject to 'improvement' as any other species of capital equipment. Among them Messrs. J. Warner's patent annular sails. John Ashton, *The History of Bread*, London 1904, p. 106. For some other 19th century improvements in windmills, Rex Wales, 'Suffolk windmills', *Newcomen Soc.*, XXII, XXIII, 1941-3.

[37] For silk, *Morning Chronicle*, 22 Nov. 1849, Macclesfield; Sir Frank Warner, *The Silk Industry*, London 1922, pp. 135, 319, 322; P.P. 1886 (c.4715) XXII 2nd Rep. Depression of Trade, Q.7454; for woollens, Rousiers, *Labour*, p. 306; J. H. Clapham, *The Woollen and Worsted Industries*, London, 1907, p. 58. For hand-looms in the Kidderminster carpet factories, W. J. Gordon, *Midland Sketches*,j london 1898, p. 104.
[38] J. W. C. Haldane, *Steam Ships and their Machinery*, London 1893, p. 000.
[39] Harry Brearley, *Steel-Makers*, London 1933, pp. 26-8. For an earlier account, John Scoffern *et. al.*, *The Useful Metals and their Alloys*, London 1857, p. 349. In the 1890s, Sheffield used 14,000 clay crucibles a week in which to fuse its steel. R. A. Hadfield, 'On the early history of Crucible steel', *Iron and Steel J.*, 1894, XLVI 324.
[40] A. B. Searle, *Refractory Materials*, London 1917, pp. 308-9.
[41] There is an account of the collier brigs by one who served with them in Walter Runciman, *Collier Brigs and their Sailors*, London 1926.
[42] F. M. L. Thompson 'Nineteenth-Century Horse Sense', *Econ. H. R.*, Feb. 1976, 2nd Ser. XXIX/I, for these and other useful horse statistics and an interesting discussion of them.
[43] Mitchell and Deane, p. 60.
[44] Samuel, 'Mineral Workers', in *Miners, Quarrymen, and Saltworkers*, 1977, p. 9 for some details of the traffic.
[45] *Morning Chronicle*, 23 Dec. 1850.
[46] C. W. Waring, 'On the application of machinery to cutting coal', *Transaction South Wales Institute of Engineers*, 1862-3, III 95.
[47] *Manchester Examiner*, 26 July 1865, p. 5, cols. 2-3; cf. also *Barnsley Chronicle*, 1 Sept. 1866; *Capital and Labour*, 15 April 1874; *Iron and Coal Trades Review*, 15 May 1872, p. 386.
[48] J. E. Williams, *The Derbyshire Miners*, London 1962, p. 174.
[49] H. J. Habakkuk, *American and British Technology in the Nineteenth Century*, Cambridge 1962, p. 200. For some of the technical obstacles encountered by the coal cutter P.P. 1873 X(313) S.C. on Coal, QQ.2198-9; Rousiers, *Labour*, pp. 121-2; P.P. 1892 XXXVI (6795-VII), R. C. on Labour, Answers to Questions, Group 'A', p. 263; A. J. Taylor 'Labour Productivity and technical innovation in the British coal industry', *Econ. H. R.*, 2nd Ser. XIV, pp. 58-62.
[50] In earlier years the hewer had been expected to do his own propping, but in the second half of the 19th century they were vigorously resisting this because it cut down their piece-rated earnings. *The Times* 5 July 1873, p. 5; *Iron and Coal Trades Review*, 13 March 1872.
[51] Mitchell and Deane, pp. 60, 119.
[52] Thompson, 'Nineteenth Century Horse Sense', Appendix, p. 80, from estimates supplied by Alan Griffin.
[53] 'Tools', *The Quarry*, April 1896, pp. 63-4.
[54] W. G. Renwick, *Marble and Marble Working*, London 1909, p. 28; *Crossing's Dartmoor Worker*, Newton Abbot 1966, pp. 70-1; J. A. Howe, *The Geology of Building Stones*, London 1910, p. 99.
[55] Interviews of the writer with David Powell, Ivor Probert, Harry Powell, Clydach Vale, Mon., Aug. 1974.
[56] William Knight, 'Granite Quarries of Aberdeenshire', *Transactions of the Highland Agricultural Society*, 1835, 2nd Ser. IV.
[57] The 'Blondin' was invented by Fyfe of Kenmay, one of the chief granite-working centres of Aberdeenshire: William Diack, 'The Scottish Granite Industry', *The World's Work*, 1903, pp. 636-7. I am grateful to Bob Duncan for this reference.
[58] Fred Bower, *Rolling Stonemason*, London 1936, pp. 40-1.
[59] Walter White, *A Londoner's Walk to Land's End*, 2nd ed., London 1861, pp. 126-30 for a vivid description.
[60] Joseph H. Collins, *Principles of Metal Mining*, London 1875, pp. 60, 63.
[61] From an autobiographical account recorded in Kenneth Hudson, *The History of English China Clays*, Newton Abbot 1969, p. 32.
[62] This paragraph is based on L. T. C. Rolt, *The Potter's Field: a History of the South Devon Clay Industry*, Newton Abbot 1974.
[63] John Grey of Dilston, 'A View of the present state of agriculture in Northumberland', *Royal Ag. Soc. J.*, 1841, II 165.
[64] *British Husbandry*, 1837, ii, p. 340.
[65] John Algernon Clarke, 'On the great level of the Fens', *Royal. Ag. Soc. J.*, 1847, VIII, pp. 92, 101, 119, and also his *Fen Sketches*, London 1852, p. 256.
[66] E. J. T. Collins, 'Harvest Technology and Labour Supply in Britain, 1790-1870', *Econ. H. R.*, 1969, 2nd Ser., XXII/3; and David Morgan 'Harvesters' in *Village Life and Labour*, London 1975. Collins article is of first-class importance for the subject of this essay.

[67] Charles Whitehead, 'Report . . . Market Garden Competition', *Royal Ag. Soc. J.*, 1879, 2nd Ser. XV, p. 842. For some estimates of labour costs on arable farms see the very interesting article by Frederic Clifford, 'The Labour Bill in Farming', *Ibid*, 2nd Ser., 1875.

[68] Charles Whitehead, 'Hops, Fruit, Vegetables', *Ibid.*, 2nd Ser., XIV, 1878, p. 750.

[69] *Ibid.*, pp. 850, 864. For potato planting near Gillingham, Trinity Coll., Cambridge, MS Diary of A. J. Munby, XVIII, 9 March 1863. I am grateful to Anna Davin for this and the other Munby market gardening references.

[70] C. W. Shaw, *The London Market Gardens*, London 1879, p. 187.

[71] Whitehead, *Royal Ag. Soc. J.*, 1879, p. 845.

[72] Whitehead, *loc cit.*, p. 846; Shaw, *op cit.*, p. 191, P.P. 1868-9 (4202.I) XIII, 2nd Rep., Emp. in Agriculture, A; Pershore, Worcs.

[73] Whitehead, *Royal Ag. Soc. J.*, 1879, p. 858, Munby Diaries, XII, 1 March 1862; XXV, 16 May 1865.

[74] For migrant pea-pickers and contract work in Essex, Glenny 'Market Gardening', *Victoria County History of Essex*, 1907, II, pp. 476-7; G. Cuttle, *The Legacy of the Rural Guardians*, Cambridge 1934, pp. 267, 273; *Toilers in London by One of the Crowd*, London 1883, p. 96, gives a description of the pea-pickers' living quarters. P.P. 1867-8 (4068.I) XVII, 1st Rep., Emp. in Agriculture, App. Pt. II, p. 359 for itinerant Irish pea-pickers at Stoke Poges, Bucks.

[75] Patrick MacGill, *Children of the Dead End*, London 1914, p. 75. For Irish potato pulling gangs in earlier times, Emp. in Agriculture, 1st Rep., P.P. 1867-8 (4068.I) XVII, c. 120, 272; d. 212, 260, 276; and pp. 385, 387, 388, 390; *ibid*, 2nd Rep., P.P. 1868-9 (4202.I) XIII, App. II. Ai, 130. For a more recent account of the Irish potato picking gangs in Scotland, *Observer*, 13 June 1971.

[76] Whitehead, *Royal Ag. Soc. J.*, 1879, pp. 841, 848; Whitehead, *Market Gardening for Farmers*, London 1880, p. 12; for a personal recollection of this kind of work at Tiptree, Essex Record Office, T/Z 25/378, Mrs. Field, 'My First Job'.

[77] Andrew Wynter, *Curiosities of Civilisation*, London 1860, p. 235; Shaw, *London Market Gardens*, pp. 159-60.

[78] Oral history: Interviews of the writer with Mrs. Annie McClough of Liverpool, April 1972.

[79] Booth MSS., B.117, p. 34.

[80] Charles Booth, *London Life and Labour*, London 1904, 1st Ser. IV, 288-9.

[81] James Jefferys, *Retail Trading in Britain, 1850-1950*, Cambridge 1954, p. 181; Janet Blackman, 'The Food Supply of an Industrial Town', *Business History*, 1962, V/1, pp. 86, 91; Robert Roberts, *The Classic Slum*, Penguin ed. pp. 88-9.

[82] J. B. Short, *The Butcher's Shop*, Oxford 1928, pp. 11, 13.

[83] For the bad meat trade, see police court prosecutions in e.g. *The Times*, 5.2.61; 28.3.61; 8.4.61; 23.8.66; 28.9.66; *Morning Chronicle*, 30.5.49; for the 'slink' butchers of Newton Heath, Manchester, Rawlinson, *Rep. to the General Board of Health . . . Newton Heath*, London 1852, pp. 28, 29-30, 45, 52-3; Holborn archives, Holborn Bd. of Works M.B. 6 Jan. 1873 for Joseph Hart, a Leicestershire dealer who made a speciality of bad meat transactions; for dealings on Salisbury Plain, Alexander Somerville, *The Whistler at the Plough*, Manchester 1851-3, I 389-90.

[84] J. Chalmers Morton, 'On Cheese-Making in Home Dairies', *Royal Ag. Soc. J.*, 2nd Ser. XI, 1875, p. 273; Hall, *Pilgrimage of English Farming*, p. 410; G. Gibbons, 'Cheddar Cheese-Making', *Journal of the Bath and West of England Society*, 1890, 4th Ser. I., p. 102; G. E. Fussell, *The English Dairy Farmer, 1500-1900*, London 1966, pp. 191, 229, 240, 242. T. D. Acland and W. Sturge, *Farming of Somerset*, London 1851, p. 53.

[85] P.P. 1864 (3381) XXI, Ann. Rep . . . British Fisheries . . . 1863, p. 14; Bob Gilding, 'Journeymen Coopers' in *The Workshop Trades; The Practical Fishmonger and Fruiterer*, London 1914-17, 1, 68; IV, 130-6; G. R. Sims, *Off the Track in London*, London 1911, pp. 49-53; Booth MSS, B.346 p. 113; Ruskin College history thesis: Tony Wailey, 'The Marsh side Shrimpers', 1974; P.P. 1898 (c.8753) XIV, Rep. Fish Curing trade, p. 11.

[86] P.R.O., HO45/9622/A 17459; William Jago, 'Modern Developments of Bread-Making' *Journ. Soc. Arts*, 27 Dec. 1889, XXXVIII 93; Andrew Low, 'Flour Milling under Free Trade' in Cox, *British Industries*, p. 256; G. J. S. Broomhall and J. H. Hubback, *Corn Trade Memories*, Liverpool 1930, pp. 40, 42, 43, 189, 244; John Burnett, *Plenty and Want*, London 1966, pp. 104-6; Marx, *Capital*, p. 232; P.P. 1862 (3027) XLVII, Rep . . . Journeymen Bakers, p. xiii; George Dodd, *The Food of London*, London 1856, p. 197; John Burnett, 'The Baking Industry in the 19th Century', *Business History*, V/2; John Ashton, *History of Bread*, p. 140; Ure, *Dictionary*, I 408; *The Miller*, 1 March 1875, p. 4; *Bakers' Record*, 28 Aug. 1869, p. 5; 25 Sept. p. 4; 9 Oct. p. 4 – for some representative complaints from the 'full-priced' end of the trade.

[87] John Burnett, *Useful Toil*, London 1974, pp. 92-5; Banbury Ref. Lib. MSS, Arthur Jonas, 'A Cherwell Boy or Banbury in the 1900s'; Samuel, 'Quarry Roughs', in *Village Life and Labour*, London 1975, p. 202. For the increase of 'small makers' specialising in halfpenny and farthing

novelties, A. E. Coombe, 'A General Chat on the Toffee Trade', *Confectionery*, 12 March 1896, p. 50; *Confectionery*, 13 April, 12 May 1896; J. D. Burn, *Commercial Enterprise and Social Progress*, London 1858, p. 130; Andrew Wynter, *Our Social Bees*, 2nd Ser., London 1866, pp. 204-5. Goldsmith's Coll., 'The Working Classes of Edinburgh, 1853; Confectioners'. So late as the 1890s G. H. Duckworth, visiting an East London sweet factory, noted that 'the steam process used in jam-making is not powerful enough for sweets', Booth MSS. B. 117 36.

[88] R. A. Church, *The Great Victorian Boom, 1850-1873*, London 1975, p. 34.

[89] E. Dobson, *Rudiments of the Art of Building*, London 1849, pp. 25, 42-3, 261. For the extraordinary elaboration expected in high-class painter's work, J. W. Facey, *Elementary Decorati n*, London 1882; Adam Hammond, *Rudiments of Practical Bricklaying*, London 1874, pp. 1, 25, 42-3, 44; Dobson, *Building*, pp. 119-23. For a vivid illustration, 'The Staircase Hand of the Old School', *The Builder*, 6 July 1878, XXXVI 693-4. F. Walker, *Brickwork*, London, 1885, pp. 58-87.

[90] Sidney and Beatrice Webb, *The Story of the King's Highway*, London 1963, p. 200.

[91] Telford's specifications for the London and Holyhead road, quoted by Thomas Hughes, *The Practice of Making and Repairing Roads*, London 1838, p. 28.

[92] Andrew Wynter, *Peeps into the Human Hive*, London 1874, II, pp. 187-8; Samuel, 'Mineral Workers', p. 83 for the competition between asphalt and granite.

[93] Booth MSS, B 149, fols. 2, 3, 7.

[94] Ronald Clark, *The Development of the English Traction Engine*, Norwich (1960), p. 167; cf. also *The Builder*, 19 Oct. 1867, XXV 767; 14 Dec. 1867, XXV 915.

[95] Quoted in *Report of the Highway and Sanitary . . . Joint Committee of the Paddington Vestry on Wood and Other Pavements*, London 1878, p. 51.

[96] *The Quarry*, May 1896; Knight, 'Aberdeenshire Granite', pp. 66-8; P.P. 1912-13, (Cd 6390) XLI, R.C. Metalliferous Mines, QQ. 8593-4; M. Powis Bale, *Stoneworking Machinery*, London 1884, p. 75.

[97] For some examples, *The Builder*, 11 May 1867, 6 July, 21 Sept. 1878, XXXVI 997. 'Church Building News', together with its appendage, 'Dissenting Church Building News' is a regular column in *The Builder* of the 1860s and 1870s, and 'Town Halls and Exchanges' became a regular heading in 1865. For the employment this gave Jude Fawley, Thomas Hardy, *Jude The Obscure*, St. Martin's paperback ed., Part 5 ch. VII.

[98] Henry Broadhurst, *The Story of His Life Told by* Himself, London 1901, pp. 28-9; *Reminiscences of a Stonemason*, pp. 89-90, 105. So late as the 1890s stone working machinery seems to have made little impact on the industry, so far as stonemasons themselves were concerned: cf. Booth MSS., A.3 fol. 279 (interview with Hancock, secretary of the Operative Stonemasons); Congress House records, United Builders and Labourers, *Trade Circular and Monthly Report*, Sept. 1895, pp. 8-9. For an early strike against machinery, interestingly by masons' labourers, *Bee-hive*, 25 Aug. 1866, p. 6, col. 3. For patents and experiments, M. Powis Bale, *Stoneworking Machinery*, London 1884; *Building News*, 6 Jan. 1865, XII 17; *Capital and Labour*, 1875, II 722; 14 Nov. 1877, IV 605. For the eventual impact of machinery on workers in the trade, *Reminiscences of a Stonemason*, p. 255.

[99] Thomas Brassey, the contractor, quoted in Terry Coleman, *The Railway Navvies*, London 1965, p. 39.

[100] *Personal recollections of English Engineers*, by a Civil Engineer, London 1868, p. 42; Coleman, *Railway Navvies*, p. 48.

[101] When a moving platform was invented by the engineer to 'supersede the necessity of thus perilling life and limb', the navvies, 'considering it to diminish their labour and wages', broke it. F. S. Williams, *Our Iron Roads*, London 1852, p. 72.

[102] P.P. 1867 (3873) XXXII, R.C. Trade Unions, Q.2546; *Free Lance*, 27 Feb. 1869, IV 66.

[103] *Reminiscences of a Stonemason*, London 1908.

[104] Carpenters claimed 'grinding money' when they were sacked from a job – i.e. a monetary equivalent for the time it took them to make their tools ship-shape. For the one hour claim, *North Yorks. Bldg. Trades Review*, 22 Jan. 1876, p. 7; *Brick, Tile and Bldrs. Gaz.*, 8 May 1888, III 336; for the two hours claim, *Builders Wkly. Reporter*, 26 Feb. 1875, p. 209; *Capital and Labour*, 5 May 1875, 8 March 1876.

[105] Davidson, *Houses*, p. 59.

[106] John Black, *Gas Fitting, A Practical Handbook*, London 1886, pp. 17-18; W. E. J. Crane, *The Sheet-Metal Worker's Guide*, London 1883, p. 22.

[107] Black, *Gas Fitting*, p. 18.

[108] For the small builder, A. Howarth and Mona Wilson, *West Ham*, London 1907, pp. 10-12; *Brick, Tile and Bldrs. Gazette*, 19 April 1888, III 298; H. J. Dyos 'The Speculative Builders . . . of Victorian London', *Victorian Studies*, Summer 1968; for the Manchester outrages, P.R.O., H.O. 45/8428; *Builders Trade Circular*, 5, 12, 19 May 1870; Richard Price 'The Other Face of Respectability', *Past and Present*, Feb. 1975.

[109] *Brick, Tile and Bldrs. Gaz.*, 19 April 1888.

[110] Rowland Kenney, *Westering*, London 1939, pp. 80-1.

[111] George R. Burnell, *Rudimentary Treatise on Limes, Cements . . . Plastering*, London 1850, for contemporary uses of lime.

[112] *Morning Chronicle*, 29 Dec. 1849.

[113] Dobson, *Building*, p. 109; Davidson, *Houses*, p. 46.

[114] C. Le Neve Foster, *Ore and Stone Mining*, London 1894, p. 546.

[115] For an excellent account of the Purbeck industry by a stoneworker, Eric Benfield, *Purbeck Shop*, Cambridge 1948. For contemporary accounts, *Morning Chronicle*, 23 Jan. 1850; C. E. Robinson, *Rambles in the Isle of Purbeck*, London 1882; Dorset County Mus., O. W. Farrer, 'The Marblers of Purbeck', *Purbeck Soc.* Papers, 1859-60, pp. 191-204.

[116] Merfyn Jones, 'Y Chwarelwyr: the Slate Quarrymen of North Wales', in *Miners Quarrymen and Saltworkers*, p. 121.

[117] *The British Clayworker*, Oct. 1898.

[118] Charles Tomlinson, *Illustrations of the Useful Arts and Manufactures*, London (1859), p. 26. E. Dobson, *Rudimentary Treatise on Bricks*, London 1850, p. 22; Tomlinson, *Illustrations*, p. 25; J. Geraint Jenkins *Traditional Country Craftsmen*, London 1965, p. 149. According to Ure, clay was tempered 'by the treading of men or oxen' and in the London district mainly by horse-ginned pug-mills: Ure, *Dictionary*, I 443.

[119] Arthur Aiken, *Illustrations of Useful Arts and Manufactures*, London 1841, pp. 19-20, 70; H. Chamberlain, 'The Manufacture of Bricks by Machinery', *Soc. of Arts J.*, 6 June 1856, IV 491-2. For an account based on oral history, Donald Young, 'Brickmaking at Broadmayne', *Proc. Dorset Natural Hist. Soc.*, 1967, LXXXIX; Dobson, *Bricks*, p. 81. Percy Dell of Cholesbury Common, Chesham, letter to Bill Keal of Princes Risborough, 29 April 1873. I am grateful to Bill Keal for collecting this account by one who worked as a hand-brickmaker in the 1920s.

[120] Chamberlain, 'Manufacture of Bricks', pp. 491-500; *Brick, Tile and Blders. Gazette*, 12 April 1887, II 295; William Johnson, *Lecture on Brick-Making Machinery and Bricks*, Leeds 1886, pp. 3, 14; *The British Clayworker*, 1893, III 214; A. Searle, *Modern Brickmaking*, London 1911, pp. 39, 424-5. For the predominance of hand brickmaking in 1870s Birmingham, A, H. Stephenson, *The Trade Associations of Birmingham Brick Masters, 1864-1933* (privately printed, 1933), pp. 83, 84, 90, 96. For the transition from hand to machine in the Nottingham district, Webb Coll., Sect. A. Vol. X, fols. 82-8.

[121] P.P. 1873 (C.745) XIX, Fact. Inspectors Reps., p. 19.

[122] Thomas Allen, *A History of Surrey*, London 1831, I 35; K. W. Gravett and E. S. Wood, 'Merstham Limeworks', *Surrey Arch. Colls.*, 1967, LXIX; Surrey Record Office, Merstham Limeworks, accounts and papers. I am grateful to Miss Gollancz, the County archivist, for bringing these papers to my notice.

[123] Dobson, *Bricks*, p. 122.

[124] *Ibid*, and Chamberlain, 'Manufacture of Bricks', p. 500.

[125] James Greenwood, 'Mr. Dodd's Dust-Yard', in *Unsentimental Journeys*, London 1867, pp. 64 sq. for a description, and engraving, of one of them. The most famous metropolitan dust heap was the one at the King's Cross end of what is today Gray's Inn Road. It was sold by an enterprising London capitalist to the Tsarist government for the rebuilding of Moscow after the great burning of 1812. G. M. L. Strauss et al., *England's Workshops*, London 1864, p. 161. *London City Mission Magazine*, 1 Nov. 1860, XXV 318-9. *Good Words*, 1 Sept. 1866, p. 645; *Ibid.*, 1879, pp. 739-42; J. Thomson and Adolphe Smith, *Street Life in London*, London 1879, pp. 94-5; 'Our Dust-Bins', *Leisure Hour*, 2 Nov. 1968. Uncle Jonathan, *Walks in and Around London*, London 1889, p. 150.

[126] P.P. 1883 (c.3516) XVIII, Rep. . . . White Lead Poisoning.

[127] P.P. 1893-4 (c.7239) XVII, White Lead Committee, pp. 7-8; QQ. 245, 250, 2527-9, 2'20, 3008.

[128] Brian Didsbury, 'Cheshire Saltworkers' in *Miners, Quarrymen and Saltworkers*, pp. 155-7; *Pottery Gazette*, 1 July 1880, IV 432; *Iron and steel J.*, 1895, LVIII 327; Apsley Pellatt, *Curiosities of Glass Making*, London 1849, pp. 61, 82; Percival Marson, *Glass and Glass Manufactures*, London 1932, p. 81; Walter Rosenheim, *Glass Manufacturing*, London 1919, pp. 107, 133.

[129] Robert Swinburne, 'The Manufacture of Glass', in ed. W. G. Armstrong, *The Industrial Resources of the Tyne, Wear & Tees*, Newcastle upon Tyne 1864, pp. 195-6; Theodore Barker, *Pilkington Brothers and the Glass Industry*, London 1960, p. 87.

[130] 'Glass and Silicates' pp. 79, 80-1, in Bevan, *Manufacturing Industries*.

[131] George Dodd, *Where Do We Get It*, London 1862, p. 248; cf. also 'Glass and its Manufacture', p. 20 in his *Curiosities of Industry*, London 1852. For other descriptions of sheet glass making at Messrs. Chance's Smethwick works, *Morning Chronicle*, 23 Dec. 1850; 'Birmingham and her Manufactures IV', in *The Leisure Hour*, 20 Jan. 1853.

[132] P.P. 1865 (3548) XX, 4th Rep. Children's Emp. Com. Q.14; *Leisure Hour*, 2 Nov. 1868; Strauss, *England's Workshops*, pp. 16-17; 4th Rep. Children's Emp. Com. QQ.82-3; *Morning Chronicle*, 23 Dec. 1850; P.P. 1899 (c.9073) XII, 3rd Rep. . . . Dangerous Trades, pp. 16-17.

[133] 4th Rep. Children's Emp. Com., QQ. 15,018-20.

[134] 'Birmingham Flint Glass Manufacture', in Timmins, p. 430; Rosenheim, *Glass*, p. 78. Cf. Webb Coll., N. of England Glass Bottlemakers, Half-Yearly Rep., Dec. 1890, p. 31, 'Under the Gas Tank system work is a lot more severe . . . . The heat is 50 degrees hotter'.

[135] 'Birmingham Flint Glass Manufacture' in Timmins, p. 531.

[136] Children's Emp. Com. Q. 12-13, 86; George Dodd, *British Manufactures*, London 1844-6, IV 61-5; Pellatt, *Curiosities*, pp. 125 sq.; Dodd, *Curiosities of Industry*, Glass, pp. 13, 15. I am grateful to Toshio Kusamitsu for helpful discussion on this point.

[137] W. C. Scoville, *Revolution in Glassmaking*, Cambridge Mass. 1948, p. 175. Charles J. Phillips, *Glass, The Miracle Worker*, New York 1941, pp. 19, 155, 191, 199. Cf. P.P. 1913 (6711) XXIII, Rep. Night Work, QQ. 842-5.

[138] Neil McKendrick, 'Josiah Wedgwood and Factory Discipline', *Historical Journal*, 1961, IV; Walter White, *All Round the Wrekin*, London 1860, p. 297; *Pottery Gazette*, 2 July 1883; *The Builder*, 19 Oct. 1878; P.P. 1863 (3170) XVIII, 1st Rep. Children's Emp. Com., pp. 33-4; P.P. 1889 (5697) XVIII, p. 90.

[139] Old Potter, *When I was a Child*, London 1903, p. 46.

[140] P.P. 1889 (c.5697) XVIII, p. 90. *Morning Chronicle*, 24 Jan. 1850; 'Through the Potteries', *Industrial World*, 2 Dec. 1892.

[141] P.P. 1861 (161) XVI, 3rd Rep. Med. Off. Privy Council, App. VI, Dr. Greenhow Rep. on Lung Diseases, p. 108; P.P. 1863 (3170) XVIII, 1st Rep. Children's Emp. Com. App. p. 2.

[142] P.P. 1893-4 (C.6894-XXIII) XXVII-I, R. C. Labour, Collet, pp. 61, 62.

[143] Arlidge, *loc cit.*; P.P. 1910 (cd 5278) XXIX, Rep. . . . Lead in Potteries, App. XXVII, p. 64.

[144] P.P. 1863 (3170) XVIII, 1st Rep. Children's Emp. Com., App. pp. 3, 7, 8, 17, 18, 21, 23.

[145] Collet, *loc. cit.* p. 62.

[146] J. R. McCulloch, *A Statistical Account of the British Empire*, London 1837, II 118; P.P. 1892 (c.6795-VI) XXXXVI, R. C. Labour, Group C II, Q.15,038; *Morning Chronicle*, 15 Nov. 1850; Ure, *Dictionary*, II 676-95, III 848-9; *The Manufacturer*, 25 April 1868, p. 27; Nuffield Coll., *Work in Bristol*, Bristol 1883, pp. 116-25; W. G. Rimmer, 'Leeds Leather Industry in the Nineteenth Century', *Thoresby Soc. Pubs.* 1960, XLVI/2; *Boot and Shoe Trade Journal*, 15 Jan. 1887; P.P. 1883 (c.3778-I), XXVIII, Local Government Bd. Rep., App. 13 'London Hide and Skin Trades'.

[147] Alexander Japp, *Industrial Curiosities*, London 1880, p. 48-50; G. Phillips Bevan, *Industrial Classes and Industrial Statistics*, London 1876, p. 147; *Hatters Gazette*, 30 April 1877; *The Hosier and Glovers' Gazette*, 1 Feb., 1 June 1884; P.P. 1868-9 (4202.I) XIII Women and Children in Agriculture, 2nd Rep., K.44-54, k. 135-85; P.P. 1865 (3483) XXVI, 7th Rep. M.O. Privy C., App. 6, Dr. Hunter's Rep. on Rural Labs., pp. 198, 253; 'The Leather Trades of Birmingham' in *Saddlers, Harness Makers and Carriage Builders Gazette*, Dec. 1873 to Jan. 1874; Thomas Middlemore, 'The Birmingham Saddlery Trade' in Timmins, pp. 474-5; W. J. Gordon, *Midland Sketches*, London 1898, p. 120.

[148] Japp, *Ind. Curiosities*, pp. 146-7.

[149] P.P. 1890 (169) XVII, SC Sweating, 5th Rep. p. xxxvl; *Morning Chronicle*, 23, 30 January 1851; R. L. Greenall, 'The History of Boot and Shoemaking at Long Buckby', *Northamptonshire Past and Present*, V/5, 437-445 is a welcome study of one of the new Northamptonshire outworking villages – one which remained a centre of the hand-sewn trade down to the end of the 19th century. Oddly it omits to mention Long Buckby's peculiar political complexion: in the 1840s it was a kind of Chartist commune, whose activities can be followed in the *Northern Star*.

[150] Walter Rose, *Fifty Years Ago*, p. 27; Joseph Severn, *The Life Story of a Phrenologist*, Hove 1929, p. 58; George Sturt, *The Wheelwright's Shop*, Cambridge 1963, p. 57.

[151] *Morning Chronicle*, 4 July, 8 Aug. 1850; Christopher Thomson, *The Autobiography of an Artisan*, London 1847, pp. 166-7, 171.

[152] *Morning Chronicle*, 29 Aug. 1850.

[153] Alfred Williams, *Life in a Railway Factory*, London 1915, p. 56; W. Mosses, *The History of the United Patternmakers Association, 1872-1922*, n.p. of pub.,1922; J. W. F. Rowe, *Wages in Theory and Practice*, London 1928, p. 93; 'Pattern Making and Capitalism', *The Socialist*, Nov. 1904 (this last is an exceedingly interesting article, written by a patternmaker and describing the way his fellows were being robbed of their skill. I am grateful to Anna Davin for the reference); Strauss, *England's Workshops*, pp. 217, 243; Wynter, *Our Social Bees*, 1st Ser., p. 220. For the new classes of coopers' industrial ware, Bob Gilding, 'Journeymen Coopers' in *The Workshop Trades*.

[154] T. A. Welton, 'Forty Years of Industrial Changes', *Trans. Manchester Stat. Soc.*, 1897, p. 168; Sir Westcott Abell, *The Shipwright's Trade*, Cambridge 1948, pp. 108-11.

155 Charles Knight, 'Localised Handicrafts in South Midlands Agricultural Districts', *British Almanac Companion*, 1861, p. 27. There is a good general account of the industry in L. J. Mayes, *The History of Chairmaking in High Wycombe*, London 1960.

156 *Morning Chronicle*, 12 Sept. 1850.

157 Oral History: conversation of the writer with Charles Connor, 10 Dec. 1972. Mr. Connor started work as a cooper in the London docks in 1912.

158 Goldsmith's Coll., Condition of the Working Classes of Edinburgh and Leith, 1853.

159 *Morning Chronicle*, 8, 22 Aug. 1850; for difficulties in assembling a kit of cabinet-making tools, *The Life and Struggles of William Lovett*, London 1876, pp. 28-30. 'A workman was not thought much of without a good chest of tools', Islington Ref. Lib., MS Diary of Henry Price, 1824-1905, p. 12. Price was a cabinet-maker who worked in the London piano industry.

160 When the premises of Messrs. Holland and Hannen, the Bloomsbury builders, were destroyed by fire in 1866, joiners' tools were valued from £4 and £5 to as much as £50 or £60 a man. *Annual Register*, August 1866, II 124. Darlington joiners in the same year estimated the average value of their tools at £20, with 'not less than 1s. 6d. a week' for replenishment and repairs, *Darlington and Stockton Times*, 23 June 1866. *Morning Chronicle*, 11 July 1850, 'Labour and the Poor, Met. Dists. LIX, Carpenters & Joiners', p. 5 is full of valuable details about tool insurance.

161 Leone Levi, *Wages and Earnings of the Working Classes*, London 1884, pp. 146-7.

162 Mosses, *Patternmakers*, p. 9; J. W. C. Haldane, *Life as an Engineer*, London 1904, p. 76. *The Autobiography of Peter Taylor*, Paisley 1903, p. 90; Webb Coll., U.K. Pattern Makers Assoc., *Monthly Reports*, May 1886.

163 *The Cabinet and Upholstery Advertiser*, 14 Dec. 1878.

164 H. J. Fryth and Henry Collins, *The Foundry Workers*, Manchester 1959, p. 44.

165 Ure, *Dictionary*, II 203.

166 *Recoll. of English Engineers*, p. 218.

167 Ure, *loc cit.*

168 For the trade union, Fyrth and Collins; for the restrictive practices, *Capital and Labour*, 22 April 1874; W. G. Riddell, *Adventures of an Obscure Victorian*, London 1932, pp. 33-6; Booth MSS. B.89, fol. 55r; P.P. 1867-8 (3980-1) XXXIX, R. C. Trade Unions, QQ. 13, 176-13,211. Ure describes the moulders as 'a very numerous class of mechanics demanding and receiving high wages'. Ure, *loc cit.*

169 S. F. Walker, 'Starting an Engineering Business II', *Practical Engineer* 1907, XXV quoted in Roderick Floud, *The British Machine-Tool Industry, 1850-1914*, Cambridge 1976, p. 50; Baines, *Industrial North*, pp. 50-1; Edward Cressy, *A Hundred Years of Mechanical Engineering*, London 1937, p. 208.

169a Floud, *Machine Tools*, pp. 42, 51, 53, 55-8, 60, 65, 67, for a helpful discussion of this point. Goldsmith's Coll., Condition of the Working Class of Edinburgh: Smiths; H. W. Strong, *Industries of North Devon*, rep. Newton Abbot 1971, p. xiii; J. E. Bellis, 'A History of G. E. Bellis & Co.', *Newcomen Soc.*, 1964-5, XXXVII; L. N. Burt, 'Watford & Papermaking', *ibid.*, 1945-7, XXV 105; A. Hawkes, 'Intensified Production', *Trans. Birmingham Assoc. of Engineers*, 1907, p. 9 for some contemporary examples. 'Speciality of manufacture is a peculiarity of all the great iron establishments in England and in many in Scotland', Kydd, Condition of the People, fol. 86.

170 Wal Hannington, *Never on Our Knees*, London 1967, pp. 28-9. This book should be warmly commended to anyone who cares about London history or who would like to read the autobiography of a fine agitator. There are brief references to the St. Pancras Ironworks in *Architects and Contractors Handbook*, London 1883, pp. 32, 39, 112, which has some drawings of its products, and Ronald H. Clark, *The . . . English Steam Wagon*, Norwich 1963, p. 58.

171 W. F. Watson, *Machines and Men. An Autobiography of an Itinerant Mechanic*, London 1935, p. 21; Wal Hannington, *The Rights of Engineers*, London 1944, p. 13; Rowe, *Wages in Theory and Practice*, pp. 106-7. For the use of grindstone as an unofficial meeting place, *Autobiography of Peter Taylor*, p. 55.

172 For a brief description of a workshop in 1870s Shoreditch see George Barnes, *From Workshop to War Cabinet*, London 1924, pp. 26-7.

173 Rowe, *Wages in Theory and Practice*, pp. 106-7; Hodgson, *Sheet Metal Work*, p. 217; *Autobiography of Peter Taylor*, pp. 90-2; Haldane, *Life as an Engineer*, p. 69.

174 James Jefferys, *The Story of the Engineers, 1800-1945*, London (1945) pp. 24, 55-6; Rowe, *Wages in Theory and Practice*, pp. 90-2, 263-6.

175 Jefferys, pp. 59, 101, 123; Rowe, pp. 90-2, 101-2; Cressy, *Mechanical Engineering*, p. 208.

176 For some examples, *Autobiography of Peter Taylor*, pp. 127, 137, 145, 166; W. G. Riddell, *The Thankless Years*, London 1948, p. 146; David Kirkwood, *My Life of Revolt*, London 1935, pp. 53, 60.

177 For the relationship of boy to adult male labour at two leading machine-builders, P.P. 1864 (3414-I) XXII, 3rd Rep. Children's Emp. Com., c.22-6.

[178] Rousiers, *Labour*, pp. 254-5.
[179] Williams, *Our Iron Roads*, p. 250.
[180] *Autobiography of Peter Taylor*, p. 105; Kenneth Hudson, *Working to Rule*, Bath 1970, pp. 38-9, 50-1, 52-3.
[181] Williams, *Our Iron Roads*, p. 190; for a splendid description of the smith's work at Swindon in later times, Williams, *Railway Factory*, pp. 82-8.
[182] *Hugh Stowell Brown, His Autobiography*, ed. W. S. Caine, London 1887.
[183] Eric N. Simons, *Lockwood & Carlisle Ltd. . . . A Chapter of Marine History* (Sheffield) 1962, pp. 10, 11.
[184] Riddell, *The Thankless Years*, p. 22.
[185] Robert Armstrong, *Rudimentary Treatise on Steam Boilers*, London 1850, p. 98; George Piggott, 'Boiler Plate Working', in Timmins, p. 97; at Laird's Birkenhead, in the early 1890s, the larger boiler plates were bent by machinery, the smaller ones by hand: James Samuelson, *Labour Saving Machinery*, London 1893, p. 26.
[186] Williams, *Our Iron Roads*, p. 230; *Industrial Lancashire*, p. 128; Armstrong, *Steam Boilers*, 1865 ed., pp. 141-2; Goldsmith's Coll., Kydd, Condition of the People, fol. 88.
[187] Williams, *Railway Factory*, p. 113.
[188] David Pollock, *Modern Shipbuilding*, London 1884, p. 136; David Pollock, *The Shipbuilding Industry*, 1905, p. 122; For the hand-riveting of gun-carriages, J. S. Jeans, *Notes on Northern Industries*, London (1878), p. 117; for its use in galvanized iron, P.P. 1864 (3414-1) XXII, 3rd Rep. Children's Emp. Com. b.166. For the relationship of boys to men in the riveting gangs, *ibid*, c. 64-73; *Illustrated Exhibitor*, 30 Aug. 1851, p. 233; Williams, *Our Iron Roads*, i. 181.
[189] Lady Bell, *At the Works*, London 1908, pp. 68-70.
[190] Ure, *Dictionary*, II 579-80; *The Engineer*, 19, 26 May 1865; Baines, *Industrial North*, p. 25; *Ironworkers Journal*, 15 April 1872, pp. 2-3; Williams, *Railway Factory*, p. 17.
[191] Sir George Head, *A Home Tour of the Manufacturing Districts*, London 1835, pp. 130-1.
[192] J. W. C. Haldane, *Steam Ships and their Machinery*, London 1893; 'Rolling Armour Plates', *The Builder*, 21 Sept. 1867.
[193] W. Graham, 'Tin and Tinplate', in Bevan, *Manufacturing Industries*, p. 168; Henry Loveridge, 'Wolverhampton Trades' in Timmins, p. 118; also *ibid*, p. 639.
[194] Ure, *Dictionary*, I 904-5; James Wilson, 'On the Manufacture of . . . Cutlery', *Soc. of Arts J.*, 11 April 1856, IV; 357-61; G. I. H. Lloyd, *The Cutlery Trades*, London 1913, p. 37.
[195] Robert K. Dent, *Old and New Birmingham* P.P. 1880, (c. 6330) XIX, Fact. Insp. Reps. pp. 27-8; W. Graham, 'Brass Founding', p. 146 in Bevan, *Manufacturing Industries;* Edward Peyton, 'The Manufacture of Iron and Brass Bedsteads', in Timmins, p. 627; P.R.O., H.O. 45/10115/B 12393K; H.O. 45/8124/28; P.P. 1862 (179) XXII, 4th Rep. Med. Off. Privy Council, App. IV, p. 144; W. Graham, *The Brassfounders Manual*, London 1868, pp. 8, 10-16, 20-1, 28. '. . . Brass-founding is not a round of . . . mechanical operations . . . it is a "craft and mystery" of the utmost variety, and of ever-increasing novelty and improvement'. Goldsmith's Coll., Thomas Cooper, 'Pictures of the People', fol. 4.
[196] *Capital*, p. 000.
[197] Norman T. Willis, *Woad in the Fens*, Lincoln 1970; H. O. Clark and Rex Wailes, 'The Preparation of Woad in England', *Newcomen Soc.* XVI, 1935-6; J. B. Hurry, *The Woad Plant and its Dye*, Oxford 1930. I am grateful to Eve Hostettler for these references.
[198] P.P. 1861 (161) XVI, 3rd Rep. M.O. Privy C., pp. 130-1; P.P. 1864 (3389) XXIV, Comm. on non-inspected Mines, pp. viii, xiv-xv.
[199] Williams, *Railway Factory*, p. 115.
[200] J. S. Wright, 'On the Employment of Women in Factories in Birmingham', *Trans. Nat. Ass. Soc. Science*, 1857, pp. 539-40.
[201] George Smith, *The Cry of the Children*, from the *Brickyards*, Leicester, 1871, p. 11; P.P. 1873 (c.745) XIX, Fact. Insp. Reps., pp. 14-15; P.P. 1866 (c.3675) XXIV, 5th Rep. Ch. Emp. Com., K,55,59. At a Tipton yard it was estimated that the girl clay-carriers, who brought clay to the moulder's table, made 300 journeys a day, travelling 17 miles and carrying an average weight of eirht tons each. They carried the clay in their arms or on the head. P.P. 1865 (3473) XX, Fact. Insp. Reps., p. 22.
[202] *Autobiography of Peter Taylor*, p. 76.
[203] P.P. 1865 (3548) XX, 4th Rep. Ch. Emp. Com., Q68; also generally *ibid*, QQ64-79 and pp. xliii-xliv.
[204] Charles Booth, *London Life and Labour*, London 1904 2nd Ser. II 136.
[205] Dodd, *Days at the Factories*.
[206] P.P. 1912-13 (Cd.6184) XXVI, Rep. . . . Shuttle-Kissing.
[207] *Morning Chronicle*, 4 March 1850.

[208] William Felkin, 'The Lace & Hosiery Trades of Nottingham', *Stat. Soc. J.* 1866, XXIX 536-8.

[209] E. J. Hobsbawm, 'The Tramping Artisan' in his *Labouring Men* is the fundamental article on this subject.

[210] Goldsmith's Coll., Samuel Kydd, 'The Condition of the People', fol. 88. 'As to the boilermakers, the numbers vary with the work. We have never less than forty or fifty, and at present we have 100', Captain Watson, Gen. Supt. Cunard Co., *Rep. . . . Unemployed in Liverpool,* Liverpool 1894; for the still greater irregularity of employment in the ship repair yards, Howell Coll., Am. Soc. of Shipwrights, 60th *Q. Rep.* Jan.-March 1897, p. 17, Jan.-March 1899, p. 1.

[211] H. J. Fyrth & Henry Collins, *The Foundry Workers, a Trade Union History,* Manchester 1958, p. 44n. For the remarkable turnover in T.U. membership partly reflecting this high level of unemployment, Howell Coll., Friendly Soc. of Iron Founders, *Ann. Rep.* 1887, p. 10; and Fyrth and Collins, p. 57.

[212] Webb Coll., Glass Bottle Makers of Yorkshire, *Q. Reps.* 12 June 1909, XXII 76. For fluctuations of employment in the flint glass trade, Rousiers, *Labour,* pp. 52-3.

[213] For a valuable and extended discussion of the theme, Habakkuk, *American and British Technology.* Habakkuk's study had provoked an extensive critical literature. Among the more accessible studies are those reprinted in ed. S. B. Saul, *Technological Change, the United States and Britain in the 19th Century,* London 1970; and ed. D. H. Aldcroft, *The Development of British Industry and Foreign Competition, 1875-1914,* London 1968.

[214] Coleman, *Railway Navvies,* p. 50.

[215] *The Builder,* 28 Sept. 1878, XXXVI 1010-11 for an early account of it.

[216] Sir Bosdin Leech, *History of The Manchester Ship Canal,* Manchester, 1907, II 22; Samuelson, *Labour-Saving Machinery,* pp. 76-7; J. Swift, 'Engineering', in Galton, *Workers,* p. 103.

[217] *Ironmonger* 18 Jan. 1879, XXI 69. For the extensive controversy on the comparative merits of English and American locks, *Ibid,* pp. 35, 43, 52-3, 74-9, 106, 133, 199, 172-4, 259-61, 323, 338; for an anticipation of it, Elihu Burritt, *Walks in the Black Country and its Green Borderland,* London 1868, p. 213.

[218] P.P. 1854-5 (O.11)L, Committee on Machinery, p. 69; Ure, *Dictionary,* II 216.

[219] John T. Day, 'The Boot and Shoe Trade' in Cox, *British Industries,* p. 239.

[220] Reference temporarily mislaid. For transatlantic contrasts of the same kind in tinplate working: J. H. Jones, *The Tinplate Industry,* London 1914, pp. 130-1.

[221] Great Exhibition of 1851, *Juries Reports,* I 428; P.P. 1854-5 (O.11)L, Committee on Machinery, p. 52.

[222] Marx, *Capital,* pp. 390-1.

[223] S. J. Chapman, *The Cotton Industry and Trade,* London 1905, p. 30; L. G. Sandberg, 'American Rings and English Mules: the Role of Economic Rationality' in Saul, *Technological Change.*

[224] Roy Church, *Great Victorian Boom,* p. 44, A. J. Taylor, 'Labour Productivity and Technical Innovation in the Coal Industry 1859-1914', *Econ. H. R.,* 1961, 2nd Ser. XIV. For the apparent slowing down of mining productivity, either because of worker-resistance or a declining proportion of face workers: R. Walters, 'Labour Productivity in the South Wales Steam-Coal Industry, 1870-1914', *Econ. H. R.,* May 1975, 2nd Ser. XXVIII/2.

[225] Hobsbawm, 'British Gas Workers' in *Labouring Men,* pp. 160-1; Webb Coll, E, A/XLI fol. 298; Will Thorne, *My Life's Battles,* London (1925), pp. 64-5.

[226] Clara Lowe, *God's Answers: A Record of Miss Macpherson's Work at Spitalfields,* London 1882, p. 25.

[227] R. A. Church, 'The British Leather Industry and Foreign Competition 1870-1894', *Econ. H. R.,* 2nd Ser. XXIV, 1971, pp. 549-50.

[228] W. Turner Berry, 'Printing' in ed. C. Singer et al, *A History of Technology,* Oxford 1958, V 713.

[229] For the workers' opposition cf. Jaffray papers British Museum Add. MSS., 57576, 57623.

[230] Bill Benfield, 'Women Bookbinders and the Bible Strike of 1849', in *The Workshop Trades.*

[231] Alan Fox, *A History of the National Union of Boot and Shoe Trade Operatives, 1874-1957,* Oxford 1958, pp. 14-15.

[232] C. P. R. Mountfield, 'The Footwear Industry of the East Midlands', *East Midlands Geographer,* June 1966, IV/1. In Leeds and Bristol riveting marked the beginning of wholesale manufacture. *Boot and Shoe Trade Journal,* 24 Oct. 1885, XVI 3 and *ibid,* 26 Jan. 1884, XIII 56-7.

[233] *Boot and Shoe Trade Journal,* 23 April 1892, XXVII 535.

[234] *Oliver's, 1869-1950*, Leicester 1950, p. 8; Northampton Ref. Lib., *Recollections of William Arnold*, pp. 20-3; for another description of the process, *Boot and Shoe Trade Journal*, 12 Feb. 1881, VII 77.

[235] Fox, *Boot and Shoe Operatives* p. 15; for some contemporary references to outworking riveters, *Boot and Shoe Trades Chronicle*, 1 Aug. 1879 (Daventry) and *ibid*, p. 22 (Leicester).

[236] *Ironmonger*, 30 Nov. 1870, XII 1010-11; 31 Oct. 1871, XIII 912.

[237] *Ironmonger*, 3 May 1879, XXI 541.

[238] Reference temporarily mislaid.

[239] Andrew Wynter, *Our Social Bees*, London 1861, p. 189. The description of earlier stages of the needle-making process is taken from Ure and from W. C. Aiken, 'Needles', pp. 106-7, in Bevan, *Manufacturing Industries*.

[240] *Pottery Gazette*, 1 April 1882, VI 343.

[241] Thomas Phipson, 'The Pin Trade' in Timmins, p. 601; S. R. H. Jones, 'Price Associations and Competition in the British Pin Industry', *Econ. H. R.*, May 1973, 2nd Ser. XXVI/2; Ure, *Dictionary*, III 458-9.

[242] C. Violet Butler, 'Pins', in *Victoria County History of Gloucestershire*, London 1907, II 207.

[243] *Pottery Gazette*, loc. cit.; Harold Owen, *The Staffordshire Potter*, London 1901, p. 311.

[244] For accessible accounts, Owen, pp. 63-71; 'An Old Potter', *When I Was a Child*, London 1903, 186-7.

[245] S. J. Thomas, 'Pottery' in ed. F. Galton, *Workers on their Industries*, London 1895, p. 191.

[246] *Pottery Gazette*, 1 Nov. 1879, p. 428; *Pottery and Glass Trades Gazette*, April 1881, V 305.

[247] James Caird, *English Farming in 1851*, London 1851, pp. 21, 84. Isaac Mead, *The Life Story of an Essex Lad*, Chelmsford 1923, pp. 32-4 for a good description of threshing barley by hand in the 1880s.

[248] Fussell, *Dairy Farming*, pp. 198-9, and David Taylor, 'The English Dairy Industry, 1860-1930', *Econ. H. R.* 2nd Ser., 1976, XXIX/4.

[249] Kenneth Hudson, *Towards Precision Shoemaking*, Newton Abbot 1968, p. 10.

[250] Clapham, *Economic History*, II 79, quoting an article by G. Richards in Trans. Inst. of Mechanical Engineers for 1885.

[251] S. & B. Webb, *Industrial Democracy*, London 1920, p. 395.

[252] P.P. 1864 (3414-I) XXII, 3rd Rep. Children's Emp. Com. p. 392.

[253] 'The Cutlery Trade of Sheffield' in Cox, *British Industries*, p. 47.

[254] *Ironmonger*, 31 Oct. 1871, p. 913; *Illustrated Exhibitor*, 23 Aug. 1851. When Messrs. Wheatman and Smith of Sheffield tried to introduce machinery into saw grinding, the men's trade union attempted to blow the works up. Frank Hill, 'Combinations in Sheffield', *Trades Societies and Strikes*, 1860, p. 582.

[255] P.P. 1863 (3170) XVIII, Children's Emp., 1st Rep. App. pp. 267, 289.

[256] William Charley, 'Flax in Ireland', *Journ. Soc. Arts*, 10 Feb. 1860, VII 202. The finer classes of Irish linen were still being made by hand in the 1870s: *The Textile Manufacturer*, 15 March 1875, p. 104.

[257] 'The Miscellaneous Trades', in Timmins, pp. 641-2.

[258] Charles Tomlinson, *Cyclopedia of the Useful Arts*, London 1866, III 190.

[259] For some examples, *The British Clayworker*, Aug. 1897, VI 112, April 1893, VII 33.

[260] Prof. Wrightson, 'Agricultural Machinery', p. 137 in Bevan, *Manufacturing Industries*.

[261] William Fairbairn, 'On ... Iron Ships ... and Riveted Joints', *Proceedings of the Royal Society*, 1872-3, XXI 260-3.

[262] Ure, *Dictionary* II 480. 'Three essential things were necessary in the manufacture of good chain cables, neither of which is cheap, viz. the best of iron, the best workmanship and the best coal'. *Mechanics Magazine*, 30 March 1860, N.S. III 209.

[263] C. S. Orwin and E. H. Whetham, *British Agriculture, 1846-1914*, London 1964, p. 348; cf. F. Clifford, 'The Labour Bill in Farming', *Royal Ag. Soc. J.*, 1875, 2nd Ser. XI 87; Fussell, *Dairy Farming*, pp. 198-9.

[264] Hackney Ref. Lib., *The Life Story of T. J. Hunt*.

[265] 'Steam power or gas power cannot be used – heavy machinery must be tabooed, for the house would not bear its vibration, were the landlord willing to allow it to be introduced – and he generally stipulates that it may not', 'London Shoe Factories', *Boot and Shoe Trades Journal*, 18 Nov. 1882, X 257.

[266] *The Builder*, 19 Oct. 1878.

[267] *Iron & Steel J.* 1895, XLVIII 326-7.

[268] *Morning Chronicle*, 4 July 1850.

[269] P.P. 1871 (440) LXII. Census of Production.

[270] For the 'bull' week in the Sheffield trades (and the 'calf' and 'cow' weeks preceding it): P.R.O., HO 45/9833/B 9744/2, *Labour News*, 28 Nov. 1874, 2 Jan. 1875; *Ironmonger*, 31 Dec. 1860, II 209, 31 Dec. 1863, V 353.

[271] W. E. Minchinton, *The British Tinplate Industry, a History*, Oxford 1957, p. 40. 'As . . . the demand did not yet exist for a standardized product there were no economic advantages to be secured by building larger works', *Ibid*, p. 35.
[272] 'On Rolling Heavy Iron', *Trans. S. Wales Inst. of Engineers*, 1860-1, II 78.
[273] Floud, *Machine Tools*, pp. 51, 55-8.
[274] Museum of Eng. Rural Life, Reading, Ransome Collection. I am grateful to Alun Howkins for this reference.
[275] Floud, *Machine Tools*, pp. 51, 55-6, 67.
[276] *Ironmonger*, 30 Sept. 1865, VII 136.
[277] *Ibid.*, 1 March 1873, XV 291; 10 Aug. 1878, XX 799.
[278] Samuel Sidney, *Rides on Railways*, London 1851, p. 96.
[279] Marx, *Capital*, 337.
[280] *Ironmonger*, 28 Feb. 1880, XXIII 298-9.
[281] W. K. V. Gale, 'Notes on the Black Country Iron Trade', *Newcomen Soc.* 1943-5, XXIV 24.
[282] J. D. Marshall & M. Davies-Shiel, *The Industrial Archaeology of the Lake Counties*, Newton Abbot, 1969, pp. 174-7; *Stephenson's New Guide to Southport*, Southport (1885), p. 92; J. M. H. Hunter, 'The . . . Location . . . of Industry in greater Nottingham', *East Mids. Geographer*, Dec. 1964, III 341.
[283] For country coopers, Bob Gilding, *The Journeymen Coopers of East London*, History Workshop, 1971.
[284] M. E. Bulkley, *Minimum Rates in the Boxmaking Industry*, London 1915, p. 4. For handmade work in paper bag making, mainly for local orders, H. L. Williams, *The Workers Industrial Index to London*, London 1881, pp. 10-11; Booth MSS. A. 17, Part A. fol. 79, B. 102 fol. 53.
[285] J. T. Sampson, 'Paper Staining', *Artisans' Reports on the Paris Exhibition of 1889*, p. 517.
[286] Gervase Wheeler, *The Choice of a Dwelling*, London 1872, p. 260. Cf. also Great Exhibition of 1851, *Juries Reports*, II 1204.
[287] Timmins, p. 66.
[288] For some early attempts, *Patent Journal*, 1847-8, IV 266; *Journ. Soc. Arts*, 30 Sept. 1859, p. 707; Ure, *Dictionary* II 202-3. For the very well sustained workers' opposition to file-cutting machinery, *Ironmonger*, 30 Aug. 1862, IV 234; 30 May 1863, V 147; 1 Oct. 1874, XVI 1233; and numerous reports during big file-cutters strike of 1866. For workers' opposition in later years, S. & B. Webb, *Industrial Democracy*, London 1920, p. 395n; Talbot Baines, *The Industrial North in the Last Decade of the Nineteenth Century*, Leeds 1928, p. 63; P.P. 1886 (C.4715) XXI R.C. Depression of Trade, QQ. 1150, 1184-1209.
[289] *Ironmonger*, 10 Aug. 1878, XX 792.
[290] For examples in copper braziery, *Ironmonger*, 15 Nov. 1879, XXII 625; 25 Sept. 1880, XXIV 338.
[291] Arthur Shadwell, *Industrial Efficiency*, London 1906, I 150.
[292] *Ironmonger*, 1 Oct. 1877, XIX 388.
[293] S. C. L. Fuller, 'Machinery in the Carriage Manufactory', *Papers Read before the Institute of British Carriage Manufacturers, 1883-1901*, p. 361.
[295] Francis Klingender, *Art and the Industrial Revolution*, Paladin ed., London 1975, p. 144. Klingender uses this excellently apposite phrase to describe Sir Matthew Digby Wyatt's official catalogue of the Exhibition.
[296] George Dodd, *Curiosities of Industry*, London 1852, 'Wood', p. 18.
[297] Strauss, *England's Workshops*, p. 186.
[298] *Illustrated Exhibitor*, 7 June 1851.
[299] *Illustrated Exhibitor*, 23 Aug. 1851.
[300] *Morning Chronicle*, 4 July 1850.
[301] Floud, *Machine Tools*, p. 49.

Nineteenth-Century Horse Sense[1]

BY F. M. L. THOMPSON

"YELLOW pages" have a good story to tell about the character of a modern consumer society; the old-fashioned trade directories of more simple societies can be almost as eloquent. London in 1902, for example, possessed 145 jobmasters, those indispensable men who hired out horses or carriages by the hour, day, week, or season; 45 of them were in a fair way of business, running branch as well as main stables; and the trade was dominated by 14 jobmasters who provided the capital city with the equivalent of an Avis or Hertz service from chains of up to 26 stables. This, no doubt, was the small change of an affluent metropolis accustomed to taking for granted the facilities of civilized existence. A highly significant moment in the development of that civilization was captured by the *London Trades Directory* of 1902, whose entries for newly started trades included one "motor spirit distiller" and one "horse sun bonnet maker".[2] The horse-drawn world reached the apogee of its refinement and specialization at the very moment when motor vehicles were becoming sufficiently numerous for businessmen to notice their existence.

It was a proper recognition of the comparative status of these two trades that the sun-hats for horses were made and marketed by a long-established City hat manufacturer with a sound reputation for bowlers, while the petrol distiller set up his works at a discreet distance, on the south side of Battersea Bridge.[3] The market for the sun-bonnets was undoubtedly large, but its extent was unknown. It included the fashionable West End carriage trade, in which the eligible horse clientele would not have exceeded 40,000.[4] But it was much larger than just that, for the sun-bonnet was not merely a dainty aristocratic whim for smart aristocratic horses, but also an eminently utilitarian device to keep flies off honest working van horses, which were even more pestered than carriage horses during long hours of patient waiting in the streets.[5] How many van horses there might be, indeed how many commercial, working, horses of all varieties there might be, no one knew. Throughout its long reign of supremacy on the road (and, to a lesser extent on the field) the horse was so much taken for granted that no one thought it important or interesting to take a horse census. It was only after 1919, when mechanical competition had already made deep inroads into the horse's grip on land transport, that the War Office, resolved on making thorough preparations to fight the last war over again, arranged for a head count of all the horses in the country. In order to gauge the role of the horse in the pre-1914 economy and in

[1] This article was delivered as a paper, under the same title, at the annual conference of the Economic History Society, Leeds, in April 1975. Some earlier thoughts on the subject may be found in my inaugural lecture, *Victorian England: The Horse-drawn Society* (1970).

[2] *Post Office London Directory for 1902* (1902), I, vii. [3] Ibid. II, 2150, 2265.

[4] W. J. Gordon, *The Horse World of London* (1893; reprinted Newton Abbot, 1971), p. 104.

[5] Oral history: recollections of Prof. F. J. Fisher communicated at Bodington Hall, Leeds, 6 April 1975.

pre-motor society, however, it is necessary to push such knowledge backwards in time. This article tries to do in retrospect the task which an efficient War Office might have done at the time in order to keep itself acquainted with the size, nature, and location of the pool on which it could draw for remounts, artillery horses, and general transport horses in time of war. The article attempts, in other words, to draw up a profile of horse employments in the transport or commercial sector, throughout the nineteenth century; and, with the addition of the more easily traced farm horses, to arrive at a census of the horse population covering the closing century of dominance, which may serve as the statistical foundation for future inquiries into the numerous economic and social ramifications of the horse sector.

I

The onset of effective mechanical competition for the horse came only in the years just before 1914, and came not from steam but from petrol. It is important to recall that the horseless carriage was so named not solely through failure of linguistic imagination, but also because the motor-car was first and foremost a horse substitute. Within that field it is well known that in Britain the pre-1914 motor-car was an aristocratic luxury, an expensive toy for the wealthy; it was not for nothing that the Duke of Bedford, properly brought up in the old school of carriage travel, kept relays of "Rolling Royces" for the journey from Woburn to London and made sure that his town chauffeur was standing by at the Welsh Harp to take over from his country chauffeur.[1] But the early motor vehicles were not exclusively for idle pleasure, being attractive substitutes for hard-working commercial horses making frequent short journeys involving much stopping and starting, and carrying the best paying cargo of all, people. The change-over was extraordinarily rapid: in 1903 London had 3,623 horse-buses and a mere 13 motor-buses; by 1913 the roles had been reversed and only 142 horse-buses remained to face 3,522 motor-buses; similarly in the cab business, there were over 11,000 hansoms and hackneys in 1903 and a solitary motor taxicab, while by 1913 more than 8,000 motor taxicabs cruised the streets and only 1,900 horse vehicles were left in the trade.[2] Outside London, country doctors formed one group which took to the motor-car in its early days, for its convenience and utility; and it may be surmised that a prime reason for the contrasting pre-1914 fortunes of the American and British motor industries was that Britain had nothing comparable to the market provided by the millions of American "horse and buggy" farmers.[3] The point is that while it is true that some cars and some horses were the playthings of the rich, forms of conspicuous consumption, it is equally

[1] Duke of Bedford, *A Silver-Plated Spoon* (1959), p. 8.

[2] H. Llewellyn Smith, ed. *New Survey of London Life and Labour* (1934), I, 180; T. C. Barker and Michael Robbins, *A History of London Transport: II, The Twentieth Century to 1970* (1974), p. 329. See also *S.C. on Transport (Metropolitan Area)* (Parl. Papers, 1919, VII), app. 18, p. 434.

[3] The annual output of horse-drawn carriages, buggies, and sulkies in the U.S.A. in 1900 was 905,000, and the output of farm waggons and trade vehicles was 570,000.—U.S. Bureau of the Census, *Historical Statistics of the U.S.: Colonial Times to 1957* (Washington, 1960), p. 416. In Britain the annual output of horse-drawn carriages was valued at £375,000 in 1907 [equivalent to 3,750 vehicles at the low average value of £100 per carriage]; the output of goods vehicles (waggons, farm carts, vans, mail vans) was valued at £457,000 from private makers or at £810,000 with the output of railway companies and public authorities included [the higher figure implies an output of 16,200 vehicles at an average value of £50 per vehicle].—*Census of Production, 1907* (P.P. 1912–13, CIX), p. 709.

true and rather more important that other cars and most horses paid for their keep, and were producer goods with vital economic functions.

The possibilities of motor competition for the horse had been of slightly more than passing concern to the Royal Commission on the Horse Breeding Industry in Ireland in 1898, and several witnesses were asked for their views on the question. Most echoed the defiant incredulity of a forthright Yorkshireman, John Major, horse-breeder from Sledmere Grange:

> I don't think for one moment that all the cycles and motor cars in the world, after the novelty is gone, will ever cause our English gentry, or their sons and daughters, to turn their backs on a beautiful typical riding or driving horse, such as we Yorkshiremen keep trying to produce, not only because we admire them so much, but also I take them to be a great safety-valve towards keeping at a distance that man who, once he enters your doors, puts his mark on all you have and refuses you to choose your own auctioneer.[1]

It may be unclear what kind of safety-valve he had in mind, though clearly there is great faith in the horse as the protector of something, maybe of daughters' virtue and fortune against unacceptable suitors, or perhaps of property against tax collectors and bailiffs. Other breeders, dealers, and transport men may not have shared the whole of Major's creed, but did express the same confidence in the superiority and invulnerability of the horse. Only one breeder was pessimistic in 1898, arguing that "the demand for hacks or utility horses kept for ordinary use" would die out before long. But he saw the bicycle, not the car, as the culprit: "I know great numbers of people who have sold their horses and ride their bicycles."[2]

Even in 1906, when motor vehicles were a good deal more familiar and more reliable, the notion that motor traction should or could be substituted for horse traction was greeted with some astonishment by the Royal Commission on London Traffic. Many expert witnesses then put this view, but its strongest and clearest advocate was an elder statesman of the City, Sir Henry Edmund Knight, who had been an alderman since 1874 and Lord Mayor in 1882. His argument was simple and compelling: that motors would save at least 25 per cent of the street space occupied by horse-drawn vehicles, and thus solve the problem of insupportable congestion and approaching traffic paralysis. His ideas were extremely radical, and included the invention of one-way street systems, bus lanes, and smokeless zones. This last was mainly in order to "get the coal traffic out of the streets", with the incidental advantage of doing away with smog, and Sir Henry brushed aside the objections of people who protested that they liked to have a fire to poke by countering: "I do not see what good it does to poke a fire. If you saved all the trouble and dirt, I think it would be very much to your advantage to give up your idea of liking to poke a fire."[3] His supreme self-confidence and willingness to use statutes and regulations to interfere with people's private

[1] *R.C. on the Horse Breeding Industry in Ireland* (P.P. 1898, XXXIII), Q. 11275.

[2] Ibid. QQ. 1353–5 (Samuel Ussher Roberts, of the Horse Breeding Association and Royal Dublin Society). Other evidence discounting any motor threat was in QQ. 10752–3 (Burdett Coutts, M.P.); 11505–7 (William Bainbridge, Secretary of the Lancashire Agricultural Society); 541 (John O'Reilly, founder of the Irish Horse Society).

[3] *R.C. on London Traffic*, II (P.P. 1906, XL), Q. 14499.

lives provoked the baronets of the Royal Commission to pepper him with hostile and mocking questions: "What—prohibiting people burning coal?" "You mean to say you would prohibit the use of horses in London?" "Do you think Parliament would ever pass a law saying that a man should be prohibited from dragging his van by a horse?" Sir Henry answered the last question by claiming "that Parliament would have no difficulty in doing that if it were shown that the streets would be very greatly relieved and the public convenience very much facilitated. Many private persons have adopted the system of their own accord; you see many motor vehicles and motor vans about."[1] A great traffic-improver and motor enthusiast, therefore, as late as 1906, did not have sufficient confidence to rest his case on superior technology and performance, and on cost advantages, which would have led users to adopt motors out of simple self-interest, but felt obliged to invoke compulsion in order to drive the horse off the streets as a public nuisance.

It is not my present purpose to pursue the horse off the streets, since that has been a twentieth-century pastime, leading gradually but remorselessly to the situation in which many people fervently desire to drive the horse-conqueror itself, the motor-car, off the streets. But it is worth remarking that shifting the horse was a slow business; the switch may have been extremely rapid in some particular uses, such as the buses, taxis, and country doctoring, but even in 1975 the occasional horse-drawn vehicle may still be seen in city streets, though these are largely kept for prestige advertisement by breweries and tobacconists, or for competition in Easter parades. It hardly needs stating that the number of farm horses in Britain exceeded the number of tractors until 1950; it is perhaps more *recherché* to suggest that in 1939 the true horse power employed on British farms was still larger than the tractor power.[2] Town-bred historians for some curious reason expect farmers always to be slow-witted and insensitive to technical and economic incentives, and may therefore be inclined to dismiss this evidence, which in any case relates to the fields rather than the streets. More telling, therefore, is the evidence that many users, more noticeably in commercial than in private road transport, clung to their horses well into the interwar years. Motor goods vehicles on British roads only just about topped the half million mark by 1939; but, as will be shown later, there were already at least that number of horse-drawn vans and carts in use in the haulage business in the 1880's.[3] That statistic is of course no more than a sighting shot, since mere aggregates of horse and motor goods vehicles are not very meaningful in the absence of information on carrying capacities, journey distances and frequencies, payload factors, and so forth. Such data, at any rate for the horse-drawn element in any comparison, may never be obtainable.

This is not to say that the evidence is devoid of all meaning: such evidence as

[1] Ibid. Q. 14494 (put by Sir George Bartley); Q. 14454 (also put by Sir George); Q. 14458 (put by Sir Joseph Dimsdale). Sir Henry in fact did suggest that bus and van traffic should be motorized by law, but "I would not interfere with the use of horses in ordinary cabs or carriages, or light vehicles generally."

[2] Ministry of Agriculture, *A Century of Agricultural Statistics: Great Britain, 1866–1966* (1968), Tables 30, 31, and 70. Although tractors were not counted in 1939, the position in that year was, in round numbers: 55,000 tractors, 649,000 horses for agricultural use.

[3] B. R. Mitchell and P. Deane, *Abstract of British Historical Statistics* (Cambridge, 1962), Table 'Transport 7', p. 230 (hereafter Transport 7); and see below, p. 72, for horse-drawn trade vehicles.

the analysis of London street traffic counted in May 1919 which showed that motor goods vehicles were over half of total goods vehicles at only one of the 12 census points set up on roads near the heart of the City where the environmental advantages of motors were at their maximum;[1] or that which showed at a head count of horses that there were at least 374,000 used in drawing goods vehicles in Britain in 1924 and 131,000 in 1934.[2] The meaning is that horse substitution took place rather rapidly where convenience, speed, and comfort were at a premium and their cost was of little consequence or could easily be covered, that is in both public and private passenger transport; but that it was very gradual where operating costs, capital costs, flexibility, and versatility had to be carefully weighed and balanced against considerations of the price elasticity of demand both for transport services and for the goods transported, that is in much of the commercial goods transport sector. The toughness of the horse in the twentieth century as the technical ability and proficiency of motor competition became ever stronger simply serves to emphasize its effortless superiority in the nineteenth century in many social and economic activities, when the steam engine was its only serious rival.

II

When the steam engine had first taken to the rails there had been lively expectations, or fears, that this marked the beginning of the downfall of the horse and the horse interest. The early railroad gentlemen, said J. R. McCulloch, "are pleased to tell us, that, by superseding the employment of horses in public conveyances, and in the regular carriage of goods, the adoption of their projects will enable 1,000,000 horses to be dispensed with; and that, as each horse consumes as much food as 8 men, it will at once provide subsistence for 8,000,000 human beings!"[3] It did not turn out so. Far from displacing horses, the railways created new and expanding demands for horse-labour. The matter was put graphically by a London commission stable-keeper, whose business had an annual flow of 1,200 horses being looked after for clients, giving evidence to the Select Committee on Horses in 1873—which originated in doubts about the capacity of the country to supply sufficient horses to meet the large and increasing demand. In answer to

[1] *S.C. on Transport (Metropolitan Area)*, 1919, Tables 7 and 7B, pp. 397–8. The 12 census points were: Blackwall Tunnel*, Blackfriars Bridge*, Euston Road*, Fleet Street, Gray's Inn Road*, Holborn*, Ludgate Hill, New Bridge Street, Queen Victoria Street, Theobalds Road, Tottenham Court Road, and Victoria Embankment. Motor trade-vehicles were 52 per cent of total trade vehicles counted on Victoria Embankment; elsewhere the proportion ranged from $13\frac{1}{2}$ per cent to $37\frac{1}{2}$ per cent. The average of the proportions, not a very meaningful figure, was $28\frac{1}{2}$ per cent motor, $71\frac{1}{2}$ per cent horse-drawn. The total number of vehicles of all kinds counted at the five points marked * was 41,492: this flow can be analysed as follows:

trade vehicles	58 per cent	horse-drawn vehicles	45 per cent
passenger vehicles	42 per cent	motor vehicles	55 per cent

This indicates that at these five points 98 per cent of the passenger vehicles were motors, while 77 per cent of the trade vehicles were horse-drawn, when the raw figures are recalculated applying the motorized percentage of trade vehicles given in Table 7B. In the country as a whole there were 62,000 motor trade-vehicles in 1919, 101,000 in 1920: probably these represented 10 to 15 per cent of the total number of trade vehicles in use.

[2] War Office, *Report on the (Fifth) Census of Horses in Great Britain, 1934* (H.M.S.O. 1935), pp. 4–5, class (v) "light and medium trade or draught horses and light and medium trotting vanners over 15 hh", plus class (vii) "other heavy horses and heavy trotting vanners".

[3] J. R. McCulloch, *Dictionary of Commerce* (1834), II, 654.

Lord Falmouth's question whether the railways had diminished or increased the number of horses used in the carrying trade, this stable-keeper replied: "I should say that it has increased it. We thought when railways first came in that we should have nothing to do, but it has not turned out so. The greater the increase of railways, the greater will be the use of the horse . . . because there is the work to be done to and fro. The horses have to work in connection with the railways; for every new railway you want fresh horses; fresh cab horses to begin with; I know one cab proprietor for instance who used to keep 60 horses, who has now 120."[1]

For the City or West End man it was perhaps natural to begin with the cabs: they were, next to the buses, the most visible and most frequently used vehicles in the business and the fashionable streets. In the early 1830's there were about 1,265 cabs in London, a number restricted by monopoly regulations; this grew to over 2,500 by the mid-1840's, in conditions of free trade; to 6,800 by 1863, and to around 11,000 from 1888 onwards. With each cab requiring two horses to keep it on the road, this was indeed a striking contribution to the growth of the city's horse population.[2] Naturally it was not every railway passenger who began or completed his journey in a cab, though perhaps a higher proportion of female journeys went that way; and probably few cabbies relied mainly on railway fares for their livelihoods. Nevertheless the growth of passenger travel occasioned by the railways was prodigious, and the twentyfold increase in the number of railway passengers between 1840 and 1870 cannot have fallen far short of a twentyfold increase in the total number of inland travellers, and it would be reasonable to expect this to have played a major part in bringing about a two-and-a-half-fold increase in cabs.[3] Similarly, although urban growth in general and urban commuting in particular plainly played a dominant causal role, railway travellers must have made a significant contribution to bus travel, and hence to the increase in the number of London buses from 620 in 1839 to nearly 1,300 in 1850, and again from 1,268 in 1890 to 3,696 in 1902.[4] In Britain as a whole there appear to have been something like 25,000 horse-buses in 1890, a number which, on the London example, could well have increased by anything between 50 and 100 per cent by 1902.[5] Since it required an average of 11 horses to keep each bus in service, allowing each pair three hours work a day and some rest days, London alone had something like 40,000 bus horses at its peak, and the country perhaps as many as 400,000.[6]

[1] *S.C. [House of Lords] on the Supply of Horses* (P.P. 1873, XIV), QQ. 3686–7.

[2] Barker and Robbins, op. cit. I, 7, 14; *S.C. on Metropolitan Local Government* (P.P. 1866, XIII), QQ. 656–7 (Sir John Thwaites, chairman of Metropolitan Board of Works); *Accounts & Papers* (P.P. 1865, XXXI), p. 462.

[3] Mitchell and Deane, op. cit. p. 225.

[4] Barker and Robbins, op. cit. I, 59, 245; *R.C. on London Traffic*, I (P.P. 1906, XL), QQ. 7418–21 (Sir Alexander Carmichael Bruce, Asst. Commissioner of the Metropolitan Police).

[5] *30th Report of the Commissioners of Inland Revenue* (P.P. 1890, XXVI), app. lvii, pointing out that the courts had just decided [*Law Reports*, 24 QBD, 172] that an omnibus plying for hire between King's Cross and Victoria was a carriage plying for hire within the definition of a hackney carriage, and hence was only liable to the annual hackney licence (15s.) instead of the carriage licence (£2 2s.) previously applied to omnibuses. Hence the reduction in carriages paying at the £2 2s. rate from 79,603 in 1889 to 51,685 in 1890, and the increase in those paying at the hackney rate from 53,160 in 1889 to 88,116 in 1890.—Ibid. p. 15. In 1902 the numbers paying the hackney rate reached 133,164, but how much of this further increase was attributable to buses is not known.—*Annual Report of the Commissioners of Inland Revenue* (P.P. 1903, XIX).

[6] *R.C. Road Traffic*, I, 1906, QQ. 25024–8 (Daniel Duff, manager of London Road Car Co.); Gordon, op. cit. p. 12. Since the population of London was about one-seventh of the total urban population of

To the railwayman, however, the railway effect was most strikingly seen in the ownership and employment of horses by the railway companies themselves. Horses were of course used directly in railway operation, mainly for shunting and train-marshalling, but also for hauling on private mineral lines, long after the perfection of the steam locomotive; how many is not known. The more important use, however, was in the collection and distribution of goods at railway terminals and stations. Some companies relied on independent carriers for this service, and the L.N.W.R. for example had a long-standing arrangement with Pickfords which dated from the earliest days of the London and Birmingham Railway; by the 1870's Baxendale, Pickford's proprietor, had 1,500 horses at work in London, chiefly on railway account.[1] Most of the major companies—the Midland, Great Eastern, Great Northern, and Great Western, for instance—kept their own horse and van fleets, with stables of 500 to 1,000 horses apiece near their London terminuses in the 1870's, studs which by 1890 had grown by a third; by that year Pickfords had 4,000 horses in their London business, which of course was not exclusively railway business.

All this represented rather more than a trifling investment for the railway companies. In London alone it was estimated that there were 6,000 railway-owned horses in 1890, valued at £60 each and with a working life of four or five years; and the necessary stabling was obviously most extensive, the Great Western for example having a large new block in South Wharf Road containing four storeys of horses from which the top layer looked out on to Paddington station roof.[2] This pattern of railway ownership of horses was repeated, on a lesser scale, in all the main provincial centres, and in some quite small places as well where particularly sharp inter-railway competition for traffic made it desirable for each company to operate its own cartage service in order to attract and capture custom.[3] No national total for railway-owned horses is available for the peak period at the turn of the century; but a century after the railways began the four main line companies still owned between them, in 1934, 14,834 horses.[4]

The road haulage business in general, not merely that segment of it which was handled by railway-owned horses, formed the central contribution of the horse to the nineteenth-century economy. W. J. Gordon was exaggerating when he claimed, in 1893, that:

> The carrying trade of these days is in the hands of the railway companies, and the carrier's horse is for all practical purposes the railway horse. Of the 84,000,000 tons of general merchandise hauled along the railroads of this island in 1890, the bulk was collected and distributed in railway vans.[5]

It was quite true that the bulk of the rail-borne merchandise could hardly escape being touched by horse; to which should be added all the coal consumed in

Britain, one might reasonably expect London buses to be about one-seventh of all buses, suggesting a British total in 1902 of about 25,000 buses, and 250,000 to 275,000 bus horses.

[1] *S.C. on Horses*, 1873, Q. 3689 (Banks). [2] Ibid. Q. 3690; Gordon, op. cit. pp. 49–60.

[3] I owe this information to G. Channon, who incorporates it in 'Pooling Agreements between the Railway Companies involved in Anglo-Scottish Traffic, 1851–69' (unpublished Ph.D. thesis, London University, 1975).

[4] *(Fifth) Census of Horses, 1934*, p. 5; this was larger than the number of horses in the army, which in 1934 was 12,605.

[5] Gordon, op. cit. p. 49.

domestic heating—estimated at 18 million tons in 1870, 31 million tons in 1890, and 38 million tons in 1913[1]—which was delivered by cart. But it was not true that all the carrying trade was in the hands of the railways: coastal shipping was still most active and extremely important on certain routes throughout this period, and some of the cargoes carried must have generated demands for road transport; canal traffic, though no doubt increasingly confined to bulk freights which could be handled at origin and destination without horse intervention, must have continued to feed some goods into the road distribution system; much goods traffic, in country areas and towns alike, never touched a railway at any point in its movements but was carried over short distances—from farms into market towns and out again to consumers, between factories and workshops within a town, and from them to retail outlets in the same town or its suburbs; and a proportion of imported consumption goods may well have moved from docks to warehouses and processing plants, and on to the final consumers, entirely by road. The railway-generated and rail-borne contribution to the internal carrying trade was, therefore, only a fraction of the total. Unluckily, it is extremely difficult even to hazard a guess at the likely size of this fraction, just as it is extremely difficult to attempt to measure the size of the pre-motor road-haulage business.

III

The taxman is the economic historian's best friend when it comes to measuring activities, and although he is a great disappointment in his lack of serious attention to nineteenth-century commercial road transport he is not totally unhelpful. An annual tax was collected on goods vehicles only during the first quarter of the nineteenth century; there was a lower rate of duty on bone-shaking taxed carts without springs, and a higher rate on carts "with springs, not metallic", and these were repealed in 1825. It seems very probable that the tax was difficult to collect and was widely evaded: there were 20,465 taxed carts known to the Inland Revenue in 1810, and 15,037 in 1824, and these have to be considered grossly insufficient for the actual needs of internal trade at the time.[2] From 1825 until the coming of motor lorries the vehicles of commercial road transport escaped taxation altogether, and there is therefore no record of their numbers.

But in 1888 the taxman was given the merest whiff of blood when Goschen, most uncharacteristically, proposed to introduce a tax on horses, vans, and wheels. His justification was that this was not a new imposition of imperial taxation at all, but was to be a species of local taxation to provide a source of revenue for the newly established county councils, which would be levied from road users and used to help pay for road upkeep, thus reducing the burden on the general body of ratepayers. It had become a canon of financial orthodoxy by the 1870's that locomotion ought not to be taxed; it might possibly be expedient to tax some forms of pleasurable or frivolous locomotion, but taxation of commercially useful and productive locomotion was a harmful and distorting imposition on the main artery of economic life. The *Economist* repeated the revealed

[1] P. Deane and W. A. Cole, *British Economic Growth, 1688–1959* (2nd edn. Cambridge, 1967), Table 55 applied to Table 54, pp. 219, 216.

[2] *Accounts & Papers* (P.P. 1830, xxv), accounts of carriage duty receipts, 1810–14 and 1825–9, pp. 226–9; *13th Report of the Commissioners of Inland Revenue* (P.P. 1870, xx), pp. 546–7, 550.

truth: "It was long ago agreed that we are all as consumers interested in having the roads maintained in a state of efficiency. Hence, these are retrograde taxes."[1] Goschen, as correct a Gladstonian as was to be found among late Victorian Chancellors, was therefore hard pressed to justify his heresy; he argued in terms which anticipated Lloyd George and his 1909 road fund:

> The cardinal point of our idea is to look at the fact that the roads are used gratis now, and that those who use a road contribute nothing to its support except by the highway rate, which falls equally on the whole body of the community ... If there are 50 ratepayers, and 40 of them use the road on foot, while 10 of them use it on horseback or on wheels, is it not fair that the 10 should pay something more as a contribution than the 40? ... quite apart from the Carriage Tax, which is a tax mainly on the more luxurious carriages—carriages used for pleasure—there is at present no tax on any other vehicles, except omnibuses and a certain number of vehicles which ply for hire. But the heaviest description of vehicles—such as railway vans, coal contractors' carts, and vehicles of that character—[An hon. Member: Brewers?] Yes; brewers' drays are a good instance. These vehicles use up the roads without paying one shilling more than anyone else for their maintenance. Well, we propose to put a duty of £1 a year upon every vehicle exceeding 10 cwt. in weight... with an exemption, which has always been made in favour of agricultural carts, carts used by farmers for the purposes of husbandry and not for haulage. Those vehicles which are propelled on roads by any other means than by horses will also be included ... We propose also, to put a very small Wheel Tax upon every vehicle.
>
> Colonel Nolan (Galway, N.): Not on carts?
>
> Goschen: Yes. We propose a duty of 2s. 6d. per wheel upon all carts over 2 cwt. [Colonel Nolan: Oh!] ... We have put it extremely low—5s. upon a two-wheeled and 10s. upon a four-wheeled cart. Gentlemen's carts and pleasure carriages will continue to pay higher duties—namely, the present Carriage Tax ... this tax of £1 10s. on the heaviest carts is not excessive, for they will still be paying less than a brougham with two horses, although causing much more wear and tear to the roads. I have had many suggestions that bicycles and tricycles should be taxed... I do not say that there is in principle any objection to such a tax ... but I do not like to multiply little taxes of this description, and I recognize the great service that has been conferred on a vast number of young men, who, by means of bicycles and tricycles, are enabled to explore the country, so I do not propose to impose such a tax as part of the scheme...
>
> I now approach with some trepidation the next licence duty ... I look to an older friend—to a tax which existed for very many years, and which was then taken off—the duty which used to be imposed upon horses of every kind, and which was repealed in 1874. The old duty was imposed on all horses, whether they were engaged in trade or otherwise ... I do not propose to tax what I may call trade horses, or horses used in agriculture; but I do propose ... to impose a tax of £1 on pleasure horses—[*Cries of* "Oh!"] [An hon. Member: Hobby horses!] I am afraid there is not the same satisfaction in regard to this £1 duty as in regard to the licence tax on heavy vehicles. Hon. Members will see that I put the cart before the horse. But let me remind hon. Members on whom this tax will fall. It will fall a good deal on what I may term the well-to-do classes. . .[2]

[1] *Economist*, XLVI, 31 March 1888, 406.

[2] *Hansard* (Commons), 3rd ser. CCCXXIV, Ways and Means—Financial Statement, 296–9, 26 March 1888.

But Goschen was altogether too serious and solemn to push through such grave taxes by cracking jokes, and altogether too much of the solid City man to be credible as a Robin Hood Chancellor out to soak the rich. He was at once roundly attacked by an equally humourless fellow Conservative, Henry Chaplin the squire of Blankney, a racing man, member of the Jockey Club, and Member for Sleaford, who momentarily chanced to have the freedom of being out of office. Casting his eye along the Treasury Bench, said Chaplin, "from the Chancellor of the Exchequer to the First Lord of the Treasury [W. H. Smith] and the row of Ministers who are sitting there tonight, I came to the conclusion that there was not a single man amongst them who knew a horse from a cow."[1] It was a damning indictment of a Conservative Government and a lament for the old agricultural base of the party, scarcely turned aside by Goschen's production of an Under Secretary to the Local Government Board, Walter Long, as the minister competent to speak on the difference between a horse and a cow.[2] The proposal to revive the horse tax and confine it to pleasure horses indeed gave ample scope for prolonged and detailed debate on the equitable definition of "pleasure", with the riding horses of farmers, country doctors, and clergymen being trotted out as deserving to be treated as non-taxable business or professional horses, in much the same way as the tax treatment of limousines is currently determined by the use to which they are put rather than by their engine size. Fascinating as this was, however, it was on the van and wheel taxes that the whole scheme came to grief.

A protest league, like an incipient road hauliers' association, sprang up almost overnight under the leadership of R. V. Causton, the carriers' and vanmen's M.P. He marshalled 40 other M.P.s, from both parties, and a collection of carriers, van owners, coal merchants, tradesmen, costermongers, wheelwrights, van builders, saddlers, harness makers, and tarpaulin makers from all parts of the country to form a Cart, Van, and Wheel Tax League with a London central committee and Anti-Wheel Tax Association branches in many provincial towns, which was so active in meetings, demonstrations, and deputations, and such a nuisance in its parliamentary pressures, that by November 1888 Goschen had been forced to withdraw his proposals.[3] Hence, despite a valiant effort, no tax statistics of horse-drawn trade vehicles were collected. Nevertheless the effort had produced some official estimates, and despite the expert criticism by the trade of their inaccuracy and insufficiency these estimates do provide a lower limit for the number of horse-drawn trade vehicles in 1888. The official estimate was that there were 300,000 trade carts with two wheels and 150,000 trade carts with four wheels in use in Great Britain in 1888; one-third of these 450,000 carts were thought to be heavy vehicles of more than 10 cwt unladen weight.[4]

The trade argued that these figures were gross understatements and that the tax burden to be passed on to the public by the carrying trade would not be the

[1] Ibid. 351, 26 March 1888; R. J. Olney, *Lincolnshire Politics, 1832–85* (Oxford, 1973), p. 164, for Chaplin's lack of humour.

[2] *Hansard*, loc. cit. 352, 26 March 1888.

[3] Highlights of the anti-wheel tax activities were reported in *The Times*, 9 April 1888, p. 8b, 23 April 1888, p. 12d, 25 April 1888, p. 8c, 10 Nov. 1888, p. 13e, 15 Nov. 1888, p. 13e; and a procession and demonstration in Hyde Park, ibid. 7 May 1888, p. 8d. *Hansard* (Commons), 3rd ser. CCCXXXI, 20, 23 Nov. 1888, 485–6, 28 Nov. 1888, and 525–6, 29 Nov. 1888, recounts the final withdrawal.

[4] *Horse Duties and Wheel Tax* (P.P. 1888, LXV), p. 285.

£300,000 a year estimated by the Treasury, but more like £1 million or even £2·4 million at the highest guess.[1] One wild protester alleged that the London carriers alone would have to pay £½ million a year, thus implying that there were well over 500,000 carts, vans, and waggons in London, which was totally absurd. Goschen countered this extravagance by telling a deputation that "I too have taken a census of the carts which are in London" which showed that there were about 60,000 trade vans and carts, including railway vans and public carriers' vehicles.[2] There is no means of knowing how reliable this "census" was, whether it was a guess or a count; certainly the London Excise officers who supplied the figure had no existing data on carts in their files from which they could have compiled it. Translated into its horse-equivalent the total of 60,000 vans for London seems to be at least on the same wavelength as W. J. Gordon, who a year or two later appears to have reckoned on about 40,000 horses employed in pulling trade vehicles, which he classified as brand new or first-hand horses, a figure which he then calmly doubled by accepting "the usual estimate that half the London horses are at least second-hand".[3] A London complement of 60,000 vans and carts was compatible with the original estimate of 450,000 for Britain as a whole. On the other hand, Goschen conceded that this original estimate had been deliberately conservative, in order not to exaggerate the amount of revenue to be expected; and the middling estimate from the trade, of revenue receipts of £1 million, equivalent to 1,500,000 trade carts and vans, may perhaps be accepted as the upper limit for the number of horse-drawn trade vehicles in 1888.[4] But such an upper limit implies, if the London weighting is at all accurate, 200,000 carts in London, a number so large that the Excise officers could hardly have failed to notice it.

A more credible upper limit may, therefore, be derived from Gordon's outside figure of 80,000 cart and van horses in London. All the contending parties appeared to agree that, on average, there was one horse to each van or cart. This may seem a niggardly ratio in view of the large number of heavy vehicles drawn by two or more horses, and the not uncommon abnormal loads such as anchors, ship's propellers, boilers, and large pieces of machinery which moved through the streets behind teams of ten or a dozen horses. On the other hand, many carriers and traders said that they kept spare carts or vans to ensure that they always had enough serviceable ones on the road, and that they in fact owned more carts than horses; the extreme case was put by a firm in the London hay

[1] *The Times*, 9 April 1888, p. 8b, 13 April 1888, p. 11f. [2] Ibid. 25 April 1888, p. 8c.

[3] Gordon, op. cit. p. 164.

[4] *Hansard* (Commons), 3rd ser. CCCXXIV, 596–7, 6 April 1888. The official estimate, and a revised estimate on the same distribution of vehicle types to yield £1 million revenue, were as follows:

	Official estimate		*Trade estimate*	
	No.	*Yield* £	*No.*	*Yield* £
Carts with 2 wheels, at 5s.	300,000	75,000	1,000,000	250,000
Carts with 4 wheels, at 10s.	150,000	75,000	500,000	250,000
Carts weighing more than 10 cwt at £1 in addition to the 5s. or 10s. duty	150,000	150,000	500,000	500,000
Total carts	450,000		1,500,000	
Total yield		300,000		1,000,000

trade, which claimed that "in order to clear trucks as they arrive at the railway depot we are obliged to keep an average of two vans to each horse we employ."[1] The 1:1 horse/cart ratio may, therefore, be accepted, though it probably does tend to underestimate the number of horses in the trade sector. At this ratio Gordon's figure implies a total of 600,000 carts and vans in Britain; a middle figure between this and the official estimate, something over half a million, may therefore seem to be about right for 1888.

This half million was obviously but a point on a trend line of commercial road vehicles and horses running through the entire horse-drawn part of British industrial and urban growth. Unfortunately it is the only remotely direct observation possible on that trend line, and the problem is to discover some acceptable way of extrapolating backwards and forwards from this uneasily fixed point. The total amount of equipment employed in road transport by public carriers and by traders in their own businesses clearly bore a direct relationship to the volume of inland commodity trade of all descriptions—long distance, short distance, intra-urban, manufacturing, wholesale trade, and retail distribution all lumped together. This in turn had a relationship to national income, though it may seem unlikely to have been a constant at the level appropriate for 1888; and the inland trade volume doubtless also had a direct relationship to the volume of goods traffic carried by rail, particularly after railway freights got into their full stride from the late 1860's onwards. Finally, one would expect there to be a relationship between the number of people employed as carmen, carriers, carters, and hauliers and the number of road vehicles; that the 1891 occupation tables suggest that there were 2·5 carts to every carter is a little disturbing if one expects to find one driver for each vehicle, but possibly only those who drove for public carriers described themselves in the census as "carriers or carters" while drivers employed by firms and businesses which ran their own fleets of carts may well have returned their occupations under those of their employers.[2] It may well be that in reality none of these three relationships was a constant. Nevertheless, in the absence of any guide to the directions in which they may have altered over time they have been calculated as constants for the period 1871–1911; and also for 1851, since it is the earliest year for which data on all three proxies are available, and out of mere curiosity for 1811 as well, from national income estimates alone. The results are given in Table 1.

The rail-based estimate for 1851 is known to be heavily biased downwards, since at that time railways were still in the early stages of making inroads into the country's inland freight business, and only in the following twenty years captured and consolidated their hold on their high and relatively stable late nineteenth-century share of total traffic; if this figure is removed from the 1851 column as obviously much too low, the remaining average suggests 188,000 commercial vehicles for 1851. For the years in which they are available before the duty was repealed in 1874 the horse-tax records generate some confidence that the cart

[1] *The Times*, 29 March 1888, p. 8a.

[2] Some backing for this assumption is provided by the 1911 Census, in which 19,943 persons were returned as "motor car drivers (not domestic), motor cab, motor van drivers", at a time when there were said to be 40,000 motor lorries in use and about 15,000 motor cabs.

Table 1. *Estimates of the Number of Commercial Road Vehicles, in '000, in Great Britain, 1811–1911*

	1811	*1851*	*1871*	*1881*	*1891*	*1901*	*1911*
1. National income (£ million)	301	523	916	1,051	1,288	1,642	2,154
Carts, vans, waggons	116	203	355	401	500	637	836
2. Goods carried by rail (million tons)	—	*c.*60	166	241	305	410	517
Carts, vans, waggons	—	106	272	393	500	671	844
3. Number of carmen, carriers, carters, and hauliers ('000)	—	68	90	145	195	311	318
Carts, vans, waggons	—	174	231	372	500	797	815
Average of three estimates of carts, vans, waggons	—	161	286	388	500	702	832

Sources:

1. Deane and Cole, op. cit. Table 37, app. III.
2. Mitchell and Deane, op. cit. Table Transport 5, Railway Capital, Mileage, Traffic, and Finances—Great Britain, 1838–1938, pp. 225–6.
3. Occupational Tables (England and Wales, and Scotland), Decennial Censuses.

estimates are of a credible order of magnitude. The horse tax was imposed in 1784, and from 1797 horses used in trade were liable to duty, but only as part of the general residual category of "all other horses and mules" which for a while also contained agricultural horses. Even after the farm horses dropped out of duty, from 1823, the tax category of "other horses" was not necessarily synonymous with "trade horses". From 1854, however, the horse-tax assessment was rearranged in schedules on the lines of the income tax, and "Schedule F: Horses used for Trade" appeared. Even this schedule was misleading, since it included horses under 13 hands—that is quite small ponies—kept for riding or drawing carriages, which were properly "pleasure" horses and not trade horses at all; and until 1870 it excluded horses kept by job- and post-masters, stage and hackney carriage proprietors, these being otherwise taxed, all of which were commercial horses and some of which were goods as distinct from passenger horses. In spite of these pitfalls there is every reason to believe that there were at least 147,000 trade horses in 1851, and, when allowances are made for the trade ponies and jobmasters' horses which were hidden in other tax classes at that date, more likely 175,000; and in 1871 there were 300,000 trade horses.[1]

[1] *13th Report of the Commissioners of Inland Revenue*, 1870, contains a history of the horse and carriage duties, pp. 550–1, and annual series of numbers taxed, 1812–68, pp. 552–5. Details for 1869–72 are in *S.C. on Horses*, 1873, app. A; and for 1869–75 in *18th Report of the Commissioners of Inland Revenue* (P.P. 1875, xx), p. 529. Returns for the years 1810–14 and 1825–9 are in *Accounts & Papers, Account of Horses Charged with Duty* (P.P. 1830, xxv), pp. 233–5; but returns for the period 1784–1810 have not yet been traced. These different sources, where they overlap, do not tally precisely, but the discrepancies are unimportant for present purposes. The problem of the "horses under 13 hh." has been met by applying the proportion indicated by the detailed return for 1868:

not exceeding 13 hh. kept for riding or drawing carriages	66,151
not exceeding 13 hh. kept for any other purpose	24,640
total not exceeding 13 hh.	90,791

For the horses kept by job- and post-masters, and stage and hackney-carriage proprietors, which were brought into the returns for the first time in 1870, it was stated that "the number of such horses at the time of the change in the law in 1869 was estimated at 75,000".—*S.C. on Horses*, 1873, app. A. One-fifth of these have been counted as "trade horses", the rest as "commercial passenger horses". *Accounts &*

It is legitimate to entertain doubts about the reliability of the horse-tax figures since it was a notoriously difficult tax to collect, and evasion either through ignorance or from forethought was particularly likely in the commercial sector. All errors, however, served to understate the true horse population; and the honest, taxed, population was very nearly adequate to draw the estimated number of commercial road vehicles in 1851 and 1871. Moreover the taxed trade horses were increasing at an annual rate of 7,000 in the five years 1865–9, and at over 4,000 a year in the final five years of the duty, 1870–4, which included the very period of acute shortage in horse supply which accompanied and followed the Franco-Prussian War and the boom of 1873, and which occasioned the 1873 Select Committee on the Supply of Horses. Such growth rates make trade horse populations of 350,000–390,000 for 1881 and 500,000 for 1891 entirely plausible.

All in all the horse data provide about as much verification of the cart estimates, for the period 1851–91 at least, as could well be expected. One or two loose ends from the trade horses remain, before it is possible to return to the horse sun-bonnets and the gentleman's or "pleasure" horse. To start with, the lone estimate of 116,000 carts and vans for 1811 may not seem quite so forlorn when the horse-tax returns reveal that in 1812, the first year for which the classification was made, there were 151,945 taxed horses described as "not wholly used in husbandry". This description is of course opaque, but since it excluded riding and carriage horses (234,288), horses let to hire (1,365), race horses (only 541 of these), farm horses over 13 hands (740,079), farm horses under 13 hands (35,239), and horses belonging to small farmers paying less than £20 a year in rent and not keeping more than two horses (24,563), it must have been very largely composed of horses employed in trade; it is an open question whether that included the passenger trade as well as the goods trade.[1] In any case, it looks as though the trade vehicles of 1812 numbered more like 100,000 to 150,000 than the 19,141 "taxed carts" which were recorded.[2]

A more serious question, perhaps, is whether there can really have been anything like 800,000 to 850,000 trade horses in 1911, and whether these could have declined to the 374,000 horses counted in 1924 as drawing trade vehicles, let alone to the 243,051 horses which were counted in the same year as being stabled in towns.[3] With real national income hardly growing at all between 1911 and

---

Papers (P.P. 1854–5, xxx), A Return of Carriages etc. in four classes, gave in Class 4 an estimate of "the Total Number of Private Waggons, Carts, Drays, and Vans, as nearly as can be ascertained" of 137,002 for Great Britain in 1854.

[1] *Accounts & Papers* (P.P. 1830, xxv), pp. 233–5. The stage coach, hackney carriage, and post horses were not subject to the ordinary horse duty, since their activities were subject to separate, and higher, rates of duty under the carriage taxes.

[2] Ibid. pp. 226–9. A more credible figure for trade vehicles can be derived from the pre-war tax returns, which suggest that in 1790 there were 119,000 "taxed carts" (two wheelers),-and 117,000 "taxed waggons" (four-wheelers).—Ibid. (P.P. 1792, xxxv), No. 763, account of produce of taxes for one year; see also ibid. Nos. 764 and 765. The "taxed waggons" included an unknown proportion of farm waggons, since any farmer with a four-wheeled waggon was liable to the annual duty on one waggon—though on no more than one, regardless of how many more he might possess; the "taxed carts", however, expressly excluded all carts used in farming.—23 Geo. III cap. 66, which imposed these duties. Hence these tax figures give comfortable support to the notion that there were about 150,000 trade vehicles at the beginning of the 1790's.

[3] (*Fifth*) *Census of Horses*, *1934*, pp. 4–5, 12.

1924, and the total volume of internal trade probably growing very little, it is reasonable to argue that the increase in motor goods vehicles in this period was chiefly a matter of horse substitution and only in very small degree a matter of catering for expanding or wholly new traffics. On this assumption the reconciliation of the trade-horse estimate for 1911 with the trade-horse count of 1924 requires that each net addition to the national fleet of motor goods-vehicles should have supplanted 2·9 horse-drawn carts, vans, or waggons; if the starting-point of this calculation, 40,000 motor goods-vehicles in 1911, is an overstatement, then the substitution factor should be correspondingly reduced.[1] Such a substitution rate at least looks plausible. If a motor lorry could not do the work of three horse-drawn vehicles, which with their complement of horses would have cost about £300 to £500 at 1911 prices, then it would have been a poor investment.[2]

IV

Gentlemen's horses, like trade horses, flourished with the increase in railway travel, and more than doubled in number between 1851 and 1901. They were expected to learn good railway manners, for although when frightened by a railway engine blowing off steam at an unaccustomed place like a level crossing the railway company was actionable for negligence, when frightened by a railway engine blowing off steam at a regular place for blowing off steam like a railway station the railway company was not guilty of negligence in failing to screen the railway from the public road.[3] Unfortunately the pleasure horses were little better behaved than the trade horses, statistically. They can be dug out of the

[1] Motor goods-vehicles 40,000 in 1911, 203,000 in 1924.—Mitchell and Deane, op. cit. Transport 7. But the work of Mr D. L. Munby of Nuffield College, which it is to be hoped he will shortly publish, suggests that there is considerable exaggeration in the early years of the series of motor vehicles which is generally used. Meanwhile it can be noted that the number of motor vehicles licensed in 1910–11 was:

motors (not hackneys)	75,617
hackney motors	31,899
motor cycles	48,857
total motors	156,373

This compares with the total of 193,000 in Transport 7 for 1911. The explanation may be that "motors (not hackneys)" included goods vehicles provided they did not ply for hire, as well as private cars, and that this has given rise to some double-counting in Transport 7. *5th Report of the Commissioners of Customs and Excise* (P.P. 1914, xvii), Tables 89 and 90 contain the detailed figures, and early motor historians may find it a more accurate and informative source than Transport 7.

[2] The only time-series of horse prices available is that of average values per head of horses imported into the U.K. 1861–5 to 1901–5.—*Agric. Statistics* (P.P. 1907, xcvii), p. 218:

	Average value per head, £		*Average value per head, £*
1861–5	25·4	1886–90	17·7
66–70	24·9	91–5	23·7
71–5	33·2	96–1900	25·9
76–80	28·2	1901–05	25·2
81–5	20·3		

Horses, however, were generally imported unbroken; the carrying trade generally bought their horses ready for work, at five or six years old. I have therefore followed Gordon for 1893 (op. cit. p. 67) in valuing trade horses at £50 per head in 1911. The value of carts, vans, and waggons varied greatly according to their specifications, and the range £50 to £120 has been assumed in the text.

[3] *Law Reports*, 14CB, NS 54, and 21 QBD 453, cited in G. H. H. Oliphant, *The Law of Horses* (1847, 5th edn. 1896).

mire of the horse-tax returns, with a certain amount of difficulty, in the period before 1874; and something can be inferred about the number of carriage horses from the annual carriage duties or licences which continued to be paid through to the 1920's. It is simply the fortune of war that after 1906 the returns of four-wheel carriages with two or more horses sometimes contained private motor cars, and the returns of two-wheel carriages, quite logically, included motor cycles.[1] They can also be approached through Goschen's abortive tax proposals of 1888, in which his projected tax of £1 a year on pleasure horses was backed by an estimate that there were 500,000 of them.[2] In all the fuss over the cart and wheel taxes this estimate of pleasure horses escaped criticism, though the unfairness of taxing the professional riding horses or pony-trap horses of clergymen and country doctors did not go unnoticed. The figure may, therefore, be accepted as a lower-bound estimate. But since there were at that time 51,000 two-horse private carriages and 363,000 one-horse private carriages, it might be thought that these alone required 414,000 horses if they were to be fully horsed; a contemporary, admittedly an Irishman and a horse breeder, spoke as if there were about 200,000 hunters among the 300 hunts in England alone; so that, should there have been as few as 600,000 pleasure horses there would still have been none to spare as ordinary riding horses.[3]

The results of these calculations, and of those for the public passenger service division of the commercial sector, are presented in Table 2 (see Appendix), which also incorporates the more familiar figures for the numbers of horses in agriculture. Even this last is not a straightforward concept, since the Agricultural Returns collected information on the number of horses kept on agricultural holdings. Not all the holdings were farms in the normal sense of the term, since they included small holdings of accommodation land of the kind which might be occupied by butchers, innkeepers, livery stable keepers, jobmasters, and the like; and not all the horses on the holdings, even those on the genuine farms, were actual farm horses. The *Returns* distinguished between "horses used for agriculture, including mares kept for breeding", and "unbroken horses kept on farms"; from 1911 there was a third category, of "other horses on agricultural holdings", but since this included saddle and carriage horses, and vanners, it has been excluded from the final tally in Table 2 in case it poses some risk of double-counting.[4]

[1] *Report of the Commissioners of Inland Revenue for 1906* (P.P. 1907, xx). Motor-cycles, which do not appear in Transport 7 until 1910, numbered about 34,000 in 1906.— *5th Report of the Commissioners of Customs and Excise*, 1914, Table 89.

[2] *Horse Duties and Wheel Tax*, 1888, p. 285.

[3] *30th Report of the Commissioners of Inland Revenue*, 1890, p. 15; figures for 1890, after buses had been dropped from the 2-horse carriage classification. *R.C. on Horse Breeding Industry in Ireland*, 1898, QQ. 418–21, 435–8 (Joseph O'Reilly). O'Reilly spoke of annual replacements of 20,000 to 30,000 hunters, implying a total stock of 200,000 to 300,000. A fashionable hunt like the Melton might have a field of 500 in the 1870's, with many taking out two or three horses each.—*S.C. on Horses*, 1873, QQ. 3363–4 (Admiral Rous).

[4] *Agricultural Output of G.B. Census of Production Act, 1906* (P.P. 1912–13, x), pp. 11–13, and *Agricultural Output of England and Wales, 1925* (P.P. 1927, xxv), p. 53, for the definitions. In 1924 there were 213,481 "other horses on farms". The horse count of that year showed 926,316 horses which were neither used for agriculture nor unbroken (under three years old).—*(Fifth) Census of Horses, 1934*, pp. 4–5. A minimum of 712,935 non-agricultural horses were, therefore, neither kept nor owned by farmers in 1924; only 243,051 of these were recorded as "stabled in towns".—Ibid. p. 12.

V

It is apparent from the 1924 figures in Table 2, which were the result of an actual head-count undertaken for the War Office as a preparation for mobilization of cavalry and military transport horses, that the pre-1914 estimates of horses in the private sector, especially in the carriage and trap division, are highly conservative and possibly seriously understate the real growth of this group between 1871 and 1901.[1] In 1924, with 474,000 private cars and 496,000 motor-cycles on the road, and with over 98 per cent of the passenger vehicles in London motorized, the horse-substitution phase of the rise of private motoring had been largely accomplished; yet there were still 52,000 carriage and trap horses, and 170,000 ponies and cobs chiefly used for trap work, keeping going in this motorized world.[2] To reconcile the 1901 and 1911 estimates for private carriage horses with the 1924 figure would require that not more than half private road transport had been motorized by 1924. This consideration alone is sufficient to authenticate the pre-1914 totals for the horse population of Britain, in Table 2, as conservative, lower-bound, estimates. The estimate for 1871, moreover, accords to within five per cent with the confident statement of Admiral Rous to the 1873 Select Committee on the Supply of Horses that "there are 2,000,000 horses in the country."[3] It is well buttressed by various earlier estimates: Middleton thought there were 1·8 million horses in 1813; Colquhoun reduced this to 1·5 million, in his 1815 *Statistical Account*, a number which McCulloch accepted for 1834; while Earl Lovelace thought in terms of two million in 1848.[4]

No contemporary guesses have been found to lend an air of respectability, or otherwise, to the later estimates. But a national herd of over three million horses was well within the breeding and importing resources of the country at the end of the nineteenth century. There were then usually about 126,000 foals under one year old kept on British farms, representing the annual births; while imports from Ireland, an important source for the maintenance of the British population, averaged 39,000 a year in 1896–1900. In these years imports from foreign countries were almost exactly balanced by exports of horses of working age, and may therefore be ignored. There was a large export of worked-out horses to Belgium and the Netherlands, running at some 29,000 a year in the late 1890's and growing to over 51,000 by 1906, but these were for slaughter and eating and from the demographic point of view are merely one of the commercial manifestations of horse deaths. The horse births available to sustain the British population were thus running at about 165,000 a year. The average annual death-rate for horses was found, on inquiry, to be 4·7 per cent in 1906; on that footing the number of young horses coming on every year was sufficient to support a population of 3·5 million.[5]

[1] The censuses of horses, of which that in 1924 was by inference the fourth since the 1934 census was described as the fifth, stemmed from the recommendations of the *Committee on the Supply of Horses for Military Purposes* (P.P. 1914–16, xxxix), p. 489.

[2] (*Fifth*) *Census of Horses, 1934*, pp. 4–5; Transport 7; and see above, p. 64, n. 1.

[3] *S.C. on Horses*, 1873, Q. 3425 (Admiral Henry J. Rous).

[4] J. Middleton, *General View of the Agriculture of Middlesex* (1813), p. 639; McCulloch, op. cit. ii, 653–4, also citing Colquhoun's estimate; Earl Lovelace, 'On the Sub-Division of Real Property . . . in France', *Jnl. Statistical Society*, ii (1848), 316.

[5] Horses under one year old from annual *Agric. Statistics;* imports from and exports to Ireland, imports

With these preliminaries out of the way, what does Table 2 signify? An expectation that horses are to be found predominantly down on the farm proves to be true, but only for the early nineteenth century and for the interwar years. By 1871 the off-farm horses were almost equal to the on-farm horses, and in the late 1890's and early 1900's definitely exceeded them. Whether all these off-farm horses were also town horses is another matter, and it probably remained the case that a majority of the horse population lived in the country at all periods. In the largest town of all, London, private horse-keeping was becoming a growing headache at least from the late 1860's because of the unpleasantness of driving in the congested streets, and because of the difficulty and expense of finding stabling, factors which were to rear their heads again in relation to cars from the 1950's with the telling difference that horses could not be left permanently parked in the streets.[1] The rate of growth of the private horse population of London therefore slowed down markedly in the late Victorian years, a development which may well have been paralleled in the larger provincial towns. The commercial horses, however, both those used in passenger services and those used in goods transport, necessarily congregated in the towns; and Table 2 suggests that these already formed the majority of off-farm horses by 1871, and grew vigorously until they formed two-thirds of this sector by 1901.

The town horses at their peak were thus over a million strong, and when allowance is made for the private horses kept in towns may well have approached 1·5 million. Each one produced three to four tons of droppings a year, which if properly made up with straw litter was the basis of 12 tons of good manure; but which if left about in the streets was just four tons of nuisance.[2] Each one contributed to traffic congestion and obstruction according to its degree; the Board of Trade produced a table of "road obstruction factors" in 1919, taking the amount of obstruction caused by a motor taxicab as par: a two-horse slow trade vehicle rated as ten, a one-horse fast trade vehicle as three, a horse-drawn bus as five, and a horse-carriage as two.[3] The enormous improvements in the urban environment—in health, in smell, and in traffic flows—stemming from the achievement of the motor in cutting down the town horses to 243,051 in 1924 and to 131,812 by 1934 are thus abundantly clear.

from and exports to foreign countries, in *Agric. Returns for 1906* (P.P. 1907, xcvii), Tables lxxx, xcvii, cxviii; horse death-rate in *Agric. Output of G.B. Census of Production Act, 1906*, 1912–13, pp. 11–13.

[1] *S.C. on Horses*, 1873, Q.20 (Joshua East, London jobmaster); Gordon, op. cit. p. 105; and see F. M. L. Thompson, *Hampstead: Building a Borough, 1650–1964* (1974), pp. 280–1.

[2] Agricultural student's notebook, *c.* 1880, in author's possession; cf. Joel A. Tarr, 'Urban pollution—many long years ago', *American Heritage* (1971), 65–6, citing American public health literature of the early twentieth century, which found an average daily output of manure per town horse of 22 lb. which is 3·58 tons a year.

[3] *S.C. on Transport (Metropolitan Area)*, 1919, p. 376. The complete table of "road obstruction factors" reads:

Passenger vehicles		*Trade vehicles*	
Electric tram	9	One horse, fast	3
Omnibus, horse	5	One horse, slow	7
Omnibus, motor	4	Two horse, fast	4
Cabs, horse	2	Two horse, slow	10
Cabs, motor	1	Light motor	1
Carriages, horse	2	Heavy motor, fast	3
Carriages, motor	1	Heavy motor, slow	5
Cycles	½	Barrows	4

In conclusion, does the study of the horse population require any further justification? Behind the large numbers lay large industries and occupations, hitherto virtually unnoticed by economic or social historians. Breeding and rearing in order to sustain the population and provide for its expansion was not, on the whole, a specialized branch of British farming, since with astronomically high infant mortality rates of around 50 per cent in the first two or three months of life and with great uncertainties about the character and prowess of the offspring, it was very risky to commit one's fortunes to horse-breeding. Southern Ireland, it is true, had a high reputation for the production of the best hunters and saddle horses in the world; hilly Wales was renowned for its ponies and cobs; and the East Riding, which in the mid-century was the home of large barouche horses and hunters, became the most-prized centre for hackneys and jobmasters' horses in the late Victorian period.[1] But aside from the occasional wealthy gentleman-farmer like Burdett Coutts with his 500-strong hackney studs at Brookfield in Renfrewshire and in Yorkshire, the pure horse farm was extremely rare even in these breeding regions, and the actual production within them took place in small lots as sidelines to general farming. All in all, however, the sale of horses to the British was worth over £1 million a year to the Irish by the 1890's; and the annual output of horses sold away from farms was reckoned to be worth £1·5 million to British farmers in 1906.[2] These of course were minute sums in relation to the value of the annual output of livestock products at this date, £95 million.

The population, however, required feeding, at a rate of something like 1·4 tons of oats or corn and 2·4 tons of hay per head per year for town horses.[3] British town horses alone would, therefore, have been consuming between 1·5 and 2 million tons of oats or its equivalent and between 2·5 and 3·5 million tons of hay a year in the 1900's, equal to the entire home output of oats and one-third of the output of hay.[4] In practice many such horses lived on imported feed, particularly maize; which was just as well, since the horses on farms had to eat too. Put in another way, it was estimated that in the 1890's and 1900's it cost about 10s. a week to feed a town horse, so that their annual feed bill was somewhere between £26 and £39 million, a sum approaching 100 per cent of the annual value of all crops sold off British farms.[5] The horse consumer market, therefore, was one of the British farmers' most important markets, all the more important because the hay component in it was naturally protected from foreign competition.

Apart from its daily needs, the horse was a not inconsiderable consumer of other products. But it is perhaps unnecessary to enumerate the smiths, farriers, saddlery and harness makers; whipmakers, stirrup, bit, and spur makers;

[1] *R.C. on Horse Breeding Industry in Ireland*, 1898, QQ. 10593–4, 10679 (Burdett Coutts, M.P.).

[2] *Agric. Returns for 1906*, Table LXXX; *Agric. Output of G.B. Census of Production Act, 1906*, 1912–13, pp. 11–13.

[3] Consumption rates are those adopted by the U.S. Board of Agriculture for town horses, for the period 1910–31.—*Historical Statistics of the U.S.: Colonial Times to 1957*, p. 262. No similar calculations were made by the British Ministry of Agriculture, but there is no reason to suppose that British eating habits differed significantly from American ones, at least in hay, oats, and corn.

[4] Output figures from *A Century of Agricultural Statistics*, pp. 112, 120.

[5] Barker and Robbins, op. cit. I, 243–4; Gordon, op. cit. p. 67, says 12s. per week for feed for carriers' horses.

wheelwrights, carriage- and coach-builders, fitters, painters, upholsterers, and trimmers; cart, van, and waggon makers; coachmen, grooms, cabmen, flymen, carmen, carriers, carters, and hauliers; horse-keepers, horse-breakers, horse-dealers, jobmasters, and livery-stable keepers—not forgetting the knackers—all of whom depended for their livelihoods on the horse, when all were overshadowed by hay as big business. The straw in the horse sun-bonnet was indeed a very small straw in the wind. If it has served no other purpose, it may have served to introduce the basic groundwork for more serious research into the backward and forward linkages of the horse in the pre-motor economy.

Bedford College, London

[*Appendix overleaf*

APPENDIX

Table 2. *Estimates of the Horse Population of Great Britain, 1811–1924**

Columns refer to the years centring on the stated year, except for column 8, which is the year of an actual census.

Numbers in '000

	1 *1811*	*2* *1851*	*3* *1871*	*4* *1881*	*5* *1891*	*6* *1901*	*7* *1911*	*8* *1924*
1. Trade horses, goods	151	161	286	388	500	702	832	374
2. Buses and trams	—	34	35	?	279	308	25	—
3. Stage carriages, post-horses, other hackneys	100	69	123	?	79	156	138	—
A. Commercial sector [1+2+3]	251	264	444	?	858	1,166	995	374
4. Private carriage horses	236	135	201	585	500	400	537	52
		114†	139‡					170
[4a. No. of private carriages	*51*	*205*	*432*	*460*	*415*	*450*	*324*	*181*]
5. Riding horses, hunters	§	28	74	§	§	200	§	260
								51
								16
B. Private or pleasure sector [4+5]	236	277	414	585	500	600	537	549
C. Horses not on farms [A+B]	487	535	858	?	1,358	1,766	1,522	923
6. Used in agriculture	800	?	940¶	987¶	1,087	1,089	1,087	753
7. On farms, unbroken	?	?	314‖	441‖	394	422	408	212
[7a. 'Other horses' on farms	?	?	?	?	?	*150*	*218*	*213*]
D. Horses on farms [6+7]	800	?	1,254	1,428	1,481	1,511	1,495	965
E. Total [C+D]	1,287	?	2,112	?	2,839	3,276	3,017	1,888

* These estimates are intended to be conservative, and will no doubt be found to have omitted several distinct categories of horse employment. One obvious omission is of pit ponies, and I am indebted to Dr A. R. Griffin of Nottingham University for the following estimates of their numbers (somewhat arbitrarily assigned by me to dates corresponding to those used in Table 2): 1851, 11,000; 1881, 25,000; 1911, 70,000; 1924, 65,000.

† Horses of farmers, clergy, and bailiffs were made subject to duty in 1854, and were estimated at 114,000.—P.P. 1870, xx; P.P. 1873, xiv. They might equally well be counted in Row 5.

‡ Horses used by farmers, clergy, surgeons, etc.—P.P. 1870, xx. § Included in Row 4.

¶ Excludes breeding mares. ‖ Includes breeding mares.

Sources:

Row 1. Table 1 above, p. 72. For 1811: horses, other than pleasure horses, "not wholly used in husbandry".—P.P. 1830, xxv.

Row 2. Horse multiplier of 10 applied to numbers of buses and trams, except for 1901.
1851: buses and stage carriages in 1853.—*Accounts & Papers*, (P.P. 1859, Sess. 1, xiv).
1871: bus horses made subject to horse duty from 1870 and the separate duty on stage carriages abolished.—*S.C. on Railway Passenger Duty* (P.P. 1876, xiii), p. vi. The resulting increase in the number of horses charged to duty was estimated by the Inland Revenue at 75,000; in fact the increase of 1870 over 1869 was 162,000, of which 4,000 represented the "normal" pre-1869 annual increase. The "abnormal" increase of 158,000 has been attributed 35,000 to buses and 123,000 to Row 3.—P.P. 1873, xiv.

1891: buses became liable to the hackney carriage rate of duty in 1890, formerly having paid the £2 2s. rate of duty for four-wheel, two-horse carriages. The increase in hackney carriages of 1890 over 1889 was 34,900; but the decrease of carriages paying the £2 2s. rate was 27,900, and this figure has been taken for the buses and trams.—P.P. 1890, XXVI.

1901: 132,000 hackney carriages paid duty in 1901.—P.P. 1902, XXII. 28,000 of these have been taken as buses and trams, but in order to reflect the 1891–1901 increase in London buses and possibly in some provincial towns (see text) a horse multiplier of 11 has been applied.

1911: there were 2,341 horse-bus drivers left in 1911, and as many conductors, compared with the 11,974 men in horse-bus service in 1901, and hence a considerable number of horse-buses were still in use; 2,300 have been assumed. It has been assumed that horse-trams, while still numerous in 1901, had become rarities by 1911.—*Census*, Occupation Tables.

Row 3. 1811: Middleton, op. cit. p. 639.

In remaining columns a horse multiplier of 2 has been applied where numbers of cabs are separately recorded.—*S.C. on Metropolitan Local Government* (P.P. 1866, XIII), Q.Q. 656–7. Where only a total of undifferentiated "hackney carriages" is recorded a horse multiplier of 1½ has been used.

1851: cabs and public passenger vehicles (excluding buses) recorded for 1853 and 1854.—*Accounts & Papers* (P.P. 1854/55, XXX). Posthorses derived from the numbers of posthorse licences issued in 1853.—P.P. 1859, Sess. 1, XIV.

1871: as for Row 2.

1891: 53,000 hackney carriages were recorded for 1889, before the entry of buses into this duty class.—P.P. 1890, XXVI.

1901: as for Row 2.

1911: 94,000 hackney carriages (horse-drawn) were recorded in 1914, of which 2,300 have been taken as buses (see Row 2). In 1911 itself there were 108,000 horse-drawn hackney carriages, which with the same deduction for buses and the same horse multiplier yields 159,000 horses.—P.P. 1914, XVII.

Row 4. 1811, 1851, and 1871: horse tax returns.—P.P. 1830, XXV; P.P. 1870, XX; P.P. 1875, XX.

1881: derived from the ratio of domestic grooms and coachmen: total pleasure horses in 1901, viz. 75,000 : 600,000. In 1881 there were 73,000 domestic grooms and coachmen; no comparable occupational category was employed in 1891.—*Census*, Occupation Tables.

1891: the estimate for Goschen's horse tax.—P.P. 1888, LXV.

1901: estimated from Row 4a, on the assumption that not more than 10 per cent of carriages did not have horses to pull them.

1911: same formula as for 1881. In 1911 there were 67,000 domestic grooms and coachmen.—*Census*, Occupation Tables.

Row 4a. Simple addition of four-wheel and two-wheel, two-horse and one-horse, carriages, without weighting, from annual carriage duty returns.—P.P. 1830, XXV; P.P. 1870, XX; P.P. 1881, XXIX; P.P. 1907, XX; P.P. 1914, XVII; P.P. 1926, XXVIII.

Row 5. 1851: described as "riding horses".—P.P. 1870, XX.

1871: four-fifths of the duty class "horses under 13 hh." assumed not to have been used in trade.

1901: O. F. Christie, *The Transition to Democracy* (1934), p. 172, citing C. E. Playne, *The Pre-War Mind in Britain* (1928).

Rows 6 and 7. 1811: horses used in husbandry and paying duty in 1812.—P.P. 1830, XXV.

1871–1911: figures from *A Century of Agricultural Statistics: Great Britain, 1866–1966* (1968), Tables 20, 70; and Mitchell and Deane, op. cit. pp. 82–3.

Row 7a. 1901–1924: saddle and carriage horses, and vanners, kept on farms but not used in agriculture: P.P. 1912–13, X; P.P. 1927, XXV.

Column 8. (*Fifth*) *Census of Horses, 1934* (1935), which repeated the figures of the (fourth) census taken in 1924.

Row 1: class (v) light and medium trade or draught horses, and light and medium trotting vanners, over 15 hh. *plus* class (vii) other heavy horses and heavy trotting vanners.

Row 4: class (iv) carriage and trap horses over 15 hh. *plus* class (ii) ponies and cobs 14–15 hh.*

Row 5: class (i) ponies 14 hh. and under* *plus* class (iii) riding horses and hunters over 15 hh. *plus* thoroughbreds.

Row 6: class (vi) horses wholly or mainly used for agricultural purposes [cf. 920,000 such horses in *Agric. Statistics* for 1924].

Row 7: horses under three years old [cf. 268,000 such horses in ibid.]

* Some of these classes may well have been drawing either goods or public passenger vehicles, so that the 1924 total for Row A is probably under- and for Row B over-stated.

Part 6

Supply and Demand, Trade Cycles

D. E. C. Eversley

9 The Home Market and Economic Growth in England, 1750–1780

1 Introduction[1]

In recent years the increasing preoccupation of economists with the problems of the developing countries has caused a systematic re-examination of the process of economic growth in general. More from a desire to use past experience as an aid towards the identification of categories for analysis than from any simple belief in the possibility of deriving universal explanations from a system of crude parallels, the economic historian of eighteenth-century England has been called upon to make his own contribution to the resumed debate. In this way he has come to rewrite, in the space of a decade, the account of a period which, in a generation from Cunningham to Mantoux and Clapham, had already received thorough treatment. In the nineteen-fifties it may have appeared sufficient to fill in the gaps left in the general outline by providing detailed accounts of individual firms and areas, or to continue the discussion of the precise role played by such categories as the rate of interest or entrepreneurship as a factor in industrial growth.

Ten years later the chain of interrelated events which we have come to call the Industrial Revolution would pass muster, in any review, only

[1] This contribution began as a paper read in Cambridge in 1962. It was then circulated under the title 'In Pudding Time' and, as the name implies, was mainly concerned with the earlier eighteenth century. This part of it was soon superseded by the appearance of A. H. John's paper of 1964. The duplicated version received detailed criticism from a number of friends whom I would like to thank for their help in revision: T. S. Ashton, W. Ashworth, T. C. Barker, J. Blackman, J. D. Chambers, A. W. Coats, E. Hobsbawm, J. D. Marshall, H. C. Pentland, S. Pollard and B. E. Supple. They are not responsible for the errors that remain despite their advice to remove them.

if it could be made to fit the categories of the era dominated by the logic of the schools of Rostow, Kuznets, and others whose main preoccupations were not with history at all, but with a world where backward nations were failing to reach the point of 'take-off into sustained growth' which marks the beginning of a true revolution—meaning a drastic change in the rate of change. The main explanations put forward under these new headings have been well reviewed by Hartwell[1] and we need not repeat them here. There is in many of these reinterpretative studies a penchant, as Hartwell noticed, to select one cause or set of causes rather than another as being the principal agent of change. Each writer, whilst acknowledging that the industrial revolution could only occur when a large number of conditions had been satisfied, feels nevertheless that he must draw attention to something which was unique in the situation of eighteenth-century England, so that one could claim, with some hindsight and following a logic not susceptible to proof, that if all other conditions had been fulfilled, there was yet one which was the crux of the matter: whereas some of the others might be superfluous, one set of social or economic conditions was indispensable.

What follows here is open to this charge: except that there are three sets of conditions which appear to be equally important, which are to some extent linked, and which, it may fairly be claimed, cannot be found in the same combination in other European countries at that time, and have perhaps not reappeared since, in the same strength, in countries struggling to escape from the fruitless cycle of beginning effort and subsequent stagnation. They experience a failure to maintain equilibrium growth rates for three sectors: home food production, domestic demand for consumer goods, and population.

2 The Thesis Stated

To lend greater precision to this contribution, and to place it in the context of material already published, it may be as well to start with the theme developed by A. H. John and others, most succinctly stated in a paper given at the Economic History Conference at Reading in 1964 and since published.[2] For our purposes, the most relevant part of this

[1] R. M. Hartwell, 'The Causes of the Industrial Revolution. An Essay in Methodology', *Econ. Hist. Rev.* 2nd ser., XVIII (1965), p. 164.

[2] A. H. John, 'Agricultural Productivity and Economic Growth in England, 1700–1760', *Jour. Econ. Hist.*, XXV (1965), pp. 19 ff.

thesis is one we readily accept: that the low food prices of the period 1730–50, due to a succession of favourable seasons and agricultural improvements, allowed large sections of the working population to feed well and yet retain a margin between income and expenditure required for minimum subsistence, a margin which could be spent, and was spent, on consumer goods, thus providing the necessary buoyancy of the home market at a time when industry was just beginning to transform itself, in some sectors, by the adoption of power, division of labour, and techniques of mass production and distribution. Let us rephrase this general argument in a series of propositions (at present only to be applied to the period before 1750) which are either implied or spelled out in John's work:

(1) Agricultural production rises relative to the existing domestic and foreign markets to a sufficient degree to allow prices to remain stable or even fall compared with earlier periods.

(2) This makes it possible for a larger number of people to work in industry and live in towns without causing famine.

(3) Population rises only slowly, at any rate not fast enough to cause food shortages, and the economically active part of the population in particular remains small enough in relation to labour demand (which may be a matter of place, skill, age or sex), at least in some areas, to keep the level of money wages high relative to food prices.[1]

(4) The workers, or some of them, still maintain their earnings, though they could satisfy minimum needs with less work.

(5) They spend their earnings in part on manufactured commodities. The price of these rises, or the producers see that they can sell more at the same price—in either case an incentive is given to increase the scope of production.

(6) Though exports take place, and their level may be rising, this additional home demand is seen as the major stable and long-term factor in promoting industrialization.

We shall have to qualify all these assertions later, but let us first of all come to the final part of John's thesis: that this state of affairs lasted only until about 1750, and that after that date there was something of a collapse due to bad weather, and failure of agricultural productivity to keep pace with accelerated population growth and urbanization, and

[1] For examples of regional shortage, see A. H. John, 'War and the English Economy', *Econ. Hist. Rev.*, 2nd ser., VII (1954–5), p. 340.

this led to the failure of the domestic market. If after 1750 industrialization continued, it must have been due to exports taking the place of home demand.

Against this view we now advance a different thesis: that the situation continued to be favourable to home demand from 1750 to at least 1780;[1] that exports before that date were neither at a sufficiently high level nor sufficiently stable to warrant the amount of investment that took place; that the rise in food prices was not drastic before 1780 in relation to incomes; and that in any case some rise in food prices does not necessarily squeeze the margin of expenditure on other products; that the period from 1750 to 1780 is crucial as the 'warming-up' period just before the take-off into sustained growth; that these thirty years laid a sufficient foundation of home consumers to enable industry to withstand the collapse of exports in the seventies; that this happened because increased real incomes tended to be spent on consumer goods rather than extra food purchases, thus still further stimulating industrial production and accelerating the movement of existing and new labour into the manufacturing sector; and that in any case the whole mechanism is to be explained not in terms of global populations, classes, or commodities, but in terms of regions and income groups.

Statistically, the period 1750–80 must be shown, if these assertions are to be proved, to have satisfied these conditions: the population of the country must have risen faster than before 1750, but not as fast as after 1780; prices of foodstuffs must have risen faster than in the period which used to be regarded as one of depression for the agricultural sector, but not as fast as during the period of war and explosive population growth; production of commodities, where this can be measured, must be shown to have grown faster than population, and faster than exports; innovations must have been publicized and adopted at a greater rate than previously, and more new enterprises started, especially of the kind catering for the home market.

This kind of examination has to be conducted principally at a quantitative level. Too much of what has appeared in the past has been of a literary kind. The hardships of the later part of the eighteenth century, which obtruded themselves into the historian's consciousness to the extent of precluding, apparently, any confidence in the possibility

[1] These breaks are artificial and have no normative meaning, but are dictated by one's precursors. For a different chronology yielding different interpretations, cf. Habakkuk and Deane, 'The Take-off in Britain', ch. 4 in W. W. Rostow (ed.), *The Economics of Take-off into Sustained Growth* (1963), p. 67.

of an extended domestic market, are the legacy of indignant pamphleteers and their twentieth-century chroniclers: both rightly concerned that we should know of the miseries suffered by a part of the population, they deflected our attention from the fortunes of another part. In the same way the opinions of contemporary political economists, though often valuable, are equally often wrong-headed and misleading. We shall try as far as possible, therefore, not to use the analysis of these writers, except where they quote facts as to prices, wages or output.

3 The Domestic Mass Market

In concentrating on the three factors mentioned, we do not dismiss others as irrelevant; indeed, it will be found that there is a good deal of overlap. For instance, many writers stress the need to accumulate capital in sufficient amounts to make industrial output rise faster than population. Some of this capital comes from sections of the population for whom levels of prices of necessaries, or the profitability of agriculture, are not of importance: but some of it also comes from the savings of those who may suffer from unfavourable price movements, and a great deal of it comes from landowners who might have had no investible funds had the price of produce remained too low to ensure a continued flow of rents. To take another category of explanations, the discussion of the significance of 'leading sectors'[1] may be as much concerned with industries serving the home market as with exports. General theories of social structure, such as Nef's view of the importance of the new attitude engendered by the Renaissance in providing both a taste for a civilized standard of living and attitudes to labour and the poor consistent with a rise of mass consumption, provide a framework for the analysis of demand.[2]

Nevertheless, the claim is here put forward that there is one central theme to all industrialization analysis: the emergence, and maintenance,

[1] *Ibid.*, p. 70. We omit here any general discussion of theories of economic growth or of particular aspects, such as the theory of effective demand. The leading recent writers on growth (Rostow, Kuznets, Gershenkron) mostly concentrate on the supply side of the implications of higher *per capita* incomes, and on savings, but not on consumption. We shall not discuss here whether any account which fails to describe the structure of demand is complete.

[2] J. U. Nef, *Cultural Foundations of Industrial Civilization* (Cambridge, 1958), esp. ch. V and VI.

of a large domestic market for mass-produced consumer goods.[1] The term 'market' must here be understood in a specific sense: specific, that is, to the time, the place, and the sector of the economy. When we say that the rapid growth of the iron industry in the second half of the eighteenth century was based on a domestic market we mean by this only that, given the state of communications, in the areas where this industry was located, and given the relative fall in prices which was associated with large-scale production, it was possible to find people with incomes and demand schedules capable of absorbing this increased output. There is, in other words, not one market any more than there is one economy. There are regions, some of which have been clearly identified, and others which are more difficult to describe, because they varied in exact delimitation for various products, and at different times.[2] Chambers's *Vale of Trent* is perhaps the classic case of the description of such a region within which food was produced, industry grew, and the products of industry were sold—almost indeed a closed system, but one which makes sense in relation to the factors operating within the economy.[3] There are also, besides the regions, recognizably separate markets, on a regional or national scale, by income groups, occupations, or fashions. The growth in the production of soap and bleach and alkalis is not due to any general adoption of habits of cleanliness, but to the emergence of some more groups, at the margin, who washed regularly, and beyond that margin, those who washed occasionally; to the increased demand of some households for white linen, relatively more people who dressed in muslin, and so on. For this purpose it is enough if a large part of that new market is purely local and even if it consists

[1] For other statements of this view, sometimes in a different context, see Hartwell, *op. cit.*, pp. 169, 177; Witt Bowden, *Industrial Society in England towards the end of the Eighteenth Century* (2nd ed., 1965), pp. 65 ff.; J. L. and B. Hammond, *The Rise of Modern Industry* (1925), p. 67; E. W. Gilboy, 'Demand as a Factor in the Industrial Revolution', in *Facts and Factors in Economic History* (Cambridge, Mass., 1932), p. 639 ff. But curiously enough, Miss Deane, whose earlier work seemed to add so much force to the thesis here stated, seems to rely firmly on exports again, as do most of the Cambridge economists and historians, in her latest work, *The First Industrial Revolution* (Cambridge, 1965), esp. ch. 4 (hereafter referred to as *First Ind. Rev.*).

[2] Cf. D. E. C. Eversley, 'Population, Economy and Society', in Glass and Eversley, *Population in History* (1965), pp. 26 ff. Some eighteenth-century economists (e.g. R. Cantillon) believed that food could rarely be transported over a distance of more than ten miles, excluding, of course, water transport.

[3] This is not to imply that there was no national economy. There was specialization, and some products peculiar to one region were sold on most national markets. Many statistical series move in exact parallel, though actual levels differ widely at any one point of time.

to some extent of those who help in the production of a commodity and their households. There is such a thing as a subsistence element in every industry.

Commodities were often classified in the past according to whether they were luxuries or necessities, and something of this distinction survives in recent discussions. It is claimed that a good deal of what was produced (or, which amounts to the same thing, bought from abroad with exports of home products) was for the wealthy sections of the community, and given great inequalities of income, these alone might account for most of the increase in production. But apart from the fact that the products in question were often far from being luxuries in any sense of the word, we need to remember that very few articles could be described with any confidence as being for the consumption of the rich only. Without giving this evidence any quantitative significance, we can cite a mass of contemporary sources alleging the prevalence of 'luxurious' habits amongst the 'poor'; a complaint, shorn of its moralizing overtones, which means nothing more than that some labourers liked tea with sugar even when both were heavily taxed; that women decked themselves out in clothes considered too good for them; and that in some cottages you might find a bit of carpet or even a piano.[1] What seems necessary for growth is that the very exceptional expenditure should become a little less so, that articles later to be described by Nassau Senior as 'decencies', half-way between luxuries and necessities, should spread through some more of the 'middling sorts of people', and that some labourers should take it into their heads so to misspend their income (according to their betters) as to go short of food and put themselves into debt for a looking-glass or a pair of gilt buckles for their Sunday shoes.

The wealthy household did, of course, play its part in increasing the scope even of the industrial market. Whilst the rich might have bought little for their personal use that was mass-produced, their households needed metal and textile manufactures, their servants needed dress and equipment and the building of their houses involved increasing consumption of glass, iron, lead, and brass.[2] To point to the consumption pattern of large and wealthy households, however, is not to weaken the theme of the mass market: inasmuch as much of what was bought was for servants, we need to look at the domestics as part of the total labour

[1] On the meaning of the relative terms used at different times, see Eversley, *Social Theories of Fertility and the Malthusian Debate* (Oxford, 1959), pp. 89 ff.

[2] Cf. Witt Bowden, *op. cit.*, p. 66.

force, and if their real incomes rose and they lived a more civilized life at the expense of their masters, that is not so different from a similar change in the living pattern of a craftsman working for the big household at one remove, or another artisan whose clientèle might be amongst the *petite bourgoisie*.

4 The Distribution of the National Income

To gain a better idea of the possible size of the new domestic markets created, we might start with some idea of national income distribution. What was the strength, in the middle of the eighteenth century, of that group of the inhabitants of the kingdom which was neither wealthy nor poor—the middle class, for the sake of brevity? Let us look again at Gregory King's famous table.[1] He assumed that there were over two and a half million people in his time in the category 'Increasing the Wealth of the Kingdom'. We can disregard the top ranks, down to the esquires, or those receiving more than £450 per annum, and we are still left with over 2.6 million people organized in households with a total income of £1 per week or more. Let us, to be on the safe side, assume that King exaggerated the incomes of the poorer freeholders and farmers, the craftsmen and artisans. If we counted only the 'persons in office', the clergy, the lawyers, the richer freeholders, the officers, etc., we still have a million people (including dependants) with incomes in the range £50–£400. Let us even exclude poor curates, pedlars, half-pay pensioners, and others unlikely to have much of a spending margin. But whichever way we look at it, we must believe that something like 10 per cent of the population were, in King's time, the nucleus of the home market. In actual fact, judging by the sort of wills we mention below, there must have been many in the class of artisans, lesser yeomen, and independant traders, publicans, and so on who were well within the 'spending group'. What the postulated growth of the home market implies is merely that this middling group should have increased in size, relatively as well as absolutely, and had a greater margin of expenditure above necessaries.

[1] As reproduced in Charles Wilson, *England's Apprenticeship* (1965), p. 239, or Deane, *First Ind. Rev.*, pp. 8–9.

Can we estimate how much they spent? If we start again from Gregory King,[1] we have a total of nearly £10 m. p.a. spent on manufactures and the products of mining and building operations, or something like £10 for the average family. But we are not concerned with a hypothetical average family. If we exclude 80 per cent of the population on the grounds of poverty, the average spending of the remaining fifth becomes £35. This sum (after allowing for food and rent) does not seem incompatible with King's view of the structure of the group in question.

Let us assume that by 1800, when the total population was about 10 million, this same group numbered 2 million, and that another 10 per cent of the population (1 million) had been added by the process of industrialization and the growth of commerce. The growth of population and the economy would then have increased the size of the consuming group from 1 to 3 million. (We realize that this is a circular argument; it is put forward here only in order to provide a framework for later calculations.) If this group had exactly the same pattern of expenditure as their forerunners in King's time, and if prices had been constant (a not impossible assumption, though the 'basket' of goods is not constant), they should between them have purchased £30 m. worth of manufactures, etc. This happens to be what Arthur Young thought the value of the market was in 1770.[2] There are a lot of ifs and buts about this sort of play with figures, yet they provide us at least with some sort of dimension for the phenomenon that has to be explained. Let us put it in another way: if we do allow that 'industrial' output trebled between the end of the seventeenth and the end of the eighteenth century, and if exports of manufactures can only account for a small proportion of the total, and if the products known to have constituted the bulk of the output in 1800 were not of the type sold to the wealthy classes, what other construction can we put on the process than that the 'middle class' of consumers expanded?

In fact, the process of structural change in the market is much more complex than this simple model suggests. First of all, we have the phenomenon of urbanization and rural specialization, which involved an unknown proportion of the population in increasing dependence on the supply of consumer goods from commercial sources as opposed to

[1] P. Deane and W. A. Cole, *British Economic Growth 1688–1959* (Cambridge, 1962), has the best analysis of King's and subsequent national income estimates. See pp. 155 ff.

[2] *Ibid.*, p. 156.

their own or their neighbours' home-made product. Examples of this sort of thing would be clothing (change from home-spun and woven), and boots, beer, eating utensils to replace the wooden home-made article, candles instead of rush-lights, furniture, fuel, and even less obvious types of expenditure, e.g. patent medicines instead of local herbs and soap in the absence of a running stream.

A much larger area for change in the expenditure pattern of any group of given real income levels arises from the change to cash wages and the urban environment. Many observers have noted the fact that urban and industrial wages were much higher than rural wages, and modern interpreters have dismissed this fact as unimportant, on the ground that the money differentials bear no relation to real wages: because in the country such a large part of total income was in kind, in food allowances, beer, rent, gardens, common rights, and so on, whereas in the towns everything had to be bought. This attitude towards the differentials tends to gloss over their significance for the structure of the market. Larger cash incomes inevitably mean more choice, and choice tends to move expenditure patterns away from the old essentials. Hence the complaints of the moralizers. In the country, so the assertions run, people had plenty to eat and a simple life.[1] If food was particularly cheap and abundant, they ate more or worked less. They neither saved, nor did they buy other things: they had all they knew or needed. But in the towns they developed a taste for new types of goods, apart from the fact that many commodities must now be purchased.[2] It was typical of the writers on the distress of the population at the end of the eighteenth century that they alleged misspending of incomes on inessential fripperies.[3] We need not ask here whether this was true or not as an explanation for distress: but we can say that if it was so, then the extension of the markets went even further downwards than we have assumed. If Ure's picture[4] of cottage equipment in industrial Lancashire has any significance, the process continued even through the worst years. We do not need to go as far as this: but the existence of the charge of misspending gives us a good lower limit to our description of the consuming classes.

1 The tradition is perpetuated. Cf. R. Postgate's Preface to the *Good Food Guide*, 1965/6 ed., p. 10.

2 Witt Bowden, *op. cit.*, p. 219.

3 This runs like a refrain through works like Sir F. M. Eden's *The State of the Poor* (1797), I, Bk. II, ch. II: 'On the Diet . . . of the Labouring Classes'. He frequently quotes from the attempts of Count Rumford to improve the pattern.

4 Andrew Ure, *The Philosophy of Manufactures* (1835).

There is a general point to add here which has some significance for total demand. If, as we assume, there is an element of re-distribution of the national income involved in the process of putting disposable cash into the hands of the ordinary consumer, we are entitled to believe that because of the higher propensity to consume attributed to the small income-earners, compared with the wealthier classes, the total effective demand for goods and services will therefore be relatively greater. It is not necessary, for the purpose of this analysis, to assume any real diminution in income for the upper ranges. There was not, as yet, much re-distributive taxation (other than through the Poor Laws). It is enough to postulate that a larger part of a rising national income should go to the middle or even lower groups to justify the assumption that the propensity to consume will rise.

Nor is this view incompatible with the idea of increased savings: as the incomes of the bourgeoisie rise, this group can both demand more consumer goods and save more. What they cannot do is to emulate the luxurious habits of the really wealthy: but then, as we try to show, luxury consumption is little help to industrialization.

5 The Flow of Income

We now consider the whole question of how any considerable body of members of the 'middle' group of economically active people could have acquired and maintained a standard of real wages[1] high enough to sustain the prolonged demand which we postulate as the core of industrialization. We need to distinguish here between short-term and long-term factors. The situation in the 1730s and 1740s is not difficult to understand. There had been considerable losses of population during the periods of high mortality (1725–30 and 1741–2) in most parts of the British Isles. There was consequently a shortage of labour in many sectors. At the same time food prices were at their lowest level since the previous century. There is no difficulty about purchasing power here. A high demand for labour results merely from the attempt to sustain the previous level of activity: unless land goes out of cultivation, there will

[1] S. Pollard, 'Investment, Consumption and the Industrial Revolution', *Econ. Hist. Rev.*, 2nd ser., XI (1958–9), p. 221, argues as many others do for a progressive decline from 1750, but this is not inconsistent with the view that the standard was higher than it had been in any previous age, or, on this interpretation, subsequently. But as we try to show, this decline is not compatible with production figures.

be some competition for labour. The effect of corn bounties has been disputed, but whether it was due to this subsidy or not, nobody now doubts that the level of agricultural output was well maintained. Exports remained steady or rose in some sectors; industrial production did not decline. (We remember that all this need not apply to every sector: it is enough that it should be true of some regions, so long as there is internal mobility. Regions of high labour demand keep up real wages to a level consistent with creating demand for consumer goods: poorer regions can only reduce this demand to zero; therefore we need only be interested in the areas of prosperity.)[1]

We now need to explain why this admittedly favourable situation continues after 1750, i.e. what long-term factors continue to operate when the exceptional demographic and climatic factors no longer apply. In the first place a closer examination of prices[2] shows that the higher prices of the period 1750–80 do not mean a return to the levels of before 1730, either in money terms or in real terms. There are many reasons for this. Although it is true that there were some bad seasons in these thirty years, the general run of meteorological observations[3] suggests that there were no excuses for famine prices. Secondly, the improvement in communications (roads, canals, coastwise shipping, river navigation) meant that it became progressively easier to iron out local shortages. (Though these had not in every case reduced total purchasing power very drastically, they must have had their effect precisely on that marginal class of consumer with which we are here most concerned, by re-distributing incomes in favour of producers and middlemen). There is now a considerable amount of evidence on the growth of a national food market during this period: so well organized in fact that prices in urban areas were often lower than in the countryside. We also know that wheat prices are not in themselves a good guide, or the only guide, to food prices as a whole. We find, for instance,[4] that malt prices do not rise as much as wheat prices. Wet seasons or very cold winters which injure one type of crop may benefit

[1] Cf. J. D. Gould, 'Agricultural Fluctuations and the English Economy in the Eighteenth Century', *Jour. Econ. Hist.*, XXII (1962), p. 318. For a good discussion of the demand for labour and its implications for wages and investment, cf. also Deane, *First Ind. Rev.*, pp. 134 ff.

[2] See below, pp. 240–3.

[3] See below, p. 244.

[4] The detailed prices of various cereals in relation to the foreign trade in these commodities need re-examination. For a good contemporary presentation, see Michael Combrune, *An Enquiry into Prices of Wheat, Malt . . . from the Year 1000 to the Year 1765* (1768), pp. 80 ff., for our period.

another.[1] Floods may increase the growth of herbage, hard frosts make the soil friable, dry summers can ripen and bring home one harvest whilst injuring beasts. The real test is whether the food situation got so desperate as to injure the demand for other goods. Many authorities also take the view that low prices are not in themselves necessary to promote high real wages—Young thought that high prices led to greater exertions to maintain a given standard.[2] That prices rose cannot be denied, but such a rise does not mean either that purchasing power fell or even that there were serious changes in demand to the detriment of the market for inessentials. No one living in the 1960s could doubt this, yet writers on the eighteenth century still tend to regard a period of rising prices as some sort of calamity. Given the conditions set out here we would argue that, as a result of the experiences of the previous decades, a large part of the population would seek to maintain in the face of increased difficulties the standards they had achieved, by combinations, by greater exertions, and if necessary by restricting fertility. And beyond that, it is still possible that they would go without some things we might consider essentials to keep the 'decencies'.

There is evidence that the price of many of these 'decencies' fell during the period under review. Others rose. There is not likely to be a universal explanation for this. On one hand, transportation costs were lowered in many instances. Mass production also played a part, as did the improved marketing methods in general. On the other hand, we need not exclude the possibility that in some years and in some markets the rise in food prices led to a temporary collapse of home consumption. If exports also suffered, this would explain the price fall. But if we then see that total production, in the long run, actually rose, the reduction in costs of production and distribution remains the only feasible explanation. The most important piece of evidence, however, will be the increased investment in productive capacity for those goods which were largely sold on the home market, and to the group of consumers we are discussing. This investment presupposes confidence in the long-term ability to purchase, and in itself, by leading to cost reductions, ensures the growth of the market.

Lastly, we need to consider the question of conscious manipulation of income levels. The whole of our hypothesis is undermined if we accept the view that wages tended to be at or very near minimum

[1] Potato prices are a case in point. Unfortunately good series are difficult to obtain but there are some in Thorold Rogers, Beveridge, etc.

[2] A Young, *Eastern Tour* . . . (1771), IV, p. 319, and passim.

subsistence levels. We say 'undermined' rather than destroyed, because the incomes of many of the groups we have described are amenable to the same pressures as wages: craftsmen, small industrial producers working on their own account, farmers, shopkeepers, professional persons, can all to some extent fix their own level of incomes according to their costs and their view of what is reasonable. But even for the daily or weekly wage-earners there is little evidence that the labour market operated in such a way as to force them to accept wage reductions even when the price of food fell. Arthur Young did not believe that prices determined wages.[1] Cannan demonstrated long ago that Adam Smith came to abandon the view that the worker was always paid on the margin of survival.[2] The changing attitudes to labour demonstrated by Coats[3], the comparative laxness shown in Poor Law administration, and the great increase in charitable provision all add up to a picture inconsistent with the view of a whole working population perpetually kept on minimum rations.[4] Gilbert's act legalized existing practices.[5] Against the barbaric cruelties of some employers (though most examples in any case date from later than 1780) must be set the many instances of industrial employers known to have provided high standards of comfort at least for their skilled workers. Company provision of housing, piped water, instruction, and amusement are the industrial equivalent of wages in kind in the agricultural sector: with the difference that in the changed environment such benevolence (or self-interest) constitutes a demand for industrial products.[6]

Once again, also, we need to point out that we are dealing with a regional model. The amount of internal migration which has been

[1] *Ibid.*, pp. 312 ff.

[2] Edwin Cannan, *Theories of Production and Distribution* . . . (3rd ed., 1917), p. 233.

[3] A. W. Coats, 'Changing Attitudes to Labour in the Mid-Eighteenth Century', *Econ. Hist. Rev.*, 2nd ser., XI (1958), pp. 35 ff. This is an important article which almost makes one take it for granted that incomes in most cases were well above subsistence level.

[4] Cf. also J. D. Chambers, 'The Vale of Trent', Suppl. to *Econ. Hist. Rev.*, No. 3 (1957), pp. 57–58. Attitudes to labour are a complex product of philosophical outlook, economic necessity, and habits of work and leisure, as Professor Chambers pointed out in one of his characteristic letters after reading my earlier paper.

[5] The motives of the Poor Law authorities were of course also mixed—many profited from their own laxity. But squires took pride in well-fed labour and tenantry and were the first to attack the industrialists for starving their workers. There is a distinct lessening, towards the end of the eighteenth century, of the number of attacks on 'luxury', meaning some comforts for the poor.

[6] Cf. J. D. Chambers and G. E. Mingay, *The Agricultural Revolution 1750–1880* (1966), p. 112.

demonstrated[1] can only be explained in terms of a differential in opportunity and real wages (though that may have been in part imaginary). It is not necessary to suppose that the whole labour force, or even the whole of the skilled labour force, continued to have this margin over subsistence: it is enough for labour to have continued to be relatively short in some sectors in order to explain the continuation or growth of the market. The rapid growth of centres like Birmingham and the Black Country, of London, of the East Midlands, and of South Wales, all provide examples of areas of almost chronic labour shortages, where the difference between industrial wages and those paid in adjoining rural areas were much larger than can be explained in terms of additions to wages in kind. Moreover, the rate of population growth by natural increase before 1780 was much too small to lead to any permanent overstocking of the market: though the balance of births over deaths noticeably increased after about 1765, this did little to add to the available labour force before 1780.

To sum up this part of the argument: starting with a relatively cheap food supply and a shortage of labour before 1750, we postulate the existence of a margin between income and expenditure on basic necessities. Between 1750 and 1780, this margin was not reduced, and in many cases, must have become larger: (i) because food prices did not rise sufficiently to cancel out the original benefit; (ii) because relative labour shortages continued in many areas and in respect of many skills; and (iii) because employers made no concerted or conscious attempt to reduce real wages to subsistence levels. In some cases, industrial paternalism increased the demand for consumer goods. As a result, the larger labour force contains more consumers for the products of industry than ever before.

6 The Extent of Exports

We shall now consider, in turn, four areas of evidence which have a bearing on the attempt to give greater precision to the general propositions we have sought to establish: the scope for exports, the industrial evidence as to the structure of the market, the agricultural sector, and the rate of population growth. Of these, exports probably present the least problem. The opinion that traditional accounts of the place of exports in the growth of the economy were exaggerated has been

[1] Deane and Cole, *op. cit.*, pp. 106 ff.

expressed at intervals ever since the cotton industry first came to prominence,[1] but on the other hand there has been a recent revival, partly through the popularity of the 'leading sector' theory, of the view that Britain's command of world markets was the real incentive to industrialization.[2] There is, however, no need for a single view of the problem. Nobody would deny that exports provided a large part of the total stimulus.[3] Moreover, most accounts are concerned with the main period of growth, after 1780, and there is no question at all that the foreign sector became increasingly more important with time. All we try to show here is that between 1750 and 1780, that is, when the foundations were laid, and many important advances made, the overseas market did not play a major role in providing the justification of expansion for the country as a whole. If for some sectors, and some areas, exports were already vitally important, this certainly was not known to those responsible for the conduct of other sectors. National preoccupations, as shown in parliamentary debates and legislation, were still with older sectors: the remnants of mercantilism sought to encourage industries which had little future. The rarity of any rational discussion of exports in the works of the economists before 1780, except in terms of the old staples or agricultural products, proves their ignorance rather than the absence of export sectors. But it also makes the operation of the leading sector rather obscure to follow.

Exports could be seen in three ways.[4] If it is true that the major part of the total increase in consumption in the country consisted of luxury foods and textiles, then exports can be seen only as stimuli to the demand for labour. We assume that the countries from which our imports originated obtained purchasing power which in turn gave employment to the producers of agricultural and industrial commodities, to workers in marine and commercial services. But even in this case the increase in foreign trade creates additional incomes at home. If,

[1] T. S. Ashton, *An Economic History of England: The Eighteenth Century* (1955), p. 63. See also the quotation there from D. Macpherson's *Annals of Commerce* (1805): 'the people of Britain being the best customers to the manufacturers and traders of Britain'. Porter's *Progress of the Nation* (1st ed., 1836) also made this point at a time when exports were much more important.

[2] For a clear expression of this view, see K. Berrill, 'International Trade and the Rate of Economic Growth', *Econ. Hist. Rev.*, 2nd ser., XII (1959–60), pp. 351 ff.

[3] In particular, critics of the home market thesis argue that the faster *rates* of foreign trade growth provided the more important stimulus, as did wars and the opportunities provided by the acquisition of colonies. But I do not see how these impulses can last if the total quantity involved remains very small.

[4] Cf. Gould, *op. cit.*, p. 318, and Deane, *First Ind. Rev.*, pp. 66 ff.

secondly, a proportion of the imports consisted of goods for the consumption of ordinary people (in the shape of tea, sugar, and tobacco), then existing additional home purchasing power has the effect of creating these imports, making possible exports in return, which then once more lead to increased incomes, thus setting up a cycle of growth. If, thirdly, a large part of imports consists of raw materials or semi-fabricates, these can only be required if the domestic demand is already in existence, from whatever class of incomes it proceeds, so that the increase in foreign trade can be seen as a link in a chain of home growth.[1]

All three elements are present, and each makes a contribution. Conversely, the demand from other countries for our own exports, regarded in isolation from the pattern of imports, must be based on changes in demand structure overseas analogous to our own: in fact, since British exports tended for a long time to be much more prosaic than her imports, it is possible to argue that the growth of the 'ordinary' consumer abroad was also of importance to us.

So far we have omitted any reference to the terms of trade. Any of the three mechanisms just described can be restated as favourable changes in the terms of trade, except where inelasticities in the supply of imports prevent this. A discussion of the movement of the terms of trade in detail, however, is outside the scope of this paper, since we are not so much concerned with the reasons for the distribution between home and foreign trade as with the facts of the case.[2] However, it may be noted in passing (although on the statistics which will be presented it cannot really be deemed a very important point in our period) that at any given level of home demand, additional exports may make it possible to reduce production costs per unit, which may further enlarge the scope of the total market without affecting the terms of trade one way or another.

What, then, are the salient facts about exports? (See Table 1.) We may base ourselves, without further discussion of the validity of the

[1] William Enfield, *Essay towards the History of Leverpool* [sic] (1773), has an analysis of the composition of the trade through that port. This shows that although exports were largely of the expected industrial consumer goods, imports were either raw materials or goods clearly for the luxury trade, with the exception of sugar and tobacco, which might have had a wider market. (I owe this information to J. D. Marshall.)

[2] For a good discussion of the significance of the terms of trade in this context, see C. Kindleberger, 'Foreign Trade and Economic Growth', *Econ. Hist. Rev.*, 2nd ser., XIV (1961–2), p. 290.

Table 1. British Exports 1740–99: Annual Average values
(all figures represent £ '000s except where otherwise stated)

	Domestic exports	All non-textiles	All iron	Nails (cwt.)	Glass	Lead	All textiles	Woollen only[1]	Cotton	Iron and cotton as % of all exports
1740–9	6,556	1,153	290	25	17	160	4,399	4,149	11	4.6
1750–9	7,748	1,521	408	32	42	151	5,805	5,311	85	6.2
1760–9	10,436	2,066	568	30	63	175	6,334	5,447	227	7.6
1770–4	10,030	2,335	766	37	74	179	6,526	5,347	239	10.0
1775–9	8,504	2,323	512	16	53	151	5,043	4,261	256	9.0
1780–9	10,190	2,848	613	17	59	175	5,710	4,250	756	13.4
1790–9	16,569	4,330	1,116	32	141	179	10,288	6,361	2,626	22.4

[1] Woollen: computed values, as in Schumpeter, Table XIII.

Source: Schumpeter, *English Overseas Trade Statistics 1693–1808.*

price levels underlying the statistics of value, on the tables published by Mrs. Schumpeter[1] and Deane and Cole.[2] The level of domestic exports achieved (annual averages) was just over £6.5 m. by 1740–9. It rose to a record of over £10 m. by 1760–9, and fell again to just over £9.2 m. by 1770–9. (Different authorities cite slightly different figures, but the effect is the same.) Taking Mrs. Schumpeter's Table II[3] (exports of English produce and manufactures only), we observe a first peak average of £9.1 m. p.a. in 1749–51, falling irregularly to a low point of £7.3 m. p.a. in 1779–81, and then climbing very steadily to reach an average of £20 m. at the turn of the century. These global figures conceal, of course, the changing structure of the export trade. In 1750 agricultural produce and fish still accounted for over £2 m. worth of exports; by 1780 this was little more than half a million pounds. Taking the principal industrial products, therefore, we expect to see a somewhat more marked trend towards increase, and Table 1 certainly shows important increases in the value of iron to a peak in 1774, followed by a fall to 1775–9, and the same is true in the case of glass, lead, and nails, with only tin increasing more or less steadily. But then all these non-textile industrial commodities together account for only about £1.5 m. p.a. round 1750, and at their peak to little more than £2 m. Textiles are clearly very much more important: at just on £5 m. p.a. *c.* 1750, rising to a plateau of over £6 m. by 1750–75, and falling again to £5 m. by 1775–9 before their spurt to the end of the century which formed the whole basis of Rostow's argument. But again these global figures must be broken down further in order to obtain the true picture of industrialization. Even at the pre-1780 peak, cotton was as yet an insignificant part of the whole: woollen exports came first, with an average value of between £5 m. and just over £5½ m., linen next with £400,000–£600,000, and cotton last with a mere £240,000, a figure which applies to the whole period 1760–80. These are all long-familiar facts, yet seen against the necessity to explain the continued growth of the whole economy from 1750 to 1780, they become significant. Above all, they

[1] E. B. Schumpeter, *English Overseas Trade Statistics, 1697–1808* (Oxford, 1960); Deane and Cole, *op. cit.*, B. R. Mitchell and P. Deane, *Abstract of British Historical Statistics* (Cambridge, 1962). Perhaps this is the place to record the view that the publication of these three volumes transformed the arena of debate on economic growth in England by providing the first widely accessible quantitative basis.

[2] Deane and Cole, *op. cit.*, ch. II. Also in Deane, *First Ind. Rev.*, pp. 60 ff., for new empirical justifications for using the figures as published by Mrs. Schumpeter, at least for comparative purposes.

[3] *op. cit.*, p. 15.

need to be set against the sort of figures which belong to the 'take-off' period, which are altogether of a different order of magnitude.[1]

We shall discuss the structure of the home market later, but at this point it is worth considering whether raw material import figures can shed any light on this situation. If we look at raw cotton imports, for which we have good quantitative series[2] we notice that by 1750 just over 2 m. lb. of cotton were imported, and that this figure was still the same by 1760. Remembering that the value of exports remained static over the next twenty years, we then note that the raw material imports rose to over 7 m. lb. before 1780 (having attained this figure already in 1777 and 1778). Unfortunately, we do not seem to have a good coefficient for the conversion of raw cotton quantities (or values) into finished values: there are too many variations in the mixtures of fibres, in the counts, and in the finishing of fabrics, including dyeing and printing, to obtain a useful conversion. But one thing stands out clearly even at this stage: if raw cotton imports, small as they were, rose more than threefold between 1750 and 1780, and if, as is claimed, the cotton industry was of some significance even by then, it looks as if the home market were of more significance than the static export market. (Another way of putting this is to point out that the ratio of cotton exports in 1795–9 to the average for 1775–9 was as 10:1 whereas raw material imports (quantities) were only 5:1, showing once more that the export markets became relatively much more important towards the end of the century.)

Naturally, if we do not deal with absolute figures but with relative increases (as historical statisticians have done since Hoffmann)[3] there are some rather startling changes to record even for the period under review. But we must beware of reading very much into these. For any one firm, or industry, clearly, a doubling of exports, from, say, £10,000 to £20,000 p.a., is an important event, stimulating investment and no doubt imitators. But such periods of sudden increase of opportunity for one or another section were familiar in English economic history since

[1] Compare Schumpeter's Tables VII, VIII, X, XII, XIV, XV, and XVI, with IX, XI, XIII, and XVII. Since the series are not continuous, the contrast between the earlier and the later period is not always quite so apparent.

[2] Schumpeter, *ibid.*, Table XVIII, p. 62. I do not know why Miss Deane (*First Ind. Rev.*, p. 60) claims that figures for raw cotton are not given in Mrs. Schumpeter's work. Surely 'Wool Cotton' means just that.

[3] Deane and Cole, *op. cit.*, p. 78; W. Hoffmann, *British Industry, 1700–1950* (Oxford, 1955), esp. Table 54, which shows clearly how calculations of percentage changes distort rather than illuminate when one is dealing with non-comparable quantities.

the Middle Ages. They seldom led to lasting growth. Therefore we must now turn to the home market to explain the real impetus given to industrial growth in the period 1750–80.

7 Organization of the Home Market

Our information regarding the shape of the internal incomes and consumption structure in Great Britain in the eighteenth century is necessarily fragmentary, and we can only follow existing authorities into this difficult field. First, we remind ourselves again that we are dealing with a series of regional economies, and that we cannot expect incomes, or expenditure, to be equally distributed in proportion to the population. Not only was total income very different, but also its distribution as between groups. Miss Deane provides us with the following rough outline[1] (see Table 2). Allowing for all the doubts which such calculations naturally arouse, we arrive at a gross national product which rises two and a half times between the end of the seventeenth century and 1770 and then doubles again (in money terms) by 1811. During this time (and here we are on safer grounds) population rises by 35 per cent from the first to the second point, and is roughly double, in 1811, what it had been 120 years earlier. In money terms, therefore, income per head more than doubles. The sector with which we are here concerned (manufactures plus trade plus transport) accounted for less than a third of the GNP *c.* 1688, and perhaps half by 1811. Exports (as here valued) were roughly £5 m. in 1688, of which certainly less than £4 m. were home industrial products. This leaves something over £10 m. as the combined turnover of industry and trade (after allowing for exports f.o.b.) for the home market, or let us say £10 per notional household per annum. By 1770 the manufacturing commercial sector is estimated as being worth £43 m., and after allowing roughly £10 m. as the value of exports of goods of British origin, we have more than £30 m. available for the home market, or perhaps £25 per household. Prices, of course, cannot be compared, as between these two earlier dates, but

[1] Based on Deane and Cole, *op. cit.*, ch. V. In *First Ind. Rev.* Miss Deane convincingly demonstrates (p. 63) that exports were worth about £2. 0*s*. 0*d*. *per capita* in 1752–3 and had risen only to £2. 15*s*. 0*d*. *per capita* in 1797–8. Although this is to be taken at constant (1796–8) prices, this still seems an insignificant addition to national income. Using a slightly different basis of years, exports price indices, and population, one could produce a static *per capita* figure.

Table 2. The Composition of the National Income 1688–1812
(millions of pounds except where otherwise stated)

	Gross national product	Of which manufactures plus distribution and transport	Of which exports	Home market	Population (millions)	Households (Pop. ÷5) (millions)	Home consumption per household	Exports per head
							£ *s.* *d.*	£ *s.* *d.*
G. King (1688)	50	15.5	5.0	10.5	5.5	1.1	9 10 0	1 0 0
A. Young (1770)	128	43.0	9.5 (domestic)	34	7.0	1.4	25 0 0	1 7 0
Colquhoun (1812)	250	130	40	90	11.0	2.2	40 0 0	3 13 0

Source: Derived from Deane and Cole, *British Economic Growth.*

These figures are to be taken only as crude approximations.

for what it is worth Schumpeter's price index[1] shows that prices, if anything, fell in this period of more than eighty years.

If we now turn to the period 1770–1811, we find that the value of the industrial/commercial sector's output has gone up to about £130 m. (being by now more than half the national product.) Exports, though they exceeded £40 m. in most years after 1806 until about 1818, when they fell again,[2] still leave a larger amount *per capita* to be consumed at home—perhaps £90 m. all told, or nearly £40 per household. Price levels have risen, of course, though probably not by 100 per cent,[3] and perhaps the most that could be said would be that domestic demand is holding its own relatively, at a time of rapidly rising population and despite the severe hardships of the war and accelerating industrialization. The implications of this for the 'standard of living' debate[4] may be worked out by others. In any case from our point of view it is not relevant. The important factor rather seems to be the increase in *per capita* consumption of home products in the time of Arthur Young. This will be independent evidence: we adduce it here to show that Miss Deane's calculations are at least consistent with our own thesis, which now needs reinforcing with proofs of a quite different kind. But we may at this stage represent the long-term trend diagrammatically.[5] This simply illustrates the widening gap between exports and total production and trade, which must represent home consumption, and we also show consumption per household.

Some allowance also has to be made for the increased need for producer goods at a time of accelerating industrialization. Investment has to come out of that slice of the GNP which is, in our calculations, a residual. What that allowance should be would depend on one's view of the probable capital/output ratios obtaining in our period. We have not got the documentary evidence to generalize on this point, but in any case the argument is not affected. It is relevant for the standard of living debate to know how much of the residual went on ordinary consumer goods, and how much on capital goods, since the latter do not enter into the calculation of current real incomes. The fact is that

[1] E. B. Schumpeter, 'English Prices and Public Finance', *Review of Economic Statistics* (1938), as quoted in Deane and Cole, and Mitchell and Deane, *op. cit.*

[2] Mitchell and Deane, *op. cit.*, p. 282.

[3] Cf. diagram at end of Deane and Cole, *op. cit.*

[4] E. J. Hobsbawm and R. M. Hartwell, 'The Standard of Living during the Industrial Revolution: A Discussion', *Econ. Hist. Rev.*, 2nd ser., XVI (1963), pp. 149 ff.

[5] Cf. Deane and Cole, *op. cit.*, p. 81, and Diagram I herein.

either kind of expenditure would creater further domestic incomes, act as a stimulus to industrial production, and form an integral part of the accelerator mechanism.

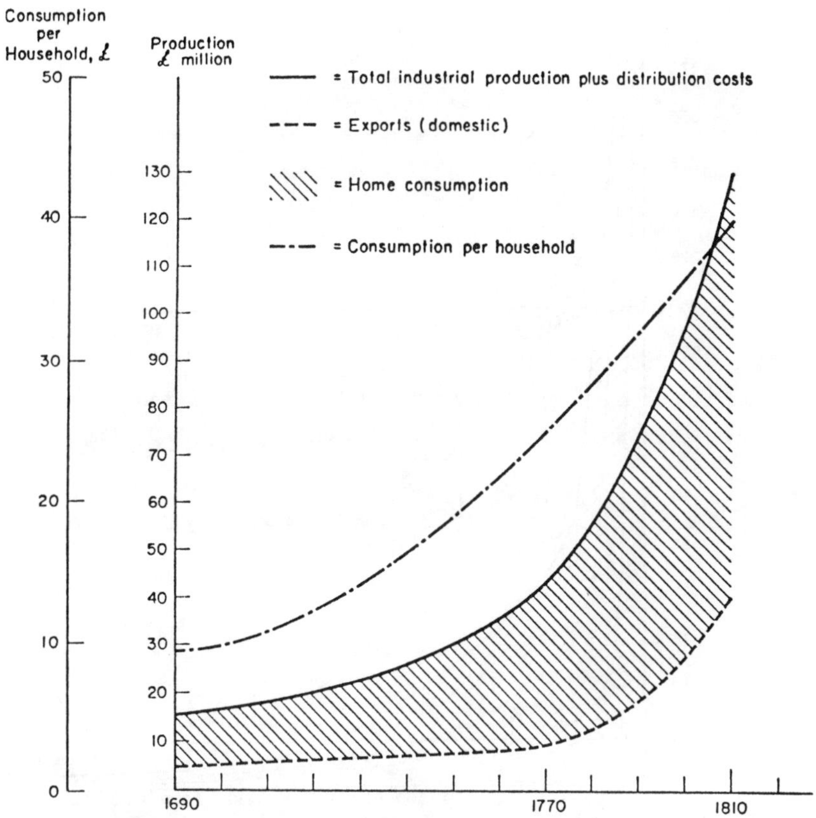

Diagram I. The Home Consumption Gap

We are now postulating the existence of a large and increasing difference between total industrial output (plus the inland trade which distributed commodities) and exports, constituting home demand. In our notional calculations we have used the idea of consumption by each household unit. Clearly, it is not possible that anything like such an average was attained by the majority of households. On the other hand, it is equally unlikely that all this extra consumption went only to the top layers of the income structure. Theoretically, of course, the top

10 per cent of all incomes could have accounted for 90 per cent of the national income. In 1770, 140,000 households could have shared £110 m. (or about £800 each), and the other 1,260,000 households shared

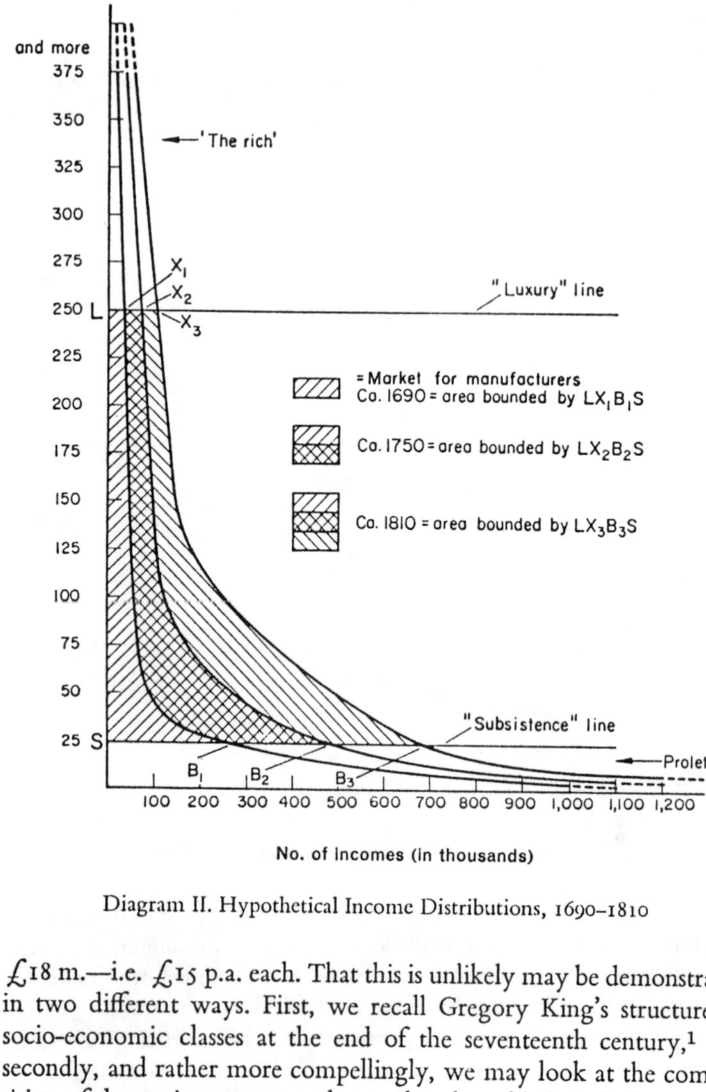

Diagram II. Hypothetical Income Distributions, 1690–1810

£18 m.—i.e. £15 p.a. each. That this is unlikely may be demonstrated in two different ways. First, we recall Gregory King's structure of socio-economic classes at the end of the seventeenth century,[1] and secondly, and rather more compellingly, we may look at the composition of domestic output to ask ourselves how likely it is that a small

[1] See above p. 213.

minority of households increased its consumption of the commodities involved on such a scale.

Let us return to the position of our group of 'middling consumers' in King's table of classes. It may be re-defined as being neither on the subsistence line nor rich. The diagram (II) shows a hypothetical income distribution before the process of creating middle incomes begins—with a few rich people and a lot of poor. Then there begins to be a bulge which we show at two successive stages. It will be argued that rising national income could benefit the wealthy greatly, and yet leave room for this new middle class, even without ever touching the poor about whom laments continue. This diagram also introduces two theoretical lines which may help to clarify the argument. The upper 'cut-off' line L–X denotes the luxury line: everything that is spent above this level is deemed to go on exotic imports, personal services, the products of goldsmiths and bookbinders. Everything below the line S–B we may term 'bread', i.e. minimum subsistence, including elementary shelter and clothing. It is in between the two lines that we find expenditure on industrial commodities, less essential foods, services, transport, etc. (This ignores savings which are likely to come more out of the area above LX than out of LXBS.) It will be seen that a doubling of the national income, with this sort of distribution, could more than double the amount spent on manufactured commodities, etc.

8 Indicators of New Activity

Let us now review the factual evidence available to support the contentions made. What indicators are there of increased activity during the period 1750–80? It would be disingenuous to begin by quoting Ashton and others for the belief that 1760 is a good starting date for the Industrial Revolution as such. But the reason for their choice was that, from their intimate knowledge of industrial records, it was clear that the quickening of the pace in so many industries occurred round about that date. One of the striking basic changes is in the rate of construction of new and improved means of communication. The history of the English canal network is too well known to require rehearsal here: suffice it to say that apart from the early Lancashire canals, the main Midland canal network was complete or under construction by 1780 and primarily served the needs of inland trade. The rise of the Potteries, to which we shall refer again later, is the prime example, but the Black

Country and Birmingham also benefited largely from these developments.[1] It is true that the Grand Trunk and the Staffordshire and Worcestershire canals, as well as the various navigations linking Liverpool to Manchester and other parts of the port's hinterland, were used to reach the sea: but this was as much connected with coastwise shipping, and the needs of the fast-growing port towns themselves, as with exports. Above all, the canals carried food for inland consumption and therefore served as a stimulus to demand since they helped to reduce the price of food just as they brought down the price of coal in places like Manchester or Birmingham.

Similarly, there were considerable improvements in road-building during the same period. The passing of the Turnpike Act of 1772 comes both as a culmination of several years of increasing pressure on parliamentary time owing to road legislation and as the start of a new spurt in building. The 1770s saw significant accelerations in road transit times, in the regularity and reliability of coach services,[2] as well as cost reductions, absolute or relative. The effect of this on the inland market is obvious, through the speed-up of mails, the increase in commercial travelling,[3] and the aid to the creation of larger regional markets.[4] At the same time, of course, the demand for roads comes from those who have reason to believe they will benefit.[5] In fact, all the arguments which Professor John adduces for the earlier epoch[6] apply *a fortiori* in our period.

Our next type of evidence comes from the history of enterprises themselves. To take only a few of the famous names in British industrial history whose firms were either founded in the period 1750–80 or first achieved prominence then: Arkwright, Boulton, Bramah, Crawshay,

[1] W. T. Jackman, *The Development of Transportation in Modern England* (2nd ed., 1962), p. 366.

[2] *Ibid.*, pp. 686 ff.

[3] Cf. G. W. Daniels, *The Early English Cotton Industry* (Manchester, 1920), pp. 64–65, for the activities of the Manchester Riders, 'The Society of Travelling Scotchmen of Bridgnorth', etc. (*c.* 1780).

[4] This is not to imply that these regional markets were created by transport. They had existed in the seventeenth century. It is rather a question of scale and regularity, reliability, predictability.

[5] This is best studied with the help of the agricultural surveys of the time of Arthur Young and Sir John Sinclair, from which one can see the sequence of promotion of turnpikes. See R. Gaut, *A History of Worcestershire Agriculture* (Worcester, 1939). The Chances of Bromsgrove (later glassmakers) helped to promote the turnpike which linked the iron centre of Dudley with the main nail-making centres, and so on.

[6] In *Jour. Econ. Hist., loc. cit.* pp. 27–28.

Dundonald, Garbett, Keir, Reynolds, Roebuck, Strutt, Sam. Walker, Wedgwood, and Wilkinson[1] all flourished in this period which is supposed to have seen a slowing down of the pace of expansion. It is the period of the most important improvements in the steam engine, of the widespread adoption of coke smelting in the iron industry, as well as perhaps the most significant in the history of technology in general.

Did these entrepreneurs then produce mainly for export—or even to any large extent? This is an important question, since if there is anything in the 'leading sector' argument, it operates surely at the personal level; that is, the stimulus to investment comes precisely through the best-known entrepreneurs in their time, even if they do not account for the largest part of the total output. Here, quite clearly, we can see the predominance of the home market. None of the men mentioned were active in woollens, the largest export industry of the time. Some of them only produced raw materials or semi-fabricates, and we cannot be certain of the ultimate destination of their output, but people like Boulton, Reynolds (Coalbrookdale), and Keir were definitely only marginal exporters. Wedgwood may serve as a prime example, because the legend of the Wedgwood exports is widespread. It is of long standing and was perhaps perpetuated by Mantoux,[2] who based his assertion on the annual export from Burslem of 550,000 'articles'. But total official export values in the 'glass and pottery group' were only of the order of £20,000 before 1780. Official valuations are much too low when it comes to fine porcelain, but this was not a significant part of total output. The number of 'pieces' involved was, in fact, eight million, so that each item is valued at less than one penny. All this can only have constituted a very small fraction of the business of the Potteries. Wedgwood's share in all this has been well described by Mr. McKendrick,[3] and we need only add here that Wedgwood did indeed count much on the prestige of his exports, such as the dinner services for foreign potentates; but they cannot have made him rich, especially since he was often not paid (or meant to be paid) by his noble clients. Wedgwood made his considerable fortune from very common articles indeed.[4] He

[1] These are taken in alphabetical order from Everett Hagen, *On the Theory of Social Change* (Homewood, Ill., 1962), pp. 305–8. His complete list is worth studying: one should construct a sort of centre of gravity of the working lives of all the characters he lists.

[2] See P. Mantoux, *The Industrial Revolution in the Eighteenth Century* (new ed., 1961), pp. 386 ff.

[3] In *Econ. Hist. Rev.* 2nd ser., XII (1959–60), pp. 408 ff.

[4] Cf. E. Meteyard, *The Life of Josiah Wedgwood* (1865), I, pp. 191 ff.

may have ostentatiously broken a plate which was not up to his declared standards, but he flooded the domestic market with 'seconds'—plates which sold at 2*s*. or 3*s*. a dozen. In any case, his chief export business did not develop until after 1775. We cite this case because it particularly illustrates the necessity of an expanding home market as a basis for exports. First of all, the home market could absorb stocks if exports ceased or diminished. Secondly, it provided the foundation for mass production, so that the cost per unit could be reduced to levels which made it feasible to export articles with a relatively unfavourable weight/value ratio, like pottery. Thirdly, it provided an outlet for the inevitable sub-standard qualities (this applies particularly to pottery, glass, and dyed and printed textiles).

At the same time, as the case of Wedgwood again illustrates, we are dealing here with commodities which typically find a ready sale in the sector to which our diagram has drawn attention: below the luxury class, above the range of mere subsistence. Glass shows a similar record. The statistics of duties paid[1] point to an annual consumption of 113,000 cwt. before 1780. That represents about 11 m. square feet of glass, worth at least £1 per cwt., and a great deal more for better varieties. But until the big upwards move in 1796 only about 10 per cent of the volume of glass produced was exported—and a good deal of this was of the best type which paid for transportation costs. Even in the days of window tax, therefore, we must reckon that 10 m. square feet of glass were sold at home.[2] A great deal of this was third, fourth, and fifth grade, as the glass makers reckoned it (only a tiny proportion of all glass was perfect: a large part of the crown glass made consisted of the 'bull's eye' which became a sought-after commodity only in the twentieth century and was considered waste in the eighteenth, yet glass makers relied on a ready sale to keep the cost of their product low). Even allowing for the building of many royal and noble palaces, we cannot doubt that the greater part of the output must have gone to our new class of consumers: those who did not stuff their windows with rags on one hand, but could not afford plate glass or clear panes for their leaded windows on the other.

[1] From T. S. Ashton, *Economic Fluctuations in England, 1700–1800* (Oxford, 1959), Appendix; Mitchell, *op. cit.*, p. 267, and T. C. Barker, *Pilkington Bros. and the Glass Industry* (1960), pp. 52–53. The statistics are difficult to interpret, but the text given here seems to follow from the figures.

[2] Dr. Marshall has calculated that 20,000 houses per annum, window tax or not, would use up that much glass by themselves, allowing a little for repairs, at £2. 0*s*. 0*d*. per house.

We have not space to pursue this argument in detail, but it can be shown that for most other commodities the story was the same. Later historians were too much under the influence of what happened in the war years to notice the break that had come just before the war. That the Midlands nailing industry was export-orientated was widely believed as the result of Thomas Attwood's impassioned appeals to the Select Committee of the House of Commons investigating the effects of the Orders in Council in 1812.[1] Yet some simple statistical checks show that the number of nailers necessary to produce the quantity exported to the United States before the embargo would not have kept more than a small fraction of those Attwood cited as unemployed working at their forges. If there were, all told, as many as he claimed, it would only mean that a very small minority was working for exports. *A fortiori*, the 1,700 tons per annum which represented the peak in the export business before 1780 would (depending on the number of nails per cwt.) keep only a few hundred domestic nailers busy. On the other hand, there is every indication that the building booms of the period[2] would by themselves account for most of the increasing output of the industries producing domestic hardware and ironmongery. Ashton describes[3] the great boom of the late 1770s; this collapsed after 1778, but the cessation of new orders would mean no immediate reduction in the output of materials required for finishing. Further, although the exports of hardware and nails decreased sharply during the American War of Independence, total production does not seem to have suffered —at least so far as the Midlands records go.

In general, such production figures as we possess for British industry show very little sign of any slackening of the growth of output during the years 1750–80, despite the fickleness of all markets. All statistics suffer from imperfections. Thus, Hulme's well-known series of figures[4] relating to iron production have been examined by Dr. Flinn,[5] and he has shown that the earlier statistics are overweighted in favour of returns from declining areas, as well as being compiled to prove specific points.[6] Unfortunately it is precisely for our period that we lack specific

[1] Cf. my more detailed account of this in *Victoria History of the Counties of England, Warwickshire* VII (City of Birmingham) (Oxford, 1963), pp. 97, 117.

[2] Ashton, *Fluctuations, op. cit.* pp. 98 ff.

[3] *Ibid.*, p. 100.

[4] E. W. Hulme, 'Statistical History of the Iron Trade of England and Wales, 1717–1750', *Trans. Newcomen Soc.*, IX (1928–9), pp. 12–35.

[5] M. W. Flinn, 'The Growth of the English Iron Industry', *Econ. Hist. Rev.*, 2nd ser. (1958–9), pp. 144 ff.

[6] Mitchell and Deane, *op. cit.*, p. 126.

estimates, but there is now little dispute about the fact that the output of pig-iron, which had been round about 20,000–25,000 tons in 1720, reached about three times that quantity before the French wars began to provide fresh stimuli, and the history of enterprises clearly shows that the greater part of that increase falls into the period between 1760 and 1780. Yet the total weight of iron products exported (not necessarily consisting of 100 per cent iron) did not exceed 21,000 tons in any year before the drop of the 1770s,[1] and was less than 10,000 tons in fifteen out of the thirty years.

If one examines the carefully compiled series now available in the *Abstract of British Historical Statistics* by Mitchell and Deane, one can only say that most of them show a more or less marked acceleration of expansion during the period 1750–80, and that where there are details, expansion continues during the export slump in the 1770s.[2] Although it would be unwise to construct a general index of production from such scanty sources, the general trend is unmistakable, and read in conjunction with the export figures, gives us a large increase in *per capita* production and consumption in the period 1750–80. (We return below to the special question of the behaviour of these series in the period 1770–80.)

Yet another way of looking at this question is to pursue the question of innovation.[3] Dr. Hartwell and others have argued cogently for the idea that the rate of innovation is a good measure of the existence of a free and progressing society. We have been warned against taking the patents of invention too seriously as a source in earlier periods,[4] but in the middle and later eighteenth century no important changes in practice occurred which would invalidate the series.[5] On this evidence alone, the number of patents sealed had been about eighty in the period 1740–9, rose to over 100 in the following decade, and to nearly 300 in the decade before 1780. This is quite in accord with what we know

[1] *Ibid.*, p. 144, Schumpeter, *op. cit.*, Tables VII and IX.

[2] Mitchell and Deane, *op. cit.* Coal, p. 110, from Newcastle and Sunderland; coal into London, p. 112; export slump, p. 120; copper, p. 156; printed goods charged with duty, p. 184; cloths from the West Riding, p. 189 (striking increase in view of fall in exports); similarly with linen stamped in Scotland, p. 200; imports of raw silk in relation to silk exports, pp. 205–9, showing a normal relationship of ten times as much raw silk and thrown silk imported as silk goods exported (even allowing for waste this is considerable, especially considering also mixtures); tallow candles, p. 262; paper, p. 263; starch, p. 264; soap, *ibid.*, etc.

[3] Hartwell, *Econ. Hist. Rev.*, XVIII, p. 175.

[4] Mitchell and Deane, *op. cit.*, p. 244; cf. also Deane, *First Ind. Rev.*, p. 128.

[5] *Ibid.*, pp. 268–9.

about industrial enterprises. As we know, many of the most important innovations were not patented when they were introduced. Nevertheless it is inconceivable that the desire to record patents is separable from the idea that the invention may be useful, or indeed from the effort to search for new methods most noticeable in a time of high economic activity.

Lastly, under this heading, we cite some evidence from the Midlands to bear out the general contention.

Our knowledge of the goods retailed in eighteenth-century England is fragmentary, but much light has been thrown on the changing pattern of consumption by the surviving inventories of wills proved in the courts of the time. No systematic analysis of these seems as yet to have been made, but there are some useful series affording scope for comparison. Thus Sedgley, in Staffordshire, a town with a good deal of mining and industry, but also significant farming until the nineteenth century, had a manorial court with limited probate jurisdiction until 1857.[1] Unfortunately wills are rare after 1730, but even by this time there is plenty of evidence that ordinary industrial workers (in this case people working on their own account) did own specimens of the products of neighbouring and even distant industries. The inventories show that few of these people were wealthy, though a single 'gentleman' of Ettingshall near-by had his personal estate proved at just under £400: the average for the eighteenth-century wills is under £100.

But even Richard Wainwright, nailer, in an inventory sworn on 27 August 1739, had the little decencies of life: fire shovel, coal hammer, toasting iron, house bellows, a copper can, wooden furniture, a tun dish, scissors, a warming pan, a brass kettle, two iron pots, one pail, a search (sieve), two old candlesticks, glass bottles and earthenware, linen of all sorts in a chamber, a pair of bedsteads, a 'coverlid', a rug, a blanket, a kneading tub, two barrels, two coffers, a box, some trenchers, pewter, a brass skimmer, a brass basting spoon, an iron flesh fork, a tin calender, and so on. A poor man indeed, worth less than £10, and indeed none who figure in the Sedgley annals had less than Richard. By 1782, John Atwood of Coseley, proved at just over £20, had a clock and case, four brass pans, table linen, and books valued at £1 5*s*. A victualler had possessions valued at £284, yet he is well below the ranks of the gentry, the big tenant farmers, or the early industrialists. Even if one does not have detailed knowledge of the exact description of his

[1] J. S. Roper, *Sedgley Probate Inventories*, privately published (Dudley, 1965), p. i. Wainwright's inventory, No. 130.

goods, one recognizes amongst them immediately, for instance, the common products of Coalbrookdale, a few miles away, which can still be seen there in the museum today. (In these particular lists the description 'Dale' occurs several times.) Predominantly, these are cast- and wrought-iron goods, brass, copper, tin products as well as wooden utensils with metal facings, handles, and so on. Increasingly, there are textiles, in the shape of blankets, linens, pillows in their ticking, rugs, curtains, and cloths.[1] There is evidence of much glass (in the case of the victualler, this has replaced pewter for drinking, though there are still plenty of pewter plates as well as china ones). Inspection of many other wills of the dioceses of Lichfield and Worcester during this century confirms the general impression that the number of people possessing goods they must have bought in their lifetime, rather than inherited, grows steadily. From the point of view of industrial production this is crucial. Mere inheritance, the passing on of wearing apparel and daily linen, of utensils and furniture, could mean, at the margin, little demand. But the new products are striking. Moreover, the total of households is clearly on the increase, especially as the number of people reaching adult age must have been a record by the last third of the eighteenth century, judging by the survivors of 1821.

Perusal of the catalogues of Birmingham firms of the second half of the eighteenth century clearly demonstrates their home-market orientation.[2] Although there were also many exporters, and a number of foreign firms had settled in the city as merchants, the surviving ledgers and correspondence demonstrate the small part played by direct overseas customers or the factor and wholesale merchants in the ports. One cannot be certain of the destination in each case, but one of the chief characteristics of the trade is in the mass of articles of relatively low value which would hardly pay for transporting to distant markets, except for the rather specialized goods used in the African trades (like guns which exploded when fired, mirrors, and beads).

In Birmingham, too, we can observe in detail the buoyancy of industrial wages, which constitute at once a proof of the increasing volume of trade even when exports are low, and provide another link in the cycle of growing supply and demand. Many commentators go on to show that these high industrial earnings also forced up agricul-

[1] Cf. Daniels, *loc. cit.*, and A. P. Wadsworth and J. de L. Mann, *The Cotton Trade and Industrial Lancashire* (Manchester 1931), pp. 238 ff., for systems of distribution.

[2] See Eversley, in *V.C.H. Warks.*, VII, pp. 91, 92, 101, 112, and *passim*.

tural wages in the vicinity[1] and thus once again added to purchasing power. We have the evidence of much contemporary unfavourable, or in some cases approving, comment to the effect that even the most ordinary people in the streets sported the Birmingham-made fashion—the buckle, and then the ornamental button, the snuff-box and the chain.

We sum up this section by repeating that what we have tried to show need only apply to a fraction of the total population. If the new standards of consumption were widespread in Birmingham, they may have been unknown amongst South Wales iron-workers or Wiltshire farm labourers. The existence of generations of consumers does not mean that there were not years when few could buy anything except basic necessities. But it seems clear that for most of the time, and for much of the country, we have no difficulty in establishing the existence of a broadly based home market.

Finally we must once again remind ourselves that the mechanism described here depends on a most delicate balance of factors. Though, on our evidence, this is how the extra purchasing power was created, and how the demands were met, there is no law of nature that this must be so, or that the sequence of events forms part of an equilibrium growth situation. At any point of time the process could have been halted. Additional labour (from spare rural capacity rather than a high rate of natural increase) might not have been forthcoming at the right time and in the right place. Innovations might not have been made or adopted to ensure that output could be raised at steady or lower prices. Increments in output of a speculative kind might not have found markets at home (thanks to a reasonable distribution of incomes and a better transport system), or abroad (thanks to expanding trading horizons). In each case the impetus would have been lost.

Nor can we list here all the 'environmental' conditions for economic growth—by now we take these for granted. An equable climate, political stability, the liberty of the individual, the 'protestant ethic', the absence of heavy expenditure on defence or royal extravagances—all the favourite pre-conditions for growth are present. This background makes the process possible, though it does not ensure it. If the social and political climate of Britain in 1750 provides the necessary conditions, home demand is the main element of the sufficient cause.

[1] Witt Bowden, *op. cit.*, p. 255; Eden, *op. cit.*, II p. 739, from Arthur Young.

9 Food Prices and Home Consumption

We need now to investigate rather more deeply the causes of this buoyancy of the home market. For the earlier period 1720–50 John ascribed it largely to the relatively favourable movements in agricultural prices.[1] To repeat the argument: it is now generally believed that the low prices of corn in particular between 1730–50 had beneficial effects on the economy as a whole: it is now doubted whether the 'depression', which had previously been seen through the eyes of those interests which suffered from low produce prices, really did exist, and that whatever purchasing power may have been lost because some farmers or landlords found food production less profitable, was more than made up by the new margin of expenditure available to the growing non-agrarian sector.[2] With all this one cannot quarrel. But then, according to this school of thought, prices rose after 1750: this benefited the farming community, but depressed the living standards of the industrial population, and this is the cause of the slump. (It is logical, of course, that if one does not think that the loss of purchasing power of farmers was significant before 1750, one allows that the return to their prosperity later had no very great positive effects, especially as they were relatively less numerous.)[3]

The exact relationship between food prices and economic growth has been systematically studied by Mr. Gould.[4] Let us here briefly summarize the conditions which need to be fulfilled for low grain prices to have positive effects on home demand (leaving out the export effects, since we have already discounted the importance of the foreign market):

(*a*) farm incomes must be a relatively unimportant part of the total national income, or a reduction in profits must leave the demand of the agrarian sector unaffected (e.g. because most of the incomes involved were those of owners or large tenants or corn merchants who even in bad years would not be at subsistence level);

(*b*) the low food prices on the wholesale markets must be passed on to the ultimate consumer, at least in large measure;

[1] John, *Jour. Econ. Hist.*, 1965, p. 32, and *passim*.

[2] Chambers and Mingay, *op. cit.*, ch. I.

[3] H. J. Habakkuk, 'The Eighteenth Century', *Econ. Hist. Rev.*, 2nd ser., VIII (1955–6), pp. 437–8.

[4] Gould, *loc. cit.*, pp. 329 ff.

(*c*) wages must be inelastic in a downward direction: i.e. not firmly tied to the prevailing price of bread;

(*d*) any savings on minimum subsistence must be expended on other commodities. (It may, of course, be that if grain is cheap there is merely an increased demand for other foodstuffs, so that their price rises and there is still no margin for other purchases.)[1]

First, we need to look at grain prices to see what, in fact, happened to them in the thirty years after 1750. In the most recent comparative series, all admittedly from the south of England,[2] we certainly see no drastic increase during this period. Wheat exceeded 40*s*. a quarter only once in the period 1730–50, and did so regularly after that date: but it only surpassed 50*s*., in at least two of the three markets, on four occasions in the next thirty years. To compare an average level of under 45*s*. with the prices of the end of the century is not justified. Barley and oats follow much the same line and wool prices are equally steady.[3] The Gilboy prices indices,[4] whilst clearly demonstrating the exceptional *baisse* of the thirties and forties, also show that the consumers' price index on the basis of 1700 = 100 did not rise above 110 before 1780. But cereals alone are clearly rising faster than other commodities, thus rather discounting the theory that lower cereal prices drive up the price of everything else.

The question of real wages has received much attention from E. E. Gilboy and more recently from E. H. Phelps-Brown and Sheila Hopkins.[5] Unfortunately most of the continuous series published are based on wages paid by institutional employers, and refer to day labourers. But an increasingly high proportion of workers (especially in textiles, coal, and the metal industries) were paid by piece rates and their earnings are hard to ascertain.[6] Moreover, even time payments were much distorted, in the eighteenth century, by wages in kind and allowances on one hand, and the truck system on the other. In any case, the

[1] Deane and Cole, *op. cit.*, pp. 62–65: a striking demonstration of the probability of steady consumption of grain per head during the eighteenth century.

[2] Mitchell and Deane, *op. cit.*, p. 486.

[3] *Ibid.*, p. 496.

[4] *Ibid.*, pp. 460 ff.

[5] E. H. Phelps-Brown and Sheila Hopkins, 'Seven Centuries of the Prices of Consumables, compared with Builders' Wage Rates', *Economica*, N.S., XXIII (1956), pp. 296 ff.

[6] Professor Ashton has kindly warned me against time rates and has given me abundant material to support his point. For some instances, see his *Economic History of England*, pp. 60–61. I still think there is *some* validity in the long series of time rates.

evidence is contradictory, so that, for instance, the Gilboy index for Lancashire shows considerable gains in real wages,[1] rising to 185 (1700 = 100) in 1769, whereas for London the record is much worse. Phelps-Brown's records based on the building trades sometimes show falls when the Gilboy index rises.[2]

For what the fragmentary evidence is worth, then, it tends to show that wages were slightly more inelastic downwards than upwards: that is to say, incomes lagged behind prices at all times, but they were not so much depressed by low prices as they were raised by high ones, and the higher prices rose towards the end of the century, the faster incomes rose, until they were tied to food by the Speenhamland system.[3] Arthur Young was perhaps the only contemporary observer who really did look at prices and incomes in various areas in detail, and it was his opinion that real wages had risen. In his average for twenty years he thought that the price of labour had risen 25 per cent, and the price of wheat and malt had fallen.[4] But these averages are misleading, as are all national indices, and as we have stated, there is no need for all real wages to have risen. We can be certain at least that some real wages rose, and that this rise was greatest in areas like Lancashire where the industrial opportunities were greatest. We must not forget that so much of what industry produces is sold back to the local workers themselves.[5] A record of wheat prices paid in Birmingham (and directly relevant to the standard of living of the urban industrial consumer, unlike some of the series quoted earlier), compiled by Dr. J. M. Martin, shows the same facts equally strikingly.[6] We have, of course, by 1780 a national grain market, with nothing like the differences in local prices which were sometimes observed before the days of the canals.[7] Prices of round about 3*s*. per bushel had been common before 1750. There were

[1] Mitchell and Deane, *op. cit.*, pp. 460 ff.

[2] The Phelps-Brown 'highs' are 1736, the late 40's to 1756, 1759–63, and 1776–82. Gilboy's Lancashire real wages are high in 1732–5, 1743–4, 1761, 1768–70, and 1776–9.

[3] See below, p. 246.

[4] Young, *op. cit.*, IV, pp. 344 ff.

[5] We remind ourselves that wages were most buoyant precisely in the growth industries and areas which had to, and did, attract migrants. Their local purchasing power is important, especially since they may be assumed to have had low dependency ratios.

[6] Unpublished private communication, but see also Dr. Martin's Ph.D. thesis, *Warwickshire and the Parliamentary Enclosure Movement* (Birmingham, 1965), which, like his earlier M.Com. thesis, throws a lot of light on the structure of the eighteenth-century grain market. See Table III.

[7] Chambers and Mingay, *op. cit.*, p. 110.

temporary shortages in 1756, 1757, and 1758, but in 1761 the price was again 3*s.* 2*d.* From then until 1781 there was a long-term rise, and something between 5*s.* and 6*s.* was the median, taking summer and winter prices together. This, if one likes, is a doubling of the level of the 1730s—but it certainly is no great increase, if any, on the prices of the early part of the century. Seen in relation to what we know of Birmingham artisans' wages, even the figure of 6*s.* probably means no reduction in purchasing power compared with the years before 1750 (which were, industrially, nothing like as booming as the later period). At any rate, Birmingham records[1] carry few complaints of corn prices, except in really acute crises like the winter of 1772. We may readily believe that during these stringent months purchases of non-necessities declined, but that is not at issue.[2]

Given, then, the fact that produce prices remained reasonable, to what may this be attributed? Generally speaking, two factors are at work: good harvests and increased efficiency. The good harvests (due mainly to good weather) are the Cleopatra's nose of the whole situation: the one completely extraneous and fortuitous factor unconnected

Table 3. Grain Prices—High and Low 1752–80 (per quarter)[3]

		Ashton/Windsor			Martin/Birmingham	
		s.	*d.*		*s.*	*d.*
	1752	42	0		36	0
High	1757	52	0		56	0
Low	1761	30	6		28	0
High	1767	58	0		64	0
Low	1769	44	6	(1770)	44	0
High	1772	52	0		57	6
Low	1779	40	0		42	8

For sources see text and notes.

[1] J. A. Langford, *A Century of Birmingham Life* (Birmingham, 1868), I, which is a conscientious summary of local news from the most important Birmingham paper.

[2] Chambers and Mingay, *op. cit.*, pp. 111–12.

[3] All these prices must be handled with great care, as there are a great many local, seasonal, and qualitative variations. The compilers of the series quoted here have been insistent on their restricted value.

with economic development, a true gift of God. The 1730s remain the most striking manifestation of this: but our later period is still one of remarkably favourable climatic conditions.[1]

The keeping of systematic weather records does not begin till much later, and again there are regional variations, but the available summaries[2] only show up the expected pattern. In 1756 there was a wet spring, and the 'wettest summer in memory of man'.[3] But this bad year was followed by five very good ones. 1762 was bad again, the next three years were good. 1766 was very wet and caused all sorts of losses apart from a bad grain harvest, 1767 was good. Then follows the one period which has given the whole second half of the century a bad name, 1768–73, but the rest of the seventies was good. These widely authenticated climatic facts are faithfully reflected in the price series, as they are in the records of declining exports of corn and subsequent imports. By 1767 rather large quantities of foreign food could be brought in at fairly short notice[4]—in that year nearly half a million quarters were brought in, a record until 1783. Since as late as 1763 exports were little less than that, we can say that in this period the available 'reserve range' of wheat alone was of the order of a million quarters. This means that good harvests are useful in keeping prices low, but they are nothing like as essential as they are in a situation of total dependence on home-grown food, and in the absence of export markets which provide an outlet for the surplus the farmer must be encouraged to grow in order to supply the home market adequately even in years of bad harvests.

Summing up on this score, we can say that although there were nine or ten wet years out of thirty, we have to characterize the supply situation as perfectly adequate. (It may be remarked that in the period 1730–50 which has figured so largely in these discussions there were also bad seasons. 1736[5] is reported on by Baker in exactly the same terms as many of the years in the early seventies, and the disaster of 1739–40 is too well known to require comment here—in its consequences on

[1] For this topic, see E. L. Jones, *Seasons and Prices* (1964).

[2] Despite Dr. Jones's work, based on T. H. Baker, *A Record of Seasons and Prices*, there is still a good deal of scope for comparative work, on the lines done by M. Leroy Ladurie in France. See also C. E. Brooks, 'The Climate of the First Half of the Eighteenth Century', *Q. Journ. of the Roy. Metereolog. Soc.*, 56 (1930), pp. 389–402, and J. Glasspoole, 'Two Centuries of Rain', *The Metereol. Magazine*, vol. 63, 1928. See also Deane, *First Ind. Rev.*, p. 228.

[3] Jones, *op. cit.*, p. 141. In general, bad weather is much better documented than good seasons.

[4] Mitchell and Deane, *op. cit.*, p. 103, and the series by Combrune, *op. cit.*

[5] Jones, *op. cit.*, p. 138.

mortality, economic activity, and foreign trade this was far worse than the difficulties of the sixties and seventies.)

As regards productivity, it cannot seriously be argued that the rate of improvement of agricultural organization and technique was slower after 1750 than before. Whether one goes by the period of activity of the pioneers (only Tull and Townsend function before 1750), by the enclosure movement,[1] which only reached its first peak in the years 1764–77, or by the direct evidence of sales from the great estates and the movement of produce by canal, it is clear that the period under discussion witnessed the beginning of the agricultural revolution which laid the foundations for the industrial one. This is as we would expect: for it is precisely the rising prices which provide the incentive, the visible growth of the market. Low prices and abundant harvests, before 1750, could not move the landlords, even with good corn bounties, in the same way as the later period when good harvests were no longer a disaster, because they clearly mean larger quantities sold at prices which were an improvement on the 1730s, and when the growth of the urban market was there for all to see.[2] (It must be remembered that before 1760 very few people thought that the population was growing at all; there were no canals, and the rate of growth even of the London market was not visible to the naked eye.)[3]

After 1750 the London market became increasingly important and, in fact, distorted the national price structure completely.[4] Hence it is not surprising that different authors vary in their assessment of the state of the market. A comparison of the Windsor and Birmingham series in Table 3 shows a wider range of oscillations in the provincial market, far from water transport, until the canal connexion was made. Hence food prices must always be scrutinized in their context before conclusions are drawn, either for the consumer or the producer.

The fact that these relatively low prices and steady supplies constituted

[1] Chambers and Mingay, *op. cit.*, p. 83.

[2] Cf. G. E. Mingay, *English Landed Society in the Eighteenth Century* (1963), ch. VIII.

[3] The Yield Ratios collected by Prof. B. H. Slicher van Bath (*A.A.G. Bijdragen*, X (1963), pp. 53 ff.), especially when compared with other series, give some fascinating but unfortunately fragmentary clues to productivity. Much remains to be done in this respect. The English series are thinnest just when we would find them most interesting, since they effectively begin only with the statistically-minded Arthur Young.

[4] Cf. F. A. Wendeborn, *A View of England towards the Close of the Eighteenth Century* (1791), I, p. 268: many things were cheaper in London than in their county of origin.

a powerful stimulus to demand did not escape contemporaries and, indeed, entered into the accounts of the classical and neo-classical economists.[1] Malthus himself in one of the characteristic passages where he reflects upon the fact that his theoretical relationships between population and subsistence cannot be tested in prudent England, attributes the rise in the standard of living of his compatriots to low food prices.[2] Although the reference there is to the period 1720–50, it must be noted that he merely points to the fact that during those years the labouring people acquired their new tastes, and the text implies that these remained until the end of the century. Only after 1790[3] is the erosion of the customary levels through price rises such that population increases at a faster rate, prudence having been diminished. The same sequence underlies the views of Alfred Marshall,[4] basing himself partly on the evidence of Eden.[5] But nowhere is there better evidence for the firm establishment of the new customary standards than in the Poor Law administration at the time of Speenhamland, when few quarrelled with the view that wheaten bread was the Englishman's birthright and that his income must be adjusted to safeguard his supplies. Had there really been a price rise of a serious kind before the wars, this standard, established for a generation and a half, would have been abandoned again.[6] In itself, of course, the survival of the wheat standard at least in the south is not evidence that there was also purchasing power for other commodities: but what we are here stressing is that Smith, Malthus, Eden, and the Poor Law authorities saw good food as the prerogative even of the lowest income groups for whom we would not claim that they were purchasers of industrial commodities on a significant scale. It is merely implicit in any more or less continuous income function that if the poorest had this, the next layers had more.[7]

[1] See Adam Smith, as quoted in Hartwell, *op. cit.*, p. 179.

[2] T. R. Malthus, *Principles of Political Economy* (1836), pp. 228–9.

[3] Malthus, *Essay on Population* (Everyman ed., 1914), II, p. 139.

[4] Alfred Marshall, *Principles of Economics* (8th ed., 1930), p. 189.

[5] Eden, *State of the Poor*, I, pp. 560–4.

[6] The foreigners always marvelled why it was that the English dog would not eat offal like his European counterparts. In the last analysis there are deeper sociological, rather than economic, explanations of why the British were able to insist on a superior diet. The connexion between political and social structure and its effect on the economy is only just beginning to be traced. Cf. Peter Laslett, *The World We have Lost* (1966), pp. 107 ff. 'Did the Peasants really Starve?' See also Deane, *First Ind. Rev.*, p. 10.

[7] But Eden did not think wheat was necessary. This would be true even where the bottom strata did not have wheat. Professor G. P. Jones has pointed out to me that at Swarthmoor Hall they also ate oaten bread. Eden continually stresses

10 The Test of 1770–80

The significant test for the validity of the thesis put forward here must come from an examination of conditions during the decade 1770–80. This, on the basis of any foreign trade or shipping figures, comes out as a bad time for British exports. This slump is largely, but not wholly, attributable to the interruption of trade with the North American (Continental) colonies, which was worth £1.5 m. less in 1776–80 than it had been ten years earlier. Other trade increased in part, or declined, largely owing to repercussions from the American war ('triangular' trade routes suffered, but Canada profited).

Recovery did not begin until 1782. In terms of growth rates the seventies must have been disastrous for the export-orientated industries, whatever they were. Volumes, where we know them, declined even faster than values, since prices were also falling.[1] Now, if it had really been true that the price rises due to the inadequacies of agricultural supplies to the domestic market had any serious repercussions for the consumer, then the total value of production should have fallen at least during those ten years, investment should have lagged, and cries of distress should have been heard. But imports did not decline to anything like the same extent (except for colonial goods, many of them for re-export to America, which naturally slumped). In the case of textile raw materials, there is an almost continuous rise of imports in the 1770's (hemp, silk, cotton, wool).[2] London used record quantities of sea

---

the regional differences (*ibid.*, I, 496–501 and elsewhere), but one should read into them no more than is warranted, i.e. that to the upper classes and most southerners (like Samuel Johnson) to eat anything except wheat was inferior, and that whoever ate barley, oats or rye must be poor. The corollary is, especially for Eden of Co. Durham, that the poor should help themselves by not asking for wheat. It was not, after all, as if the principles of Speenhamland were known only in the southern counties: only that the catastrophe of the wartime prices was less noticeable in the more adaptable northern areas.

[1] There were a few exceptions. Cf. Schumpeter, *op. cit.*, Table VIII, p. 24.

[2] The 1777 level of raw wool imports was a record until 1785. When linen exports were at their lowest level for thirty years (from England, in 1781), Scottish linen stamped for sale rose to above 15 million yards for the first time (Mitchell and Deane, *op. cit.*, pp. 200–1). In 1775 the lowest amount of silk manufactures was exported since 1745, and the largest amount of raw silk ever was imported the following year. Cotton re-exports existed, but were negligible.

coal.[1] Sales of copper ore in Cornwall and Devon[2] reached record levels in the mid-1770s—a particularly important commodity in judging the level of activity in the new Midlands industries.

For many other domestic consumption figures, the same general picture emerges. Beer consumption,[3] if we can judge by the quantity of strong beer charged with duty, went up in total and probably remained level *per capita*: malt remained steady, but hops increased considerably.[4] Spirit gallonage went up by over 10 per cent, though, of course, it did not reach pre-1750 levels.[5] Candles, paper, starch, and soap—all these common consumer goods which come to our knowledge through the duties paid on them climb steadily, in total and *per capita*, throughout the 1770s. Lastly, the acceleration on the rate of applications for patents is noticeable in the 1770s.

In industrial history this decade is also pre-eminent for the rate of foundation of new enterprises. It was a period of great activity at Coalbrookdale,[6] culminating in the building of the Iron Bridge which epitomizes innovation, willingness to invest, and the response to an active inland market. The Grand Trunk Canal was built. In 1771 Arkwright settled at Cromford, and by 1779 his mills there had reached the state that made them an object of pilgrimage. In fact, the great majority of his factories in all parts of the Midlands and the North-West originated in this period. The first Watt steam engines were erected in Cornwall at the end of the decade when the local mine-owners were unable to fulfil demands for copper ore from existing levels. All these selections are to some extent arbitrary, but they serve to reinforce the general point.

If export industries had produced 'leading sector' effects before 1770, industrialization should have been slowed down. Instead of this there was visible acceleration just when, according to Professor John's thesis, the home market should have become much more difficult through rising prices and faster population growth. Since the evidence for reduced exports in the period 1770–9 is incontrovertible, and since pro-

[1] Mitchell and Deane, *op. cit.*, p. 112.

[2] *Ibid.*, p. 157.

[3] *Ibid.*, p. 251.

[4] *Ibid.*, pp. 247–251. The whole of the brewing statistics are best explained in P. Mathias, *The Brewing Industry in England* (Cambridge, 1959), esp. Tables 36 and 41. Beer was definitely getting stronger, hence the malt/hop discrepancy. This in itself should be a sign of affluence!

[5] Mitchell and Deane, *op. cit.*, pp. 255–6.

[6] A. Raistrick, *Dynasty of Ironfounders* (1953), pp. 91 ff., for a good picture of the period.

duction and innovation also flourished, we may take this as additional proof of our main thesis, that the purchasing power of broad sections of the home market was still increasing at a beneficial rate.

11 The Rate of Population Growth

We have been at pains to stress the rising or at least stable purchasing power of the home market in the period 1750–80. The question naturally arises: what was the rate of population growth, and the distribution of the increase, that permitted this state of affairs? And why did the 'Malthusian' effects not make themselves felt at this time? In other words, why did England not produce a Malthus twenty years earlier?

The short theoretical answer is that population growth at this time was 'right'—neither so small as to cause shortages of labour, or of demand for goods and services, nor so large as to reduce real wages, create labour surpluses and destroy the basis of demand. Such an answer is not incompatible with the existence of pockets of excessive or insufficent growth, with distress in some areas and great prosperity in others. Without the benefit of the results of recalculations of British population figures in the eighteenth century now being undertaken, we may base ourselves on the tables published by Deane, Cole and Mitchell as affording at any rate a consistent outline. The Cambridge calculations,[1] based essentially still on Rickman but corrected in the light of later work such as Brownlee's, give a rate of natural increase of only about one per thousand for the period 1701–50. This is, in fact, composed of a number of short-term rates: growth early in the century, stagnation or even recession 1720–30, and then accelerated growth to 1750. For the period 1750–80 with which we are here concerned the growth rate is given as 6.8 per thousand, which is very similar to that which obtained since 1730. Only after 1780 do growth rates rise to above 1 per cent to a maximum of over $1\frac{1}{2}$ per cent in the early nineteenth century.

Mr. Colin Clark[2] has furnished us with an excellent model of eighteenth- and early nineteenth-century population growth based on an observed rate of natural increase just quoted, and consistent with such vital rates as we do possess. He has based himself (see Table 4) on

[1] Deane and Cole, *op. cit.*, p. 115.
[2] *Population Growth and Land use,* London 1967, p. 56.

Table 4. Demographic Model of Eighteenth-Century Population Growth (after Colin Clark)

Age group	Female survivors of 1,000 total births based on Swedish mortality of		Expected[b] percentage married	Marital[c] age-specific fertility	Age-specific fertility all women[d]	Offspring at survival rates of	
	1757–63[a]	1816–40	not widowed			1757–63	1816–40
15–19	1,448	1,731	2	.43	.01	14	17
20–24	1,391	1,683	30	.43	.13	181	219
25–29	1,338	1,627	60	.41	.25	334	406
30–34	1,247	1,558	70	.37	.26	324	405
35–39	1,202	1,487	75	.30	.22	265	327
40–44	1,130	1,400	70	.15	.10	113	140
45–49	1,049	1,308	65	.02	.01	10	13
Total fertility					4.90		
Gross reproduction rate					2.38		
Net reproduction rate						1.241	1.527
Average rate of population growth % per year						0.7	1.4

[a] Only general survival rate given: ratios of female to general rate assumed the same as in 1816–40.
[b] Based on 1851 data for England and Wales, but raised slightly because of nineteenth-century rates of male emigration having probably been higher than eighteenth century.
[c] Henry's general estimate for eighteenth- and nineteenth-century Europe (15–19 added).
[d] Product of two previous columns.

the Swedish life tables of the two periods in order to give some precision to his calculations, but certainly his assumptions fit the British pattern. It will be seen that he assumes only a third of the women marrying before the age of 25, total marital fertility of just under five children per married woman, and a rate of infant and child mortality reducing a gross reproduction rate of 2.38 to a net 1.241 in the middle of the eighteenth century and 1.527 early in the nineteenth. Whilst we would not assert at this stage that these rates are correct in detail, it is useful to adduce them to show that there is a satisfactory growth model which fits in with the sort of hypotheses here advanced.

These figures are significant. The 1750–80 rate is the nearest of all those given to that obtaining in many west European countries today. It is well below current Dutch or North American rates, which approximate more to those ruling in England round 1800. But the significant feature is that a growth rate of under 1 per cent, slowly increasing, is much more favourable to economic growth than that of the euphemistically-termed developing nations of Asia and Latin America, where 2, 3, and even 4 per cent per annum is still common. A rate of 0.7 per cent is manageable in an economy where capital resources are not yet easily mobilized, communications still in need of much improvement, and agricultural output not yet rising at the sort of speed which mechanization made possible much later. On the other hand, the total increase of one and a half million people between 1750 and 1780 was probably as high as that observed in any previous period of a hundred years, and it provided a useful labour force where it was needed. If the whole country increased by under 25 per cent in thirty years, Lancashire increased by a third, the West Riding by half, Warwickshire by just under 40 per cent.[1] But Devonshire stagnated. There is no question that the increase was absorbed: according to Miss Deane's calculations, the industrial counties gained more people by migration than by natural increase. Poverty may have been a problem, unemployment as such was unknown.

Given the differential in agricultural and industrial wages previously mentioned, which in the case of Lancashire and Warwickshire certainly meant an increase in real wages, we can say that the shift of population into the industrializing and urban counties meant an increase in consumption per head. In any case, it must be stressed again, all migration from country to town means a stimulus to industrial production, for however much the old self-sufficient community may have changed

[1] Deane and Cole, *op. cit.*, p. 103.

even before 1750 to a system of craft specialization, it was only with the concentration in towns that the possibility of mass production and distribution became feasible, and only with canals that prices of supplies to distant markets bore some relation to the cost of production.[1]

What, then, were the factors which kept population growth to such a convenient level? There are basically two types of causes involved in every such growth rate: involuntary and voluntary. The involuntary factor in the holding down of the population increase is mainly the failure of mortality to fall any further. As we now know, the main reasons for such reduction of mortality as was experienced in the eighteenth century was not the increase in medical or surgical skill, but the non-recurrence of periodic epidemics on the scale known before 1740, which sometimes wiped out the gains of a previous generation. This has been explained in detail elsewhere,[2] and we do not here enter into discussions of the relative merits of smallpox inoculation and environmental factors. Infant mortality, in particular, was probably not reduced at all, or by only very little. Adult mortality from the common endemic diseases remained at a high level, and did not improve until the nineteenth century.[3] But because after 1740 there were no great crises of mortality (defined as periods when deaths exceeded births over any considerable area, i.e. excluding London or Liverpool, where this may have happened at intervals until much later), marital unions had a better chance of continuing unbroken. This in turn had an effect on fertility. We do not yet know whether there was also an increase in age-specific fertility (i.e. the number of children born per thousand women of any given age group), but it is likely that there was, because the proportion of first marriages to all marriages increased in a period of smaller mortality in the middle age groups. This would be so even if there were no reduction in still births, or in the premature involuntary ending of pregnancies through infection and accident. We can only note at this stage that mortality declined, but not as fast as it did later and that therefore the rate of increase in population growth is kept down on this account alone.

Secondly, there are the voluntary causes. What Malthus and others

[1] See above p. 217.

[2] Cf. Eversley, 'Population, Economy and Society', in Glass and Eversley, *op. cit.*, pp. 53 ff., and P. E. Razzell, 'Population Change in Eighteenth Century England. A Re-Interpretation', *Econ. Hist. Rev.*, 2nd ser., XVIII (1965), pp. 312 ff.

[3] T. McKeown and R. G. Record, 'Reasons for the Decline of Mortality in England and Wales during the Nineteenth Century', *Population Studies*, XVI (1962), p. 94.

tried to convey in the passages previously mentioned was that prudence was being exercised in England precisely on account of the standard of living already attained. Malthus meant by this late marriage, and certainly there is no sign (contrary to previous assumptions) that there was a fall in the age at marriage in the eighteenth century. When the Registrar-General first reported, the age at first marriage was where it was everywhere in the 'western' system.[1] Malthus only told us what happened in the most improvident strata of society, and systematized his fears about the future, no doubt influenced by the effects of wartime rises and the spread of the Speenhamland system. His picture is not to be taken as representative of England in the second half of the eighteenth century.

Late marriage has important economic connotations quite apart from its effect on the rate of population growth. Firstly, it helps to raise savings, especially where the middle classes are concerned. Provided they are not hoarded, savings do not undermine the market, but divert expenditure to capital goods and thus help to stimulate some types of industry (e.g. shipbuilding) as well as laying the foundations for future mass production. Secondly, late marriage was designed to, and probably did, ensure a standard of living far above the breadline: in other words, households thus founded were more likely to be significant consumers than those begun too early. Whilst there is no doubt about the limitation of fertility by late marriage, we have no evidence as to whether there was widespread avoidance of marriage. By the time there are accurate data on the number of persons dying unmarried, it certainly does not look as if celibacy was an important factor. The possibility cannot be ruled out in some areas, especially those from which there was sex-biased migration.

We are on still more difficult ground when it comes to the voluntary limitation of births within marriage. We can no longer doubt that there was a possibility of this, after Dr. Wrigley's demonstration of the course of fertility at Colyton.[2] We already know that the French began to restrict births even before the Revolution. There is increasingly important material relating to abortions and the use of prolonged lactation to avoid conception. Such data as we possess up till now certainly does not suggest that birth intervals between live births much in excess of

[1] J. Hajnal, 'European Marriage Patterns in Perspective', in Glass and Eversley, *op. cit.*, pp. 101 ff.

[2] E. A. Wrigley, 'Family Limitation in Pre-Industrial England', *Econ. Hist. Rev.*, 2nd ser., XIX (1966), pp. 82 ff.

twenty-seven months were normal, especially where the mother was under 30 years of age, but they do occur, and whilst many of these cases no doubt result from spontaneous abortion or simple failure to conceive, it is possible that they indicate deliberate limitation—whether by abstention, *coitus interruptus*, the use of simple appliances, or deliberate abortion.

At any rate, the crude birth rates which have been collected in English case studies so far[1] are nowhere near those which would have ruled if all marriages which were contracted had lasted for an average of fifteen years (which on present mortality evidence they should have done) and had produced the biological maximum compatible with a year's breast feeding, i.e. a child roughly every two years, or seven per marriage. The highest marital fertility rates we have encountered for this period are only just on five children per marriage (there were few sterile couples). More significantly, many calculations have produced averages of only just over three children per marriage.

Therefore it cannot be ruled out that there was some limitation. Now, it will be recalled that we are dealing in this essay with only one section of the total population: the artisans, the tradesmen, the more substantial farmers, the engineers and clerks. They are the new middle-income group, they are also the rising part of the population. They are precisely those of whom nineteenth-century writers claimed that they were most prone to marry late and carefully and later to limit their families.[2] They were the nascent *bourgeoisie*. We therefore do not have to postulate that either the common labourers or the wealthy sections limited their families in one way or another. If we assume, as we do here, that between 20 and 30 per cent of the population were affected by the new movements, and if that section limited its families to the extent of one child less than the unrestricted average would have been, this would alone account for a 5 per cent reduction of total fertility, or a birth rate about two per thousand less than it might have been. This is highly probable and consistent with all the facts we have observed.

To sum up this part of our argument. We have said that population grew at a natural rate of seven per thousand. Death rates were down on previous totals to the extent that epidemics failed to occur, but still

[1] For a bibliography of such studies, cf. Lynda Ovenall in ed. Wrigley, *An Introduction to English Historical Demography* (1966), pp. 249 ff., e.g. Chambers for Nottingham, Sogner for Shropshire, Eversley for Worcestershire. There are others in preparation showing the same pattern (Martin for Warwickshire in his Ph.D. thesis, *op. cit.*).

[2] Eversley, *Social Theories of Fertility*, ch. IV, sections 2 and 3.

remained well above the level of the early nineteenth century. The birth rate fell in later years, as population grew, but fertility had not yet reached its maximum by any means. Statistically, it seems probable that the difference between the natural increase rate of 1750–80 and that prevailing two generations later is accounted for in equal measure by the falls in mortality and the rise in fertility still to come.[1]

If growth rates had been twice as high as they were (i.e. had the population grown by 3 m. instead of 1½ m.), and especially if that additional growth had been due to higher fertility and lower infant mortality, it is highly likely that the standard of living of the population would have fallen. More young children mean higher dependency ratios, to use a term of modern growth analysis: more mouths to feed and fewer hands to feed them, to use Malthusian language. Larger families are, even today, poor consumers except for basic necessities. The small family (especially if the mother works) buys more luxuries, remains in better health, saves more. There can be little doubt that the equilibrium of economic growth in this period is connected with a highly convenient population growth rate. Could there have been a causal connexion? The answer is probably positive. After 1740, for the first time, more people live in a non-catastrophic world. Food, shelter, and clothing became a certain expectation. Calamitous deaths become rarer. Opportunities for upward movement become more frequent. Under these circumstances, prudence about marriage, and restriction of families, go with saving, geographical and social mobility, and reduced mortality. Reduced mortality results from better environmental conditions, the more regular supplies of necessities, and the incipient improvements in drainage, water supplies, and the rest of the familiar though overstressed catalogue of wonders of the Industrial Revolution.[2]

12 The General Model and Conclusions

We have now set out at length the elements in the socio-economic situation which lead, in our view, to the conclusion that the years from 1750–1780 form an essential part of the preparation for full-scale

[1] This cannot be wholly gauged by looking at the crude (i.e. non-specific) birth- and death-rates, which depend too much on rapidly changing age and sex structures.

[2] This is not meant to imply any disrespect to Ashton's views in *The Industrial Revolution*—these things have some validity, though not, in the light of most recent work, as much as we all once believed.

industrialization. Without doubting the significance of the earlier period, we have tried to show that the impetus of these thirty years was even more important because it consolidated the advance gained, perpetuated the new structure of the home market and encouraged the entrepreneurs of the time to take up innovations and to invest vastly larger sums of capital.

In this situation the family, the unit of consumers and producers, plays a basic role. The backbone of the industrial State is that part of the labour force which in our diagram of increasing real incomes is contained in the central growing bulge. Let us classify the conditions for its survival.

(1) It must have an assured income above the subsistence level.
(2) It must be mobile to be available wherever skilled labour and management are required.
(3) It must be allowed to keep its rising real wages, which must not be eroded by its own actions (e.g. too many children), the State (taxation), or their employers (by adjusting wages to living costs when prices fall), or by other parts of the population (by the arrival of a surplus of labour through natural increase or immigration).

On the other hand, the following need *not* apply:

(1) There is no need for the disappearance of a proletariat, or the cessation of its growth. In fact, as cheap labour it performs regrettable useful functions so long as the proportion of have-nots and buy-nots does not increase too fast.
(2) There is no need for this class of consumer/producers to exist everywhere. Within a regional economy, it is enough that there should be growing pockets of such people.
(3) They need not always be consumers: the system will tolerate years of stringency due to higher food prices, temporary closure of some markets, or redundancy for some skills.
(4) Rising real wages are not incompatible with rising money wages, nor high money wages with cheap products.

On balance, of course, the number and proportion of active consumers must grow continually, especially so as to provide a good market for consumer durables and the construction industries. Guessing from the statistics already quoted, it would not be impossible to postulate that in 1750 something like 15 per cent of the population had family incomes in the range between £50 and £400 p.a., and that by 1780 the proportion had risen to 20 per cent or even 25 per cent, and in absolute numbers this would mean an extra 150,000 households which functioned as consumers for the range of goods we have described. On the totals available from industry and commerce for the home market (see Table 2), this would mean that each of these households (over 300,000 of them) might be able to spend £30 per annum on industrial output and commercial services. This would account for £10 m. of the total (which, as we saw, might be computed at £30 m. in 1770). This would leave £10 m. for the much smaller number of really large incomes (30,000 households at £300 each) and another £10 m. to be shared by the residual 1,200,000 poorer households, at £8 each, or 3*s*. per week. In case this seems high, this would comprise not only consumer durables and bricks, glass, nails, etc., for housing, but also candles, all clothes, tools, and so on. In each case, obviously, there is to be included the proportion attributable to each notional household of public expenditure (Poor Law, street commissioners, defence) and private investment (agricultural implements, machines, means of transport). There is no implication that the 3*s*. would have to be found in cash, though there is no reason why this would not have been possible as an average.

Clearly this model of the national income structure (see diagram III) is purely hypothetical. It is presented here merely to show a *feasible* basis for the kind of economic growth we observe. It has been arrived at, in many ways, as a residual, by the rejection of the alternative hypotheses which were, basically, either that exports were responsible for the major part of the increased total demand, or that it came only from the wealthy households and their immediate retainers and suppliers. It would have been absurd to attempt to distribute the whole of the extra purchasing power equally through the whole population.

Lastly, since we have postulated that the whole structure varied from region to region, we must allow for the group in question to have been either much larger or much richer in some areas, and smaller or poorer in others. And since we have said that the sums involved are long-term averages, it will follow that at times locally or nationally both the group and its income were smaller and at others larger. Therefore we must not, as the Hammonds once did and as their successors do to this day, take the lamentations of pamphleteers like David Davies, or even

Sir F. M. Eden[1] at times, at their face value, but allow that what they said was true of its time and its place, and the people of whom they wrote.

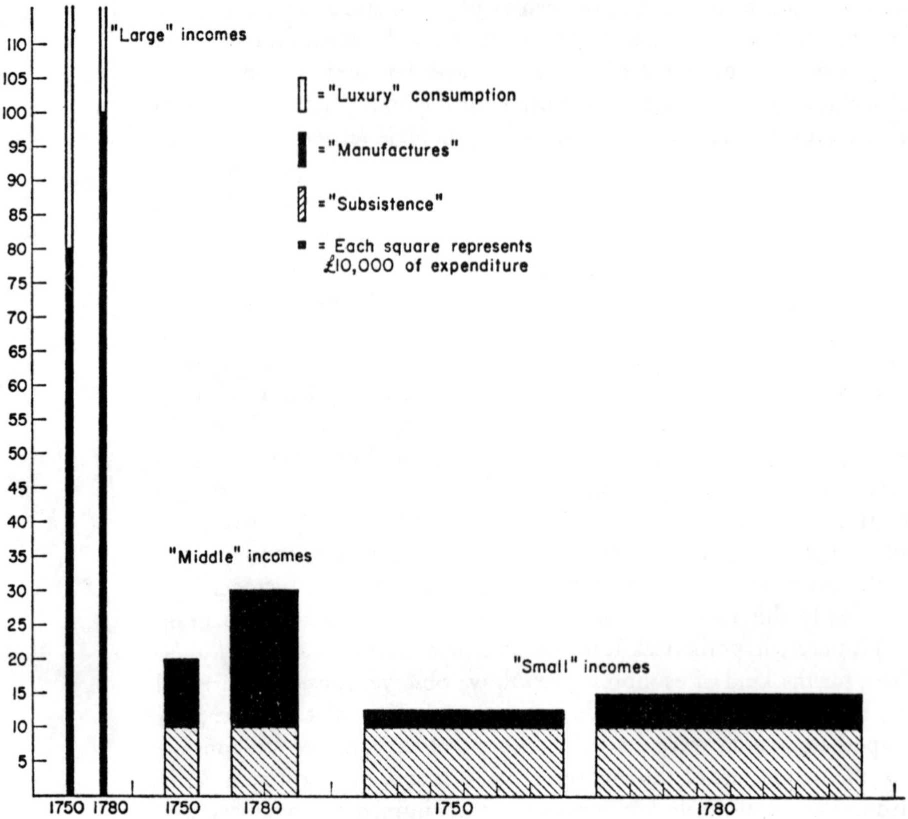

Diagram III. Household Income and Expenditure, 1750 and 1780

For it is our final conclusion that if the condition of the worst-afflicted section of our industrial and agricultural proletariat in the early nineteenth century, especially in the northern cities and some of the backward rural counties, had been the common fate of all the country's working population, industrialization could not have taken place. The

[1] David Davies, *The Case of Labourers in Husbandry Stated and Considered* (1795) is a case of special pleading; Eden is not, except that anyone writing at the time is bound to overstate the dismal aspect of the changes of the last years of the century which many considered to be a long-term trend.

incentive to investment and the increase in the scale of operations must have come from an increasingly broadly based and in the long run continuous and expanding home market. The foundation of the industrial revolution, in this model, is laid by the home sales of articles of everyday life to a section of the labour force which was neither very poor nor very rich, but wielded effective purchasing power in the face of many difficulties. Had food shortages, labour displacement by mechanization, or the subsistence theory of wages won the day, there would have been no industry. The remarkable (and probably unique) feature of the British industrial and commercial scene was the existence of this free, mobile, prudent section of the population which was able to assert some choice in the pattern of its expenditures sufficiently often to provide the stimulus which at least in our period neither foreign markets nor luxury consumers could have given. With some variations therefore, we conclude that the causes operative in the period 1730–50 continued to exercise much influence for the next thirty years, and that it is during this period that the most important foundations of the industrial State were laid.[1]

[1] This does not exhaust the catalogue of possible analyses. A paper of mine published in 1960, 'Population and Economic Growth in England before the Take-off' (First International Conference of Economic History, Stockholm, 1960, *Contributions and Communications*, The Hague, 1960, pp. 457 ff.), contained something of a shopping list of economic and demographic phenomena which awaited further investigation, having been mentioned in passing in the literature. Most of them still lack a researcher. Finally, the whole of the approach of this paper has been deliberately kept to the field of agrarian and industrial economics and demography, under the strong influence of the man to whom this volume is dedicated. One cannot rule out the possibility, however, that a more strongly sociological approach may also prove fruitful.

DEMAND AS A FACTOR IN THE INDUSTRIAL REVOLUTION

ELIZABETH WATERMAN GILBOY

I

IN THE field of economic history as well as that of economic theory there has been a tendency to overemphasize the factor of supply. Precisely as the classical economists were inclined to accept demand as given and constant, most of the economic historians of the nineteenth century concerned themselves with a detailed analysis of changes in the technique of production, the decay and expansion of certain industries, the effects of power machinery upon production, and so on. Little attention has been paid to changes in the nature of demand, even to the undoubted extension of demand, and especially is it true that the mechanism by which these changes occurred has been overlooked. Labor, as well, has been considered rather as a factor in production than as the major portion of the consuming public. When demand has been touched upon at all, it was usually dealt with in vague and general terms, with reference to Adam Smith's theory of the extension of the market. Exceptions occur, of course. Mrs. Knowles, Hobson, and a few others were well aware of the significance of developments of the demand factor with respect to the Industrial Revolution. But even they made little attempt to analyze the structure of internal demand; rather they stressed the increase of commerce, and the opening of new markets in undeveloped countries.

Obviously the factory system with its complicated industrial mechanism cannot function profitably without a large and growing demand ready and willing to absorb its products as fast as they are produced. The factory form of organization was not new. It had existed in ancient Rome, again in mediaeval Florence, and at various times in England and on the Continent, previous to the Industrial Revolution. Before the eighteenth century, however, the factory had never become typical. It existed sporadically to provide a few articles of luxury for the upper classes. The thesis of this essay is, indeed, that the factory could not become typical until demand had been extended and had become sufficiently flexible throughout the entire population to consume the products of large-scale industry. In other words, the Industrial Revolution presupposes a concomitant development and extension of consumption.[1]

The "extension of the market" school, as those writers who emphasize the phrase of Adam Smith might be called, do not go so far as to examine consumption standards. Why should new countries made accessible to English goods by the expansion of commerce and the greater facilities of transportation have been eager to buy the output of the factories? Why, again, should the home population suddenly afford a market for a greater number of products? These questions have been unanswered for the most part.

Lewinski, for example, explains the origin of the Industrial Revolution in Belgium by the typical "extension of the market" approach. The growth of the market for

[1] I am greatly indebted to Professor Gay for pointing out to me the importance of the analysis of consumption in connection with the Industrial Revolution. He is, in fact, the only person known to me who has stated the above hypothesis definitely and clearly, though not in print. I owe to him the impetus which started me upon the investigation of demand, not only in connection with the Industrial Revolution, but in general.

Belgian goods was due to the growth of population within the country, primarily, and to the extension of the external market resulting from political alliances, first with France and later with Holland.[1] He places emphasis upon the increase in population as indicating an increase in demand which forced industry to adopt new production methods, largely because of the action of diminishing returns.[2] It is difficult to see why an increase of population as such should necessarily lead to an increase in demand. The result might just as plausibly be a decrease in the standard of living of the population. The pressure of population upon subsistence, reducing the standard of living of the majority of the population to a bare minimum of existence, is a doctrine which has been familiarized by Malthus and Ricardo. And it has been a fact in the case of China, and, until recently, of India. If population increase were in itself a sufficient reason for increasing demand, calling forth more highly developed methods of production, the Industrial Revolution ought to be in an advanced stage in both India and China. Increase in population cannot by itself increase demand except in the case of a people with a standard of living so firmly intrenched that any lowering of the standard is implacably resisted. Then the increase in numbers may well serve as a stimulus to the discovery of profitable means of improving the technique of production. Undoubtedly this is what occurred in Belgium, although it is not brought out by Lewinski. Even when population increase does increase demand, it is not usually the most important aspect of the change in demand and probably would not alone lead to extensive industrial reorganization. An expansion of the standard of living of the population, as well as the growth of new wants, is necessary.

[1] *L'Evolution Industrielle de la Belgique* (1911).
[2] *Ibid.*, p. 58.

Mrs. Knowles and the Hammonds seem to have had some such process in mind when they selected commerce as the cause of the Industrial Revolution in England.[1] They point to the new commodities introduced into general consumption as a result of seventeenth- and eighteenth-century commerce, and admit that an increase in the general demand of the population took place at the same time. The process by which this widespread increase in demand occurred, however, is not analyzed; and it may be pointed out in passing that the Hammonds' implicit admission of a rise in the standard of living of the population as a whole in eighteenth-century England is somewhat inconsistent with their theory of increasing misery of the working classes after 1750.

An unusual emphasis upon the significance of demand is to be found in Hobson's *Evolution of Modern Capitalism.* His fourth essential condition for the development of capitalism is "the existence of large, accessible markets *with populations willing and economically able to consume the products of capitalist industry.*"[2] If these markets do not have populations willing and able to consume industrial products on a large scale, the type of business made widespread by the Industrial Revolution cannot exist. Hobson is aware of this and also makes consumption an integral part of his theory of over-saving as the cause of business cycles. That theory cannot be examined here, but it is of some interest to note Hobson's stress upon the part played by demand. It is made apparent again in his treatment of wages, in which he examines the justification of high wages from the point of view of the consumption of the workers and of the community as a whole.

Pillai comes very near to the point of view of the present

[1] L. Knowles, *The Industrial and Commercial Revolutions in Great Britain . . .* (1924); T. and B. Hammond, *The Rise of Modern Capitalism* (1925).

[2] See p. 2.

paper in the initial pages of his study of the Industrial Revolution in India. He contends that the extremely slow penetration of modern industrialism into India is due to the difficulties of changing the standards of consumption of the general population.[1] Probably there are still other authors besides those mentioned here who have recognized the significance of demand. None of them, however, pay any attention to the process by which demand changes. Hobson comes nearest to analyzing the mechanism by which changes in demand occur, but even he deals with the term in a rather vague way and links his analysis to an idealistic scheme of social progress.

It is therefore the task of the present essay, in so far as it can be done within the limits of a brief paper, to sketch the relation between changes in demand and changes in the technique of production. The theoretical argument will be illustrated principally, but not entirely, by the situation in England in the eighteenth century, at the time when the radical industrial changes which are classed under the heading of the Industrial Revolution were taking place.

II

The theory is simple. It may be briefly stated thus: necessary concomitants of the growth of large-scale production, and especially of its initial stages, are (1) changes in the shape of the demand schedule, to use Marshallian terminology,[2] or in the nature of demand, of the various

[1] *Economic Conditions in India* (1925).

[2] For the benefit of the general reader not versed in Marshallian economics, it may be explained that "demand schedule" is the term for a hypothetical list of prices and quantities, expressing the relation between the number of units of a commodity which an individual or a group *would* buy and the price per unit, at any one time under given conditions. Mathematically, price is usually expressed as a variable function of quantity. The standard of living of a person or group may be looked upon as composed of a series of interrelated schedules of this sort, all of which may change in shape and position, affecting other schedules in so doing.

layers of consumption within the general population; (2) a shift in the demand schedules of the group, or an increase in demand; (3) the introduction of new wants; and (4) mobility of individuals within and between the various classes of the population. In other words, the tastes of the population are changed both in nature and in quantity, new commodities are incorporated into the consumption of the group, and there is considerable shifting between the members of the various groups within the community. If any one of these factors is present, the others are likely to emerge. They are all interrelated. They mark a society which is socially unstable, in which standards of living are changing, and in which class lines are not clearly drawn.

A society in which the standards of consumption of a majority of the population are fixed and stable, wherein people are contented with what they have and desire nothing more, scarcely provides fertile soil for the sowing of industrial changes. Even an increase in population may have little effect in stimulating methods of production. If diminishing returns set in, the excess population may emigrate or die of starvation or plague, or war may ensue. New production methods *may* be invented, especially with a net population increase, but as a rule a complete overturn of the social structure is needed as well. The continued social stability of mediaeval life is a case in point; then the standard of living of each class was fixed by custom and tradition, and centuries elapsed before this stability could be broken down.

In order that a shift in the demand schedule may occur, individuals must be able to buy more units of a commodity at the same price, or the same amount of the commodity at a higher price. In other words, the entire schedule must shift upwards, indicating a greater buying power. There must therefore be an increase in the real income of the population. Real wages must rise, either through a decrease in prices or through a rise in money income, so that the majority of the population have more to spend.

The spread of new wants throughout the majority of the population is one of the most active forces leading to changes in standards of consumption. The introduction of new commodities leads people possessed of an economic surplus to try this and that, and finally to include many new articles in their customary standard of life.[1] If the economic surplus is not at hand, supposed necessaries may be neglected in order to obtain novelties, or individuals may be motivated to work harder in order to obtain the economic surplus. The mechanism by which these new wants are incorporated into all classes involves interclass competition, not only in consumption but in production: there must be sufficient mobility within the groups so that some individuals, at least, actually move from one group to another, while many more are actuated by a belief that they can do so.

The process is something like this. New articles are made available by commerce and taken up as a fashion by the wealthy and leisured members of society. The process may end there. If the various classes are not sufficiently closely related so that interclass competition in consumption is possible, if the general population have neither the desire nor the means to experiment with these new commodities, they will end as they began, articles of luxury for the rich. Commerce in the Middle Ages was largely in such luxuries; it was not until the seventeenth and eighteenth centuries that the ordinary individual would or could begin to include such goods in his own consumption. It

[1] See Hazel Kyrk, *Theory of Consumption*, for a statement of the existence of an economic surplus, and an available supply of varied economic goods as a stimulus to consumption change.

did not occur to the average villein or serf to desire spices or jewels or silken robes. He expected to live and die in his own stratum, content with a simple diet and coarse and unornamented clothing. Envy of the upper classes, closely followed by an attempt to imitate them, what Defoe called "apeing one's betters," is a comparatively modern phenomenon. The removal of political, legal, and economic restrictions upon the lower classes, the breakdown of feudal and manorial customs, were essential to its development.

As Tawney observes:

> The tendency and direction of the forces released by the Industrial Revolution, if that phrase is still to be retained, are not open to question. They were those described by Sir Henry Maine, when he wrote of "the beneficent private war which makes one man strive to climb on the shoulders of another and remain there." Compared with that of most earlier periods, the economic system which it created was fluid and elastic.[1]

The fluid and elastic system was not entirely created by the Industrial Revolution. It existed prior to the development of the Industrial Revolution — or such a great economic change could not have happened when and where it did — and was then enhanced in turn by the industrial development.

It is part of Tawney's criticism of the modern economic and social order to deplore "that beneficent struggle." He would substitute social and collective goals for those of individual gain in wealth and power. It is extremely dubious whether modern industrial society could have begun or could continue without the type of individualistic competition which he most decries. It may be limited and controlled, the rules of the game may be made more stringent and applicable to all, but it must be there. Ethically it

[1] R. H. Tawney, *Equality* (1930), p. 30.

may be undesirable; but economically it is essential to the development of modern industry.

Once society is sufficiently mobile, the luxury articles of the rich will seep down through the different social ranks, and perhaps end by becoming a necessity for all. By this time the upper classes will have taken up some other article of fashion. A contemporary described the process well:

> In England the several ranks of men slide into each other almost imperceptibly; and a spirit of equality runs through every part of the constitution. Hence arises a strong emulation in all the several stations and conditions to vie with each other; and a perpetual restless ambition in each of the inferior ranks to raise themselves to the level of those immediately above them. In such a state as this fashion must have an uncontrolled sway. And a fashionable luxury must spread through it like a contagion.[1]

Although this sort of thing occurred prior to the eighteenth century, it did not take place in a sufficiently general fashion to cause widespread comment. But at that time, eighteenth-century pamphleteers in England burst forth into a lamentation upon luxury in general and particularly the luxury of the lower classes. More will be said about this later.

Still another aspect of the problem exists which is not without interest. The qualitative and quantitative changes in demand noted above may serve as a stimulus to production in another way. So far we have only mentioned the fact that demand must increase and change in order to absorb the output of industry. It may also be true that changing consumption standards will act as an impetus to labor on the side of production. A transition has to be made from labor which is independent and which works in its own home, at its own time, to labor which works under strict discipline in a factory. Usually it is assumed that

[1] [T. Forster], *An Enquiry into the Causes of the Present High Price of Provisions* (London, 1767), p. 41.

labor was forced by circumstances to make this change. Certainly it is true that in England, at any rate, the acceleration of the enclosure movement, dispossessing agricultural labor in many districts, occurred conveniently about the time when masses of unskilled labor were needed in the new factories. It is true, as well, that the state disposed of paupers by transferring them by the cartload from parish workhouses to factories.[1] What could labor do but submit?

On the other hand, many laborers may have been motivated by the desire for increased consumption of material goods, whetted by the taste for luxury already acquired. They may have been willing to submit to the routine and dependent régime of the factory because wages there were higher and steadier. There is reason to think that this was true in the north of England, at least. Redford sees in this fact one explanation of the steady and wave-like migration of labor from the country-side to the town. He goes so far as to state that the difference in standard of living, which was in favor of the industrial town, was a motivating force in impelling labor to the industrial centers.[2]

Changes in demand or consumption, therefore, may be looked upon not only as necessary adjuncts to industrial change in the sense of providing a continually expanding and varying market, but as stimuli on the production side, too. As Malthus put it:

> It is not the most pleasant employment to spend eight hours a day in a counting house. Nor will it be submitted to after the common necessaries and conveniences of life are obtained, unless adequate motives are presented to the mind of the man of business. Among these motives

[1] See Unwin, *Samuel Oldknow and the Arkwrights* (1924). Also cf. W. E. Rappard, *La Revolution industrielle . . . en Suisse* (1914), p. 231. The factory workers of the printed-cloth industry were recruited from the poorest elements in the population.

[2] A. Redford, *Labour Migration in England* (1924), p. 60.

is undoubtedly the desire of advancing his rank, and contending with the landlord in the enjoyment of leisure, as well as of foreign and domestic luxuries.[1]

Miss Kyrk stated the same idea more universally when she said, "For the sake of gains as consumers, individuals consent to more hazardous and less interesting work as workers."[2] In India the inverse has been found to be true; that is, when laborers have satisfied their fairly simple wants, which can be done in three or four days' work a week, they do not appear at the factory.[3] What is to make them continue labor which is disagreeable to them, if their material wants are satisfied? As may be imagined, the establishment of the factory system in India has been attended with many difficulties.

Of course, there have always been those who argued that necessity was the only spur to industry and efficiency. This view was particularly popular among the eighteenth-century English mercantilists whose aim was to keep labor cost, in the sense of wages, as low as possible in order that England might outsell competitors in world markets. Sir William Temple expressed their views perfectly when he said that "the only way to make them [the laborers] temperate and industrious, is to lay them under the necessity of labouring all the time they can spare from meals and sleep, in order to procure the necessaries of life."[4] Quotations of this sort could be multiplied without end. The question of necessity versus luxury as a motivating force to increase the efficiency of labor and the product of industry engendered a hot philosophical controversy between the mercantilists and their opponents. In more recent times the theory opposite to that of mercantilism has become

[1] *Principles of Political Economy* (2nd ed., 1835), p. 403.
[2] *Op. cit.*, p. 65.
[3] Pillai, *op. cit.*, chap. ix.
[4] *A Vindication of Commerce and the Arts* (London, 1786), p. 534.

more popular. The economy of high wages is a doctrine of which one hears a good deal.

Theoretically, then, it is possible to conclude that far-reaching and widespread industrial changes cannot occur except in a society in which demand and consumption standards are undergoing swift and radical readjustment. Such a society is characterized by mobility between classes, the introduction of new commodities leading to the development of new wants, and a rise in real income of the people as a whole. Let us examine briefly the origin of the Industrial Revolution in the light of these ideas.

III

The social and economic milieu of eighteenth-century England formed the background for the first appearance of the Industrial Revolution. From a country predominantly rural, with the greater part of industry carried on in the homes of the workers under various forms of the putting-out system, it was transformed by the end of the century into a country of growing industrialism with factories supplanting the home as the producing unit. It is unnecessary to describe the nature and process of this industrial change. It has been done by many economic historians and by no one more competently than by Mantoux. It may at least be hazarded, although the data are scarce and contradictory, that the standard of living of the general population was increasing at the same time.

The available wage data indicate that the wages of common day labor were certainly increasing in the districts most closely connected with economic expansion.[1] In

[1] E. W. Gilboy, "Wages in Eighteenth Century England," *Journal of Economic and Business History*, vol. ii, no. 4 (August, 1930), pp. 603–629. The statistical material for the analysis of wages and prices summarized above is given in more detail in this paper.

the north of England (the wage material is for the North and West Ridings of Yorkshire, and Lancashire) the money wages of unskilled labor had risen from a median of 9d. per day in 1700 to 1s. 9d. per day by 1790. During the same period the journeyman's daily wage had gone up from 1s. to 2s. 3d. The rise was especially rapid after 1760. Within ten years the laborer's rate went from 1s. to 1s. 6d., the journeyman's from 1s. 6d. to 2s.; whereas it had taken the first fifty years of the century for the rates to go from 9d. to 1s. and from 1s. to 1s. 6d., respectively. This upward swing of wages appeared as characteristically in the agricultural districts of the North Riding as in the industrial sections of Lancashire and the West Riding. Real wages, as measured in terms of the laborers' most important article of diet, oat bread, appear to have risen, too, by the end of the century. Other evidence as to the living conditions of the laboring classes gleaned from local records and descriptions of laboring life show a real improvement towards the end of the century, at least for the class of labor under discussion.[1] The balance of contemporary observation points in that direction.

A similar rise in real wages took place in London and the metropolitan district surrounding it. The rise is not as con-

[1] See A. P. Wadsworth and J. de L. Mann, *The Cotton Trade and Industrial Lancashire 1600–1780* (1931), bk. iv. Mr. Wadsworth's evidence as to the condition of the workers in the cotton industry, especially of the weavers, is conflicting. There are no comparable wage figures for various parts of the century. Arthur Young's figures for 1769 indicate that some of the wages were even lower than day-laboring rates, although the check weavers' were somewhat higher. The formation of the early weavers' societies, and their activities in trying to raise wages and improve working conditions, were the result of recurring periods of want due to bad harvests, wars, etc. Mr. Wadsworth's material leans towards the depressing side, although he cites some data as to the increase of luxury among the working population. He does not commit himself, however, on the course of the standard of living of the weavers during the century. Probably the question must be left open until wages are specifically investigated.

sistent nor as great as in the northern districts, but the median daily rate for the common laborer increased from 1s. 7d. in 1700 to 2s. in 1787. At the same period the wages of craftsmen went from 2s. 6d. to 3s. 2d. Most of this rise occurred before 1750. Wheat prices — again measuring by the chief food item of the laborers' budget — were falling slightly during the first part of the century, and rising during the second half. Subsidiary evidence as to the food, clothing, and general living conditions of the workers provide grounds for assuming a net gain in consumption by the end of the century. The same conclusion, it may be noted, was reached by Mrs. George, purely on the basis of pamphlet and general descriptive material.[1]

The data for the west of England are the least satisfactory of the sections investigated. There is no foundation, at any rate, for a conclusion as to rising consumption standards in this district. The median daily wage rate of common labor fluctuated in Gloucestershire around 1s. from 1736 to 1787, the only period for which figures were obtained. Craftsmen's wages were approximately 1s. 6d. per day for most of the period. Barley prices, representing the principal diet of the laborers, showed, if anything, a slight trend upwards, with extraordinarily high prices in the years of bad harvests, which appear to have afflicted the west with especial distress. Food riots were prevalent all over the country in the 'fifties and 'sixties, but struck the western counties with particular force.[2]

The apparent stability or, possibly, decline in the standard of living of the Gloucestershire laborer is of interest in connection with a suggestion noted in the previous section. The decay of the woolen industry of the west during the eighteenth century is well known. It was sup-

[1] M. D. George, *London Life in the Eighteenth Century* (1924).
[2] See Arthur Young, *Tour to the Southern Counties* (1772), p. 340.

planted by the rising textile industry of the north. The introduction of the factory system occasioned riots against machines wherever they were introduced. In the north these uprisings against the mechanism of modern industry eventually subsided. In the west, however, the workers refused to use the new machines and continued to oppose their introduction. Possibly there is a connection between the continued opposition of labor to machinery, the decline of the woolen industry of the west, and the low and unchanging standard of living characteristic of the working classes. It might be suggested that the impetus of a changing material standard was lacking, and consequently laboring-class opposition to mechanized labor in a factory could not be overcome. At least, it is plausible to suggest that this may have been a contributing factor to the decay of the industry, which depended, of course, on many other factors as well.

It is unwise to generalize about England as a whole. Distinct regional differences in the real and money wages of common labor in at least three sections of the country have been found. There may be other regional divergences; the study of wages in other trades may tell a different story. Contemporary literature, however, indicates that the situation of London and the north was more typical of the country than that of the west. Eighteenth-century social and political pamphlets are full of allusions to the growth of luxury among the lower classes. Some wrote to deplore it, others to praise it, but most writers noted its existence. Even those who stressed the darker side of working-class life were likely to let slip observations upon the inclusion of tea, wheat bread, sugar, and other luxuries of the day in the laborer's diet. On the basis of this pamphlet material alone, Lecky, an astute and eminent historian, concluded that the standard of living of the general population had risen during the century.[1] The amount of contemporary comment upon the growth of luxury is so great that it cannot be entirely ignored by the least optimistic of investigators. Some basis of fact must have existed.

It is clear that tea, sugar, and wheat bread became incorporated into the general working-class diet during the century. Meat was eaten more frequently. Cotton and silk clothes were worn by the poor as well as by the rich. More money was spent on recreation, in the form of frequenting fairs, interludes, etc.[2] Conditions in some of the new industrial towns offered better and more sanitary living conditions for the average worker than had been known before.[3] We may conclude tentatively that the consumption of the average laborer was more varied, and greater in amount by the end of the century than at the beginning.

The evidence on interclass mobility and competition is fully as difficult to evaluate. Instances occur in which laborers rose to become managers and even owners of factories. Redford quotes examples of apprentices rising to the position of manager of a mill, and the Reverend William MacRitchie, a contemporary observer, writing of opportunities for advancement in the Sheffield cutlery works, comments, doubtless with exaggeration, on "people rising every day from nothing to eminence, by dint of industry." [4] Wadsworth remarks of Lancashire: "In contrast with the regulated industry of France, or with the sixteenth century ideal of an ordered distributive state in which each

[1] *History of England in the Eighteenth Century* (1883), vol. vi, p. 186: "These complaints of growing extravagance in the industrial classes were too common in the latter half of the eighteenth century not to rest on some real foundation."

[2] Cf. George, Botsford, and many others. Evidence as to the popularity of amusements comes from Quarter Sessions Records, which record the efforts of the Justices of the Peace to decrease the number of these entertainments as injurious to the morals of the lower classes. Middlesex County Records are full of such indictments.

[3] Cf. the description of Mellor in Unwin, *op. cit.*; also Redford, *op. cit.*, p. 30.

[4] *Diary of a Tour through Great Britain in 1795* (1897), p. 67.

class performed its appropriate functions, this Manchester society possessed variety and spontaneity. But it appears almost rigid and bound by custom in comparison with working class life half a century later. . . ."[1] His own evidence throws little light on the rise of the weaver from a small master to a manufacturer. The later eighteenth century, however, appears to have been one of those periods in which the growth of new enterprise, taking place at an unusually rapid rate, so upset the customary equilibrium of social and economic arrangements that individuals were able to transcend class barriers.

Contemporaries expostulated upon competition in consumption between classes. It was complained that servants could not be told from their mistresses; that class distinction was disappearing from the country-side; and that the lower classes no longer knew their position in life.[2] For instance, Davis remarked that "a fondness for Dress may be said to be the folly of the age, and it is to be lamented, that it has nearly destroyed those becoming marks whereby the several classes of society were formerly distinguished."[3] The phenomenon was especially noticed in urban districts, in London more than anywhere else, where servants provided the link between the habits of consumption of the upper and lower classes. Down through each rank in the social scale conspicuous consumption of all kinds was imitated. Fielding describes the process as follows: "Thus while the Nobleman will emulate the Grandeur of a Prince the Gentleman will aspire to the proper state of the Nobleman, the Tradesman steps from behind his Counter into the vacant Place of the Gentleman. . . ."[4]

[1] *Op. cit.*, p. 325.

[2] For example, Defoe, *Giving Alms No Charity* (1704); the *Great Law of Subordination* (1724).

[3] *Friendly Advice to Industrious and Frugal Persons* (1817), p. 23.

[4] *Enquiry into . . . Increase of Robbers* (1753), p. 4.

It cannot be doubted in general that eighteenth-century England exhibited signs of social instability. How inclusive as to region and class of worker the increase of standard of living, how extensive the interclass competition in consumption and production, and how great the luxury of the laborers all remain to be investigated more thoroughly. The undoubted distress of certain periods, certain sections, and certain classes of workers cannot be disregarded. The fate of the hand-loom weavers, in both England and Switzerland, is a depressing tale. Nor can the evils of the early factories be overlooked. On the other hand, these unfavorable elements must not be overemphasized as they frequently have been in the past. For the present, it may be assumed that the balance was in favor of an increase of the standard of living.

The initiation of the Industrial Revolution in other countries appears to have been marked with some of the same characteristics. The Swiss weavers were very prosperous in the eighteenth century and formed the basis for the political agitation of the early 1800's. Rappard quotes a contemporary to the effect that "it was relatively easy to rise from the rank of manual worker to that of head of the industry."[1] In contrast, he is inclined to stress the misery of the early factories after the manner of the Hammonds. Children and beggars were the laboring force of many of the first factories, in degrading conditions. In this case, why did migration occur from the country to the cities, if the result were only misery? Figures are given to show that factories were supplied with very little labor from the home district. Visitors, too, commented favorably on the living conditions of the Swiss working classes, despite a lower level of wages than in other countries. Rappard attributes this to foreign optimism and appar-

[1] *Op. cit.*, p. 242.

ently disregards the fact that the standard of living may have risen. Still, he cites contemporaries on the instability and discontent of the early factor workers, their ambition for material well-being, and their going from factory to factory in search of better conditions. These conditions are not dissimilar to those which we have seen to characterize eighteenth-century England.

Pillai attributes the slow development of the factory system in India in large part to the low standard of living of the workers, and their strict caste organization which makes competition between classes unthinkable. He gives some evidence — not very complete — of a rise in wages of the lowest class of workers and of the breakdown of the caste system coincident with the growth of the factory system. The willingness of workers to labor more steadily, in the hope of acquiring the articles of comfort and luxury creeping in through the influence of foreigners, appears to be increasing.[1]

The United States imported the Industrial Revolution in a late stage and developed industrially in an exceptionally short period. It may be observed briefly that American society has been noted since its origin for its lack of class distinction, and for mobility between the indistinct class groups which did grow up; and for its general restlessness and striving for material comfort. America, in fact, exhibited in an intensified form the conditions out of which the Industrial Revolution grew in England.

In conclusion, we may say that demand is an important factor in facilitating the occurrence of industrial change, on the basis of a priori reasoning. The importance of stimulating demand is, indeed, extensively recognized by modern producers. Whereas in the early stages of the Indus-

[1] *Op. cit.*, p. 135.

trial Revolution demand was, if anything, in the van of production, today the reverse is true. Over-expanded industries resort to high-pressure salesmanship of the most far-fetched nature in order to increase a demand already existent, or to arouse it where there is none.

The continued investigation of demand in relation to the Industrial Revolution is to be urged. Elaborate statistical researches cannot be expected. It would certainly be impossible to concoct even the most dubious "statistical" demand curve from eighteenth-century data. It is possible, however, to obtain data on wages and the prices of the main foodstuffs and to indicate crudely the course of real wages. Contemporary comment may be compared and sifted in order to obtain a net balance of opinion. The use of eighteenth-century English sources has led to the conclusion that changes in demand played an important part in the initiation of the Industrial Revolution. Changing consumption standards, the increase of population and shifting of individuals from class to class, and a rise in real income provided a stimulus to the expansion of industry which must not be underestimated. Although the mistake must not be made of exaggerating, in turn, the influence of demand, its significant relation to changes in production should be clearly recognized.

JOSIAH WEDGWOOD: AN EIGHTEENTH-CENTURY ENTREPRENEUR IN SALESMANSHIP AND MARKETING TECHNIQUES [1]

By N. McKENDRICK

When Josiah Wedgwood was born in 1730, the Staffordshire potters sold their wares almost solely in Staffordshire. Their goods found their sale in the local market towns,[2] and occasionally, carried by pedlars and hawkers, they reached further afield [3]—to Leicester, Liverpool and Manchester. To sell in London was rare,[4] to sell abroad virtually unknown.[5] Yet by 1795 Wedgwood had broken through this local trade of fairs and pedlars to an international market based on elegant showrooms and ambassadorial connexions; he had become the Queen's potter and sold to every regal house in Europe. His wares were known in China, India and America. Other potters had prospered but Wedgwood had flourished above all others. Born the thirteenth son of a mediocre potter with only the promise—and a promise never fulfilled—of a £20 inheritance, he died in 1795 worth £500,000 and the owner of one of the finest industrial concerns in England. His name was known all over the world. It had become a force in industry, commerce, science and politics. It dominated the potting industry. Men no longer spoke of 'common pewter' but of 'common Wedgwood'.[6]

Such fabulous success is not easily explained. It certainly cannot be explained in terms of Wedgwood's gifts alone. For Wedgwood was fortunate in the period in which he lived. Born poor into the squalor and dirt of a peasant industry,[7] one might have thought him unlucky. Superficially he was. The ware was still crude, the market still local, the roads almost impassable, and the workmen as likely to go drinking and wenching as to appear at work. Worse conditions for industrial expansion might seem difficult to imagine. But in all this were the signs of improvement. The technical discoveries of Astbury, Booth and Whieldon had opened up new opportunities for expansion and improvement; steam power was soon to open up more. Wesley was leading men to more methodical lives, Brindley and Bridgewater were building their canals, and agitation over the state of the roads had already started. When

[1] I should like to record here my thanks to Mr. N. G. Annan, Prof. M. M. Postan, Dr J. H. Plumb and Mr C. H. Wilson for having read this article and made many helpful suggestions. I am also greatly indebted to Josiah Wedgwood & Sons Ltd. Barlaston, Stoke-on-Trent, for permission to quote from the manuscripts in the Wedgwood Museum (subsequently referred to as WMSS), and to Mr Tom Lyth, the curator, for his generous help. The manuscripts have been collected from different sources and are catalogued accordingly. I have adopted the following abbreviations — E for Etruria, L for Liverpool, and L.H.P. for Leith Hill Place (this last collection is uncatalogued). The Mosley papers, also at Barlaston, I refer to in full (they are also uncatalogued). Dubious dates or dates relying on internal evidence are given in brackets.

[2] *Pitt Agricultural Survey*, pp. 2–3, 166, for list of the 24 markets.

[3] T. Whieldon, Memorandum Book, c. 1740–52, p. 78. An unusually distant order, 'Mr Green at Hovingham, Eylsham (sic), Norfolk. Aug. 11'.

[4] *Ibid.* p. 81, 'For Miss Ferney . . . directed to Capt. Blake in Surrey St. in the Strand'.

[5] The occasional piece often found its way abroad, but the Staffordshire potters never sold there in quantity.

[6] The *Black Dwarf*, 17 September 1817.

[7] Josiah Wedgwood, *An Address to the Young Inhabitants of the Pottery* (Etruria, 27 March, 1783), p. 21.

Josiah was serving his apprenticeship, such movements were only in their infancy, but with each year they gathered strength and support. He still had to fight reaction. But in the 1760's he found allies he would have looked for in vain in the 1730's. Moreover the demand for earthenware was steadily growing. Tea-drinking —rapidly becoming a national characteristic—and beer-drinking —already established as such—were both increasing. These, and the more fashionable drinks of coffee and hot chocolate, greatly increased the demand. Further, the growth of incomes, the shift of tastes, particularly of the 'middle ranks', and the expansion of overseas trade provided market opportunities in constantly mounting numbers.[1] But most important of all, the rise in population represented a vast and growing market with ever-expanding needs. It was Staffordshire that satisfied them. For plate was too expensive, pewter too scarce, and porcelain too fragile to compete with the versatile pot. In these conditions the potteries were bound to prosper.

The reasons why Wedgwood prospered above all others have proved more elusive. Most historians have argued that his discoveries—green glaze, creamware, jasper and black basalt—won him technical supremacy over his rivals; and that his factory organization and division of labour—his stated desire 'to make such machines of the Men as cannot err'[2] — confirmed his superior quality. But this alone is not sufficient to explain his supremacy. For his inventions were quickly copied and his quality easily reproduced. They won him immediate attention but they could not keep it unless he could afford to sell his ware more cheaply than his rivals. This historians have cheerfully assumed. The statement by Professor Ashton that 'it was by intensifying the division of labour that Wedgwood brought about the reduction of cost which enabled his pottery to find markets in all parts of Britain, and also of Europe and America' [3] is merely the most recent and most authoritative of a long line of such views—Meteyard,[4] Jewitt,[5] Church,[6] Smiles,[7] Burton,[8] and Trevelyan [9] all produce the same argument. They note the efficiency of Wedgwood's factory system, his avoidance of waste, the drop in breakages through the use of canals, the cheapening of transport charges because of canals and turnpike roads, and conclude that Wedgwood's wares were obviously cheaper than his rivals. Unfortunately they were not. His goods were always considerably more expensive than those of his fellow potters: he regularly sold his goods at double the normal prices,[10] not infrequently at three times as high, and he reduced them only when he wished to reap the rewards of bigger sales on a product that he had already made popular and fashionable at a high price,[11] or when he thought the margin between his prices and those of the rest of the pottery had become too great. In 1778, for instance, he introduced a cheaper teapot to cut down the huge price gap which had arisen between his prices and those of

[1] Asa Briggs, *The Age of Improvement* (1959), p. 28.
[2] WMSS. E. 18265–25. J(osiah) W(edgwood) to T(homas) B(entley) (Subsequently referred to as JW and TB), 9 October 1769.
[3] T. S. Ashton, *The Industrial Revolution* (Oxford, 1949), p. 81.
[4] Eliza Meteyard, *The Life of Josiah Wedgwood*, 2 Vols. (1865–6).
[5] Llewelyn Jewitt, *The Wedgwoods* (1865).
[6] A. H. Church, *Josiah Wedgwood* (1894).
[7] Samuel Smiles, *Life of Wedgwood* (1894).
[8] William Burton, *Josiah Wedgwood and his Pottery* (1922).
[9] G. M. Trevelyan, *The Social History of England* (1944).
[10] WMSS. E. 18457–25. JW to TB, 14 April 1773, and WMSS. L.H.P. MSS. JW to TB, 21 & 22 April 1771.
[11] WMSS. E. 18392–25. JW to TB, 23 August 1772.

Palmer and Neale, a rival local firm, writing, 'Mr. Palmer sells his three sizes of black fluted teapots at 18/– the long doz^{ns}. that is @ 9^{d}. 1/– & 18^{d}. Per pot which we sell at 50 or 60/–!'.[1]

There are ample reasons why the usual explanations did not apply. Canals, for instance, may have cheapened his goods, but they cheapened all other potters' goods as well; and though division of labour made for cheap production, the cost of experiments and the many failures they automatically entailed,[2] the expence of commissions to artists,[3] and the relatively high wages that Wedgwood paid,[4] more than cancelled this out. But more important than this was Wedgwood's decision not to compete with his rivals in price. It was never his practice nor his intention to sell cheaply. As he wrote towards the end of his life, 'it has always been my aim to improve the quality of the articles of my manufacture, rather than to lower their price',[5] and, more important than his statements,[6] his price lists fully confirm this. His selling policy relied on quality and above all on fashionable appeal, and Wedgwood believed that high prices had an integral part to play in such a policy, writing 'a great price is at first necessary to make the vases esteemed *Ornaments for Palaces*'.[7] He did not charge his pottery at what it was worth [8] but at what the nobility would pay for it.[9]

Some idea of how this policy developed can be gained from a letter he wrote to Bentley in 1771. Faced with a mounting stock he was overjoyed at the prospect of a large order [10] from Russia: 'This Russ^{n}. trade comes very opportunely for the useful ware, & may prevent me lower^{g}. the prices here, though it may be expedient to lower the prices of the Tableplates to 4/– Per doz in London, as our people are lowering them to 2/3 or 2s here. Mr Baddeley who makes the best ware of any of the Potters here, & an Ovenfull of it Per Diem has led the way, & the rest must follow, unless he can be prevail'd upon to raise it again, which is not at all probable, though we are to see him tomorrow, about a doz^{n}. of us, for that purpose ... Mr Baddeley has reduc'd the prices of the dishes to the prices of whitestone, ... In short the *General trade* seems to me to be going to ruin on the gallop—large stocks on hand both in London & the country, & little demand. The Potters seem sensible of their situation, & are quite in a Pannick for their trade, & indeed I think with great reason, for *low prices* must beget a *low quality* in the manufacture, which will beget *contempt*, which will beget *neglect*, & disuse, and there is an end of the trade. But if any one Warehouse, distinguish'd from the rest, will continue to keep up the quality of the Manufacture, or improve it, that House may perhaps *keep up*

[1] WMSS. E. 18814–25. JW to TB, 25 February 1778.
[2] Wedgwood fired over 10,000 pieces of jasper before he achieved perfection.
[3] They included artists of the stature of John Flaxman, George Stubbs, and William Hackwood. c.f. Neil McKendrick, 'Josiah Wedgwood and George Stubbs', *History Today*, VII, No. 8 (August 1957), 514.
[4] To deal with this in detail is beyond the scope of this article.
[5] WMSS. E. 8636–10. JW to Mr. Charles Twigg, 18 June 1787.
[6] The letter which he wrote to Lord Paget (E. 18895–25. June 1 1779) saying that he wished 'his profits rather to arise from a large consumption, than from a high price with a diminished sale' which is quoted by Ralph M. Hower, 'The Wedgwoods — Ten generations of Potters', *Journal of Economic and Business History*, No. 2 (February 1932), 306, is an exception which is not convincing in face of the mass of contradictory evidence in Wedgwood's letters to Bentley, e.g. E. 18407–25. 19 September 1772, and E. 18770–25. 10 July 1777.
[7] WMSS. E. 18392–25. JW to TB, 23 August 1772.
[8] I.e. a price based on the cost of production.
[9] WMSS. E. 18307–25. JW to TB, 4 June 1770.
[10] It amounted to some £4000.

'*its prices*,[1] & the *general evil*, will work a *particular* good to that house, & they may continue to sell *Queens ware at the usual prices*, when the rest of the trade can scarcely give it away. This seems to be all the chance we have, & we must double our dilligence here to give it effect. The same Idea may be applied to Ornaments, & the crisis in which a foreign vent for our goods will be the most singular service to us, is, whilst the General Manufacture is *degradeing*, & the particular one *improving* 'till the difference is sufficiently apparent to strike the most common purchacers; & that crisis seems now to be at hand, which I am very sorry for, but it seems to me inevitable; for I am certain the Potters cannot afford to work their goods in a Masterly manner, & sell them at the prices they now do, & they will probably go lower still.'[2] He held the same view in 1773 when 'the whole of the Pottery' [3] agreed to lower their prices a further twenty per cent. For though he anxiously asked, 'Do you think we can stand our ground in London @ 5/ P (doz) for plates, when everybody around us will be selling @ 2/6 & 3/–?',[4] and discussed the possibility of having two prices, he eventually decided against cuts of any kind, writing 'We must endeavour to make our goods better if possible —other people will be going worse, and thereby our distinction will be more evident'.[5]

In taking this decision Wedgwood committed himself to new methods of selling his ware, for he not only decided on high prices, but also determined on large sales to a widespread market. He had quickly realised that at the prices he charged quality alone would be sufficient only to win for him a limited and specialized market, and to confine his sales to a small and exclusive class. Moreover in the eighteenth century his improvements and inventions did not remain his monopoly for long. They were copied and reproduced—cheaply and in quantity. Every new invention that Wedgwood produced—green glaze, creamware, black basalt and jasper—was quickly copied [6]; every new idea—jasper cameos, intaglios and seals, tea trays, snuffboxes and knife-handles—was eagerly taken up; every new design—Etruscan painting, the Portland Vase and Flaxman's modelling—was avidly reproduced. And in every case the reproductions were cheaper. Even William Adams, perhaps the finest potter amongst Wedgwood's rivals, whose products equalled if they did not surpass Josiah's,[7] could undercut his prices by 20 per cent.[8] The result was inevitable. Those customers, who had been attracted by his novelty and his quality, reluctantly but nonetheless surely left him for cheaper makers, writing like James Abernethy, 'I imagined that you was the only person that printed that sort of ware—but it seems that there are others that put up with smaller profits'.[9]

It was clearly not by novelty and originality alone that Wedgwood held his custom, nor was it solely by high quality. For his novelty did not survive for

[1] My italics.

[2] WMSS. L. H. P. JW to TB, 21 & 22 April 1771.

[3] WMSS. E. 18457–25. JW to TB, 14 April 1773.

[4] WMSS. *ibid.*

[5] WMSS. *ibid.* The other potters' prices were by now '⅓ *of our price*'.

[6] Creamware for instance — in its improved form virtually his own creation — was being made by 1784 by 25 potters in Burslem and Newcastle alone. c.f. Bailey's Western Dictionary for 1784, W. Mankowitz & R. G. Haggar, *Concise Encyclopaedia of English Pottery & Porcelain* (1957), pp. 268–270.

[7] *Ibid.* p. 226, Turner's wares were 'frequently equal in quality' to Wedgwoods', and W. B. Honey describes Adams's jasper as 'quite equal' to that of Wedgwood.

[8] Mankowitz & Haggar, *op. cit.* p. 4.

[9] WMSS. E. 30554–5. J. Abernethy to JW, 2 October 1763.

long, and his high quality was not unrivalled. They played an integral part in his sales policy, but they are not in themselves sufficient explanation of his success. He had the good sense to realise that he was not likely to invent pottery superior to his creamware, his black basalt, or his jasper. Having once achieved perfection in production, he must achieve perfection in sales and distribution. It was clear that Wedgwood must either cut his prices as his rivals did in the cut-throat race for the custom of an expanding market, or seek some new distinction to mark off his wares from the rest of the pottery. He chose the latter course, and it is with these new methods that this article is mainly concerned: how Wedgwood won a world-wide demand, and how he invented the means of satisfying it.

* * *

Partly he did this by the capture of the world of fashion. For although Wedgwood had complete confidence in his wares—writing, 'wherever my wares find their way, they will command the first trade'—he also realised that '*Fashion* is infinitely superior to *merit* in many respects, and it is plain from a thousand instances that if you have a favourite child you wish the public to fondle & take notice of, you have only to make choice of proper sponcers (sic)'.[1] The sponsors he aimed to win for his pottery were the monarchy, the nobility, and the art connoisseurs —in fact, the leaders of fashion. He quickly realised that to make pots for the Queen of England was admirable advertisement. To become the Queen's Potter and to win the right to sell common earthenware as Queen's ware, was even better. As Wedgwood wrote: 'the demand for this s^d^. *Creamcolour*, alias, *Queensware*, ... still increases. It is really amazing how rapidly the use of it has spread over the whole Globe, & how universally it is liked. How much of this general use, & estimation, is owing to the mode of its introduction—& how much to its real utility & beauty? are questions in which we may be a good deal interested for the government of our future Conduct. The reasons are too obvious to be longer dwelt upon. For instance, if a Royal, or Noble introduction be as necessary to the sale of an Article of Luxury, as real Elegance & beauty, then the Manufacturer, if he consults his own inter^{t}. will bestow as much pains, & expence too, if necessary, in gaining the former of these advantages, as he wo^{d}. in bestowing the latter.'[2] Wedgwood was not a man to fail to consult his own interests. He took immediate action.

That Wedgwood sought such patronage has been categorically denied. Miss Meteyard, for instance, wrote in tones of hushed approval, 'we have seen Mr Wedgwood working silently onwards ... unsolicitous of patronage ... having laboured to invest the articles produced by his hand with an excellence and taste hitherto unknown, he left the natural results to their own time and place of fulfilment'.[3] She closes in defiance—and in capital letters—'IT WAS PATRONAGE WHICH SOUGHT THE GREAT POTTER: NOT THE GREAT POTTER PATRONAGE'.[4] It is an eloquent defence but unfortunately grossly untrue.[5] That Wedgwood did seek such patronage is indisputable.

[1] WMSS. E. 18898–26. JW to TB, 19 June 1779.

[2] WMSS. E. 18167–25. JW to TB, 17 September 1767.

[3] Eliza Meteyard, *op. cit.* I. 368–9.

[4] *Ibid.*

[5] Apart from being untrue of the whole of Wedgwood's life, this statement also distorts the origins of the order for the Queen's tea set. The order was offered to many potters before Wedgwood saw the potential value of accepting it.

He went to endless trouble and expense to win the royal favour—the famous green and gold tea set was followed up by a box of patterns and a creamware dinner service, and by 1768 he was advertising his '*Royal Patronage*' in the St. James's Chronicle and 'in that morning paper which is mostly taken by the people of fashion' [1] to broadcast the opening of his new rooms. He did not let their support languish for want of attention, constantly urging Bentley, 'that a little push farther might still be made with due decorum'.[2] In December 1770 Her Royal Highness the Princess Dowager was being waited upon,[3] and in 1771 he was scheming to become 'Potter to *His* Majesty' and 'Potter to the Prince of Wales'.[4] Nor did he neglect the younger members of the royal family, and by 1790 he had won the title of 'Potter to her Majesty & their Royal Highnesses the Duke of York & Albany & the Duke of Clarence'.[5] He did not hesitate to exploit it to the full, writing to congratulate his partner on his efforts with the Queen, 'you have sown the seeds of a plentifull & rich harvest, which we shall reap in due time ... Their majestys are very good indeed! I hope we shall not lose their favour, & promise ourselves the greatest advantage from such Royal Patronage, & the very peculiar attention they are pleased to bestow upon our productions. It was a good hint you gave them ... I hope it will work, & have its proper effect'.[6] On every bill head, every order form and every advertisement his titles were proudly displayed.[7] For he was confident that if he had the patronage of the great, he would have the custom of the world.

Having tapped or attempted to tap all the sources of royal patronage, he next broached the nobility and gentry. For he wished 'if possible (to) do in this as we have done in other things—begin at the *Head* first, & then proceed to the inferior members'.[8] Convinced of the value of a fashionable reception for his goods, he went to great trouble and expense to achieve it. Though he fully realised the cost, interruptions and poor immediate returns of special individual orders,[9] or 'Uniques' [10] as he called them, he willingly accepted expensive and difficult commissions. Other potters fought shy of such projects, Wedgwood and Bentley accepted every challenge. They welcomed commissions from Queen Charlotte for a specially designed teaset which all other potters had refused,[11] from George Stubbs for huge stoneware plaques of great technical difficulty,[12] and from Catherine the Great for a table service requiring 1282 pieces and over a thousand original paintings.[13] Strictly uneconomical in themselves, the advertising value of these productions was huge.[14] In the same

[1] WMSS. L. 17666–96. JW to Mr Cox, 13 June 1768.

[2] WMSS. L. H. P. JW to TB, 8 July 1771.

[3] WMSS. E. 18334–25. JW to TB, 24 December 1770. Wedgwood was quite satisfied despite the small order because 'tis good to have an opening, & to be known, the former may increace (sic), & the latter cannot hurt us'.

[4] WMSS. L. H. P. JW to TB, 8 July 1771.

[5] WMSS. E 1066–2. Printed Bill Head from Wedgwood. 24 February 1790.

[6] WMSS. L. H. P. JW to TB, 7 September 1771.

[7] WMSS. E. 18341–25. E. 1066–2. E. 18504–25. and many others.

[8] WMSS. L. H. P. JW to TB, 2 September 1771.

[9] WMSS. E. 18283–25. JW to TB, 10 January 1770. 'Defend me from particular orders', also c.f. WMSS. E. 18269–25. JW to TB, 19 November 1769.

[10] *Ibid.*

[11] WMSS. E. 18073–25. JW to John Wedgwood. Postmark 17 June 1765. He says he received the order 'because nobody else wo[d]. undertake it'.

[12] WMSS. E. 18785–25. JW to TB, 18 October 1777 and many other references.

[13] WMSS. E. 18450–25. JW to TB, Postmark 23 March 1773. c.f. Dr G. C. Williamson, *The Imperial Russian Dinner Service* (1909), pp. 65–102.

[14] WMSS. E. 18498–25. JW to TB, 14 November 1773.

way, though on a lesser scale, he made pebble ware for Sir George Young,[1] cameo heads of the sons of Mrs Crewe,[2] and printed ware for Lady Isabella Stanley.[3] For as Bentley wrote of the last, 'Tho' this is a trifling matter we must please these great Friends who are warm Patrons of this Manufacture'.[3] All of these orders were 'uniques'—they could never go into general production. They were made entirely for their advertising value, to win the patronage of the court and courtly circles; the friendship of the architects and the artistic world; the favour of the fashionable aristocracy and the gentry; and—of course—the future custom of them all.

By appealing to the fashionable cry for antiquities, by pandering to their requirements, by asking their advice and accepting their smallest orders, by flattery and attention, Wedgwood hoped to monopolize the aristocratic market, and thus win for his wares a special distinction, a social *cachet* which would filter through to all classes of society. Everything was done to attract this aristocratic attention. A special display room was built [4] to beguile the fashionable company which Josiah drew after him to Etruria; [5] steps were taken to make the London showrooms attractive 'to the ladies',[6] and to keep the common folk out; [7] he was even prepared to adjust his prices downwards so that they could be paid genteelly, writing to his partner, 'I think what you charge 34/– should ... be ... a Guinea & a half, 34 is so odd a sum there is no paying it *Genteely* ...'.[8] Once attracted, everything was done to keep such attention. The good will of Wedgwood patrons never withered from neglect. Sir George Strickland was asked for advice on getting models from Rome; [9] Sir William Hamilton was asked for advice on gilding; [10] they were complimented by the reproduction of their country houses on the great Russian service; [11] and great care was taken to flatter them by giving them first sight of any new discovery.[12] The first Etruscan vases, for instance, were shown before they were put on sale to 'Sir Watkin Williams Wynn, Mrs Chetwynd,[13] Lord Bessborough, Earl of Stamford, Duke of Northumberland, Duke of Marlborough, Lord Percy, Lord Carlisle St. James's Place, Earl of Dartmouth, Lord Clanbrazill, Lord Torrington, Mr Harbord Harbord'.[14] These were the nucleus of an aristocratic claque that did Wedgwood untold good. They praised his ware,[15] they advertised it,[16] they bought it,[17] and they took their friends to buy it.[18] Wedgwood had

[1] WMSS. E. 18269–25. JW to TB, 19 November 1769.
[2] WMSS. L. H. P. JW to TB, 2 September 1771.
[3] WMSS. E. 622–1. TB to JW, 21 June 1769.
[4] WMSS. L. H. P. (a fragment). JW to TB, 27 July 1771.
[5] WMSS. E. 18878–26. JW to TB, 25 February 1779.
[6] WMSS. E. 18149–25. JW to TB, postmark 1 June 1767.
[7] WMSS. *ibid.* 'For you well know that .... my present sett of Customers... will not mix with the rest of the World...'
[8] WMSS. E. 18271–25. JW to TB, 1 December 1769.
[9] WMSS. L. H. P. JW to TB, 7 September 1771.
[10] WMSS. E. 18365–25. JW to TB, 11 April 1772.
[11] WMSS. E. 18498–25. JW to TB, 14 November 1773. An action designed to 'rivet them more firmly to our interests.' For list of views, see G. C. Williamson, *op. cit.* pp. 59–90.
[12] WMSS. E. 18274–25. JW to TB. 9 December 1769. Also E. 18273–25. Sarah Wedgwood to JW, 6 December 1769.
[13] Mrs Chetwynd was their connection with the palace.
[14] WMSS. E. 18274–25. JW to TB, 9 December 1769.
[15] WMSS. L. H. P. JW to TB, 7 September 1771.
[16] WMSS. E. 18367–25. JW to TB, 18 April 1772.
[17] WMSS. Innumerable examples, e.g. E. 30857–5 and E. 30859–5.
[18] WMSS. E. 18505–25. JW to TB, 6 December 1773. Lady Littleton, for example, 'makes a point of' taking her friends to Wedgwood's showrooms.

no scruples about exploiting their friendship and their praise. In 1776, for instance, by artful flattery he carefully prepared the ground for his new Bas-relief vases at the next season's sale, writing to Bentley, 'Sir William Hambleton, our very good Friend is in Town—Suppose you shew him some of the Vases, & a few other Connoisieurs (sic) not only to have their advice, but to have the advantage of their puffing them off against the next Spring, as they will, by being consulted, and flatter'd agreeably, as you know how, consider themselves as a sort of parties in the affair, & act accordingly'.[1] In the small, interconnected, gossip-ridden world of the English aristocracy in the eighteenth century, such introductions were vital, for even a very few sales could have an important effect.

For the lead of the aristocracy was quickly followed by other classes. Fashions spread rapidly and they spread downwards. But they needed a lead. As Wedgwood put it, 'Few Ladies, you know, dare venture at anything out of the common stile (sic) 'till authoris'd by their betters—by the Ladies of superior spirit who set the ton'.[2] Wedgwood fully realised the value of such a lead, and made the most of it by giving his pottery the name of its patron; Queensware, Royal pattern, Russian pattern, Bedford, Oxford and Chetwynd vases [3] for instance. He went further than this with some. For he was not afraid to anticipate this patronage and to give his wares its beneficent sanction before it was bestowed. When he wished to give a new cheap line in flowerpots a good send off he wrote to Bentley, 'they want a name—a name has a wonderful effect I assure you—Suppose you present the Dṣ of Devonshire with a Set & beg leave to call them Devonshire flowerpots. You smile—Well call them Mecklenberg [4] —or—or—what you please so you will but let them have a name'.[5]

Once committed to this policy of reliance on the support of the great, Wedgwood had to attend to every dictate of fashion. He could not afford to let Wedgwood ware become unfashionable. He combined with Matthew Boulton to satisfy the demand for ormolu mounted pottery; [6] he banished gilding from his vases—and gilders from his workrooms—at the command of Sir William Hamilton and an unresponsive market; [7] he made black teapots to show off to better advantage the current vogue for white hands; and he clothed naked figures that were 'too warm' [8] for English taste with voluminous draperies, writing firmly to John Flaxman that to cover 'the nudities ... with leaves ... is not enough—for none either male or female, of the present generation, will take or apply them as furniture, if the figures are naked'.[9]

To the rage for the antique and the excitement over Herculaneum Wedgwood gave special attention. It was vital that he should. For tired of the late

[1] WMSS. E. 18693–25. JW to TB, 12 September 1776.

[2] WMSS. E. 18766–25. JW to TB, 21 June 1777.

[3] E. Meteyard, *op. cit.* II, 68. denies this, writing 'These were no vulgar appelations given to flatter a patron or to insure sales; but simply showed' from whose possessions the vases had been modelled. In some cases this was no doubt true (not that JW would have neglected the advertising value in any case), but it was certainly not true of the Duchess of Devonshire, Queen Charlotte, or Mrs Chetwynd.

[4] The brother of Queen Charlotte.

[5] WMSS. E. 18811–25. JW to TB, 9 February 1778.

[6] WMSS. E. 18193–25. JW to TB, 15 March 1768.

[7] WMSS. E. 18365–25. JW to TB, 11 April 1772.

[8] WMSS. E. 18278–25. JW to TB, 28 December 1769, and E. 18523–25. JW to TB, 13 March 1774.

[9] Arthur Lane, *Wedgwood Bicentenary Exhibition, 1759–1959* (introd.) (1959), p. 6. Wedgwood was by no means consistent on this point and I hope to deal with it more fully elsewhere. C.f. also Wedgwood to Joseph Wright of Derby, WMSS. E. 672–1, and many others.

Baroque and Rococo extravagances of the middle decades of the century, the world of fashion had flocked to acclaim the new discoveries at Naples. The proliferating decoration, the exuberant colours, and the universal gilding of Rococo were banished; the splendours of baroque became distasteful; the intricacies of *chinoiserie* lost their favour. The demand was for purity, simplicity and antiquity. The Grand Tour had done much to prepare the ground in England.[1] Familiarized with the ancients for the first time, hordes of English 'Milords' [2] returned from the continent demanding the pure, the correct, the scientific art as they chose to call it. Before long the neo-classical reigned supreme, and a ready sale awaited the first potter to produce a pleasing neo-classical style. Here was the perfect market for Wedgwood to exploit. He was not the man to ignore it. He changed his style and became the prophet of the new art form. It was to this realization of the possibilities of neo-classicism,[3] whilst his rivals still busied themselves with what he called 'a dazzleing profusion of riches & ornament',[4] that Wedgwood owed much of his success. For it meant that he was fully established as the favourite of the world of fashion. He had first use of a market 'randy for antique'.[5]

Wedgwood did everything he could to promote and to serve the new fashion. He based his vases on the urns and amphorae of the ancients, he decorated them with classical swags and garlands, he reproduced their cameo medallions and reclining figures. He invented new glazes to suit these designs and revived encaustic painting to decorate them. He named his new factory 'Etruria' and inscribed on its first products the words 'Artes Etruriae Renascuntur'. To clinch his position as leader of the new fashion he sought out the famous Barberini vase as the final test of his technical skill. At first his efforts were in vain. Lady Portland, like an ecstatic squirrel with a unique nut, had secreted it away amongst her other treasures, and would show it to none but her closest friends. But her death gave Josiah his chance, and his reproduction of the vase caught the imagination of the whole continent. Every detail of the mythology behind the vase was eagerly discussed and Wedgwood's name circulated through every European court.

Moreover, Wedgwood wanted his wares to play the part in contemporary art that the statues and ceramics of the ancients had played in all previous centuries, to become in fact part of the works of art of the future. With this end in view he commissioned Wright of Derby to paint his ware,[6] and invited Romney to use his wares as background material when in want of ornaments,[7] whilst in the family portrait by Stubbs, although of an equestrian nature and in a completely rural setting, a large Wedgwood and Bentley vase found a place by Josiah's side. In encouraging this attitude Wedgwood discovered one of the most sophisticated advertising techniques of the century—for the fact that his wares alone appeared on the canvases of such famous artists was bound to excite attention.[8]

[1] It is interesting to note that Wedgwood's classical products did not sell well in Russia — beyond the reach of the Grand Tour and the new fashion.

[2] Gibbon was told that there were 40,000 Englishmen on the continent in 1785.

[3] C. H. Wilson, 'The Entrepreneur in the Industrial Revolution in Britain,' *Explorations in Entrepreneurial History*, VII, No. 3 (Feb. 1955), 137.

[4] WMSS. E. 18365–25. JW to TB, 11 April 1772.

[5] Philip Larkin, *The Less Deceived* (Hessle, 1955), p. 27.

[6] WMSS. E. 18834–25. JW to TB. Endorsed by TB, 'Should have been dated 5 May 1778.'

[7] Wedgwood Correspondence, XL (1110), 6. John Ryland's Library, Manchester.

[8] For more detailed discussion of this practice c.f. Neil McKendrick, 'Josiah Wedgwood and George Stubbs,' *History Today*, VII, No. 8 (August 1957), 508–9.

It also helped to win the favour and support of the artists and the connoisseurs. How highly Wedgwood rated this support can be seen from a discussion with 'Athenian' Stewart about whether they would gain or lose by competition with Matthew Boulton of Soho.[1] 'We agreed that those customers who were more fond of show & glitter, than fine forms, & the appearance of antiquity wo^{d}. buy Soho Vases, And that all who could feel the effects of a fine outline & had any veneration for Antiquity wo^{d}. be with us.—But these we are afraid wo^{d}. be a minority; a third class we therefore call'd in to our aid, compos'd of such as wo^{d}. *of themselves* choose shewy, rich & gawdy (sic) things, but who wo^{d}. be *over ruled by their betters* in the choice of their ornaments as well as (in) other matters; who wo^{d}. do as their *architects*, or whoever they depended upon in matters of taste directed them; & with this reinforcement we thought Etruria stood a pretty good chance with any competitor'.[2] It was this belief in the selling power of fashion and the support of the art world which led Wedgwood to spend so much time in gaining the approbation of the connoisseurs, the artists and the architects. He had no intention of relying on merit alone to sell his goods, he sought out patrons and sponsors to reinforce that appeal. Just as he felt that his flowerpots would sell more if they were called 'Duchess of Devonshire flowerpots', and his creamware more if called Queensware, so he longed for Brown,[3] Wyatt [4] and the brothers Adam [5] to lead the architects in the use of his chimney pieces, and for Stubbs to lead the way in the use of Wedgwood plaques.[6] And he was right to do so. He backed the leaders of fashion in the belief that the rest of society would follow—and they did.

The struggle that Wedgwood had to sell his magnificent jasper tablets—generally accepted to be amongst the finest things he ever made, and now amongst the most expensive—illustrates the importance of this patronage. For the lack of it damned these tablets. Some were sold,[7] but they never sold well. For the fashionable architects refused to support them. Wedgwood and Bentley did everything they could to win them round: Wedgwood assiduously cultivated the friendship of 'Capability' Brown,[8] Bentley advocated their use to Adam,[9] and was urged by Wedgwood to call on Wyatt and 'try if it is not possible to root up his prejudices & make him a friend to our jaspers. If we could by any means gain over two or three of the *current architects* the business would be done'.[10] Their high quality alone could sell only a few. They needed proper sponsors. For 'If you are lucky in them no matter what the brat is, black, brown or fair, its fortune is made. We were really unfortunate in the introduction of our jasper into public notice, that we could not prevail upon the architects to be godfathers to our child. Instead of taking it by the hand, & giving it to their benediction, they have cursed the poor infant by bell, book & candle, & it must have a hard struggle to support itself, & rise from under

1 They were considering opening a London showroom in the Adelphi.

2 WMSS. E. 18335–25. JW to TB, 24 & 26 December 1770.

3 WMSS. E. 18147–25. JW to TB, 23 May 1767. He wrote on meeting 'the famous Brown... He may be of much service to me, & I shall not neglect what chance has thrown into my way'.

4 WMSS. E. 18855–26. JW to TB, 16 October 1778.

5 WMSS. E. 18394–25. JW to TB, 30 August 1772. '...it is very much in Mr Adam's power to introduce our things into use'.

6 Neil McKendrick, *op. cit.* p. 508.

7 Sir John Wrottesley, Sir Laurence Dundass and Lady Bagot bought them.

8 WMSS. E. 18853–26. JW to TB, 6 October 1778.

9 WMSS. E. 18394–25. JW to TB, 30 August 1772.

10 WMSS. E. 18855–26. JW to TB, 16 October 1778.

their maledictions'.[1] For once Wedgwood and Bentley's marketing techniques had failed. Their salesmanship had drawn a blank.

This was, however, an exception and serves only to illustrate the importance of Wedgwood's methods and the very real influence of that fashionable support which he so ardently courted. For by these methods Wedgwood had won the patronage of the court, the aristocracy, the artists and the cognoscenti. In doing so he had gained the favour of a powerful social catalyst. For in the smaller, more closely knit society of the European nobility of the eighteenth century, these patrons, these '*lines, channels & connections*' [2] as Wedgwood called them, were of vital importance. They led the fashion. They encouraged imitation. They spread the Wedgwood name abroad and sent presents of his ware: Horace Walpole bought it [3] and wrote to his widely scattered friends about it; [4] Mrs Crewe sent a dessert service to the Countess of Zinzindorf in Vienna; [5] 'the Duke of Richmond ... made a present of a pair of vases ... to the Duke of Leinster who was in Raptures with them'; [6] and so on. Wedgwood did not let the matter rest there. He had no hesitation in exploiting this patronage. When he heard of 'a violent *Vase madness* breaking out amongst' [7] the Irish, Wedgwood wrote in haste to Bentley; 'This disorder sho[d]. be cherish'd in some way or other, or our rivals may step in before us. We have many Irish friends who are both able & willing to recomm[d]. us, but they must be applied to for that purpose ... Ld. Bessboro' you know can do a great deal for us with his friends on the otherside(of) the Water by a letter of recommendation or otherwise as he may think proper. You are to visit him soon—the rest will occur to you. The Duke of Richmond has many & virtuous friends in Ireland. We are looking over the English Peerage to find out *lines, channels & connections*—will you look over the Irish Peerage with the same view—I need not tell you how much will depend upon a *proper* & *noble* introduction. This, with a fine assortment of Vases & a Trusty & *adequate* Agent will ensure us success in the conquest of our sister kingdom'.[8]

* * *

These were the more subtle advertising techniques of Josiah Wedgwood. They assured him a favourable reception for his wares in London and in the country houses of the rich. They stimulated interest and made his products known even in the provinces. They formed the basis of his sales policy—but only the basis. He had to use more direct methods to force home his advantage and exploit the position he had won for himself. Warehouses, showrooms, exhibitions, trademarks, new standards of display, puffing articles, straightforward advertisement, free carriage, and travelling salesmen; all of these played their part in Wedgwood and Bentley's marketing campaign.

Wedgwood was quick to realize the value of a warehouse in London. For high quality goods he needed a market accustomed to 'fine prices'. He was

1 WMSS. E. 18898–26. JW to TB, 19 June 1779.
2 WMSS. E. 18314–25. JW to TB, 2 August 1770.
3 *Catalogue of the Contents of Strawberry Hill* (1842), pp. 130, 131, 179, 180, 181.
4 *Letters of Horace Walpole*. ed. Mrs Paget Toynbee (Oxford, 1903–5), IX, 305; X, 282; XI, 172. also c.f. E. Meteyard, *op. cit.* II, 72. It must be admitted that Walpole was not always admiring.
5 WMSS. E. 18350–25. JW to TB, 17 September 1771.
6 WMSS. E. 18314–25. JW to TB, 2 August 1770.
7 WMSS. *ibid.*
8 WMSS. E. 18314–25. JW to TB, 2 August 1770. c.f. E. Meteyard, *op. cit.* II, 176–7.

not likely to find it in the annual market fairs of Staffordshire—the time-honoured *entrepôt* of their county's pots—nor amongst the country folk who haggled over their wares straight from the crateman's back or the hawker's basket, and to whom expense was the controlling factor in deciding their custom. A London warehouse would give him direct access to the fashionable clientele he aimed at and an opening in what was still the major distributing centre for the wholesale trade of the country.[1]

He first opened a warehouse there as early as 1765 and it soon became an integral part of his sales organization. It gave him the opportunity to put into action some of his most creative ideas. For apart from its success in the wholesale trade,[2] Wedgwood quickly reinforced its position by developing a vigorous retail trade in London. In two years his trade had outgrown his rooms in Grosvenor Square, and he was writing to Bentley, 'We must have an Elegant, extensive & Conven(ien)t shewroom',[3] and discussing the merits of different sites. Pall Mall was thought to be too accessible to the common folk, for he wanted space for more exciting methods of display [4] rather than for accommodation of the general public.[5] He planned to have a great display of his wares set out in services as for a meal 'in order to *do the needfull* with the Ladys in the neatest, genteelest, and best method. The same, or indeed a much greater variety of setts of Vases sho^{d}. decorate the Walls, & both these articles may, every few days be so alter'd, revers'd, & transform'd as to render the whole a new scene, Even to the same Company, every time they shall bring their friends to visit us'.

'I need not tell you the many good effects this must produce, when business, & amusement can be made to go hand in hand. Every new show, Exhibition or rarity soon grows stale in London, & is no longer regarded, after the first sight, unless utility, or some such variety as I have hinted at above continues to recommend it to their notice ... I have done something of the sort since I came to Town & find the immediate good effects of it. The first two days after the alteration we sold three complete setts of Vases at 2 & 3 Guineas a sett, besides many pairs of them, *which Vases had been in my Rooms 6–8 and some of them 12 months & wanted nothing but arrang(e)ment to sell them.*'[6] (My italics). It is clear from this that Wedgwood anticipated the most modern ideas of effective display—after nearly two hundred years retail potters use almost exactly the same layout to show off their wares.

He even anticipated a rudimentary self service scheme, for he planned to have his slightly inferior goods priced according to their quality, and displayed 'in one of the best places of your lower Shop, where people can come at them, & *serve themselves*'.[7] (My italics). Further, he laid out his tiles in patterns to show their full variety; [8] he placed his cheap vases on a separate range of shelves; [9] and to give his customers a greater sense of the rarity of his goods, he strictly limited the number of jaspers on display in his rooms at any given time.[10] To

[1] C. R. Fay, *Great Britain from Adam Smith to the present day. An economic and social survey* (1928), p. 132.

[2] With the development of the canal system and the growth of turnpike trusts its importance to Wedgwood's wholesale trade naturally declined, though it was still vital for foreign dealers.

[3] WMSS. E. 18147–25. JW to TB, 23 May 1767.

[4] WMSS. E. 18711–25. JW to TB, 4 November 1776.

[5] WMSS. E. 18149–25. JW to TB, postmark 1 June (1767).

[6] WMSS. E. 18149–25. JW to TB, 1 June 1767.

[7] WMSS. L. 17677–96. JW to William Cox, 7 April 1769.

[8] WMSS. E. 18711–25. JW to TB, 4 November 1776.

[9] WMSS. E. 18364–25. JW to TB, 6 April 1772.

[10] WMSS. E. 18802–25. JW to TB, 15 December 1777.

shield the delicate sensibility of his patrons' tastes he proposed 'a Curtain immediately for your Pebble ware shelves, which you may open or shut, inlarge or diminish the shew of gilding as you find your customers affected'.[1] For their entertainment he provided pattern books in all his warehouses as 'they will be looked over by our customers here, & they will often gèt us orders, & be pretty amusem^{t}. for the Ladies when they are waiting, wch. is often the case as there are som(e) times four or five diff^{t}. companys, & I need not tell you, that *it will be our interest* to amuse, & divert, & please, and astonish, nay, & even to ravish the Ladies ...'.[2]

His success was immediate. His account books, his lists of visitors and contemporary comment all record the constant streams of fashionable callers. As early as 1769 he was taking £100 a week [3] in cash sales alone at his London rooms, in addition to numerous orders. His men had to work night and day [4] to satisfy the demand and the crowds of visitors showed no signs of abating.[5] Wedgwood's, in fact, became one of the most fashionable meeting places in London.[6] It is not surprising, therefore, that Boulton and Fothergill in Pall Mall,[7] Josiah Spode in Fore Street, Cripplegate and then at the more genteel Portugal Street, Lincoln's Inn Fields,[8] and finally Minton [9] followed Wedgwood's lead and established warehouses and showrooms in London. For a fashionable appeal in London had vital influence even in the depths of the provinces. The woman in Newcastle-upon-Tyne who insisted on a dinner service of 'Arabesque Border' before her local shopkeeper had even heard of it, wanted it because it was 'much used in Lond^{o}. at present', and she steadfastly 'declin'd taking any till she had seen that pattern'.[10]

To encourage this outward spread of fashion and to speed it on its way, Wedgwood set up warehouses and showrooms at Bath,[11] Liverpool [12] and Dublin [13] in addition to the showrooms at Etruria [14] and Great Newport Street. The effect on the Liverpool potters of Wedgwood's competition can be seen from a contemporary's comment. The Liverpool local historian Enfield wrote in 1774, 'English porcelain, in imitation of foreign China, has long been

[1] WMSS. E. 18365–25. JW to TB, 11 April 1772.
[2] WMSS. E. 18232–25. JW to TB, February 1769. My italics.
[3] WMSS. E. 30857–5 and E. 30859–5. Peter Swift to JW, 18 and 25 March 1769.
[4] WMSS. E. 18230–25. JW to TB, 15 February 1769.
[5] WMSS. E. 30857–5 and E. 30859–5. Peter Swift to JW, 18 and 25 March 1769.
[6] D. Marshall 'London and the Life of the Town' in *Johnson's England*, ed. A. S. Turbeville (Oxford, 1933), I, 187. Miss Marshall quotes Lord Townshend on 'Squire Hanger', a beau and a macaroni,

'At Tattersall's, Wedgwood's, and eke the Rehearsal,
'Then straightway at Betty's he's sure to converse all;
'At Arthur's you meet him, and the mall in a sweat,
'At Kensington Garden's he's posted vidette'.

'Tattersall's' was the famous place for sportsmen, 'the Rehearsal' refers to the Opera House in the Haymarket, 'Betty's' was the famous fruitshop in St. James's Street. I have not been able to trace 'Arthur's'.
[7] WMSS. E. 18261–25. JW to TB, 27 September 1769.
[8] Arthur Hayden, *Spode and his Successors* (1951), pp. 20–22.
[9] Minton-Senhouse MS. and Minton Account Sales, quoted by Dr J. Thomas in his unpublished Ph. D. thesis presented to London University in 1934, 'The Economic Development of the North Staffordshire Potteries since 1730, with special reference to the Industrial Revolution', p. 815.
[10] WMSS. E. 1192–2. Joseph Harris of Newcastle-upon-Tyne to JW, 1780.
[11] WMSS. Numerous letters to Mr Ward from JW; e.g. E. 4428–6 to E. 4651–6.
[12] WMSS. Numerous letters to Mr Boardman from JW; e.g. E. –8 and E. –9.
[13] WMSS. Numerous letters to Mr Brock from JW; e.g. E. 3880–5 to E. 3908–5.
[14] WMSS. L.H.P. (a fragment), JW to TB, 27 July 1771.

manufactured in this town; and formerly with success. But of late this branch has been much upon decline, partly because the Leverpool (sic) artists have not kept pace in their improvements with some others in the same way; but chiefly because the Staffordshire ware has had and still continues to have so general a demand, as almost to supersede the use of other porcelain. The great perfection to which this art, both in works of utility and of ornament and taste, is carried at the modern Etruria, under the direction of those ingenious artists, Messrs Wedgwood & Bentley, at the same time that it is highly serviceable to the public and reflects great honour on our country, must be unfavourable to other manufactories of a similar kind'.[1]

It was on the London showrooms, however, that Wedgwood lavished most of his attention. By judicious use of shows and exhibitions he kept up his London sales [2] and advertised his more spectacular productions. These were carefully stage managed. Great care was taken in timing the openings,[3] and new goods were held back to increase their effect. As Wedgwood wrote to Bentley: 'Your shew will be vastly superior to anything your good Princes & Customers have hitherto seen. I am going up-on a large scale with our Models &c which is one reason why you have so few new things just now, but I hope to bring the whole in compass for your next Winters shew and ASTONISH THE WORLD ALL AT ONCE, For I hate piddleing you know'.[4] Winter, summer, spring and autumn sales were bolstered up by the occasional exhibition. Anything they made for the Queen, for instance, was automatically exhibited [5] before it was delivered, with reproductions [6] on sale to press home their advantage after the show had ended. But the most influential exhibition of all was that of the Russian service for Catherine the Great in 1774. Its display, Wedgwood thought, 'would bring an immence (sic) number of People of Fashion into our Rooms—Wo^d^. fully complete our notoriety to the whole Island, & help us greatly, no doubt, in the sale of our goods, both usefull and ornamental—It wo^d^. confirm the consequence we have attain'd, & increase it, by shewing that we are employ'd in a much higher scale than other Manufacturers. We should shew that we have paid many comp^ts^. to our Friends & Customers, & thereby rivet them the more firmly to our interests...'.[7]

Nothing was spared. For Wedgwood was determined to make the most of the opportunity. New rooms were taken; [8] the public—or rather the 'Nobility & Gentry'—informed that admittance was by ticket only; [9] and ample advertisement planned.[10] The success of the show was certain. Regarded as one of the most popular sights in London, it was visited by Queen Charlotte and by her brother His Royal Highness Prince Ernest of Mecklenburg,[11] and by the King and Queen of Sweden, and day after day for over a month the

[1] Dr William Enfield, *A History of Liverpool* (1774), p. 90, quoted in Knowles Boney, *Liverpool Porcelain of the 18th Century and its makers* (1957), p. 7.

[2] WMSS. E. 18853–26. JW to TB, 6 October 1778.

[3] WMSS. E. 18696–25. JW to (TB), September 1776.

[4] WMSS. E. 18614–25. JW to TB, 6 August 1775.

[5] WMSS. E. 18350–25. JW to TB, 17 September 1771.

[6] WMSS. uncatalogued. JW to TB, 17 October 1771.

[7] WMSS. E. 18498–25. JW to TB, 14 November 1773. No. 2 (i.e. 2nd letter from JW to TB that day).

[8] Portland House, Greek Street, Soho. First mentioned as 'our new Rooms' on 31 May 1774.

[9] WMSS. Draft of advertisement, May 30 1774. c.f. Dr. G. C. Williamson, *op. cit.* p. 33.

[10] *Ibid.* Planned for the front page of *Public Advertiser* & *Gazeteer*. (Miss Meteyard claims that it appeared in these and *St. James's Chronicle* but it can only be traced to *Public Advertiser* for 8 June 1774.)

[11] WMSS. E. 18547–25. JW to TB, 15 & 16 July 1774.

fashionable world thronged the rooms and blocked the street with their carriages.[1] Wedgwood had ensured its success by his choice of subject alone, for almost all of those whose country seats were represented on the service trekked from their distant homes to see the exhibition.[2] The last ounce of publicity was wrung out of it, by displaying duplicates of the service in the showroom at Etruria, and painting others 'without the Frog' for a continued display at Greek Street.[3] With this exhibition he had aroused and exploited the imagination of the fashionable world. He was equally capable of harnessing the emotion of the rest of society to serve his own ends.

No public event—Chatham dying,[4] Wesley preaching,[5] or Keppel pleading [6] —lacked its commercial opportunities for Wedgwood. As early as 1766 he wrote to Bentley, 'What do you think of sending Mr Pitt upon Crockery ware to America. A Quantity might certainly be sold there now & some advantage made of the American prejudice in favour of that great Man'.[7] Similarly when Admiral Keppel was tried by court martial and, amidst great enthusiasm, acquitted, Wedgwood wrote at once for a picture to copy, regretting that he had not 'had it a month since, and advertis'd it for pictures, bracelets, rings, seals &c.' [8] Exasperated by the delay he wrote that their travelling salesman 'says he could sell *thousands* of Keppels at any price. Oh Keppel Keppel—Why will not you send me a Keppel. I am perswaded (sic) if we had our wits about us as we ought to have had 2 or 3 months since we might have sold £1000 worth of this gentleman's head in various ways, & I am perswaded it would still be worth while to disperse them every way in our power'.[9] For the same purpose the rise of Methodism, the Slave Trade controversy, and the Peace with France were all given ceramic expression: Wesley, printed in black by Sadler and Green, on a Wedgwood teapot; [10] slavery on the famous jasper medallion of the kneeling slave, asking 'Am I not a man and a brother?' [11]; the Peace Treaty on a jasper plaque specially commissioned by Josiah from Flaxman.[12] Other contemporary figures much in the public eye—Garrick, Dr Johnson, Priestley, Mrs Siddons, Captain Cook and many others [13]—joined Wedgwood's series of famous heads: Greeks, Romans, Poets, Painters, Scientists, Historians, Actors and Politicians.[14] Made up into 'Historical Cabinets' [15] these heads found a ready sale. One alone proved abortive—the Popes. They were tried but sold poorly. They lacked sales appeal, for as Wedgwood explained 'nobody now a days troubles their head about his Holiness or his Predecessors'.[16]

Wedgwood also used newspaper advertisement—in London, provincial

[1] *Diary of Mrs Delaney*, 7 June 1774. Quoted G. C. Williamson, *op. cit.* pp. 34–5.
[2] Dr G. C. Williamson, *op. cit.* The list of views (1282 in all), pp. 55–91.
[3] WMSS. E. 18540–25. JW to TB, 20 June 1774.
[4] WMSS. E. 18840–25. JW to TB. 30 June & 1 July 1778.
[5] Donald C. Towner, *English Cream Coloured Earthenware* (1957), Plate 85(b).
[6] Keppel had been accused by Sir Hugh Pallister, his 2nd in command. He was acquitted on 11 February 1779 and received the thanks of both Houses.
[7] WMSS. E. 18123–25. JW to TB, 18 July 1766.
[8] WMSS. E. 18878–26. JW to TB, 25 February 1779.
[9] WMSS. E. 18880–26. JW to TB, 1 March 1779.
[10] Donald C. Towner, *op. cit.* Plate 85(b).
[11] WMSS. E. 19002–26. JW to Dr Erasmus Darwin, July 1789. A Copy.
[12] WMSS. E. 30193–2. JW to John Flaxman, 2 November 1786.
[13] Wolf Mankowitz, *Wedgwood* (1953), catalogues for 1779 and 1787, pp. 203–75.
[14] WMSS. E. 18657–25. William Cox to TB, 24 February 1776.
[15] WMSS. E. 18433–25. JW to TB, 2 January 1773.
[16] WMSS. *ibid.* Wedgwood was careful to avoid certain political implications, however, and refused to reproduce certain heads. e.g. E. 18772–25. JW to TB, 19 July 1777.

and even continental papers. This part of his marketing programme has received little attention, historians in general preferring to quote his occasional refusal rather than his more general acceptance of this medium. His remark, 'I wo^{d}. much rather not advertise at all if you think the sales are in such a way as to do without it...' [1] clearly indicates a certain reluctance. But this can be explained. It was due to his temporary fear of further attentions from 'Antipuffado' [2]—an anonymous opponent of 'that monstrous blast of puffery' [3] which eighteenth century manufacturers used to advertise their goods. This method itself—articles pretending impartiality but in fact praising certain goods—seems to have grown out of the widespread desire of big firms to avoid direct advertisement. They shrank from what Wedgwood called 'blowing my own trumpet' [4] and preferred to get others to do it for them. The company they would have to keep must also have discouraged them, for advertisements were mainly from petty traders, hawkers, quacks, local shopkeepers, and other more dubious professions. The Queen's Potter was naturally not keen to share a column with battling women [5] and fighting cocks,[5] nor eager to offer his services alongside those of a prostitute [5] or a gigolo,[5] a wet nurse [5] or a bug killer [5]—even though the latter claimed to serve the same monarch and be the oldest in the land. Wedgwood felt the same initial aversion to using travelling salesmen because it savoured of hawking.[6]

But whatever his feelings, a study of Wedgwood's letters and of contemporary newspapers makes it quite clear that he conquered them. Certain forms of advertisement he would never countenance. He banned his showrooms from using handbills, writing 'We have hitherto appeared in a very different light to common Shopkeepers, but this step (in my opinion) will sink us exceedingly ... I own myself alarm'd ... it being a mode of advertisement I never approv'd of ...'.[7] But there were many other forms which, when his stock began to mount, he was quick to use, writing 'This seems to point out advertiseing (sic) ... All trifling objections vanish before a real necessity'.[8] His faith in the value of advertisement is further borne out by his belief that Cooper and Duburk failed in Amsterdam because they did not make 'a fair experim^{t}. what advertising &c would do'.[9] And it is conclusively proved by the numerous occasions on which he used it. He advertised his ware,[10] his warehouse,[11] and his agents; [12] he advertised his Royal patronage and the support of the nobility; [13] he marked his ware and he advertised that mark.[14] He even organized the trial over encaustic painting in London for the sake of advertisement, writing to Bentley, 'May not this affair furnish us with a good excuse for advertiseing away at a great rate?' [15] Furthermore he proposed to publish prints of the pieces of

[1] WMSS. L. H. P. JW to TB, 13 February 1771.
[2] WMSS. L. H. P. JW to TB, 11 February 1771.
[3] E. S. Turner, *The Shocking History of Advertising* (1952), ch. II, *passim.*
[4] WMSS. E. 19001–26. JW to Dr Erasmus Darwin. Endorsed. 28 June 1789. (Copy).
[5] Turner, *op. cit.* pp. 28–48 and *passim.* Turner makes no mention of Wedgwood and Bentley.
[6] WMSS. E. 18827–25. JW to TB, 16 April 1778.
[7] WMSS. E. 18427–25. JW to TB, 7 December 1772.
[8] WMSS. L. H. P. JW to TB, 16 February 1771.
[9] WMSS. E. 18616–25. JW to TB, 10 August 1775.
[10] WMSS. E. 18341–25. JW to TB, 17 February 1771.
[11] WMSS. E. 18563–25. JW to TB, 10 November 1774.
[12] WMSS. E. 18504–25. JW to TB, 2 December 1773. In this case the agent was Brett.
[13] WMSS. *ibid.*
[14] WMSS. E. 18469–25. JW to TB, 7 June 1773. 'It will be absolutely necessary for us to mark them, & advertise that mark'.
[15] WMSS. F. 18325–25. JW to TB, 13 October 1770.

furniture into which Wedgwood ware had been introduced. A step which he believed 'would give sanction, & notoriety to our productions to such a degree, perhaps, as we have at present no idea of. I would put these Nos. into the common mode of sale in all the shops, & in our own Warehouses every where'.[1]

He did not neglect to keep up a steady stream of flattering articles in the press. Some of these occurred in the natural course of events. By its own fine quality, and the judicious attention of its makers, Wedgwood's wares had many admirers amongst the literary connoisseurs and won periodic praise for them in the daily news-sheets. But Wedgwood did not rely on this alone. He speeded up the process and augmented it. Although, for instance, he received two unsolicited puffs [2] in August and September 1770, by October he was writing to Bentley, 'There is a most famous puff for Boulton & Fothergill in the St. James's Chronicle of the 9th & for Mr Cox likewise, How the Author could have the assurance to leave us out I cannot conceive. Pray get another article in the next paper to complete the Triumverate'.[3] The attacks on this puffing technique, by Antipuffado, excited such attention that Wedgwood and Bentley discussed exploiting it for their own ends. For having realized that exaggerated abuse could be as effective in publicity as praise—one of the more advanced advertising ideas—they discussed methods of provoking their anonymous attacker to strike again: 'But should not we seem a little nettled & provoked to induce him to take up his pen again, for if he thinks his writeing is of service to us, he will certainly be silent. You mention his letter as a foundation for my advertiseing—How wod. you introduce the mention of it into an advertisement?' [4] After much discussion this idea was eventually rejected, but it shows an awareness of advertising techniques far ahead of their time. They were always conscious of the value of propaganda, and they were not above suggesting to the King and their customers that there was no hope of obtaining more of the vital ingredients for their jasper. 'This idea will give limits, a boundary to the quantity which your customers will be ready to conceive may be made of these bassreliefs, which otherwise would be gems indeed. They want nothing but *age* & *scarcity* to make them worth any price you could ask for them.'[5] He could not give them age but he did his best to imply that they were scarce. It is interesting to note that Wedgwood suggested to Bentley that he should burn this letter.

A study of their advertisements reveals a number of interesting developments in their selling policy. From a copy of his first,[6] it is clear that he had decided to pay the cost of carriage on his goods to London, even though this would mean a loss of £500 a year in his profits.[7] Of even greater importance is the way this policy developed in the advertisement outlined to Bentley in 1771 when poor sales demanded 'that some additional mode of sale be thought of or our dead stock will soon grow enormous'.[8] In this,[9] free carriage to London is extended to part payment—and a very considerable part—to any place in

1 WMSS. E. 18518–25. JW to TB, 20 February 1774.
2 WMSS. E. 18323–25. JW to TB, 1 September 1770, one in the *Gazette* and another in *Lloyds*. Another in the *Daily Advertiser*, L. H. P. JW to TB, 21 January 1771.
3 WMSS. E. 18325–25. JW to TB, 13 October 1770. Later JW denied that he ever advertised without affixing his name. L. H. P. JW to TB, 11 February 1771.
4 WMSS. L. H. P. JW to TB, 12 February 1771.
5 WMSS. E. 18802–25. JW to TB, 15 December 1777.
6 WMSS. E. 18230–25. JW to Sarah Wedgwood, February 1769.
7 WMSS. E. 18191–25. JW to TB, 3 March 1768.
8 WMSS. E. 18293–25. JW to TB, 18 April 1770.
9 WMSS. E. 18341–25. JW to TB, 17 February 1771.

England. In addition he offered the first recorded example of a satisfaction-or-money-back policy. Not only is this the first of its kind to be discovered in Europe or America but it antedates John Wanamaker—who is normally given the credit for this innovation—by nearly a century.[1]

Advertisement alone, however, was not sufficient fully to exploit the English market. As Wedgwood said, 'It seems absolutely necessary for the increase of our sales ... that some means must be unremittingly made use of to awake, and keep up the attention of the world to the fine things we are making & doing for them'.[2] He felt that his rival Voyez sold his wretched seals 'by mere dint of application to the buyers',[3] and so he went to work himself armed with pattern boxes, catalogues and samples. This was so successful that he extended it, and in 1777 he took the momentous decision to make his wares known throughout the country by personal introduction in the shape of travelling salesmen, and a crude and primitive version of the modern commercial traveller or sales representative can be seen in the proposals drawn up in October of that year between Wedgwood and John Brownbill.[4] Despite early difficulties Wedgwood persevered and by 1787 there were three such travellers on the road,[5] and by 1790 a book of rules and travellers' procedure, called the Travellers' Book,[6] had been drawn up. In it the record of their sales and their expenses bears ample testimony to their success.[7]

By such means Wedgwood broke through to a national market. By novelty, quality and fashionable appeal he won the favour of London and the notice of the provinces; with sales, exhibitions, and spectacular productions—all well advertised—he publicised this support; and with warehouses, salesmen and free carriages he invented the means of satisfying that demand. Having made his ware desirable, he had made it accessible.

* * *

The capture of the English market was not enough to satisfy Wedgwood. He longed to serve the whole world from Etruria, and constantly scanned the commercial horizon for new markets. No country—Mexico, Turkey, not even China—was too distant for him to contemplate with excitement. No obstacle—Russia's taste, Spain's hostility, or Portugal's prohibition—was too great for him to hope to overcome it. Difficulties served only as a challenge to his ambition. France—home of European porcelain, centre of rococo elegance, and safe behind a high tariff wall—was the greatest challenge of all. Even the thought of it inspired Wedgwood. 'And do you really think that we may make a *complete conquest* of France? Conquer France in Burslem? My blood moves quicker, I feel my strength increase for the contest—Assist me my friend, & the victory is our own ... we will fashion our porcelain after their own hearts, & captivate them with the elegance & simplicity of the ancients'.[8]

[1] Ralph M. Hower, 'The Wedgwoods — Ten Generations of Potters', *Journal of Economic and Business History*, IV, No. 2 (February 1932), 305.

[2] WMSS. E. 18880–26. JW to TB, 1 March 1779.

[3] WMSS. E. 18507–25. JW to TB, 10 December 1773.

[4] WMSS. E. 18784–25. JW to TB, 17 October 1777.

[5] WMSS. Byerley, Howorth and Brownbill.

[6] WMSS. L. 23571. Travellers' Book, c. 1793.

[7] WMSS. *ibid.* In 10 days in June 1793, the expenses amounted to £2.9.10½ (added up wrongly by the traveller to £2.9.10) and the sales to £101.3.2. I hope to publish further details of Wedgwood's travellers elsewhere.

[8] WMSS. E. 18252–25. JW to TB, 13 September 1769.

Necessity as well as ambition led Wedgwood and Bentley to seek new outlets for their products. They needed a larger market to move their stock, to exploit the capabilities of their production machine, and to swallow old lines which had exhausted their selling power in England. In the early seventies, when sales were slack, Wedgwood wrote 'we must either find some new markets or ... turn off some of our hands'.[1] The stock was too large, '& nothing but a *foreign* market ... will ever keep it within any tolerable bounds'.[2] He determined that 'Every *Gentle* & *Decent* push should be made to have our things *seen* & *sold* at Foreign Markets. If we drop, or do not *hitt off* such opportunities our selves we cannot expect other People to be so (in) attentive to them, & our trade will decline & wither, or flourish & expand itself, in proportion as these little turns & opportunitys are neglected or made the most of'.[3]

Wedgwood seized on the slightest hint of an opening into a new market. Merely reading in Lady Mary Wortley Montague's *Letters* of the Turks' taste for pots of perfume in the numerous arches around their rooms, filled him with lust for the Turkish market. It was a purely ceramic lust, however, for he wrote, 'Let who will take the Sultanas if I could get at these delightful little nitches, & furnish them, is all I covet in Turkey at present'.[4] This casual reference conjured up a whole range of commercial possibilities to Wedgwood and he was convinced that 'if we had a clever Ambassador there som(e) thing might be done'.[5] His desire for such a contact is easy to appreciate, for the diplomatic service—though no-one has realized this before—had proved one of the most fruitful channels of entry into foreign trade.

It was yet another way in which he exploited the favour of the aristocracy and his connections with the Establishment. They had already ensured a favourable reception for his goods in England. Their influence was not unfelt even on the Continent, but it required something more than this to penetrate fully the European market. When offered on the open market through the normal channels of merchant and middleman, the high quality of Wedgwood's products earned them immediate attention, but their price worked against them. Many lay idle as dead stock, some were returned as too expensive. They required a 'proper & noble introduction' such as he had contrived for them in England to overcome this drawback. What better introduction to the heart of European courts and their fashionable attendants could be devised than through her Majesty's ambassadors?

Wedgwood realized that they were naturally keen to raise the prestige of their country, and by flattery and presents he rapidly won their allegiance. 'Suppose we were to make S^{r}. W^{m}. Hamilton a present of an Etruscan tablet... it would be the best introduction they could have in the country where he resides'.[6] His confidence in such introductions was such that he had once written, 'The Russians must have Etruscan, & Grecian vases about the 19th Century. I fear they will not be ripe for them much sooner, unless our good friend S^{r}. W^{m}. Hamilton should go Ambassador thither & prepare a hot bed to bring these Northern plants to Maturity before their *natural* time'.[7] Everywhere such introductions proved invaluable, and through the agency of ambassadors, envoys, consuls and plenipotentiaries, Wedgwood's wares

[1] WMSS. L. H. P. JW to TB, 10 April 1771.
[2] WMSS. L. H. P. JW to TB, 11 February (postmark 14 March) 1771.
[3] WMSS. E. 18384–25. JW to TB, 5 August 1772.
[4] WMSS. E. 18407–25. JW to TB, 19 & 20 September 1772.
[5] *Ibid.*
[6] WMSS. E. 18855–26. JW to TB, 16 October 1778.
[7] WMSS. E. 18367–25. JW to TB, 18 April 1772.

entered—with no trouble and little expense—the courts of Russia, Poland, Portugal, Spain, Denmark, Sweden, the Netherlands, Turkey, Naples, Turin and even into China. Such a catalogue of services is impressive. But it is by no means complete. For these men were magnificent evangelizing agents for Wedgwood's ware. Each representative did more than introduce Wedgwood into one country. Ambassadors are peripatetic beings and like malaria-carrying mosquitos they carried Wedgwood's name abroad,[1] to convert the world to what Wedgwood called 'the true belief—(a belief) in our tablets',[2] vases and multifarious productions.

Wedgwood alone amongst the Staffordshire potters enjoyed these favours, and the honour of such attention was not lost on his customers. When the Portland vase was first successfully copied it was introduced to the courts of Europe in the finest possible style through Wedgwood's ambassadorial connections.[3]

Those connections were, however, only one of the methods used by Wedgwood to break through to an international market. In the export trade no less than in England, the process of marketing pottery underwent a great change. His general sales policy was the same. He was determined on superior quality rather than cheap production to sell his wares. He was also determined to keep his prices high. From the beginning, therefore, as in England he was committed to a policy of interesting the rich and exciting the favour of the fashionable. Once more he relied on court circles to publicize the unusual quality of his wares by buying the most outstanding pieces. He knew well enough that if it was bought by kings, it would be bought by their courtiers, and once fashionable at court it would be bought by the gentry, and so on down the social scale. The ambassadors had set these wheels in motion. But more than this was required. For there were many competitors for the European market. Firms such as Boulton and Fothergill were as alive to its possibilities as Wedgwood and Bentley, and they were not squeamish in their compliments. Occasionally they stole a march even on Wedgwood, as in 1776, when Josiah wrote in anguish to Bentley, 'They are now preparing a complimentary Group with a proper Inscription, upon the death of the Grand Duchess. You see they have carried *into execution* what we have only *talked about*, and will profit by it, so surely as Princes love flattery'.[4]

Moreover, they had to make their goods easily accessible to classes outside the court circles. There was no smooth ambassadorial introduction to the minor nobilities of Europe. They had to resort to cruder methods—they proposed to send a thousand parcels containing £20,000 worth of pottery,[5] to deluge Europe with earthenware, for it seemed 'the only mode in which our goods can get into such Familys'.[6] As Wedgwood wrote excitedly to Bentley: 'This object is great indeed, and my general idea upon it is to close heartily with it to *the utmost verge of prudence* or *rather beyond* [7] ... I think we sho^{d}. not sell

[1] Men like Sir Robert Liston who bought over £238 worth of Wedgwood ware whilst he was at Madrid and Stockholm, and later visited Washington, Batavia and Constantinople on diplomatic missions, c.f. D. B. Horn, *British Diplomatic Representatives, 1689–1789* (1932), pp. 138, 144, and *Concise D.N.B.* p. 782(a).

[2] WMSS. E. 18863–26. JW to TB, 22 November 1778.

[3] WMSS. Mosley MSS. JW to T. Byerley, July 1790, containing a transcript of Lord Aukland's letter to Wedgwood.

[4] WMSS. E. 18684–25. JW to TB, 14 July 1776.

[5] WMSS. L. H. P. JW to TB, 26 October 1771. 1000 £20 parcels.

[6] WMSS. L. H. P. JW to TB, 2 November 1771.

[7] My italics.

all to Italy and neglect the other Princes in Germany & elsewhere who are waiting with so much impatience for their turns to be served with our fine things—unless you think it better to send all to one place at a time that one Agent may first do the business in Italy, then in Germany and so on to Spain, Mexico, Indostan, China, Nova Zembla and the L^{d}. knows where.' [1] Germany was, in fact, the first to be tried. It was a great risk. But it came off. Wedgwood did not propose to repeat it. Only rising stocks, and the exhaustion of all other efforts to move them, justified such storm trooper methods. It was an exceptional technique and similar only to Wedgwood's flooding of Francfurt with specially prepared goods in 1790 at the coronation of Leopold as Emperor. The goods he prepared were in celebration of the coronation and of Leopold's life.[2] For such objects he could hope for only a temporary sale, and his intention was to advertise as much as to sell. They were designed to display his goods in the most spectacular fashion to the great congregation of European nobility that gathered to watch the coronation, and to the huge crowds that swarmed in their wake, in the hope that 'the remembrance of our fine things will be implanted with sufficient force upon their minds' [3] for them never to forget them. The Portland vase was displayed there for the same purpose. For it was not the Francfurt market that Wedgwood was aiming at—he had harnessed that before—but the market of the whole of Europe. This concentration of goods at Francfurt was like throwing a pebble into a pond and Wedgwood was more interested in the ripples than the splash. For slowly the fashionable crowds would disperse and with them would go Wedgwood's cameos, carried as seals on the bellies of Polish noblemen like Prince Czartoriskie, or worn as lockets at the throats of Portuguese princesses like the Marchioness of Pombal, to startle distant families by the brilliance of their colour and the sharpness of their modelling, and to win orders by their novelty from every corner of Europe.

Having won the notice and the custom of the nobility, Wedgwood wished to proceed lower in the social scale. 'The Great People have had these Vases in their Palaces long enough for them to be seen and admired by the Middling Class of People, which class we know are vastly, I had almost said, infinitely superior in number to the Great, and though a *great price* was, I believe, at first necessary to make the vases esteemed *Ornament for Palaces*, that reason no longer exists. Their character is established, and the middling People w^{d}. probably by (sic) quantitys of them at a reduced price'.[4] Simply by cheapening goods which he had already made fashionable Wedgwood immediately opened up a great new market. But it was not a method which he relished, and his wares were still far from cheap. To win this class completely he had to appeal to the differences in its interests as well as in its purse. It clearly required different marketing techniques from those used to seduce the upper classes. The mass of the population was socially inaccessible to ambassadors, too numerous for individual parcels, and too insignificant to be flattered by reproduction. But if Wedgwood could not appeal to their vanity, he found an admirable substitute in their loyalty. He made cameo medallions of their monarchs, writing to Bentley, 'I hope to make some ... use of his C(atholic)

1 WMSS. L. H. P. JW to TB, 26 October 1771.

2 WMSS. E. 19010–26. 'Invoice of the Ornamental Ware shipped by JW & Co, to Fran(c) furt S/M.' 11 September 1790.

3 WMSS. Mosley MSS. JW to JW, II, 3 September 1790.

4 WMSS. E. 18392–25. JW to TB, 23 August 1772.

Majesty in the Spanish Trade—*if the subjects are fond of their King*'.[1] He exploited not only their loyalty to the crown but their patriotism, their pride in their national heroes, writing, 'People will give more for *their own Heads*, or the *Heads in fashion*, than for any other subjects, & buy abundantly more of them ... We should select the proper Heads for the different European Markets ... and this Plan will certainly increase our wholesale business'.[2] Their faith was equally skilfully exploited: the Popes for Italy and Spain,[3] the saints for South America,[4] Mohammed for Turkey.[5] Buddha alone of the better known gods seems to have been neglected—presumably for economic reasons.

To the varying fashions and different tastes of his foreign buyers he gave his detailed attention. For France, for instance, where the rococo wonders of the mid-century were far from dead, Wedgwood produced ormolu mounted pottery to meet the prevailing fashion.[6] Though in Russia he dumped his old goods 'much seen or blown upon', he also produced a special pot for them alone [7] and sent them 'shewy, tawdry, cheap things, cover'd all over with colors (sic)' [8] because they thought cream ware ugly. For hot climates which shared this aversion, he made 'green & Gold ware' because 'they do not like *pale, colourless ware*'.[9] To America, adjudged not ripe for expensive things at present he sent mainly cheap goods and seconds, whilst for Turkey he invented a whole new range of goods to suit its exotic fancy.[10] Nor did he neglect the minor details of national habit—cups in the Saxon fashion were made for Germany; and small coffee cups, as was their custom, for the Venetians.[11]

By these means Wedgwood had created an enormous demand for his ware both ornamental and useful. The upper classes bought both, but mainly the expensive ornamental wares, and in imitation of their social superiors the lower classes bought the useful. He had achieved this success by wide and sweeping changes in the potters' marketing techniques. He had, however, a further contribution to make. He radically altered their methods of distribution. He built canals, promoted turnpike trusts and developed a sales organization of his own. His part in the promotion of turnpikes and canals was vital to the development of Staffordshire for 'they were the basis of the prosperity of the Potteries'.[12] This aspect of his work is too well known to require repetition here. His attempt to break away from the middleman in the distribution of his goods has, however, been only slightly touched upon by other historians.

He had dealt since 1769 through middlemen such as Boulton and Fothergill, Bentley and Boardman, Hume and Walmesley, Edmund Radcliffe and a host of others abroad. But, vital as their service was to most potters, Wedgwood

[1] WMSS. E. 18669–25. JW to TB, 15 May 1776.
[2] WMSS. E. 18679–25. JW to TB, 2 July 1776.
[3] WMSS. L. 10137–12. 'A List of orders for Mr Walmesley, Deans Gate, Manchester'. 30 January 1775. 'Saints &c may answer at this market, try to provide some... & send a sett of the Popes... or a few loose ones'.
[4] WMSS. E. 18561–25. JW to TB, 5 November 1774. '...some articles sho[d]. be made on purpose for this trade relative to their Religion... Crucifixes, Saints &c'.
[5] WMSS. E. 18522–25. JW to TB, 8 March 1774. '...proper subjects for the Faithfull amongst the Musselmen'.
[6] WMSS. E. 18193–25. JW to TB, 15 March 1786.
[7] The 'black & yellow'.
[8] WMSS. E. 18487–25. JW to TB, 14 August 1773.
[9] WMSS. E. 18500–25. JW to TB, postmark 22 November 1773.
[10] WMSS. E. 18444–35. JW to TB, 4 & 6 March 1773.
[11] WMSS. E. 31191–1. TB to JW, 18 October 1776.
[12] J. H. Plumb, *England in the Eighteenth Century* (1950), p. 147.

was rapidly outgrowing his reliance on them. More and more merchants, attracted by Wedgwood's name and reputation, were writing to him personally in order to get more favourable terms.[1] Naturally Wedgwood was keen to accept their advances and dispense with the middlemen and their profit-devouring commissions, and he knew that they would 'leave us whenever they can buy 6d P doz cheaper. I would therefore wish us to have a correspondence of our own, independent of any set of men whomsoever, both at home and abroad, with the Merch[ts]. & with the Shops. We can make any quantity, & the only P—t we can now have is to make them *perfect* & *disperse* them. The former shall have my best attention here & I shall lose no opportunity of assisting in the latter as occasions may offer'.[2]

Although his reputation attracted many buyers, Wedgwood did not rely on his name alone to overcome the many difficulties—distance, language and tariff prohibitions—which foreign merchants had to face. He sought them out with pattern boxes,[3] and catalogues in translation;[4] tempted them with discounts, reductions and special terms for the first order;[5] and eased their problem of delivery by establishing foreign warehouses like those in Dublin, Paris and Amsterdam, and employing foreign agents like Veldhuyson and Perregaux. For the further comfort of his foreign buyers he employed French, German, Italian and Dutch-speaking clerks and answered their letters in their native tongue. Ample testimony to his success and the increasing momentum of commercial development he brought about can be found in his account books. And an analysis of his foreign correspondence reveals the constant expansion of foreign orders. He received his first order from Amsterdam in 1764; from St. Petersburg and Brunswick in 1769; from Dublin in 1771; from Naples in 1773; from Dessau, Leipsig and Paris in 1774; Bonn, Dresden, Dunkirk, Leghorn, Malaga, Rotterdam, Trieste and Venice in 1775; Goa in 1776; Moscow and Nice in 1777; Ostend, Rome and Vienna in 1781; Geneva in 1782; Antwerp, Brescia, Cadiz, Hamburg, Ratisbon and Stuttgart in 1783; Brussels, Genoa, Lisbon, and Palermo in 1784; Dorpat, Marseilles, Stockholm, Strassburg in 1785; Basle, Bilbao, Bologna and Madrid in 1786; Danzig, Rouen, Turin in 1787; Ancona, Berne, Oslo, Lübeck, Mittau, Nuremburg, Parma, Riga, Udine in 1788; Boulogne, Darmstadt, Douai, Mainz, Mannheim, Milan, in 1789; Göttingen, Regensburg, Tournai in 1790; Ansbach and Copenhagen in 1791; Cologne and Memmingen in 1793.[6]

[1] WMSS. E. 5077–7. Conrad Wilhelm Krause of Brunswick ('Brounschwyk') to JW. 15 February 1771. Krause had 'several Times recieved by Hands of my Frinds (*sic*) Goods from your Fabric' but now he wished to 'Negociate Direct'. There are many similar examples.

[2] WMSS. E. 18473–25. JW to TB, postmark 21 June (1773).

[3] WMSS. E. 18501–25. JW to TB, 21 November 1773. 'We shall want some hundreds of small dishes to send abroad as patterns the next spring....'.

[4] First in French in 1773 (E. 18501–25), then German and Italian in 1774 (E. 18518–25 and E. 18524–25), and finally Dutch and Russian in the same year (E. 18527–25). They seem to have been a new idea, at least to the Potteries because incredibly elaborate steps were taken to keep the illustrated ones secret in order 'to get the start one season at least'.

[5] The elasticity of JW's attitude to discounts is fully illustrated in his dealings with Messrs. James Jackson & Co of St. Petersburg. L.H.P. 1771.

[6] This list is meant to give an impression of the rapid spread of JW's exports rather than to be a complete list. Many of the dates may have to be adjusted forwards, and he is known to have been dealing with the Hague, Metz, Limoges, Zurich, Lausanne, Bordeaux, Eperney, Bayruth, St. Amand, Florence, Gothenburg, Konigsberg, Oporto, Archangel, Warsaw, Bremen and Messina by 1790, but I have not yet established when he secured the first order. This list is culled from the whole range of WMSS. but more especially from E. 609–1 to 30210–1; E. 835–2 to 1954–2; E. 2742–4 to 3282–4; E. 3724–5 to 1090–5; E. 4321–6 to 31129–6. c.f. Hower; *op. cit.* p. 309.

One final point requires attention. For no account of Wedgwood's marketing activities would be complete without some mention of the part he played in organizing the potters, appealing to ambassadors and exploiting his noble connections to bring pressure to bear upon the formation of economic policy and the government's attitude to import restrictions and prohibitions. This aspect of his career is more germane to Wedgwood's political activities and as such is beyond the scope of this article, but it is necessary to point out here that by his action he influenced the Government in its formulation of the Irish Treaty of 1785, initiated an attempt to lift the Swedish prohibition on English earthenware in 1789, and led the potters in their efforts to secure favourable commercial treaties with Portugal in 1785, and, most important of all, with France in 1787. Some idea of the effects this could have on the potters' market can be judged by a comparison of the earthenware exports to France in 1785 and 1789. In 1785 they totalled £641; four years later they amounted to £7,920.[1] Yet again Wedgwood had penetrated a market which had defied all previous English potters.

Such was his success that he had in the words of Faujas de Saint Fond, 'created a commerce so active and so universal, that in travelling from Paris to St Petersburg, from Amsterdam to the farthest point of Sweden, from Dunkirk to the southern extremity of France, one is served at every inn from English earthenware. The same fine article adorns the tables of Spain, Portugal, & Italy, and it provides the cargoes of ships to the East Indies, the West Indies and America'.[2] In Poland in 1783, it was announced that 'His Majesty (Stanislas Augustus) wishing to put an end to the considerable loss in currency caused by purchases of tableware manufactured in England, [has] established ... at great expense, a pottery at the Belvedere palace'.[3] Even the great European factories—Sèvres, Meissen, Vienna, Furstenburg, Paris and Doccia had to follow the humble Staffordshire potters and reproduce Wedgwood's designs.

If I have laboured this point it is to show to what lengths Wedgwood was prepared to go to sell his wares, to show what detailed attention he lavished on his customer's requirements and to show how misguided is the accepted and often repeated view that Wedgwood and Bentley 'were in fact too absorbed in the creation of beauty to be overmindful of the means and methods of its dissemination'.[4] Nothing could be further from the truth. In fact, far from such delightful indifference to sales and such unselfish devotion to beauty, Wedgwood was quite prepared to reproduce ugly objects if his customers wanted them, writing, 'I have a very small vase which was dug out of Herculaneum ... I do not see any beauty in it but will make something of it if Sir William Farringdon wishes it'.[5] Moreover, when his orders exceeded his output, he answered the demand by supplying ware which he had bought from other potters—potters like Lowe, Astbury, Meir, Garner, Turner, Heath, Brown and Malkin and many others [6]—whose products were usually cheaper imitations of Wedgwood, often much below his standard of production, which

[1] G. Villiers & John Baring, *Final Report of the Commercial Relations between France and Great Britain* (Parl. Report, 1834), p. 87.

[2] Faujas de St. Fond, *Voyage en Angleterre, en Ecosse et aux Iles Hébrides*, I, 112.

[3] Witold Kula: *Szkice o manufakturach w Polsce, XVIII wieku* (Warsaw 1956), I, 304, quoting from Pamiztnik... 1783 [ed. Świtkowski]. I am indebted to my colleague, Dr L. R. Lewitter, of Christ's College, Cambridge, for this reference.

[4] Meteyard: *op. cit.* I, 368–9.

[5] WMSS. E. 18271–25. JW to TB, 1 December 1769.

[6] WMSS. E. 4840–6 to 5062–6. And various other scattered references.

could never have sold in quantity without the aid of Wedgwood's marketing organization. In the 1780's when the supply constantly lagged behind the demand, Wedgwood was forced to buy in quantity from other potters—he bought £4,500's worth from George Neunburg alone in the first six months of 1784.[1] Nothing displays better the importance of Wedgwood's salesmanship than this period. For so fashionable had his name become and so popular his wares, that he could sell at a higher price what his rivals could not sell at all.

Despite constant complaints of high prices, slow delivery, bad packing and inadequately made-up orders, the retail merchants had to deal with Wedgwood in preference to any other potter. For their customers—both foreign and English, both humble and aristocratic—knew of Wedgwood ware, knew that the English queen, the Russian empress and countless foreign and native aristocrats used it, and they were determined to have those pieces of Wedgwood which they could afford. Patterns seen in the London showrooms were insisted on by ambitious hostesses in the provinces; Catherine the Great's service seen in St. Petersburg persuaded Muscovite nobles to order similar sets; heads of the Popes in jasper spread Wedgwood's name through Italy, Spain and South America; and the Queen of Portugal in cameo proved irresistible to the people of Lisbon. Medallions of the notables of Germany, Holland, France, Poland, America, Sweden, Denmark and Turkey served a similar purpose there. Once they reached these distant parts their excellence proved their own advertisement.[2]

In fact the methods of distribution suited to the peasant craft stage of the potteries had proved totally inadequate to dispose of the growing production of Etruria. And Wedgwood had completely transformed them.[3] The impact of the Industrial Revolution in the potteries had an inevitable effect on the attitude of the potters to marketing their goods. It called for new methods of salesmanship and new centres for display. To succeed the potter needed merchant partners,[4] foreign agents, salerooms, warehouses, travelling salesmen, catalogues and trained linguists to deal with the increasingly technical problems of foreign trade. He also needed improved transport and more favourable commercial agreements. It was Wedgwood who provided them, and gave the lead to the others.

Dr Thomas allows such changes to signify 'a commercial revolution in the disposal and dispersion of their goods as real and disturbing as the productive changes which occurred inside their industrial organisation and as far reaching as the Communication and Transport revolutions which occurred outside

[1] WMSS. L. 1788 to 1789. c.f. Hower, *op. cit.* p. 301.

[2] Neil McKendrick, 'Josiah Wedgwood: Industrial Pioneer', *The House of Whitbread* (Winter issue, 1957–58), p. 11.

[3] He had realized that the difficulties involved in buying from Etruria might have discouraged all but the most ardent, 'it will only be a few who have the disorder very strong upon them who will be at the trouble of procuring them at such a distance'. WMSS. E. 18318–25. JW to TB, 20 August 1770.

[4] Other potters teamed up with merchants — Josiah Spode II with William Copeland, the successful London tea merchant in 1824, Thomas Minton with William Pownall, the Liverpool merchant in 1793 — but Wedgwood's association with Bentley which began in 1769 was one of the earliest and most successful of the great eighteenth-century 'inventor and entrepreneur' partnerships, rivalling that between Boulton and Watt, Brindley and Bridgewater or Roebuck and Black. Owing to the disappearance of all but fragments of Bentley's correspondence it is difficult to do full justice to his part in the partnership in an article of this kind. His contribution can, however, be disentangled from a close study of Wedgwood's replies and I hope to show elsewhere how important it was.

their factories',[1] and dates its completion as 1850. But although the other potters did not experience such a revolution until that date, there can be little doubt that Wedgwood had initiated all the most important changes by 1790. Yet no aspect of Josiah Wedgwood's life has been so neglected as his impact on the commercial techniques of the eighteenth century. Few are more important.

For it was by such methods that a local craft became a national industry and served an international market. In 1775 Wedgwood had hoped to 'ASTONISH THE WORLD ALL AT ONCE';[2] what he expressed as a hope in 1775, he had accomplished as a fact in 1795. His ware was in universal demand. Admired by the Emperors of China, Russia and Germany; praised by scientists of the calibre of Priestley, Watt and Black; and painted by artists as fashionable as Stubbs, Romney and Wright of Derby, it was acclaimed by art, science and society. And—which was more important for Wedgwood—it was equally acclaimed by the public. For it was from his huge sales of his common useful ware—seals, buttons, inkpots, tableware and the like—that Wedgwood drew his greatest reward from his commercial campaign. The servant's hall was quick to follow its mistress's lead, and Wedgwood's accounts consistently return a higher percentage of sales and takings in his useful ware than in his ornamental: even in fashionable Bath, the proportion was 60 to 40.[3] It is therefore in the fading lists of outstanding accounts and amongst the neglected bundles of everyday orders that the true picture of Wedgwood's universal appeal and widespread success is to be found. They record the ambitions of the chef of the Yacht Inn in Cheshire who hoped to found his gastronomic reputation on Wedgwood's creamware; the taste for Wedgwood shared by a German professor at Brunswick and a bachelor don at Cambridge; the popularity of Wedgwood in a lonely military garrison in Quebec; and the purchase of Wedgwood by Edward Gibbon whilst writing his great history in Lausanne. These and many others bought it: Spanish ambassadors, Indian colonists, Bohemian nobles, Bristol chemists, Oxford colleges, Lancashire merchants and Sicilian monarchs. By superb reproduction and the exercise of his vivid entrepreneurial imagination Josiah Wedgwood had achieved his purpose. He was what he wished to be: 'Vase Maker General to the Universe'.[4]

Gonville & Caius College, Cambridge

1 Dr J. Thomas, *op. cit.* p. 771 *et seq.*
2 WMSS. E. 18614–25. JW to TB, 6 August 1775.
3 WMSS. E. 4428–6 to 4651–6. Returns to Wedgwood of the takings in the Bath salerooms.
4 WMSS. E. 18232–25. JW to TB, February 1769.

Joel Mokyr

Demand vs. Supply in the Industrial Revolution

The intelligent application of the formal principles [of economic theory] is . . . chiefly significant negatively rather than positively, for showing what is "wrong" rather than what is "right."

Frank Knight (1940)

In the process of what could best be called "making sense of the Industrial Revolution," few articles have been more influential than Gilboy's eloquent plea to view demand as an equal partner in bringing about the most profound economic change in human history.[1] The notion that *both* sides of the demand and supply equation come into play in the explanation of the crucial questions, such as "why England first" or "why the eighteenth century," has made sense to more than a generation of economic historians better trained in handling Marshall's scissors than Occam's razor. The appearance of Keynes's *General Theory* a few years later lent additional support to the notion that demand was somehow important.[2]

The present paper reexamines what may be termed the "Gilboy thesis." The concept of demand itself is too vague without additional clarification. What I shall therefore attempt is to reformulate the Gilboy thesis in consistent and testable form.

Journal of Economic History, Vol. XXXVII, No. 4 (December 1977).

I would like to express my thanks to Donald N. McCloskey, on whose very elastic supply of knowledge in economics, history, and econography I have liberally drawn. Others who have been of great help with comments and suggestions include Pekka Ahtiala, Reuven Brenner, Karl de Schweinitz, Stefano Fenoaltea, Jonathan Hughes, Eric L. Jones, Glenn C. Loury, Jacob Metzer, F. Michael Scherer, and Richard O. Zerbe.

[1] Elizabeth Waterman Gilboy, "Demand as a Factor in the Industrial Revolution," in A. H. Cole, ed., *Facts and Factors in Economic History* (1932); rpt. in R. M. Hartwell, ed., *The Causes of the Industrial Revolution in England* (London, 1967), pp. 121-38.

[2] Very explicit reiterations of Gilboy's thesis can be found, for example, in David Landes, *The Unbound Prometheus* (Cambridge, 1969), p. 46; D. E. C. Eversley, "The Home Market and Economic Growth in England, 1750-1780," in E. L. Jones and G. E. Mingay, eds., *Land, Labour, and Population in the Industrial Revolution* (London, 1967), pp. 206-59; Phyllis Deane, *The First Industrial Revolution* (Cambridge, 1969), p. 34; A. H. John, "Agricultural Productivity and Economic Growth in England," in E. L. Jones, ed., *Agriculture and Economic Growth in England, 1650-1815* (London, 1967), pp. 172-93. A recent restatement is in François Crouzet, "Western Europe and Great Britain: 'Catching Up' in the First Half of the Nineteenth Century," in A. J. Youngson, ed., *Economic Development in the Long Run* (New York, 1967), pp. 113-14. A variation on the same theme is in Nicholas Georgescu-Roegen, *The Entropy Law and the Economic Process* (Cambridge, Mass., 1971), pp. 246-47. Doubts concerning the demand hypothesis were sounded by W. W. Rostow, *How It All Began* (New York, 1975), pp. 14, 129, 172-73, and in a slightly different context by Donald N. McCloskey, "Did Victorian Britain Fail?" *Economic History Review*, 23 (Aug. 1970), 446-59.

I. DID THE DEMAND CURVE SHIFT?

It is possible to interpret the Gilboy thesis as a shift in aggregate demand, or as a shift in the market demand curve for industrial goods. Leaving the former for section III, we first consider whether the Industrial Revolution could be triggered by an outward shift of the industrial demand curve.

In her original article, Gilboy maintained that

> The factory could not become typical until demand had been extended . . . throughout the entire population to consume the products of large scale industry. . . . In order that a shift in the demand schedule may occur, individuals must be able to buy more units of a commodity at the same price, or the same amount of the commodity at a higher price . . . the entire schedule must shift upward, indicating a greater buying power.[3]

In this simple form the demand thesis is based on circular reasoning. A shift *of* the demand curve for manufactured goods can occur only if income rises, the price of nonmanufactured goods falls, or if a change in tastes occurs. Ruling out the latter for the moment, the shift in the demand curve must be caused by a rise in real income, and can therefore not serve at the same time as an explanation of it.

It is transparent that if a shift in the demand curve for industrial goods is to be used for the explanation of the rise of industrial output, the shift in the demand curve must be caused by factors other than the rise of output itself. Three alternative theories have been presented in the literature in this context, namely agricultural growth, expansion of foreign demand, and population growth.[4] It is worthwhile to examine these theories in more detail. The first and in many ways the most attractive hypothesis is agricultural progress. A decline in prices of agricultural goods will lead to an increase in the demand for nonagricultural goods if, ceteris paribus, the demand for agricultural goods is inelastic, as is usually assumed. While unimpeachable on a priori grounds, the theory runs into a timing dilemma: whereas agriculture prices fell in the first part of the eighteenth century, they started to rise after 1750, a trend that persisted until the end of the Napoleonic Wars both in Great Britain and on the Continent.[5] More-

[3] Gilboy, "Demand," pp. 122-26.

[4] W. A. Cole, "Eighteenth-Century Economic Growth Revisited," *Explorations in Economic History*, 10 (Summer 1973), 327-48.

[5] The evidence for France is summarized in C. E. Labrousse, *Esquisse du mouvement des prix et des revenus en France au XVIIIe siècle* (Paris, 1933), pts. II, III, IV, esp. pp. 137-66. For the nineteenth century, too, the timing does not work. The period usually associated with the most rapid economic growth in France (1840-70) was also one of rising wheat prices. See ibid., p. 141, and B. R. Mitchell, *European Historical Statistics* (New York, 1975), p. 742.

over, it is not always sufficiently emphasized that there should be a decline in the price of agricultural goods *relative* to the price of nonagricultural goods. The movements of relative prices are less clear-cut.

What was the contribution of falling agricultural prices to the rise in demand for nonagricultural goods? The order of magnitude of the impact of changing relative prices on demand can be estimated approximately following a procedure outlined in a recent study by R. A. Ippolito.[6] Since the price elasticities as well as the actual movement of prices are subject to large margins of error, the procedure followed has been to estimate a lower and an upper bound of the expansion of industrial demand due to the fall in relative prices.[7] The upper bound of the estimate is 26.1 percent for the entire period 1750-1850, or .26 percent annually, while the lower bound is −.25 percent annually.[8] The average rate of growth of individual output for the entire period can be estimated conservatively at 3 percent annually. The changes in relative prices thus account for at best 8 percent of the industrial expansion, while it is quite possible that their net effect was on the whole negative.

Some efforts have been made to rescue the argument by maintaining that rising agricultural prices due to an exogenous decline in

[6] Richard A. Ippolito, "The Effect of the 'Agricultural Depression' on Industrial Demand In England, 1730-1750," *Economica*, 42 (Aug. 1975), 298-312. Ippolito's overall conclusion is that the agricultural depression of 1730-50 did not constitute a major source of demand for industrial goods either. It should be emphasized that the estimates do not truly reflect the contribution of the agricultural revolution (i.e., shifts of the supply curve of agricultural goods) to the industrial revolution. This would only be the case if the supply curve of agricultural goods were perfectly elastic (as is assumed by Ippolito) or if demand were stationary. Neither of these assumptions seems plausible; hence, what is measured is the impact of price changes only, without further identification of their source.

[7] Let the demand function have the form $Q_M(D) = Y^a P_A{}^b P_M{}^c$. This means that we can write the relationship between the demand for industrial goods and the relative price $P = P_A/P_M$ as

$$\frac{\Delta Q_M(D)}{Q_M(D)} = \left[\frac{\Delta P}{P} + 1\right]^b - 1$$

An approximate value of b can be estimated by using the so-called Cournot relationship between the own and the cross elasticities of demand:

$$\gamma_A \epsilon_A + (1 - \gamma_A) b = -\gamma_A$$

ϵ_A is the own price elasticity of demand for agricultural goods and γ_A is the share of agriculture in total output. A reasonable estimate of γ_A would put it at around $\frac{1}{3}$. The value of ϵ_A is unknown, but is generally believed to be less than one in absolute value. Setting it equal to −.3 and −.7, we obtain estimates of b between −.35 and −.15.

[8] The relative prices underlying the calculations were obtained and computed from a number of series which overlap to some extent, or which had to be converted to relative prices using some assumptions about weights. The procedure followed was in each case to choose the extreme estimates, thus obtaining upper and lower bounds.

supply (for example, harvest failures) could, under certain conditions, stimulate industrial demand. The mechanism supposedly responsible for this is that a bad harvest raises agricultural prices. Since the elasticity of demand for farm products is usually thought to be less than unity, farm income would rise. If the farmers have a very high propensity to spend on manufactured goods, a harvest failure could allegedly cause a net rise in the demand for nonagricultural goods.[9] In a simple general equilibrium framework it can be proven that if both goods are normal, this cannot occur. Only if agricultural goods—taken as a whole, including both consumption and investment goods—are inferior goods (that is, the income elasticity for them is negative), demand for nonagricultural goods may (but does not have to) expand as a result of harvest failures. For all practical purposes, this version of the demand thesis can thus be ruled out. The proof of the theorem is relegated to Appendix A.

Eversley has maintained that the post-1750 rise in agricultural prices was not caused by a decrease in supply but by a rise in demand.[10] The absence of detailed output data precludes a direct test of this argument, but the work of Eric Jones and others demonstrates that agricultural productivity followed an upward secular trend throughout the eighteenth century, including the period 1750-1780. But it does not help to explain the growth in industrial production. If the increased demand for agricultural goods came at the expense of industrial goods, say through a change in tastes or income distribution, it runs directly counter to the Gilboy thesis. If it did not come at the expense of the industrial goods, where did it come from? Surely in this case rising aggregate income must be the underlying mechanism. To the extent that a higher aggregate income resulted from population growth, it will be dealt with below. But if higher income per capita is the source of the expansion of demand we are back in the same circular argument as before.[11]

[9] David S. Landes, "The Statistical Study of French Crises," this JOURNAL, 10 (Nov. 1950), 200-01. Phyllis Deane and W. A. Cole, *British Economic Growth, 1688-1959* (2nd ed.; Cambridge, 1969), p. 93. John D. Post, "A Study in Meteorological and Trade Cycle History: The Economic Crisis Following the Napoleonic Wars," this JOURNAL, 34 (1974), 338-39.

[10] Eversley, "The Home Market," pp. 240-46.

[11] A general discussion of the impact of agricultural fluctuations on the English economy is in J. D. Gould, "Agricultural Fluctuations and the English Economy in the Eighteenth Century," this JOURNAL, 22 (Sept. 1962), 313-33. Gould's attempts to explain the price movement by the operation and later repeal of the Corn Laws in the eighteenth century cannot be viewed as definitive—the export bounties were repealed in 1773, and it is hard to see how that could help raising prices.

A more convincing rationalization of the apparent paradox is provided by Eric Jones.

> It is unlikely that higher food prices after 1750 ever seriously eroded the expansion of the home market for consumer goods by eliminating the margin for such spending among the working population. Agriculture constantly demanded more labour. The "golden age" of the labourer in the second quarter of the eighteenth century apparently engrained in him tastes for manufactured goods which he was willing to work harder thereafter to gratify.[12]

In other words, there seems to have been a shift of the labor supply curve due to a reduced leisure preference. Following this argument, it can indeed be maintained that demand factors mattered insofar as the supply of labor, the demand for leisure, and the demand for goods are simultaneously determined. If there was an increase in the "demand for income," economic growth would occur, but only at the expense of leisure.[13] Possibly this change in attitudes may have been of importance in the initial stages of the Industrial Revolution but it can hardly account for sustained economic growth. It should be added that others have attributed the lengthening of the labor day to different factors such as coercion (for example, Marx and the Hammonds) or improved diets (Freudenberger and Cummins).

The second exogenous source of demand which allegedly was a necessary condition for rapid industrial growth in Britain in the last third of the century—and by implication Belgium, Switzerland, and France in the half century following Waterloo—was exports.[14] It is important, however, to phrase the question sharply; it is not sufficient to state that a "leading" nation had an export sector which pulled the entire economy behind it. In the case of early industrial Europe, if export demand was a main determinant of the patterns of industrialization and growth, it is necessary to show that the successful industrial nations were somehow favored over other potential sellers for whom the expansion of world demand did not lead to rapid industrialization. Moreover, as Kindleberger has pointed out, the

[12] Eric Jones, "Agriculture and Economic Growth: Economic Change," in *Agriculture and the Industrial Revolution* (Oxford, 1974), pp. 116-17.

[13] According to Freudenberger the effective labor input per worker doubled during the second half of the eighteenth century. Cf. Herman Freudenberger, "Das Arbeitsjahr," in Ingomar Bog et al., eds., *Wirtschaftliche und Soziale Strukturen im saekularen Wandel* (Hannover, 1974), pp. 307-20.

[14] Walter E. Minchinton, ed., *The Growth of English Overseas Trade* (London, 1969), ed.'s intro., pp. 36-52. H. J. Habakkuk and Phyllis Deane, "The Take-off in Britain," in Walt Whitman Rostow, ed., *The Economics of Take-off into Sustained Growth* (London, 1962), pp. 77-78.

coincidence between exports and growth is not self-evident. It is necessary to specify in which way foreign expansion brings about growth over and above the simple gains of trade.[15]

The issue of export as a critical component of industrial demand is complicated by the continuous improvements in the productivity of international trade. These had the effect of reducing the transactions and transportation costs, which led to increased international specialization along the lines of comparative advantage. The proportion of national product exported thus rose, and at the same time national income itself grew as the gains from international trade were realized. It could thus be argued that the ultimate causes of growth were changes on the supply side, even though from the point of view of the economy in question the changes were perceived as shifts of the demand curve. More substantial than this taxonomy is the objection that exports may have increased as a result of supply shifts which caused prices to fall, resulting in rapidly growing exports. Unless the shifts in demand and supply curves are separated it is impossible to attribute rising exports to growing foreign demand.

How important was export in the demand for British industrial output? A well known remark in McPherson's *Annals of Commerce* dismisses the importance of exports altogether, estimating the ratio of exports to home consumption at 1:32.[16] This number is obviously too low, but it illustrates the fact that some contemporaries were sceptical of the preponderance of exports in the expansion of industrial demand. The available data support that impression. Schlote's index of the relation between total overseas trade and industrial product displays a striking stability in the period in which export is supposed to have led industrialization. The index which moves slightly above 50 (1913 = 100) in the first half of the eighteenth century, leaps to 65 in the 1760s, but averages only 54 for the rest of the century, and 55 for

[15] Charles P. Kindleberger, *Economic Growth in France and Britain, 1851-1950* (New York, 1964), pp. 264-66. A longer list of possible links between exports and growth is suggested by Richard E. Caves, "Export-led Growth and the New Economic History," in Jagdish N. Bhagwati, ed., *Trade, Balance of Payments, and Growth* (Amsterdam, 1971), pp. 433-37. Some of the mechanisms suggested by Caves work in the wrong direction, however, and other nexuses seem far from easy to test in the case of Europe's industrialization, 1750-1850. For instance, does an expansion in exports stimulate and create ex nihilo entrepreneurial talent and initiative, or does it simply divert those resources away from alternative uses? Do increased exports stimulate the formation of overhead capital (e.g., harbor facilities), or should this be viewed as an additional cost imposed on the export-oriented economy?

[16] Minchinton, *English Overseas Trade*, p. 38. Deane and Cole, *British Economic Growth*, p. 42. T. S. Ashton, *An Economic History of England: The Eighteenth Century* (London, 1972), p. 63.

the period between the Napoleonic Wars and 1850.[17] Eversley stresses the powerful counterexample provided by the events of the 1770s: while exports collapsed, industrial output kept growing.[18] The aggregate data for the eighteenth century may be deficient but data for individual goods display the same features.[19] Even in the case of cotton in Great Britain it appears that the relation between exports and total output is not very pronounced. One crude test of the export-led hypothesis would expect to find a positive correlation between the rate of growth of cotton industry and the proportion of total output exported. But the correlation coefficient between the quinquennial rates of growth of cotton goods production (approximated by raw cotton consumption) and the proportion of total cotton goods exported is not statistically significant for the period 1750-1829. Turning to the nineteenth century we have access to somewhat more aggregated data, which are inconsistent with an "export pull" or staple theory of growth. Total domestic exports as a percentage of national income fell from 18 percent in 1801 to 11 percent in 1841, and then rose to 14 percent in 1851. The corresponding ratios of exports to industrial output are 76 percent, 32 percent, and 42 percent.[20] Had export demand been the one factor that singled out Great Britain as the economy most suitable to industrialization, one should observe that the role of exports increased when industrial growth was fastest, that is, after the Napoleonic Wars. Quite clearly, this version of the Gilboy thesis fails the test.[21]

Furthermore, it is clear that the British Empire, where expansions in demand would benefit British manufacturing more than other countries, was a small proportion of total industrial demand. From the Peace of Paris on, the share of the colonies fluctuates above 35 percent of total exports, although the data for 1793-1814 are not easy to interpret. After Waterloo the British Empire accounted for less than 30 percent of exports. In 1846-1850, total exports to India were

[17] Werner Schlote, *British Overseas Trade from 1700 to the 1930's* (Oxford, 1952), p. 51.

[18] Eversley, "The Home Market," pp. 247-49.

[19] Exports of woolen goods remained virtually unchanged between 1730-39 and 1780-89, while output of woolen cloth milled in West Riding grew by 396 percent (broadcloth, 1735-85) and 97 percent (narrow cloth, 1739-85). Exports of tin grew by 53 percent (1730/39-80/89), while output increased by 86 percent. Between 1750 and 1790 exports of coal rose by 60 percent while coal output more than doubled.

[20] Deane and Cole, *British Economic Growth*, p. 166. Mitchell and Deane, *Abstract*, pp. 282-83.

[21] A similar conclusion has been reached recently by Paul Bairoch, who states flatly that the industrial revolution reached quite an advanced stage in England before seeking foreign outlets. Cf. Paul Bairoch, "Geographical Structure and Trade Balance of European Foreign Trade from 1800 to 1970," *Journal of European Economic History*, 3 (Winter 1974), 569.

only 87 percent of total exports to Germany and only 30 percent higher than exports to the Low Countries. The demand exerted by colonial markets is thus not a very persuasive explanation of why Britain became the "workshop of the world."[22]

The third source of demand cited as important to the industrial revolution is population growth.[23] England's rapid population growth is supposed to have created a large market for industrial commodities, while France's slow population growth is viewed as a cause of retardation. As a matter of economic logic it simply is false that population growth, all other things held equal, will invariably increase the demand for industrial goods. Demand, after all, depends on consumers' income, not merely their numbers. Population growth will increase the number of consumers but decrease income per capita due to diminishing returns. The net effect is indeterminate. It can be shown that in general the following proposition holds: all other things equal, a rise in population will lead to an increase in the demand for any good if and only if the income elasticity of the demand function of that good is less than the reciprocal of the elasticity of non-labor in the production function. The proof of this proposition is straightforward and is deferred to Appendix B. The implications are quite striking. Assume that the economy was sufficiently competitive to approximate the elasticity of non-labor by its share in national income. Deane and Cole estimate the share of labor in national income in 1801 at about .45, which implies that all goods with an income elasticity of 1.8 and higher would experience a decline in demand as a result of population growth.[24] It can also be seen from eq. (6) in Appendix B that if we assume that the income elasticity for agricultural goods was about .5, which implies that the income elasticity for nonagricultural goods was about 1.25, a 1 percent increase in population would lead only to a .31 percent increase in the demand for nonagricultural goods as a whole. Thus population growth *alone* increased demand for nonagricultural goods by 14 percent between 1751 and 1801 and by another 29

[22] The colonial market version of the demand hypothesis is even less convincing for other European economies. Spain, Portugal, and the Netherlands—all late industrializers—had access to substantial colonial markets. Belgium and Switzerland, on the other hand, were confined to the demand structure dictated by the competitive world market, and yet underwent an industrial revolution before 1850. Ireland could have benefitted from the enormous demand exerted by Great Britain and its colonial empire, yet Ireland failed to industrialize.

[23] Deane, *First Industrial Revolution*, p. 34. Brian Murphy, *A History of the British Economy, 1740-1970* (London, 1971), p. 333. Gilboy herself expressed doubts about the importance of population growth by itself.

[24] Deane and Cole, *British Economic Growth*, p. 255. King's estimates for 1688 indicate that the share of labor income was only 39 percent, which would make the critical income elasticity equal to 1.64.

percent between 1801 and 1851. The increment of population increased demand by *less than 10 percent* of total output growth for the first half of the nineteenth century, and probably by even less than that for the period 1751-1801. Even if it is assumed that population growth was fully exogenous, its significance in generating the demand for increased industrial production was marginal.

It thus appears that cost-reducing and factor-increasing changes occupy the center of the stage: supply rules supreme. Technological change, capital accumulation, improvements in organization and attitudes, all made it possible to produce food, clothing, pots, and toys cheaper and better. But the sceptical reader may still be unconvinced. Will these goods be sold? Will it be "possible to find people with income and demand schedules capable of absorbing this increased output?" Is it not true that "the growth of industry has to be explained in terms of its markets, the reasons why increasing quantities of its products could be sold?"[25] Contrary to the first intuition, when considered in a competitive, multiproduct economy these questions are meaningless. As John Stuart Mill put it: "The demand for commodities determines in what particular branch of production the labour and capital shall be employed; it determines the *direction* of the labour, but not the more or less of the labour itself, or of the maintenance or payment of that labour."[26] Nothing has since been added to the body of economic theory that could refute this view—at least as long as the economy is in full employment. After all, the decisions on how, what, and how much to produce are made by the firm. And the individual firm always faces a demand curve which is elastic, irrespective of the demand elasticity for the industry as a whole.[27] An inelastic demand curve facing an *industry* will not affect the impact of a cost-reducing innovation on the rate of growth of the economy. In this case the income effects of the falling prices will simply be siphoned off to other industries.

II. INDUCED TECHNOLOGICAL CHANGE AND ECONOMIES OF SCALE

One interpretation of the Gilboy thesis maintains that the supply curve shifts outward as a result of increases in demand. It is to this issue we now turn.

[25] Eversley, "The Home Market," p. 211. Ralph Davis, *The Rise of the Atlantic Economies* (London, 1973), p. 304.

[26] John Stuart Mill, *Principles of Political Economy*, ed. W. J. Ashley (London, 1929), pp. 79, 87. Emphasis in original.

[27] This is true even in the case of a monopoly in which the industry and the firm are the same. A monopolist will always operate on the elastic segment of the demand curve facing him.

The idea that technological change is demand-induced is far from new, though its full impact on economic theory has not been felt until relatively recently.[28] The precise meaning of demand-induced innovation is more difficult to formalize if we keep in mind that historians have to deal—often simultaneously—with both process and product innovation. In any event, the theory that the intensity of technological change depends on demand factors is different from the Kennedy concept of "induced innovation." The latter shows how market conditions affect the location of an economy on a *given* "innovation possibility frontier." But for the Gilboy thesis in its dynamic version to be true, it is necessary that demand conditions should be capable of causing a shift *of* the frontier outward. This implicitly assumes that there is a "market" for inventions.[29] The demand for technological progress becomes essentially a derived demand, dependent on the demand for the final good. An increase in the demand for the consumption good will generate more inventions precisely in the same way as it will generate increased employment of other inputs. But in the present context that hardly resolves the dilemma first encountered in section I: whence the initial shift in demand?

Moreover, formidable doubts have been expressed concerning the "market" for inventions.[30] If such a market existed at all, it was plagued by at least three sources of market failure, namely the preponderance of externalities, the uncertainty surrounding all stages of innovative activity, and the fact that often new knowledge rapidly becomes a public good.[31] Most damaging, however, to the view that

[28] The seminal article is Charles Kennedy, "Induced Bias in Innovation and the Theory of Distribution," *Economic Journal*, 74 (Sept. 1964), 149-55. For a review of the ensuing debate, see Paul A. David, "Labor Scarcity and the Problem of Technological Practice and Progress in Nineteenth Century America," in *Technical Choice, Innovation, and Economic Growth* (Cambridge, 1975).

[29] The classic work is Jacob Schmookler, *Invention and Economic Growth* (Cambridge, 1966), esp. pp. 88-103, 202-09. Samuel Lilley, "Technological Progress and the Industrial Revolution," in Carlo M. Cipolla, ed., *The Fontana Economic History of Europe* (London, 1973), vol. 3. See also, for example, Landes, *The Unbound Prometheus*, pp. 77, 137. H. J. Habakkuk, "The Historical Experience of the Basic Conditions of Economic Progress," in L. H. Dupriez, ed., *Economic Progress* (Louvain, 1955), pp. 150-51. The debate is admirably summed up in A. E. Musson, ed., *Science, Technology and Economic Growth* (London, 1972), ed.'s intro.

[30] M. Ishaq Nadiri, "Some Approaches to the Theory and Measurement of Total Factor Productivity," *Journal of Economic Literature*, 8 (Dec. 1970), 1148-49 and references cited there.

[31] The list of inventors who failed to capitalize on their major inventions is very long, with Cort, Crompton, Goodyear, Whitney, W. Kelly, Lenoir, and Ericsson being some of the most notorious examples. Schiff has shown that there is little evidence to support the hypothesis that a national patent system makes a significant difference in the rate of industrialization. Cf. Eric Schiff, *Industrialization without National Patents* (Princeton, 1971). For a sceptical evaluation of

the "quantity" of technological progress will respond to shifts on the demand side is that the price on the vertical axis of this supply and demand model is not precisely specified. Since each invention is, by definition, produced only once, the producer has no firm basis on which to estimate his returns. It is often argued, moreover, that financial gain is only a minor consideration for many inventors: scholarly achievement, desire to improve, and pure love of inventing are often cited as prime motives of inventors.[32] This does not imply in itself that inventors are unaware of or indifferent to the economic needs of the society they live in. But a distinct possibility of market failure exists all the same. The material needs of society as reflected by the structure of market prices, and the same needs as viewed by an inventor in search of fame or satisfaction can diverge enormously. Often the full potentialities of major inventions were not recognized initially by the progenitors.[33]

Furthermore, as Rosenberg has recently stressed, it is misleading to assume that the supply of new technology was very elastic so that demand conditions set the "output" of technological change.[34] It is tautological that cost-reducing innovations will be sought after and adopted irrespective of the size of the market; competition forces firms to minimize costs. But is there strong evidence that an increase in demand will result in an increase in inventive activity? Schmooklers results seem to bear this out, but his dependent variable is patenting, not invention nor adoption of an innovation.[35] In fact, there is some evidence indicating increased efforts on technological improvement in times of reduced demand.[36] While this line of rea-

the impact of the patent system in Britain, see T. S. Ashton, *The Industrial Revolution, 1760-1830* (New York, 1964), p. 11.

[32] Joseph Rossman, *The Psychology of the Inventor* (Washington, 1931), p. 152.

[33] Edison, for example, "failed to understand the future of his invention [the phonograph] as part of the world of art and entertainment. He thought of it mainly [as a] modern dictaphone, [or] a speaking family album, to preserve the speeches of great statesmen, to teach languages . . . ." See J. R. T. Hughes, *The Vital Few* (London, 1973), p. 175. The divergence between the social and the private rates of return on innovations does not seem to have vanished over time. For a recent study see Edwin Mansfield et al., "Social and Private Rates of Return from Industrial Innovation," *Quarterly Journal of Economics*, 91 (May 1977), 221-40.

[34] Nathan Rosenberg, "Science, Invention and Economic Growth," *Economic Journal*, 84 (March 1974), 90-108.

[35] For critiques of Schmookler's findings see, e.g., Musson, *Science and Industry*, pp. 25-29, and John Jewkes, David Sawers, and Richard Stillerman, *The Sources of Invention* (2nd ed.; New York, 1969), pp. 210-11.

[36] See esp. Nathan Rosenberg, "The Direction of Technological Change: Inducement Mechanisms and Focusing Devices," *Economic Development and Cultural Change*, 18 (Oct. 1969), 23, and sources quoted there. P. Schöller, "La Transformation économique de la Belgique de 1832 à 1844," *Bulletin de l'Institut des Sciences Economiques*, 14 (Dec. 1948), 585. William

soning has its problems as well, and the entire debate is highly speculative, it underlines the weakness of a theory that associates technological change with increasing (or simply "large") demand. Even if the supply of new knowledge were highly responsive to changes in demand, the basic problem of the *primum mobile* remains. If technological change occurred when "demand conditions were ripe" (as Lilley has put it), what changed in this respect in the late eighteenth century? As was shown in section I, most of the traditional sources of demand expansion were of negligible size.

If necessity was indeed the mother of invention, surely the conception was no case of parthenogenesis. A frequently employed notion in this context is that of "bottlenecks" or "challenge and response." To the extent that what is meant here is technology's "responding" to demand, it has been discussed above. But there is a more sophisticated interpretation of this idea in terms of "technological bottlenecks."[37] Some find it useful to distinguish between external bottlenecks which pass through the market mechanism, and internal imbalances which take place at the level of the firm, but this is immaterial for the present purpose. The essence of the argument can be sketched as follows. Suppose a firm or an industry produces a good X by means of two perfectly complementary processes, *a* and *b*. The factors employed are L_a and K_a in process *a* and L_b and K_b in process *b*. Total initial output is thus given as $X_0 = \min\,[a\,(L_a,\,K_a),\,b\,(L_b,K_b)]$. Now suppose that a technological breakthrough occurs which increases the productivity of both factors in process *a* by a factor of $1 + \alpha$. Due to the strict complementarity of the two processes, however, output will not rise initially at all; a bottleneck in process *b* has emerged. Obviously any innovation that will raise the productivity of the factors in *b* by a factor of $1 + \alpha$ or less will be translated directly in an equiproportional growth in output. One could surmise that in this case the incentive to innovate would be especially high, whether the bottleneck is external or internal. But one should not forget that an alternative solution exists to the "imbalance" created, namely the reallocation of labor and capital from process *a* to process *b*. In the present example, simple calculations show the reallocated amount to be $\frac{\alpha L_a L_b}{L_a + (1 + \alpha)L_b}$ for labor and a similar quantity for capital.

Brown, "Innovation in the Machine Tool Industry," *Quarterly Journal of Economics*, 71 (Aug. 1957), 406-25.

[37] David, "Labor Scarcity," pp. 82-83. J. R. T. Hughes, "Foreign Trade and Balanced Growth: The Historical Framework," *American Economic Review*, 49 (May 1959), 335-36. Nathan Rosenberg, "The Direction," pp. 4-11.

When will a "bottleneck" result in a technological breakthrough and when will it result in a reallocation of factors? Unfortunately we have no good theory that will predict this. It seems reasonable that the ultimate result will depend on the supply side of technology, that is, whether and at what cost the existing stock of scientific and technical knowledge is capable of solving the bottleneck.[38] In addition, it is likely to depend equally on the cost and time involved in reshuffling the factors of production. As reallocation is, by comparison, a relatively continuous process, whereas inventions are more or less discrete events, it is reasonable to suggest that the more time passes without a technical resolution, the more actual reshuffling of resources will take place, gradually lessening the extraordinary payoff of a technical breakthrough. Many cases can be cited in which apparent "bottlenecks" were solved by reallocation of resources before technological changes altered the required input ratios again. Those "bottlenecks" that were resolved by a spectacular masterstroke such as Eli Whitney's or William Perkin's are likely to receive better coverage in the literature than resource reallocations, which tend to be more gradual and evolutionary. An example of the latter is the production of energy in the second half of the nineteenth century. Output of coal increased at an annual rate of 2.5 percent in Great Britain between 1854 and 1911, and even faster in the U.S. and Germany. The increased demand for energy generated by rapid growth created a bottleneck in coal mining, which was highly labor intensive. Yet there were few major technological breakthroughs in coal mining, especially in England.[39] The winning of coal continued to depend on arduous manual labor under exhausting and perilous conditions. The "bottleneck" was resolved by a reallocation of resources.[40] The proportion of male workers in mining and quarrying in Great Britain rose consistently from 4.2 percent in 1841 to 9.2 per-

[38] A good illustration is the case of steel. It has been argued that Henry Bessemer's attention was first directed to steel when a cast iron cannon was unable to fire a new projectile he devised, and that the Bessemer process was therefore a result of a "technical imbalance." It is noteworthy that a much more costly imbalance existed in constructing railroad tracks from wrought iron, which had to be replaced frequently. Experiments conducted in England indicated that steel rails could outlast wrought iron rails by a factor of seven. Fogel has estimated that the distribution of the wear of wrought iron rails was such that 55 percent of all rails laid wore out in the first ten years. Since the life of equipment was much longer and the life of the area-clearing infinite, the inability to produce cheap steel constituted a severe imbalance—which for the first forty years of the railroad was solved by factor reallocation, not by the invention of cheap steel.

[39] See, e.g., T. K. Derry and T. I. Williams, *A Short History of Technology* (Oxford, 1960), pp. 473-74. A. J. Taylor, "Labor Productivity and Technological Innovation in the British Coal Industry, 1850-1914," *Economic History Review*, 14 (Aug. 1960), 58.

[40] Taylor, "Labor Productivity," pp. 62-64.

cent in 1911. In the U.S. the proportion of miners in the labor force rose from 1.2 percent to 2.8 percent between 1850 and 1910; in Germany the proportion rose from .9 percent to 2.8 percent in the same period. About two thirds of all mine workers were employed in coal or lignite mines. Another example is cotton, often cited as the prime case of an industry in which technological change occurred in "compulsive sequences." It is rather arbitrary to cite the technological solutions of the "bottlenecks" in dyeing, carding, weaving, and spinning, while ignoring that similar problems existed in the planting and picking of cotton at one end of the process, and the tailoring of clothes at the other end. Bottlenecks here were clearly resolved by reallocation, not invention.

Repeating the words "challenge and response" endlessly constitutes no more of a theory of technological change than Oscar Wilde's parrot's "supply and demand" constituted a theory of prices. As long as challenges occur without a forthcoming response, while other major developments occur without discernible stimuli, the "bottleneck theory" ought to be treated with utmost caution. In any event, the "challenge and response" or "compulsive sequence" mechanism is a weak defense of the Gilboy thesis. After all, even if technology did respond positively to the emergence of "bottlenecks," such a pattern of technological progress is essentially a description of the precise operation of the supply side rather than a "link" between supply and demand.

More powerful in the present context is the argument that views the new technology as a shift in cost curves in such a way as to reduce average cost only at a given level of output; that is, the new technique is subject to economies of scale. The Industrial Revolution implied a manifold increase in plant size in manufacturing and transportation. Central power sources, a more sophisticated division of labor, and efficient supervision and discipline imposed on factory workers, brought about an unprecedented increase in firm scale. Hence demand conditions *could* have been important in determining where and when the shift to the new technique occurred, and whether its full potentialities were exploited. What is necessary for this line of reasoning is that the optimal plant size is large with respect to the extent of the market. More precisely, the crucial variable is the ratio of the horizontal distance of the demand curve from the vertical axis at the minimum cost price to the output at which this minimum cost is attained. If this ratio is less than one, the scale economies are not fully realized on account of insufficient demand. Even if the ratio is larger than unity but still small, insufficient competition may slow down growth by reducing overall efficiency. The question whether demand factors could have operated in this fashion is particularly complicated because Europe was, comparatively speaking, a well integrated economy, so that economies of scale could lead to specialization of large economies in goods in which increasing returns were important. In other words, if country A produced goods under conditions of economies to scale, country B would experience growth too if it could import the goods from A at a cheaper price.

It is not an easy task to substantiate the case for increasing returns in manufacturing anywhere before, say, 1870.[41] Rigorous empirical tests of microeconomic data for early European industry are unavailable. But most econometric studies for American data seem to indicate that the hypothesis of no increasing returns cannot be rejected.[42] There are simpler ways to measure increasing returns than estimating production functions. An early study by G. T. Jones assigned a very modest role to scale economies in Britain for the period 1850-1920, and there is no reason to believe that before 1850 the situation was any different.[43] The alleged existence of economies to scale in the Lancashire cotton industry is criticized in a recent article by Gatrell who concludes that "size in itself guaranteed neither efficiency in good times nor viability in bad . . . one may be most impressed by the ability of small, not to say middle-sized, units to exploit their opportunities in an increasingly competitive industry."[44] A suggestive but not entirely unambiguous procedure is to argue that if there had been industries in which potential scale economies were present but in part

[41] Even today, there is considerable difficulty in interpreting the evidence whether national market size conveys a clear-cut advantage or not, although it is likely that in some industries small nations fail to achieve scale economies or sufficient competition. Cf. F. M. Scherer, *Industrial Market Structure and Economic Performance* (Chicago, 1970), pp. 93-95.

[42] Paul A. David, "Learning by Doing and Tariff Protection: A Reconsideration of the Case of the Ante-bellum United States Cotton Textile Industry," in *Technical Choice*, pp. 142-43. See also Thomas J. Weiss, "Economies of Scale in Nineteenth Century Economic Growth," (Summary of Research Workshop), this JOURNAL, 36 (March 1976), 39-41.

[43] G. T. Jones, *Increasing Returns* (Cambridge, 1933). Jones's results are upper bound estimates of the importance of scale economies since he is unable to separate scale effects from technological change. Still, he finds an "observed elasticity" (i.e., the proportional growth of output divided by the proportional fall in price) of about 5 for the Lancashire cotton industry between 1850 and 1870. This implies a degree of homogeneity of 1.25, assuming a Cobb-Douglass production function. Sandberg has disputed the accuracy of Jones's index and insists that costs in the British cotton industry fell substantially between 1885 and 1914. He attributes this decline in costs to technological change, however, not to scale economies. Cf. Lars G. Sandberg, *Lancashire in Decline* (Columbus, 1974), pp. 93-119, 131-33. See also J. D. Gould, *Economic Growth in History* (London, 1972), pp. 229-35.

[44] V. A. C. Gatrell, "Labour, Power, and the Size of Firms in Lancashire Cotton in the Second Quarter of the Nineteenth Century," *Economic History Review*, 30 (Feb. 1977), 125.

unrealized due to inadequate demand, a process of concentration should have taken place in these industries, possibly resulting eventually in the emergence of "natural" monopolies. There is very little evidence for such phenomena anywhere in Europe before 1850—and certainly none before 1760. One—admittedly rough—indication is simply to look at the number and size of firms active in manufacturing. In 1834, there were 1134 cotton mills in Britain, which rose to 1932 in 1850, stabilizing around 2500 in the 1860s.[45] In Ghent there were 29 cotton spinning firms in 1817 (a crisis year), which rose to 48 firms in 1826 and 78 in 1839.[46] In France, Lévy-Leboyer has observed the proliferation of small firms in the textile industry.[47] In the iron industry, similarly, there is little evidence of concentration; as late as the mid–1850s British iron works were small and dispersed.[48]

It seems unwarranted to rely on scale economies to rescue the Gilboy hypothesis. Scale economies were prominent on the level of the firm, and yet could be relatively insignificant for the economy as a whole if most firms operated on the horizontal segments of their cost curves. The most significant exception to this rule is inland transportation, especially canals and railroads. Here a combination of large fixed costs and non-tradeability would be consistent with the argument that in the absence of a high level of demand there would have been much less reduction of transportation costs. Some gains might have been made on account of increased interfirm specialization, but evidence for such vertical disintegration is not strong. Stigler's rash assertion that England's early start was because "as the largest economy in the world it could carry specialization further than any other country" not only puts the cart before the horses, but has no basis in

[45] A summary of these figures, collected from various Parliamentary Papers, is provided in Mark Blaug, "The Productivity of Capital in the Lancashire Cotton Industry during the Nineteenth Century," *Economic History Review*, 13 (April 1961), 379. The average size of the 1105 cotton manufacturers surveyed in the Horner Report of 1841 was 175 workers, with the median size almost exactly 100 workers. Cf. Gatrell, "Labour," p. 98.

[46] H. Coppejans-Desmedt, "De Statistieken van E. C. Van Der Meersch over de Katoenindustrie in Oost Vlaanderen," *Bulletin de la Commission Royale d'Histoire*, 128 (1962), pp. 121-81. Xavier Heuschling, *Essai sur la statistique générale de la Belgique* (2nd ed.; Brussels, 1841), p. 96n.

[47] Maurice Lévy-Leboyer, *Les banques européenes et l'industrialisation internationale dans la première moitié du XIXe siècle* (Paris, 1964), pp. 170-71.

[48] See, e.g., A. Birch, *The Economic History of the British Iron and Steel Industry* (London, 1967), p. 205. Duncan Burn, *The Economic History of Steelmaking, 1867-1939* (Cambridge, 1940), pp. 191, 194-95. The number of firms engaged in iron production mentioned in the Coal Commission report was 342, of which 187 engaged in pig iron production. Cf. Shinichiro Kurimoto, "A Statistical Arrangement of the Royal Commission on Coal in 1871," *Quarterly Review, Nara Prefectural College*, 22 (Aug. 1974), 65-93. Kurimoto concludes that "small scale firms had not only overwhelming weights in number but also excellent activity in operating equipments [*sic*]."

the evidence.[49] An alternative approach proposed by Burnet maintains that the supply curve is continuously downward sloping, so that the entire growth process is to be viewed as an explosive disequilibrium process. In this way, he suggests, one could lend theoretical legitimacy to Rostow's take-off.[50] But Burnet's theory does not distinguish sufficiently between historical and Marshallian supply curves. The "historical" supply curve contains technological progress, resource discoveries, and capital accumulation which bring about shifts *of* the Marshallian supply curve. Burnet's bold statement that "the entrepreneur lucky enough to discover a virgin field of consumer demand can look forward to a golden age of self generating growth" is thus largely based on a misconception, although local gains are of course to be expected.

It appears that the preponderance of the idea that economies of scale were somehow crucial in spite of the absence of evidence is caused by a misunderstanding. In many cases an invention was followed by additional, gradual progress which resolved relatively minor but vital bugs in the application of the new technique.[51] These subsequent improvements are often mistaken as increasing returns—shifts *of* the supply curve are likely to be identified erroneously as points lying *on* a given supply curve when they follow a major invention as aftershocks.[52] It is possible, however, that the opportunities for "learning by doing" type of technological progress were far larger in those industries for which demand was comparatively elastic, and that the learning process was chiefly determined by the quantities produced. "Learning by doing," it appears, is the most convincing prima facie nexus between the structure of demand and the rate of growth of the economy. Whether the correlation between demand elasticity and "learning by doing" potentialities actually existed is an empirical issue.[53]

[49] George Stigler, "The Division of Labor Is Limited by the Extent of the Market," rpt. in W. Breit and H. M. Hochman, eds., *Readings in Microeconomics* (New York, 1968), p. 158.

[50] I. D. Burnet, "An Interpretation of Take-off," *Economic Record*, 48 (Sept. 1972), 424-28. Rostow himself is lukewarm about this attempt to formalize his take-off, and notes correctly that Burnet's falling supply curve has to level off somewhere. Cf. Rostow, *How It All Began*, p. 141.

[51] Nathan Rosenberg, "Factors Affecting the Diffusion of Technology," *Explorations in Economic History*, 10 (Fall 1972), esp. pp. 10-14.

[52] An interesting case in point is the debate about the adoption of the reaper in ante-bellum midwest grain farming. See Alan L. Olmstead, "The Mechanization of Reaping and Mowing in American Agriculture, 1833-1870," this JOURNAL, 35 (June 1975), 327-52.

[53] The one case that has been investigated in this respect is the New England cotton industry. Robert Zevin has found that demand was indeed elastic (the elasticity being between −2 and −3 in the 1820s, falling to about −1.5 after 1833). Paul David has found evidence for "learning by doing," although he views the learning more as a function of time than of

We return to economies of scale. As Allyn Young noted, it would be wasteful to make a hammer to drive a single nail. But one is hard pressed to come up with many examples in which the number of nails was so small to obviate the purchase of at least one hammer. And even when this was the case—could nations not borrow each other's hammers? Perhaps Argentina was too small to found a special school for railroad engineers. For that very reason the Argentine railroad system was built by British engineers and the trains hauled by British locomotives. The Belgian textile industry in the 1840s imported its largest machines, while producing most of the smaller machines itself. The average capacity of a domestically produced steam engine in the East Flanders industry was about 50 percent smaller than that of imported engines. In Verviers the capacity of imported engines was more than three times larger than that of domestic ones.[54] The one Belgian machine manufacturer who produced largely for export markets, Cockerill's works in Seraing, made machines with a capacity of 145 horsepower. His competitors, who worked largely for the domestic market, made much smaller machines—an average of 20 horsepower.[55] Today, Belgium is perhaps too small to build its own civilian airplanes—so they buy them in the United States. Luxembourg, which is even smaller, may not find it profitable to operate its own airlines—so the *Luxembourgeois* fly Sabena.

III. A MACROECONOMIC ANALYSIS

An alternative interpretation of the Gilboy thesis maintains that an expansion of demand for industrial goods does not necessarily have to come at the expense of other goods. This would be the case if the preindustrial economy had large amounts of underutilized resources. Indeed, without unemployed reserves of factors, economic expansion initiated by demand would have run into difficulties.[56] If there were large reservoirs of involuntarily unemployed labor, increases in demand would set into motion a multiplier mechanism, which, enforced by induced investment, could have led to the Industrial Revolution.[57]

accumulated output. Cf. Robert B. Zevin, "The Growth of Cotton Textile Production after 1815," in R. W. Fogel and S. L. Engerman, eds., *The Reinterpretation of American Economic History* (New York. 1971), p. 135, and David, "Learning by Doing," p. 167.

[54] Royaume de Belgique, Ministère des Travaux Publics, *Statistique de la Belgique: Mines, usines minéralurgiques, machines à vapeur, 1839-44* (Brussels, 1846), p. 48.

[55] Lévy-Leboyer, *Les banques*, p. 361.

[56] The same point is made by McCloskey, "Did Victorian Britain Fail?" p. 455.

[57] See, for example, Murphy, *A History*, p. 387; Habakkuk, "The Historical Experience," p. 153; Jan de Vries, *The Economy of Europe in an Age of Crisis, 1600-1750* (Cambridge, 1976), pp. 177, 241. The assumption that the early industrial economy was subject to serious Keyne-

This view is reflected in the famous but infelicitous note on the final page of Hicks's *Value and Capital* that "the whole Industrial Revolution of the last two hundred years has been nothing but a vast secular boom, largely induced by the unparalleled rise in population." It is well known that unemployment and pauperism were widespread in preindustrial societies, and Western Europe was no exception.[58] But unemployment is by no means the same as underutilized resources. The latter would only be the case if aggregate demand was insufficient so that people willing and able to work could not find employment. In a preindustrial or early industrial economy a large amount of "natural" or frictional unemployment is to be expected.[59]

Decisive evidence of whether eighteenth- and early nineteenth-century Europe was in fact in something like Hansen's secular stagnation or Joan Robinson's "limping" golden age is not easily obtained. It is a difficult problem to distinguish in a meaningful way between voluntary and involuntary unemployment of employable workers. But there is one body of evidence which requires scrupulous attention in this context, namely the writings of the so-called mercantilist or "pre-Adamite" school. Between William Potter's *The Key to Wealth* (1650) and James Steuart's *An Inquiry into the Principles of Political Economy* (1767), a long list of political economists expressed opinions which all contain an element of Keynes's theory of aggregate demand. It is unmistakable that many of these writers thought that unemployment could be remedied by an exogenous expansion of effective demand. Keynes cited with enthusiasm Bernard de Mandeville's *Fable of the Bees*, and one could provide much more sophisticated and elaborate illustrations from such eminent writers as William Petty, Nicholas Barbon, and George Berkeley. There can be no doubt that from many points of view these writers should be viewed as precursors of Keynes.[60]

---

sian unemployment is made also by J. L. Anderson, "A Measure of the Effect of British Public Finance, 1793-1815," *Economic History Review*, 27 (Nov. 1974), 610-19.

[58] See esp. D. C. Coleman, "Labour in the English Economy of the Seventeenth Century," *Economic History Review*, 8 (Dec. 1955), 280-95.

[59] For a view that attributes "unemployment" to dietary inadequacies, see Herman Freudenberger and Gaylord Cummins, "Health, Work, and Leisure Before the Industrial Revolution," *Explorations in Economic History*, 13 (Jan. 1976), 1-12.

[60] See, for example, William D. Grampp, "The Liberal Elements in English Mercantilism," *Quarterly Journal of Economics*, 66 (Nov. 1952), 465-501. Douglas Vickers, *Studies in the Theory of Money, 1690-1776* (Philadelphia, 1959). T. W. Hutchison, "Berkeley's *Querist* and its Place in the Economic Thought of the Eighteenth Century," *British Journal for the Philosophy of Science*, 4 (May 1953), 52-77. N. G. Pauling, "The Employment Problem in Pre-Classical English Economic Thought," *Economic Record*, 26 (June 1951), 52-65. S. R. Sen, *The Economics of Sir James Steuart* (Cambridge, Mass., 1957).

But even if we accept that these writers thought that the unemployment they were witnessing was of the Keynesian type the issue is not settled. How can we be certain that they actually saw what they thought they saw? Heckscher, Blaug, and Coleman among others have pointed out that there are powerful alternative explanations to the observed unemployment in the pre-industrial economy.[61] One is a high preference for leisure resulting in a backward bending supply curve, which could account for much of the unemployment observed by contemporaries.[62] High seasonal variance in the demand for labor combined with high adjustment and transportation costs led to widespread seasonal unemployment.[63] The importance of this factor appears to have been diminishing with the increasing availability of nonagricultural employment in agrarian communities.[64]

[61] Eli Heckscher, *Mercantilism* (2nd rev. ed.; London, 1955), vol. II, pp. 340-58. Mark Blaug, *Economic Theory in Retrospect* (Homewood, Ill., 1968), p. 15. Coleman, "Labour," p. 289.

[62] I. D. S. Ward, "George Berkeley: Precursor of Keynes or Moral Economist on Underdevelopment," *Journal of Political Economy*, 68 (Feb. 1959), 31-40; Landes, *Unbound Prometheus*, p. 59.

[63] Seasonal unemployment struck agriculture and industry alike, since weather conditions affected water and wind mills, road conditions, bleach fields, and so on. John Law, in his *Money and Trade* (1705), seemed to think that seasonal unemployment could get as high as 50%. A 1752 pamphlet cited by Mantoux places the proportion of time in which journeymen tailors were unemployed at about 40%. (Cf. Paul Mantoux, *The Industrial Revolution in the Eighteenth Century* [rev. ed.; New York and Evanston, Ill., 1961], p. 71n.) In the mid-eighteenth century construction workers were idle at least "four or five months in the year." R. Campbell, *The London Tradesman* (1747), pp. 103-04; rpt. in M. Dorothy George, ed., *English Social Life in the Eighteenth Century* (London, 1923), pp. 32-33. Seasonal unemployment was particularly severe where rural industry was absent and the crops not diversified, particularly in Ireland. As late as 1836 G. C. Lewis noted that two thirds of the Irish work force was not employed all year round, and that this irregularity was the true cause of poverty in Ireland. Cf. G. C. Lewis, *On Local Disturbances in Ireland* (London, 1836), p. 312. Other examples are cited by T. S. Ashton, *An Economic History*, pp. 202-03; idem., *Economic Fluctuations in England, 1700-1800* (Oxford, 1959), p. 6. Wages fluctuated seasonally, so that it is not easy to distinguish between voluntary unemployment (i.e., consumption of leisure) and involuntary seasonal unemployment.

[64] A possible explanation of the widely observed phenomenon of "unemployment" in preindustrial Europe could be built on the idea of structural unemployment. If the marginal productivity of labor was lower than some accepted minimum of subsistence (possibly zero), it follows that people willing to work cannot find employment. For a precise formulation of this idea see R. Eckaus, "The Factor-proportions Problem in Underdeveloped Areas," in A. N. Agarwala and S. P. Singh, *The Economics of Underdevelopment* (London, 1958). Under these circumstances a change in the composition of demand could reduce unemployment if demand shifted toward comparatively more labor-intensive goods, thus increasing the total demand for labor. William Petty's recommendation to employ idle workers to "build a useless pyramid upon Salisbury Plain, bring the stones at Stonehenge to Towerhill or the like" sounds reasonable on this background. It is quite clear, however, that such compositional effects were becoming rapidly less important as the Industrial Revolution proceeded and industrial production was becoming more capital-intensive. The argument could, however, go a long way in explaining the rapid expansion of domestic industry before 1750, which was much more labor-intensive than factory production.

The above is not meant to imply that shocks and fluctuations did not have a profound impact on the level of economic activity. Harvest failures, fluctuations in exports, political upheaval, tariffs, and wars caused widespread distress and unemployment.[65] But as a *long-run* description of a normal state of affairs, involuntary unemployment in preindustrial Europe seems a dubious proposition. One reason is that in a barter economy Keynesian unemployment cannot occur. While Europe and North America were gradually becoming more monetized in the seventeenth and eighteenth centuries, they still contained large pockets of barter exchange. Moreover, in the seventeenth and eighteenth centuries, both in England and in Europe, long-term investment in capital goods was typically financed out of own funds (family loans, retained profits). This implies that investment cannot exceed *ex ante* savings, but as Heckscher pointed out, the "pure" savers who did not plow their savings back into their business, had relatively few alternatives to hoarding. As capital markets improved, the gap between savings and investment narrowed. Heckscher "tests" this hypothesis by looking at secular price movements in the eighteenth and nineteenth centuries.[66] The tests are suggestive but not definitive, since the supply of money is not held constant.

One possible answer to the attempts to search for long-run Keynesian unemployment in pre-modern Europe is provided in the *General Theory*. Keynes's rather cavalier description of the pre-modern economy has it that:

> It is impossible to study the notions to which the mercantilists were led by their actual experiences, without perceiving that there has been a chronic tendency throughout human history for the propensity to save to be stronger than the inducement to invest. The weakness of the inducement to invest has been at all times the key to the economic problem. Today the explanation of the weakness of this inducement may chiefly lie in the extent of existing accumulations; whereas, formerly, risks and hazards of all kinds may have played a larger part.[67]

What Keynes overlooks is that in order to have continuous involuntary unemployment, hoarding has to exceed dishoarding over prolonged periods of time. In other words, a secular upward trend in the demand for money should be discerned. Why would such a trend

[65] See, for example, Joel Mokyr and N. Eugene Savin, "Stagflation in Historical Perspective: The Napoleonic Wars Revisited," in Paul Uselding, ed., *Research in Economic History*, vol. 1 (1976), pp. 198-259.

[66] Heckscher, *Mercantilism*, vol. II, pp. 348-54.

[67] John Maynard Keynes, *The General Theory of Employment, Interest, and Money* (New York, 1936), pp. 347-48.

occur? Keynes's explanation of "risks and hazards" explains the absolute level of the demand, but it is hard to argue that alternative assets were becoming *gradually riskier* than money. But deflationary pressures could result also if the economy was growing or monetizing, however slowly, and the consequent increment in the demand for money exceeded the growth in the money supply. Such deflationary pressures could have produced unemployment if the price level did not adjust sufficiently.

The absence of unemployment or aggregate output data exclude any direct testing of this hypothesis, but some suggestive facts cast doubt on it. First, if it is true that the demand for money for any reason rose faster than its supply, it is reasonable to suppose that interest rates would have been subject to upward pressure. Available data do not support this hypothesis: on the whole, interest rates declined during the seventeenth century while in the eighteenth the downward trend tended to be obscured by wartime borrowing after 1740.[68] Secondly, the evidence indicates that the supply of high-powered money started to grow at an accelerated rate at some point after 1680.[69] Equally important was the emergence of nonmetallic money as a means of exchange. Here one should count as money not only the notes issued by the bank of Stockholm, the Bank of England, and the American colonies, but also bills of exchange which were increasingly made negotiable. The latter, especially, provided a true source of "inside money," making the money supply more responsive

[68] De Vries, *The Economy of Europe*, p. 211; Sidney Homer, A *History of Interest Rates* (New Brunswick, N.J., 1963), pp. 133-43, 155-80.

[69] The best indicator is still total output of the silver and gold mines in Mexico and South America collected by Von Humboldt and published in 1809 in his *Essai politique sur le Royaume de Nouvelle-Espagne*. These figures were refined and completed by Adolf Soetbeer, *Edelmetallproduktion und Wertverhältnis zwischen Gold und Silber seit der Entdeckung Amerikas bis zur Gegenwart* (Gotha, 1879). Soetbeer's figures show a marked acceleration of bullion output after 1680. Moreover, if his figures are corrected in the ways suggested by W. Lexis, the acceleration is even more marked. Cf. Wilhelm Lexis, "Beitrage zur Statistik der Edelmetalle," *Jahrbücher für Nationalökonomie und Statistik*, 32 (1879), pp. 361-417. The results of applying Lexis's critique to Soetbeer's figures are provided below:

Total World Annual Output of Bullion
(in millions of piastres)

Year	*Soetbeer estimate*	*revised Soetbeer estimate*	*Year*	*Soetbeer estimate*	*revised Soetbeer estimate*
1601-20	99.9	81.9	1701-20	99.8	89.5
1621-40	94.0	76.9	1721-40	130.8	120.5
1641-60	90.4	74.1	1741-60	164.6	155.3
1661-80	86.5	71.2	1761-80	175.3	162.4
1681-1700	91.6	76.4	1781-1800	207.8	192.4

to the needs of the economy.[70] Thirdly, if considerable involuntary unemployment had existed in preindustrial and early industrial economies, it should be expected that sudden expansions of aggregate demand should have affected output and employment in a significant way. It seems, however, that in the one test case for which evidence exists, the Napoleonic Wars, this was not the case.[71] It may thus be inferred tentatively that preindustrial and early industrial economies were as a rule *on* their transformation curves. This is not to say that in no sense were there any underutilized resources in preindustrial Europe which could be brought into productive activity during the industrialization process. Obvious examples are a better allocation of resources due to increased efficiency of the market mechanism or the formation of additional factors of production complementary to labor. But such movements should be viewed properly as supply and not as demand related phenomena.

IV. CONCLUSIONS

The intention of this paper has been to examine the Gilboy hypothesis in every possible interpretation, and to decide whether we can assign an important role to demand factors in the explanation of the industrial revolution. Few of the various alternative interpretations withstand the scrutiny of a priori reasoning or empirical tests. The old schoolboy view of the industrial revolution as a "wave of gadgets" may not be so far off the mark after all, provided we allow for "more" as well as for "better" gadgets, and we include abstract improvements such as organizational change, changes in workers' attitudes, and so forth, as "gadgets" in a wider sense.

If demand was not a "factor," what exactly was its place in the industrial revolution? To start with, we observe that any supply shift will affect the economy in direct proportion to the proportional size of the industry affected relative to the economy as a whole. For a once and for all fall in costs the demand structure only matters for determining the composition of the increment in national income but not its size. If the supply shift is a continuous process, however, the shape of the demand curve does matter, because it determines the future

[70] Fernand Braudel and Frank Spooner, "Prices in Europe from 1450 to 1750," in *Cambridge Economic History of Europe*, vol. IV (Cambridge, 1967), pp. 386-87. It is worth noting that Braudel and Spooner claim, albeit without precise evidence, that not only the supply of money rose considerably in the eighteenth century, but the demand for money in fact declined. Cf. ibid., p. 450.

[71] For details, see Mokyr and Savin, "Stagflation in Historical Perspective," pp. 210-23.

pattern of the relative size of the industry in question in the economy. To be more precise: if the own price elasticity is more than unity in absolute value it can be shown that (1) the overall impact of a constant rate of cost reduction will increase over time, and (2) the spillover effects for all other goods taken together are negative. This casts a peculiar light on the role of so-called leading sectors which are supposed to grow due to a very elastic demand curve. Moreover, as output expands, the economy will move down into the inelastic segment of the demand curve. At that point the continuous fall in costs will result in increased demand for all goods (that is, for the good in question and for all other goods taken together), but this impact will slowly peter out over time. Something similar to this process happened in the market for textiles between 1760 and 1860. The elastic demand caused output to grow very rapidly as a response to initial price reductions, so that subsequent price reductions were applied to a much wider base.

Secondly, there is the stability of demand. While the absolute level of demand for industrial goods cannot be used to explain the timing and speed of the industrialization process, heavy fluctuations in demand had an adverse effect on growth, due to the substantial costs of resource reallocation, the acquisition of new information, and so forth. Wars, revolutions, blockades, tariffs, harvest failures, and other unanticipated catastrophes inhibited growth not so much through the first moments of the demand function parameters as through the second moments.

A third way in which demand related factors could have been important in determining the speed and timing of the industrialization process is through their determination of the intersectoral terms of trade. As a consequence of a change in the terms of trade income distribution may change. For example, income may be redistributed from industrial workers and capitalists to landowners. It is, in fact, possible that industrialization could lead to "immiseration" of the modern sector in a way that is analogous to the well-known possibility of "immiseration" of a country increasing its exports. Note, however, that "immiserizing growth" can occur at the level of the economy but not at the level of the entire world. Similarly in a closed economy, it can occur at the level of a sector, but the economy as a whole is better off. Still, if the modern sector has a higher savings propensity or lower risk averseness than the other sector, such a worsening of the terms of trade may affect the rate of growth.[72]

[72] Joel Mokyr, *Industrialization in the Low Countries, 1795-1850* (New Haven, 1976), ch. 7.

To summarize, the traditional notion that supply and demand were somehow symmetric in the industrialization process is unfounded. The determination of "when," "where," and "how fast" are to be sought first and foremost in supply, not demand related processes. Statisticians warn that error can take the form of excessive credulity ("type II") as well as excessive scepticism ("type I"). The wide and uncritical acceptance of the Gilboy thesis is an example of a type II error which has crept into our textbooks and journal articles. It is hoped that the present essay will administer to the demand theory a dose of well-deserved compensating scepticism.

Northwestern University

Appendix A

Assume an economy with the following characteristics: (1) It produces two goods, agricultural goods (A) and nonagricultural goods (M). (2) The two goods are consumed by farmers and nonfarmers. (3) The demand for agricultural goods has a price elasticity of less than one in absolute value for both the farmer's and the nonfarmer's demand functions. (4) Both goods are normal; that is, all income elasticities are positive. (5) The supply of manufactured goods is a function of relative prices, but (6) the supply of agricultural goods depends in the short run only on weather conditions. Proposition: An exogenous decline in agricultural goods due to harvest failures will always result in a fall in the demand for nonagricultural goods.

Proof

There are four different demand functions in this system. Let the subscript 1 denote the demand of farmers for agricultural goods, 2 the demand of nonfarmers for agricultural goods, 3 the demand of farmers for nonagricultural goods, and 4 the demand of nonfarmers for nonagricultural goods. The demand system can thus be written as

$$Q_i(D) = f_i(Y'_j, P_A, P_M) \qquad \begin{array}{l} i = 1,2,3,4 \\ j = A \text{ for } i = 1,3 \\ \quad\;\; M \text{ for } i = 2,4 \end{array} \qquad (1)$$

where Y'_j (j = A,M) is nominal income, P_A and P_M the money prices of agricultural and nonagricultural goods, and Q(D) means the quantity demanded. Assuming that all demand functions are homogeneous of degree zero, we divide everything by P_M, thus using nonagricultural goods as our numéraire. Thus

$$Q_i(D) = f_i(Y_j, P, 1) \qquad \begin{array}{l} i = 1,2,3,4 \\ j = A \text{ for } i = 1,3 \\ \quad\;\; M \text{ for } i = 2,4 \end{array} \qquad (2)$$

where Y_j is income in terms of M-goods and $P = P_A/P_M$.

The price elasticities of the functions described by eq. (2) will be denoted by b_i, i = 1 . . . 4. Note that by assumption $-1 < b_1, b_2 < 0$. The *market* demand functions have demand elasticities which are weighted averages of the sectoral elasticities. Let θ_i (i = A,M) denote the proportion of good i consumed by the producing sector. Then

$$\delta = \theta_A b_1 + (1 - \theta_A) b_2 \tag{3}$$

is the overall price elasticity for agricultural goods with respect to P, and

$$\beta = (1 - \theta_M) b_3 + \theta_M b_4 \tag{4}$$

is the overall price elasticity for nonagricultural goods with respect to P, and should be interpreted as a cross elasticity. Finally we note that

$$Y_A = PA \tag{5}$$

denotes agricultural income, where A is exogenous, and

$$Y_M = M(P) \tag{6}$$

denotes industrial income. M is a function of P, that is, the supply of M-goods can have a positive price elasticity. This is expressed in γ, the elasticity of M with respect to P which is negative (since P contains P_M in its denominator), since some agricultural goods are used as inputs into manufacturing.

Let us now define $D = Q_3(D) + Q_4(D)$. We have to show that the elasticity of D with respect to A is positive under the assumptions stated. Using the chain rule, we obtain

$$\frac{dD}{dA} = \frac{\partial f_3}{\partial Y_A}\left[\frac{dP}{dA}A + P\right] + \frac{\partial f_3}{\partial P}\frac{dP}{dA} + \frac{\partial f_4}{\partial M}\frac{dM}{dP}\frac{dP}{dA} + \frac{\partial f_4}{\partial P}\frac{dP}{dA} \tag{7}$$

which can be written in terms of elasticities

$$\frac{dD}{dA}\frac{A}{D} = a_3(1 - \theta_M)\left[\frac{1}{\delta} + 1\right] + b_3\frac{1}{\delta}(1 - \theta_M) + a_4\gamma\frac{1}{\delta}\theta_M + b_4\frac{1}{\delta}\theta_M \tag{8}$$

where δ, b_3, and b_4 are as defined in eqs. (3) and (4) and the a's are the corresponding income elasticities.[73]

Assume that the theorem is false, that is, eq. (8) is negative. By multiplying the right-hand side of (8) by δ, it can be seen that (8) will be negative if

$$a_3(1 - \theta_M)[1 + \delta] + b_3(1 - \theta_M) + a_4\theta_M\gamma + b_4\theta_M > 0 \text{ (since } \delta < 0) \tag{9}$$

[73] The parameter δ is not precisely a demand elasticity. Rather, it is the elasticity of an envelope curve of shifting demand curves. An exogenous decline of A causes an initial rise in P as the economy moves up along the demand curve. In a general equilibrium context, however, δ incorporates the secondary effect produced by the decline in M (from eq. 6), causing a leftward shift in the demand for agricultural goods, partially offsetting the rise in P.

From eq. (4) we substitute into (9)

$$a_3(1 - \theta_M)[1 + \delta] + \beta + a_4\theta_M\gamma > 0. \tag{10}$$

The relationship between δ and β can be expressed by using the Cournot aggregation relation.

$$\beta = -(\delta + 1)\frac{PA}{M} \tag{11}$$

Substituting (11) into (10) we obtain

$$(1 + \delta)\left[a_3(1 - \theta_M) - \frac{PA}{M}\right] + a_4\theta_M\gamma > 0 \tag{12}$$

Since γ is negative and $\delta > -1$, this implies

$$a_3 > \frac{PA}{M}\frac{1}{1 - \theta_M} = \frac{PA}{Q_3} \equiv \frac{Y_A}{Q_3} \tag{13}$$

using the definition $I - \theta_M \equiv \dfrac{Q_3}{M}$.

Eq. (13) can be written as

$$\frac{\partial Q_3}{\partial Y_A}\frac{Y_A}{Q_3} > \frac{Y_A}{Q_3} \tag{14}$$

which implies

$$\frac{\partial Q_3}{\partial Y_A} > 1 \tag{15}$$

Eq. (15) says that the farmers' marginal propensity to consume manufacturing goods exceeds unity. That means that if their incomes increase by one penny, they will increase their spending on manufacturing goods by more than one penny; in other words, their consumption of agricultural goods will be reduced as their income rises. But we assumed that all goods were normal, and hence there is a contradiction. This completes the proof.

Appendix B

Proposition

All other things equal, population growth will increase the demand of those goods the income elasticity of which is *lower* than the reciprocal of the elasticity of non-labor in the production function.

Proof

Let total national income be denoted as

$$Y = f(L) \quad f' > o,\ f'' < o \tag{1}$$

Let y be income per capita, Y/L, and let the demand for any good by one individual be

$$q(D) = g(y) \tag{2}$$

Total demand is thus

$$Q(D) = Lg(y) \tag{3}$$

Differentiating (3) with respect to L yields

$$\frac{dQ(D)}{dL} = g(y) + L\left(g'(y)\,\frac{dy}{dL}\right) \tag{4}$$

By using the chain rule and simplifying we obtain

$$\frac{dQ(D)}{dL} = g(y) + g'(y)(f' - y) \tag{5}$$

which can be written in elasticity form

$$\frac{dQ(D)}{dL}\,\frac{L}{Q(D)} = 1 - g'(y - f')\,\frac{L}{Q} = 1 - \frac{(y - f')}{y}\,\frac{g'(y)y}{g} \tag{6}$$

which will be positive if

$$\frac{y}{y - f'} > \frac{g'(y)y}{g} \tag{7}$$

The left-hand side of (7) is the reciprocal of the share of non-labor in national income while the right-hand side is the income elasticity of the good under consideration. This completes the proof.

W. W. Rostow

Business Cycles, Harvests, and Politics: 1790-1850[1]

You cannot get them to talk of politics when they are well employed.—William Mathews, 1833.

I have observed during the whole time I have been in Lanarkshire that any rise in the rate of discount at the Bank of England has been immediately, or at least shortly, followed by an increase both of crime and of civil suits, and if it continues long, of mortality and typhus fever. So much so, that, as I am an official member of the prison board and of most of the Charities, I have always made it a rule to say . . . "Gentlemen, the Bank of England have raised their discounts, you had better immediately take measures for enlarging the prison accommodations and for extending the infirmaries and Poor houses."—A. Alison, Sheriff of Lanarkshire, 1848.

I

HISTORIANS of every shade of bias admit the importance of the influence of economic situations on political and social events. The weight attached to economic factors or, more precisely, the mechanism of their action is, however, by no means settled. The most familiar relation that has hitherto been emphasized in the years 1790-1850 links the mechanical inventions of the late eighteenth century to the growth of the factory system and to the consequent rise of a large urban proletariat and a powerful middle class. From these relations, which are essentially sociological, efforts have been made to explain the political forces that produced the Reform Bill of 1832, the Chartist Movement, the repeal of the Corn Laws. In the realm of cultural history, the rise to dominance of a philosophy of individualism and a cult of romanticism have been linked to the same forces, with ramifications in economic doctrine, religion, architecture, and poetry. Lectures on "The Industrial Revolution" may well be a part of courses on the romantic poets or on British foreign policy. Such attempts at interconnection represent a long-run analysis of economic influences. For many purposes, especially where economic influences operate at several removes, that sort of generalization is adequate.

The experiences of the western world since 1918 and, more especially, since 1929 have impressed observers, however, with the tremendous im-

[1] This article is an outgrowth of work done by the author on the study of "The Growth and Fluctuations of the British Economy, 1790-1850," under the direction of Dr. Arthur D. Gayer.

pact of short-period economic forces. Changes in social structure, in political atmosphere and policy, and in intellectual attitudes can be more or less directly traced to the depression after 1929. While it is true that these trends (and the causes of depression as well) have a long history, reaching back, at least, to 1873, their timing, their intensity, and their unique character are closely connected with recent short-run developments. Many historians have taken account of this type of influence, but rarely systematically.[2]

From 1790 to 1850 there were at least three major economic forces that contributed, at intervals, to British social and political unrest: cyclical unemployment, fluctuations in domestic harvests, and technological unemployment. The latter, by itself, was not likely to produce major disturbances; nor can it be sharply distinguished from cyclical unemployment. The underemployment of hand-loom weavers was, admittedly, an important element in the Luddite and Chartist movements; and the resentment of the hand-loom weavers against the introduction of machinery often gave a peculiar character to the activity of wider groups. The most serious unrest, however, was a product of cyclical depression and high food prices.

II

Good harvests with resulting low grain prices were calculated to call forth complaint from the landholders and from tenants burdened with fixed rent payments. The demand for grain was sufficiently inelastic to bring a decline in gross income when good harvests caused a sharp fall in price; and, until 1832 at least, the agrarian interest was disproportionately represented in Parliament. The simplest short-period, economic-political relation is that between the wheat price and the Corn Laws.

The Corn Laws were altered principally in the following years: 1791, 1804, 1815, 1822, 1828, 1842. Repeal came, of course, in 1846. A glance at the annual average wheat price reveals the principal setting for these amendments.

The wheat price fell from 56s. per quarter in May 1790 to 42s. in October 1791. The movement continued to a low point of 37.0s. in May 1792. The harvest of 1791 was "one of great abundance" and, under the

[2] Of the studies covering the first half of the nineteenth century G. D. H. Cole's *Short History of the British Working Class Movement, 1789-1927,* makes perhaps the most consistent use of a framework of business fluctuations and price movements. See, for example, I, 79, 88, 105, 177-188. But in describing strikes, and even phases of political development, many opportunities for relating the data to short-period fluctuations are missed.

prevailing corn law, the fall in price was sufficient to cause the ports to be closed to foreign grain:[3] "but the low price was productive, as usual, of complaint on the part of the landed interest, and was the occasion of a fresh corn bill." The inadequate harvests of the following years, however, kept the wheat price at or above 50s., and the Act of 1791 was not called into operation.

The catastrophic fall in the wheat price, from 154s. per quarter in March 1801 to 50s. in February 1804, produced similar, though even more violent complaint. The Corn Laws were again modified. The area under cultivation had, of course, been greatly expanded between 1793 and 1804. A succession of abundant harvests brought forth an unparalleled supply of wheat.[4] The import limit was raised to 63s.[5] The cutting off of Baltic supplies resulting from the resumption of war, together with bad harvests kept the wheat price above the new minimum until 1815. Like the Act of 1791, that of 1804 was never operative.

From August 1812, when the wheat price was 152s., to January 1816 a steady decline took place. The break-up of the Continental System and, finally, the return to peace, as well as good harvests, caused this fall.[6] Despite considerable opposition[7] it was judged that only an 80s. import limit could protect the capital newly invested in agriculture.

Until the close of 1818 the wheat price remained above 80s., aided largely by inadequate harvests on the Continent and considerable exports from Britain to France.[8] But good harvests then brought on a decline to 39s. at the close of 1822. The period of severe agricultural distress has colored the whole view of British agriculture in the three decades after the Napoleonic Wars.[9] In 1822, however, the Government was trapped

[3] T. Tooke, *A History of Prices* (1838), I, 81. This act provided for a duty of 24s. 3d. if the price was under 50s.; 2s. 6d. at or above 50s. and under 54s.; 6d. at or above 54s., and a bounty of 5s. on exports at a price under 44s. *See also* W. Smart, *Economic Annals of the Nineteenth Century* (1917), I, 90-92; and D. Macpherson, *Annals of Commerce* (1805), IV, 219-220.

[4] T.Tooke, *A History of Prices* (1838), I, 237-239.

[5] Under the Corn Law of 1804 a duty of 24s. 3d. per quarter was imposed when the wheat price was under 63s.; and 2s. 6d. per quarter, when at or above that rate and under 66s.; and 6d. when above 66s. The lowest price reached between 1804 and 1815 was 66s., in November 1807.

[6] Tooke, *op. cit.*, I, 322-325; II, 2-4; also *Annual Register*, 1814, 219.

[7] T. Doubleday, *A Financial History of England* (1859), 227, notes that the protection of soldiers was required for members of the House of Commons, on the passage of the 1815 Corn Law, such was "the fury of the people."

[8] Tooke, *op. cit.*, II, 16-18.

[9] For a modification of the traditional view of unrelieved depression, see G. E. Fussell and M. Compton, "Agricultural Adjustments after the Napoleonic Wars," *Economic History*, February 1939.

between the farmers' petitions and the opposition to further protection from labor, commercial, and industrial interests. At Peterloo, three years before, "No Corn Laws" had appeared on the banners. Unlike the position in 1815 the decline in agricultural prices (1818-1822) was accompanied by a decline in import prices and nonagricultural domestic prices. The Corn Law of 1822 modified only slightly the terms of the Act of 1815.[10] "The farmers had asked for bread and gotten a stone"; but there were others, too, asking for bread between 1818 and 1822.

From the second quarter of 1823 to the last quarter of 1828 the price of wheat hovered between 50s. and 70s. per quarter. Although the farmers were far from content, their relative position was, with respect to profits, probably no worse than that of the manufacturer or exporter; after the crisis of 1825, in fact, it was probably better. And those who sought a reduction in agricultural protection were victorious in the Corn Law of 1828.[11] It is probable that the growing prestige of free-trade ideas and the parliamentary influence of industrial and mercantile groups played some part in moderating the 1815 bill.[12] There was, however, an immediate economic basis reflected crudely in the movement of relative prices in the twenties:[13]

	Domestic Price Index	*Import Price Index*	*Wheat Price* (s. per quarter)
1823	97	99	52
1827	106	82	56

It is clear that, from 1823 to 1827, the wheat price did not share the net fall experienced in most other markets. After 1825, a peak year in general prosperity, this disparity was especially felt; and it was in the post-crisis atmosphere that antagonism to the Corn Law of 1815 developed.[14]

[10] *See* Smart, II, 117-118, for a detailed account of this bill which, like those of 1791 and 1804, was never really operative. A feature of its terms was the application of a sliding scale of duties when the price rose above 70s.

[11] Smart, *op. cit.*, II, 439. The sliding scale introduced in 1828, after two years of controversy, ranged from a duty of 34s. 8d. when the home price was 52s., to 1s., when the home price was 73s. In fact this amounted to only a very slight modification on the preceding bill, although the trend against agricultural protection had clearly set in.

[12] Their long-term pressures were symbolized, perhaps, in the person of William Huskisson.

[13] Nonagricultural domestic commodities show, on the whole, a greater net decline than the domestic index, which includes wheat and a variety of other agricultural commodities heavily weighted.

[14] *See* Tooke, *op. cit.*, II, 136, on food prices (1826-1827), relatively high "in the distressed state of the manufacturing population."

Until the last quarter of 1832 the wheat price remained well above 50s. and, although the farmers never ceased to complain, their position was not desperate. The three following years (1833-1836), however, brought abundant harvests, low prices, and extensive parliamentary investigations. Although the pressure for further protection increased, no action was taken by a Parliament in which anti-agrarian interests had been materially strengthened by the Reform Bill of 1832.[15]

At the close of 1836 the wheat price again rose suddenly as the harvest of that year appeared inadequate.[16] Chronically bad yields kept the price abnormally high until 1842. In this period the anti-Corn Law forces crystallized outside of Parliament, deriving additional strength from the generally depressed state of industry and the high level of unemployment, especially after 1839. This protracted pressure on real wages helped bring about the Whig tariff reforms of 1841 : but "Corn duties they left where they were, crying over their shoulders as they were being pushed out of office that a reasonable fixed duty . . . was the right thing."[17] Peel was elected on the issue of the sliding scale, and in 1842 his modifications of the Corn Law of 1828 consisted in lowering the maximum duty and in making the sliding scale less steep.[18] He himself believed this arrangement to be a considerable reduction in protection, and it was put forward as such. There is no doubt that the high food prices and depression in the previous few years had, by 1842, helped to discredit the whole argument for agricultural protection and for tariffs generally.

The role of the Irish famine in the suspension and, ultimately, in the repeal of the Corn Laws is a familiar short-period influence, as is also the tangled and dramatic political story of 1846. It is probable in fact, that strictly economic considerations played a somewhat lesser part in the final repeal than in some of the earlier modifications. In 1845 and the first three quarters of 1846 the domestic wheat price ranged between 45s. and 59s. A few years before it had been over 70s. The Irish famine might

[15] The Poor Law of 1834, however, may be considered, in part, as a concession to the hard-pressed land owners. It was in the agricultural districts that the burden of poor rates, under the old system, was most severely felt.

[16] Tooke, *op. cit.*, II, 157-158.

[17] J. H. Clapham, *An Economic History of Modern Britain* (1926), I, 497. The Whigs went out of office in 1841, to be succeeded by Peel and the Tories, until 1846.

[18] W. Page, *Commerce and Industry*, I, 128-130. The new scale started with wheat at 50s. paying 20s. duty, diminished to 1s., when the price was 73s. (as in the scale of 1828). There were two "rests" in the scale, wheat from 52s. to 54s. paying a duty of 18s. a quarter and wheat between 66s. and 68s. paying duty of 8s. This, it was hoped, would check the speculations of corn dealers, who had been tempted, by the steepness of the scale of 1828, to hold back corn in the hope of getting higher prices.

have been dealt with by extraordinary measures short of actual repeal. The rise in the wheat price (to a peak of 93s. in June) in 1847, however, would almost certainly have ended agricultural protection then, if its end had not been accomplished earlier.

This account is not meant, of course, to deny the long-period factors making for a reduction in agricultural protection: the growth of population, the accelerated industrialization of Britain, the widening political power of the urban middle classes and their free-trade doctrines. But it is clear that the timing of the events leading up to repeal were closely connected with the British harvests and other short-run factors influencing the absolute and relative level of the prices of agricultural products.

III

A. *The Speenhamland System, 1795.* The years 1794 and 1795 saw some industrial recovery in Great Britain, from the depression of 1793. A more powerful force, however, affecting labor's position, was a rise in foodstuff prices, due primarily to bad harvests.[19] The wheat price was 43.2s. per quarter in January 1792, 108s. in August 1795. Cost-of-living indexes reveal this rise graphically.

	Gilboy-Boody (cost of living in London)	*Silberling* (wholesale prices of foodstuffs, etc.)
1793	148	106
1794	168	110
1795	179	130

Although money wages rose, there seems little doubt that they rose "in a very inadequate proportion to the increased price of the necessaries of life."[20] There was widespread evidence of physical distress,[21] and the wage-subsidy scheme for out-of-door relief was instituted, much in the tradition of the Elizabethan poor laws.

B. *The Combination Acts, 1799 and 1800.* From the last quarter of 1796 until about the middle of 1799 the price of wheat and the cost of living remained moderately low. Despite the brief but severe depression of 1797, these were, internally, years of relative peace. The wheat price

[19] Tooke, *op. cit.*, I, 181-183, and 187.

[20] Tooke, *op. cit.*, I, 185-186.

[21] Tooke, *op. cit.*, I, 225-226. It is perhaps worth noting that the situation in 1795-1796 had direct and immediate influence on the ideas of Thomas Malthus, Frederick Eden, and Thomas Paine, as well as many others.

(which, at the end of 1798, was down to 48s.) then rose to a peak of 154s. in March 1801. The cost-of-living indexes move as follows:

	Gilboy-Boody	*Silberling*
1798	165	121
1799	229	143
1800	252	170
1801	190	174

The government acted by offering bounties on grain imports, by sending agents to the Baltic ports, and by encouraging the process of enclosure. In 1799-1801, in general, the working classes were fairly well employed. Even the crisis in the Hamburg trade, in 1799, did not induce a prolonged deflation.[22] Under these circumstances the workers had considerable market leverage in contracting money-wage bargains.[23] In the attempt to maintain their real wage, at a time of rapidly rising costs of living, the men resorted to various types of combinations. Even agricultural workers banded together in certain areas, notably in Norfolk.

Although the combination movement was very much the outgrowth of a particular short-period situation, and although the typical expression of discontent was the local bread riot or strike, the unrest was, at times, successfully linked with republican ideas. The Corresponding Societies, particularly, attempted to shape and unify the general dissatisfaction around the current liberal platform. With the memory of the French Revolution fresh in mind the Government acted to repress the Corresponding Societies and the Combinations. The Acts of 1799 and 1800, reënforcing existing legislation,[24] made illegal all collective working-class activity except the guild functions of the Friendly Societies.

C. *The Repeal of the Combination Acts, 1824 and 1825.* From 1820 through the early months of 1825 a fairly continuous increase in output and employment occurred. In the latter stages of the boom prices rose, relieving manufacturers briefly from the chronic downward pressure

[22] Willard Thorp's *Business Annals* (1926), 152, seriously misrepresents the position of British industry from 1798 to 1801. Each of those years is headed "depression," although output was almost certainly increasing and unemployment was low. This is not meant, of course, to deny the reality of the difficulties due to high living costs, from 1799 to 1801.

[23] For the role of a prosperity demand for labor in the unrest of these years see Macpherson, *op. cit.*, IV, 475 and 500. Strikes for higher wages were, in this as in later periods, a familiar characteristic of the latter stages of business expansion.

[24] For the position of the Combination Acts of 1799-1800, as part of a long tradition of limitation, *see* D. George, "The Combination Acts Reconsidered," *Economic History*, 1926.

that had existed since 1814. But the period 1820-1824 saw a coincidence of increased output and a sagging price level. Foodstuff prices, too, were fairly low. It is not surprising, then, that in the four years after Peterloo British labor was relatively peaceful.

The fact of increasing prosperity, too, made it possible for the industrialists and the Government to afford a greater tolerance:[25] "Exceptional measures of repression were allowed gradually to lapse; the activities of spies were relaxed; and the law was set less freely in motion against working-class attempts at combination." It was in this atmosphere, early in 1824, that Place and Hume maneuvered the repeal of the Combination Acts.

The repeal immediately brought into the open the trade unions which had been operating under cover in the previous two decades; and it encouraged the formation of many others. A wave of strikes broke out and, in the following year, an aroused Parliament seriously limited the easy-going terms of the Act of 1824.

There can be little doubt that the strikes of 1824-1825 can, in some measure, be attributed to the repeal of the Combination Acts. Two other factors, however, were then operating. In the first place, in the latter half of 1824, the boom was suddenly accelerated, pushing the major British industries close to full employment. Enormous exports to South America and to the United States, as well as widespread internal enterprise, created a typical, late prosperity situation. Strikes for higher wages would, normally, be expected.

This tendency was accentuated by a second factor, a sudden rise in living costs:[26]

	Tucker (London artisans)	*Silberling*
1823	124	111
1824	126	113
1825	137	128

In 1825, with a confidence born of some five years of increasing employment, the unions instituted numerous strikes for higher wages—in the cotton, wool, coal, iron, building, and other trades.[27] At about the

[25] Cole, *op. cit.*, I, 88.

[26] For an account of the inadequate harvest which largely explains this rise, *see* Tooke, *op. cit.*, II, 132-135.

[27] For a detailed account, from contemporary sources of strikes in 1824-1825, *see* Smart, *op. cit.*, II, 232-233, and 306-313.

middle of 1825, however, the business cycle turned downward; and, although the strikes continued for some time, by the end of the year "combination . . . was knocked on the head. Bradford weavers and combers went back to work at the old wages . . . so did the Renfrewshire colliers."[28] In the bitter industrial conflicts that continued into 1826, labor was no longer on the offensive, but attempting to preserve wage rates in the face of a declining industrial demand.[29]

The repeal of the Combination Acts is properly regarded as an expression of the general trend toward laissez faire, paralleled, in the twenties, by the Huskisson tariff reforms. Hume, in Parliament, presented the measure in such a light. Both Hume and Place regarded the unions as illiberal institutions brought into being by the repressive action of the Government; and they looked forward to their disappearance with the repeal of the Combination Acts.[30] Nevertheless, the tolerant action of Parliament in 1824 was directly connected with the previous years of prosperity; the violence of the strikes of 1824-1825 was largely the outgrowth of the situation in the labor market on either side of the cyclical turning point, accentuated by rising costs of living. It is possible, too, that the intemperance of the reaction of the Government in withdrawing, in 1825, a large part of the freedom granted in the previous year, may be linked to the change in the industrial outlook which occurred in that year. With commodity prices falling, and disillusion setting in with respect to the newly floated Latin American mining issues, the doctrine of laissez faire, as applied to labor organization, seemed somewhat more empty than in 1824.

D. *The Factory Act of 1847.* Factory acts in the first half of the nineteenth century were passed in 1802, 1819, 1833, 1842, and 1847. Each arose in a unique political setting; but each saw a similar combination of humanitarian and anti-industrial groups arrayed against the manufacturers. Within the ranks of the manufacturers there were, of course, notable exceptions: men like Peel and Whitbread and Owen, to whom the conditions of factory labor were ethically outrageous and who believed that shorter hours and better conditions meant great efficiency and profits. In parliamentary debate, humanitarian arguments led to

---

[28] Smart, *op. cit.*, II, 312-313.

[29] *See Annual Register*, 1826, Chronicle, 67, 109-112, 115, 149-151.

[30] J. L. and B. Hammond, *The Town Laborer*, 134-136; Page, *op. cit.*, I, 77, quotes Place as follows: "combinations will soon cease to exist. Men have been kept together for long periods only by the oppression of the laws; these being repealed, combinations will lose the matter which cements them into masses and they will fall to pieces."

rebuttal based on pleas for the freedom of the individual, or originating in attacks on state paternalism. To these the manufacturer would often add the claim that shorter hours meant a serious reduction or even the destruction of the existing margin of profit, and the loss of foreign markets.

There is, however, probably some significance in the fact that these acts were all passed at, or close to, a low point in cyclical fluctuations. The years 1819 and 1842 are such troughs in general business conditions, while 1833 and 1847 were also generally depressed years (the troughs were in 1832 and 1848).[81] To a limited extent the children and women working in the factories and mines were competing with the men available for the jobs. At a time of severe cyclical unemployment it would be natural, then, that the men should complain, and attempt to oust their competitors or to limit their working time. A major driving force behind the movement which led to the Act of 1833, for example, was, "the hope of absorbing men who are 'hanging on the trade idle.' "[82]

In the case of the ten-hour bill of 1847 the role of depression is even clearer. From 1845 onward unemployment was steadily increasing. In 1844 a ten-hour bill was defeated; in 1847 it was quietly passed:

> *The Times,* in a leading article on the following day, said it was not to be imagined that there had been any considerable degree of conversion on the subject. The argument stood very much where it had done in 1844, and had, in fact, been almost exhausted in that memorable struggle. The absence of fierce opposition was attributed in a large measure to the fact that the chief argument of the opponents—namely, that the country could not spare the last two hours of industry—could not be brought forward in 1847 without inviting its own refutation, for so great was the depression of trade that the mill owners found it impossible to keep their mills working for so long as ten hours.[83]

From the side of the workers, too, a distinctly nonhumanitarian factor can be detected in the ten-hour agitation. There seems to be little question that the labor unions viewed the measure as a means of restricting the labor supply and maintaining wage rates at a time of serious depression.

[81] The Factory Act of 1802, which dealt only with the pauper children, was probably inspired almost exclusively by religious and humanitarian motives.

[82] Quoted by Clapham, *op. cit.,* I, 573-574, from First Report of the Factory Commissioners (1833), XX, 849. *See also* Frederick Engels, *The Condition of the Working Class in England in 1844* (1892), 134-148, on the competition between the men and their wives and children.

[83] *See* B. L. Hutchins and A. Harrison, *A History of Factory Legislation* (1903), 70, with reference to the *Times,* May 4, 1847.

At an early stage of the ten-hour agitation (December 1841) in a period of severe unemployment, Fielden was reported as having said:[34]

It is "the duty of individuals to curtail the quantity of production when there is an over-abundant supply of the article they produce rather than increase it and reduce wages." He considered that "a reduction of hours of labour from twelve to ten would have this tendency," and was therefore desirable, as they had already "got mills and machinery to produce more than they could find a vent for at a remunerating price."

To some extent, then, the ten-hour bill was passed because unemployment existed and because it was believed by some to be a recovery measure. In markets other than that for labor the restriction of supply, in an effort to maintain prices, was a typical depression phenomenon in these years.

The long-run economic and social influences reflected in the debates are perhaps more familiar:[35]

In Parliament, the factory question, from this time down to 1847, was really a part of the wider struggle between the agricultural landlords and the manufacturers over the repeal of the Corn Laws. "The Tories were taunted with the condition of the labourers in the fields, and they retorted by tales of the condition of the operatives in factories. The manufacturers rejoined by asking, if they were so anxious to benefit the workman, why did they not, by repealing the Corn Laws, cheapen his bread. The landlords and the mill owners each reproached the other with exercising the virtues of humanity at other people's expense."

This is not to deny, of course, that sincere humanitarians worked within the ten-hour movement; nor does it underrate the importance of the strange political battle which led, finally, to the passage of the 1847 bill. But it is clearly a case where the short-run position of the economic system—the degree of unemployment—played a part in determining the moment of its ultimate acceptance.

IV

Testifying before the Committee on Manufacturers (1832) William Mathews, Staffordshire iron manufacturer, was asked:

9991. Do you conceive that the depression of trade in late years has had any effect in producing . . . discontent?—Very great. 9992. Do you think

[34] Hutchins and Harrison, *op. cit.*, 63-64*n*.
[35] Hutchins and Harrison, *op. cit.*, 61-62, quoting Morley's *Life of Cobden*, II, 300.

the working classes of Staffordshire ever show political discontent so long as they are doing well in their particular trade?—Not at all; you cannot get them to talk of politics so long as they are well employed. 9993. Do you think any man could create discontent among them so long as they were doing well?—It is utterly impossible.

The converse of the dictum—"you cannot get them to talk of politics when they are well employed"—is not to be generalized without reservation; but within this period it serves to explain the political unrest of such years as 1811-1812, 1816, 1819, 1826, 1830-1832, 1837, 1839-1842, and 1847-1848. In each of these cases a fairly direct connection can be traced between unemployment and mass dissatisfaction. In the case of the reform bill, for example, to which Mathews's questioners were referring, there can be little doubt that the intense depression of the latter months of 1831 and the early months of 1832 contributed significantly to the pressures that led to its passage.

The activities of the Chartists, covering more than a decade, offer an interesting, if somewhat crude confirmation of this thesis. General business conditions reached a peak in 1836; 1837 was a year of severe depression; some recovery followed through 1838, to a second peak early in 1839. From the latter months of 1839 to the end of 1842 Britain suffered almost unbroken depression, exacerbated in its effects on the working classes by bad harvests. A recovery set in during 1843 which culminated in a peak in 1845. Business activity then declined to a low point in 1848. The phases of the most important Chartist activity occurred within severe depression; its temporary, but almost complete disappearance in 1843-1845 coincides with the prosperity of those years.[36]

The three focal points of Chartist activity came in 1839, 1842, and 1848. Beginning in 1837 the movement gradually grew to the point where a petition boasting some million and a quarter signatures was presented to Parliament in 1839.[37] The failure of this petition, and the Government's prosecution of the leaders, caused a temporary stagnation; but in 1842 a petition containing almost three and a third million names was placed in the hands of the Government.[38] In that year, too, the Chartists helped lead a series of bitter strikes, marked by extensive sabotage.

The wheat price fell, with the advent of a promising harvest, in the summer of 1842, and in the following year recovery was under way.

[36] This relationship has been traced in detail by Preston W. Slosson, *The Decline of the Chartist Movement* (1916), 115-137.
[37] *Ibid.*, 60.
[38] *Ibid.*, 61.

For the Chartists "a long period of discouragement and inactivity followed," until the return of depression in 1846.[39] Still another petition went to Parliament in that year, and throughout 1847 the strength of the movement increased. Early in 1848 there were large meetings in the principal cities, climaxed by the presentation of the signatures of what purported to be almost six million British men and women.[40] Threats of direct action, however, failed to materialize, and, in the following years, prosperity and inadequate leadership brought the movement to an end.

The demand for universal suffrage at this period in British history patently had roots deeper than cyclical unemployment. The chronically depressed position of the hand-loom weavers, many of whom were Chartists, was also more than a cyclical problem. Yet, apparently, depression was required before the political doctrines of the Chartist leaders could command wide or effective support.

These examples by no means exhaust the possibility of tracing important links between short-run economic fluctuations and political and social events in the years 1790-1850. There are innumerable other cases available and those traced here deserve more detailed analysis. Even these brief summaries, however, reveal the manner in which cyclical fluctuations and cost-of-living movements served to detonate and to give expression to the familiar underlying trends. They should also emphasize the distinctive economic, social, and political atmosphere of each year, or even different parts of the same year. The use of long-run conceptions like "the growth of the free-trade movement," or the "development of working-class organizations," or "the industrial revolution" tends to blur this type of distinction. A necessary, but by no means sufficient requirement for a thorough interrelation of economic and other factors is a knowledge of fluctuations in general business activity and in costs of living.

Columbia University

A Note on the Statistics

The indexes of imported and domestic prices used here have recently been constructed in the course of a study of "The Growth and Fluctuations of the

[39] *Ibid.*, 78 and 94.

[40] On close examination it was later estimated that about two million bona fide signatures were attached to this final petition.

British Economy, 1790-1850," under the direction of Dr. Arthur D. Gayer. The Gilboy-Boody index appears in "The Cost of Living and Real Wages in Eighteenth Century England," *Review of Economic Statistics,* August 1936. The Silberling index is from "British Prices and Business Cycles, 1779-1850," *Review of Economic Statistics,* 1923. The Tucker index is from "Real Wages of Artisans in London, 1729-1935," *Journal of the American Statistical Society,* March 1936.

The business-cycle index is an outgrowth of an extensive history of business fluctuations (also a part of Dr. Gayer's study). Each year is rated from 0 (deep depression) to 5 (major peak). The ratings represent judgments made on the basis of all available statistical evidence, as well as a large body of qualitative material. In addition, an index of business activity was constructed containing some half dozen of the more significant series. To present that index as a cyclical indicator would have required the abstraction of secular trends from each of the series; the index misrepresented business fluctuations at several points, in the light of wider information. It was thus decided to use the semidescriptive chart presented here. It is similar in many respects to those compiled from Thorp's *Annals* by Gottfried Haberler (*Prosperity and Depression*). The wheat price is an annual average, derived from weekly quotations (1790-1834) in the *London Gazette* and (after 1834) monthly quotations in the *Gentleman's Magazine.*

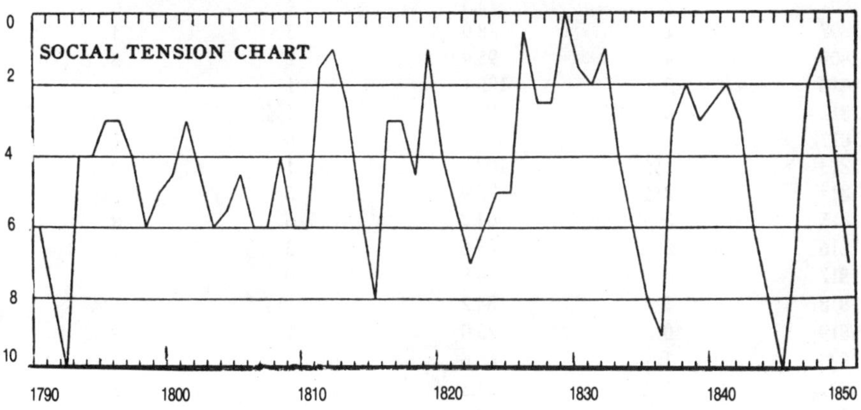

The so-called Social Tension Chart is of even more imaginative construction than the business-cycle index. The wheat price was first reduced to a 0 to 5 scale: 0 representing a year of abnormally high wheat price (and thus, like a 0 year in business activity, "high social tension"), 5 representing a year of abnormally low wheat price. This abstraction of the wheat price was then added to the business-cycle pattern, and the total plotted inversely to a high wheat price and/or estimated severe unemployment thus tending to raise the level of the chart. This method makes the quite arbitrary judgment

"SOCIAL TENSION" AND ITS COMPONENTS

	Business-Cycle Pattern	*Wheat Price* (actual annual average: s. per quarter)	*Wheat Price* (abstracted; inverse)	*"Social Tension"**
1790	3	50.5	3	6
1791	4	45.0	4	8
1792	5	41.2	5	10
1793	0	47.7	4	4
1794	1	51.8	3	4
1795	2½	74.1	½	3
1796	3	77.1	0	3
1797	0	52.8	4	4
1798	1	50.2	5	6
1799	3	67.6	2	5
1800	4	113.7	½	4½
1801	3	119.0	0	3
1802	5	67.2	4	9
1803	1	56.6	5	6
1804	1½	59.8	4	5½
1805	2½	87.5	2	4½
1806	3	79.0	3	6
1807	2	73.3	4	6
1808	1	78.9	3	4
1809	4	95.4	2	6
1810	5	106.1	1	6
1811	0	94.6	1½	1½
1812	1	125.2	0	1
1813	1½	108.5	1	2½
1814	2½	73.9	3	5½
1815	3	64.2	5	8
1816	0	75.6	3	3
1817	3	94.8	0	3
1818	5	84.2	½	5½
1819	0	73.0	1	1
1820	1	65.6	3	4
1821	1½	54.3	4	5½
1822	2	43.2	5	7
1823	3	51.9	3	6
1824	4	62.1	1	5
1825	5	66.7	0	5
1826	0	56.9	½	½
1827	1½	55.9	1	2½
1828	2	60.4	½	2½
1829	0	66.0	0	0
1830	1½	64.4	0	1½

"SOCIAL TENSION" AND ITS COMPONENTS—*concluded*

	Business-Cycle Pattern	*Wheat Price* (actual annual average: s. per quarter)	*Wheat Price* (abstracted; inverse)	*"Social Tension"**
1831	2	66.3	0	2
1832	0	58.7	1	1
1833	1	53.0	3	4
1834	2	46.3	4	6
1835	3	39.4	5	8
1836	5	48.4	4	9
1837	0	55.6	3	3
1838	1	64.3	1	2
1839	3	70.6	0	3
1840	2	66.2	½	2½
1841	1	64.3	1	2
1842	0	57.2	3	3
1843	1	50.2	5	6
1844	3	51.1	5	8
1845	5	50.9	5	10
1846	4	54.6	3	7
1847	2	69.3	0	2
1848	0	50.5	1	1
1849	1	44.2	3	4
1850	2	41.8	5	7

* "Low social tension" is indicated by high figures in this index; this series is designed to be plotted inversely.

that cyclical unemployment and high food prices were equally responsible for unrest. Low wheat prices, of course, affected farmers unfavorably and probably to a lesser extent, agricultural workers. At best, then, the Social Tension Chart summarizes influences operating on the industrial working classes.

It is perhaps unnecessary to emphasize the essentially approximate and descriptive nature of these calculations. They represent, however, a useful summary of a considerable body of evidence, and the results conform fairly well to qualitative political and social data. Intervals of "high social tension" bred known symptoms of unrest, which, in many cases, expressed themselves in important legislation or in the activities of the Luddites, the Chartists, and other groups: intervals of "low social tension" saw these movements fade from sight, although the low wheat prices that helped create them often brought the agricultural interests clamoring to Parliament.

Acknowledgements

The editors and publishers wish to thank the following for permission to use copyright material:

R. C. Allen, 'The growth of labor productivity in modern English agriculture', *Explorations in Economic History,* 25 (1988), pp. 117–46. Copyright © 1988 by Academic Press, Inc., by permission of Academic Press, Inc.;

E. J. T. Collins, 'Harvest technology and labour supply in Britain, 1790-1870', *Economic History Review,* 2nd ser., XXII (1969),, pp.453–73, by permission of the author;

D. E. C. Eversley, 'The home market and economic growth in England, 1750–1780', in E. L. Jones and G. E. Mingay, eds., *Land, labour and population in the industrial revolution* (1967), pp. 206-59, Edward Arnold, by permission of Hodder & Stoughton Ltd;

E. Gilboy, 'Demand as a factor in the industrial revolution', in A. H. Cole et al., eds., *Facts and factors in economic history: articles by former students of Edwin Francis Gay* (Cambridge, Mass., 1932), pp. 620–39, by permission of Harvard University Press;

E. L. Jones, 'Agriculture and economic growth in England, 1660–1750: agricultural change', *Journal of Economic History,* XXV (1965), pp. 1-18, by permission of Cambridge University Press;

D. S. Landes, 'Technological change and industrial development in western Europe, 1750–1914', *Cambridge economic history of Europe,* VI (Cambridge, 1965), pp. 274–310, by permission of Cambridge University Press;

N. McKendrick, 'Josiah Wedgwood: an eighteenth-century entrepreneur in salesmanship and marketing techniques', *Economic History Review,* 2nd ser., XII (1960), pp. 408–33, by permission of the author;

P. Mathias, 'Who unbound Prometheus? Science and technical change, 1600–1800', in *idem,* ed., *Science and society, 1600–1900* (Cambridge, 1972), pp. 54–80, by permission of Cambridge University Press;

J. Mokyr, 'Demand versus supply in the industrial revolution', *Journal of Economic History,* XXXVII (1977), pp. 981–1008, by permission of Cambridge University Press;

A. E. Musson and E. Robinson, 'Science and industry in the late eighteenth century', *Economic History Review,* 2nd ser., XIII (1960), pp. 222–44, by permission of the authors;

W. W. Rostow, 'Business cycles, harvests and politics, 1790–1850', *Journal of Economic History,* I (1941), pp. 206–21, by permission of Cambridge University Press;

R. Samuel, 'The workshop of the world: steam power and hand technology in mid-Victorian Britain', *History Workshop Journal,* 3 (1977), pp. 6–72, by permission of History Workshop;

F. M. L. Thompson, 'Nineteenth-century horse sense', *Economic History Review*, 2nd ser., XXIX (1976), pp. 60–81, by permission of the author;

F. M. L. Thompson, 'The second agricultural revolution, 1815–1880', *Economic History Review*, 2nd ser., XXI (1968), pp. 62–77, by permission of the author;

E. A. Wrigley, 'The supply of raw materials in the industrial revolution', *Economic History Review*, 2nd ser., XV (1962), pp. 1–16, by permission of the author.

Every effort has been made to trace all the copyright holders but if any have been inadvertently overlooked the publishers will be pleased to make the necessary arrangement at the first opportunity.